Springer Optimization and Its Applications

Volume 226

Aims and Scope

Optimization has continued to expand in all directions at an astonishing rate. New algorithmic and theoretical techniques are continually developing and the diffusion into other disciplines is proceeding at a rapid pace, with a spot light on machine learning, artificial intelligence, and quantum computing. Our knowledge of all aspects of the field has grown even more profound. At the same time, one of the most striking trends in optimization is the constantly increasing emphasis on the interdisciplinary nature of the field. Optimization has been a basic tool in areas not limited to applied mathematics, engineering, medicine, economics, computer science, operations research, and other sciences.

The series **Springer Optimization and Its Applications (SOIA)** aims to publish state-of-the-art expository works (monographs, contributed volumes, textbooks, handbooks) that focus on theory, methods, and applications of optimization. Topics covered include, but are not limited to, nonlinear optimization, combinatorial optimization, continuous optimization, stochastic optimization, Bayesian optimization, optimal control, discrete optimization, multi-objective optimization, and more. New to the series portfolio include Works at the intersection of optimization and machine learning, artificial intelligence, and quantum computing.

Volumes from this series are indexed by Web of Science, zbMATH, Mathematical Reviews, and SCOPUS.

Boris Goldengorin
Editor

Theory, Algorithms, and Experiments in Applied Optimization

In Honor of the 70th Birthday of Panos M. Pardalos

Springer

Editor
Boris Goldengorin
Center for Applied Optimization
University of Florida
Gainesville, FL, USA

ISSN 1931-6828 ISSN 1931-6836 (electronic)
Springer Optimization and Its Applications
ISBN 978-3-031-91356-3 ISBN 978-3-031-91357-0 (eBook)
https://doi.org/10.1007/978-3-031-91357-0

This Springer imprint is published by the registered company Springer Nature Switzerland AG
The registered company address is: Gewerbestrasse 11, 6330 Cham, Switzerland

Preface

This book is dedicated to the anniversary of my friend, colleague and co-author Professor Panos M. Pardalos. I will not follow tradition and tell you how old Panos is, when we met and why our scientific, philosophical and purely human interests resonated. Initially, too many colleagues, former students of Professor Pardalos, and now famous professors in many prestigious universities, public and private companies responded to the invitation to submit an article. In the end, I selected only 25 articles from 35 submitted, after strict peer review of which only 19 articles remained, comprising about 500 pages. My decision not to classify the accepted papers by general topics, and group them into chapters, is motivated not only by their diversity, but also by their unique scientific, methodological and practical individuality including usefulness both for junior university students and for advanced professionals. As a result, the articles in the book are arranged in alphabetical order by the name of the first author in each article. The papers included in this volume cover a wide range of Professor Panos Pardalos's research interests from deterministic and stochastic global optimization to integer (combinatorial) optimization, including solutions to many modern applied problems from planning theory and its applications, support vector machines, data mining and data analysis, supply chain optimization, Pareto optimality, game and equilibrium theory, vehicle routing problem and its variations, energy systems, financial and stock markets, optimal control with the key notion optimization.

A thorough review of published articles has made this book useful to a wide range of readers, from undergraduate students taking their first steps to professionals in fields such as optimization in operations research, resource management, computer science, data analysis in sports and finance, and many other applied optimization problems.

University of Florida
Gainesville, FL, USA
February 2025

Boris Goldengorin

Contents

A Case Study on Stochastic Security-Constrained Unit Commitment for Power Systems Models in Baden-Württemberg

John Alasdair Warwicker, Luc Janecke, and Steffen Rebennack

1 Introduction

Unit commitment (UC) concerns optimization models for the scheduling and dispatch of power generators subject to cost minimization [24]. The addition of operational constraints, such as transmission lines or contingency requirements (i.e. safety margins on the output), transforms the problem into the security-constrained UC model (SCUC) [8, 30]; both models are known to be NP-hard. As power systems face growing challenges from renewable energy variability and demand fluctuations, stochastic SCUC (S-SCUC) offers a robust solution for managing these complexities [31]. S-SCUC is essential for modern power systems as it incorporates uncertainty into the scheduling of generating units through a probabilistic approach, considering various future scenarios. This method enables more informed and adaptive decision-making, enhancing system reliability by allowing generation schedules to adapt to real-time changes and unexpected events [40].

S-SCUC models seek to optimize the commitment and dispatch of generating units; balance fuel costs, start-up, and shutdown expenses; and reserve provisioning while adhering to operational constraints, leading to significant cost efficiencies and effective resource management. It is particularly beneficial for integrating renewable energy sources such as wind and solar, managing their variability and uncertainty by optimizing schedules to reduce curtailment and maximize utilization, thus

J. A. Warwicker · S. Rebennack (✉)
Institute of Operations Research, Karlsruhe Institute of Technology, Karlsruhe, Germany
e-mail: john.warwicker@kit.edu; steffen.rebennack@kit.edu

L. Janecke
Institute for Automation and Applied Informatics, Karlsruhe Institute of Technology, Karlsruhe, Germany
e-mail: luc.janecke@student.kit.edu

B. Goldengorin (ed.), *Theory, Algorithms, and Experiments in Applied Optimization*, Springer Optimization and Its Applications 226,
https://doi.org/10.1007/978-3-031-91357-0_1

supporting environmental sustainability goals by lowering emissions and aligning with green energy targets. Additionally, S-SCUC aids in strategic planning and investment decisions by providing insights into the value of different generation and storage technologies under uncertain conditions and improves market efficiency by enabling better bidding strategies and contributing to price stability.

Incorporating uncertainty into SCUC is fundamental for effectively managing the complexities of modern power systems, particularly with the variability of renewable energy sources including wind and solar. S-SCUC models this variability to plan for fluctuations in renewable generation, ensuring sufficient backup generation or storage to maintain system balance, thereby reducing the need for curtailment and maximizing renewable energy use while enhancing both economic efficiency and environmental sustainability. Additionally, S-SCUC addresses demand fluctuations influenced by weather, economic activities, and consumer behaviour, enabling more accurate generation scheduling and preventing inefficiencies related to under- or over-commitment, ensuring the system can meet high-demand periods without relying on costly peaking plants or risking blackouts.

S-SCUC also manages fuel price volatility by optimizing fuel procurement and dynamically adjusting generation schedules, minimizing operational costs and maintaining cost-effective operations amid changing market conditions. Furthermore, SCUC accounts for generator availability uncertainties due to unexpected outages or maintenance needs, creating resilient generation schedules that ensure adequate reserve requirements and enhance system reliability. This flexibility supports strategic bidding and operational planning, improving overall market efficiency and stability. By modelling potential grid contingencies like line failures or extreme weather events, S-SCUC enhances grid stability and security through effective contingency planning. Additionally, S-SCUC aids long-term planning and investment by providing insights into the performance and value of different technologies under uncertain conditions, supporting infrastructure development and capacity expansion decisions. This comprehensive approach makes S-SCUC essential for achieving reliable, cost-effective, and sustainable power system operations in an evolving energy landscape.

The rest of this paper is structured as follows: In Sect. 2, we give an overview of the state-of-the-art and the different solution techniques proposed for the S-SCUC problem. In Sect. 3, an optimization model of the energy system in Baden-Württemberg is presented in detail, and computations are carried out, comparing stochastic and deterministic SCUCs in terms of performance. Section 4 presents an analysis and discussion of the results, and we finish with conclusions and outlook.

2 Stochastic Security-Constrained Unit Commitment

In addressing the complexities of modern power systems, the development and application of S-SCUC models have proven essential. Numerous S-SCUC models have been presented, each differing based on the constraints they consider.

2.1 Formulations and Constraints

Mathematically, the SCUC problem is characterized as a large-scale mixed-integer programming (MIP) problem, featuring numerous binary, continuous, and discrete control variables, along with a set of constraints [1]. The SCUC models can be categorized based on the constraints they consider.

The SCUC models described in [9, 10] integrate AC network constraints for secure and economical power generation scheduling. Using Benders decomposition, they separate unit commitment and network security checks, iteratively resolving violations to account for real and reactive power interactions. The model in [21] focuses on reconfigurable transmission networks to enhance wind power integration, leveraging network reconfiguration to minimize costs related to energy, spinning reserve, wind curtailment, and load shedding while adhering to transmission constraints.

On the other hand, several SCUC models employ stochastic methods to handle uncertainties in power system operations. The models described in [2, 40] use Monte Carlo simulations and scenario aggregation techniques to manage uncertainties such as random outages and load forecasting errors. These models optimize fuel resource allocation and emission allowances, enhancing the reliability and economic efficiency of long-term operations amid uncertain conditions. In addition to that, the model in [4] focuses on integrating wind power into reconfigurable transmission networks, minimizing costs related to energy, spinning reserve, wind curtailment, and load shedding while considering transmission constraints.

Other notable approaches include the SCUC model in [38], which incorporates wind power volatility through a two-stage optimization process. This method ensures robust unit commitment and dispatch solutions by addressing both the intermittency and the volatility of wind power generation. Furthermore, the early stochastic SCUC model presented in [29] integrates uncertainties such as renewable generation variability and load fluctuations into the unit commitment process. By utilizing scenario-based optimization and Benders decomposition, this model aims to minimize total system costs while ensuring reliability and accommodating operational constraints.

SCUC models traditionally use deterministic approaches, but the inclusion of uncertainties leads to more realistic results. Incorporating uncertainties in SCUC models enhances their accuracy and reliability. By considering various scenarios, stochastic SCUC models optimize generation schedules and reserve allocations, reducing overall costs. This approach prevents the conservative over-allocation of reserves typical in deterministic models, leading to more cost-effective operations [40]. Stochastic models account for potential system states, including equipment outages and load variations. This results in more robust scheduling decisions that maintain system reliability even under unexpected conditions. For example, the models implicitly determine reserve requirements based on specific commitment schedules for each scenario [2]. Locational Marginal Prices (LMPs) derived from stochastic models reflect the impact of uncertainties, providing more accurate economic signals for market participants. These prices are crucial for planning and investment decisions, ensuring that the economic incentives align with the actual risks and uncertainties faced by the system [20, 40]. For instance, a study on a large-scale power system showed that stochastic models led to better fuel consumption strategies and reduced emissions compared to deterministic models. The use of multiple scenarios allowed for more adaptive and flexible decision-making, enhancing both economic and environmental outcomes [1].

In general, the inclusion of uncertainties in SCUC models results in more realistic and reliable outcomes. Stochastic approaches provide significant improvements in economic efficiency, reliability, and market accuracy by considering a wide range of potential future states. These models represent a crucial advancement over traditional deterministic methods, offering a more robust framework for power system operation and planning.

2.2 *Solution Methods*

In typical UC models (and extensions thereof), the objective function seeks to minimize the total system cost [29]. The following methods are typically used within S-SCUC models with such objectives:

1. Scenario generation and reduction;
2. Benders decomposition;
3. Correct and preventative actions.

Such methods are important to handle uncertainties associated with power systems through robust scenario generation and reduction, decomposition techniques, or dynamic adjustments [16, 27, 28].

2.2.1 Scenario Generation and Reduction

This approach generates a large number of scenarios representing possible power outputs through Monte Carlo simulation. The generated scenarios are based on the forecasted power with a specified standard deviation to capture the volatility of renewable power. To reduce the computational burden of Monte Carlo simulations, Latin Hypercube Sampling (LHS) is employed. LHS is a stratified sampling method that ensures each sample is the only one in each axis-aligned hyperplane, thus reducing the variance of the simulation and providing a more accurate approximation with fewer samples [38]. After generating scenarios, scenario reduction techniques are applied to eliminate those with very low probabilities and aggregate similar scenarios. This is done to manage computational complexity while maintaining an accurate representation of possible outcomes.

These methods are advantageous because they maintain statistical accuracy while simplifying the problem through scenario reduction, which focuses on the most representative scenarios. However, Monte Carlo simulations can be computationally intensive and may not scale well for large systems, while LHS requires careful implementation to ensure accuracy. Additionally, scenario reduction might discard low-probability, high-impact scenarios, potentially compromising the robustness of the solutions.

2.2.2 Benders Decomposition

When using Benders decomposition [3, 32], the SCUC problem is decomposed into a master problem and several subproblems to manage computational complexity effectively. The master problem is a unit commitment problem which includes the objective function and constraints relating to power balance, reserve requirements, and unit operational limits and is solved by either MIP approaches or Lagrangian relaxations. The first subproblem is a feasibility check, which confirms if the current solution to the master problem can accommodate renewable power volatility in individual scenarios. The second ensures the solution meets network security constraints. If such violations are detected in either case, Benders cuts are generated and added to the master problem.

This method is flexible, allowing for various optimization techniques to solve the master unit commitment problem, and includes focused security checks to ensure solutions meet all operational constraints and accommodate wind power volatility. Despite these advantages, Benders decomposition involves multiple iterations between the master problem and subproblems, which can be time-consuming. Coordinating these iterations effectively can be complex and computationally demanding, and the iterative process may converge slowly, especially when many iterations are required to approximate the subproblems accurately (e.g. due to dual degeneracy). Hence, advanced solution techniques, such as stochastic dual dynamic programming, have been suggested [11, 41].

We note that while Benders decomposition as a method was introduced for large-scale programs with linear subproblems [3], enhancements exist to consider, for example, convex [14] and non-convex subproblems [12], as well as binary [5] and non-convex master problems [36].

2.2.3 Corrective and Preventive Actions

Corrective actions refer to real-time adjustments made to the dispatch of non-wind units to handle deviations in wind power output from the forecasted values. Such actions ensure the system remains balanced in spite of renewable power fluctuations. Preventative actions, on the other hand, are planned day-ahead adjustments to the commitment and dispatch of non-renewable units, considering possible renewable power scenarios. These actions help in pre-emptively addressing potential violations due to wind power volatility.

In general, these actions enhance real-time flexibility and reliability by providing the ability to make adjustments in response to unforeseen deviations in wind power output and ensuring the system is prepared for expected variations. These actions contribute to more reliable system operations by anticipating and addressing potential issues. However, frequent real-time adjustments can lead to higher operational costs, and both corrective and preventive actions depend heavily on the accuracy of initial forecasts, which may not always be precise. Additionally, preventive actions may result in unnecessary costs if the forecasted scenarios do not materialize as expected.

3 Methodology for Optimization Model

In the remainder, we implement an S-SCUC model to address the uncertainties in power generation using real data from Baden-Württemberg. We model the security constraint through a safety margin on the outputted power, and we use scenario generation through a Monte Carlo simulation due to its generality and flexible probabilistic framework. Although this may be impractical for very large systems and numerous scenarios, we avoid this by narrowing the scope of the study to a specific use case. In particular, extensive research was carried out in order to find or calculate suitable and realistic input values for the model parameters with the aim of providing a realistic framework rather than a theoretical one.

3.1 Assumptions and Simplifications

In order to simplify the model (and due to the lack of available data), transmission lines were not taken into account; therefore, the security constraint was simply a requirement for surplus reserve in case of increased demand. Further, we do not consider renewable generation as typical UC models focus mostly on thermal and hydropower plants. We refer the interested reader to, for example, [35, 38] for related models. Distributed generation is implicitly modelled in the demand uncertainty, as it just increases the residual load fluctuations. For pumped storage hydropower, the reservoir level was not considered as the time horizon is only 1 day. To simplify the model, we did not consider ramp-up and ramp-down rates, minimum downtime, or maximum uptime. Finally, for the generator data, whenever no specific values could be obtained for a certain type with and without combined heat and power (CHP), the program neglects the potential coupling and assumes no CHP for all generators of that type. Due to these simplifications, we did not require scenario reduction techniques as the computational complexity was not a bottleneck.

In terms of uncertainty, this work focuses mainly on stochastic scenario creation techniques in terms of demand data. Other sources of uncertainty, such as generator outages, were not modelled due to the chosen 1-day time discretizations; in this case, the downtimes would be fixed within a single period. Further, we consider the deterministic equivalent of a two-stage stochastic programming problem and do not take recourse actions into consideration; we leave this for future work.

3.2 *Input Data and Acquisition*

The SCUC model relies on input data regarding generators and the electricity demand. While some input data could be relatively easily transformed from publicly available sources, other data proved complicated to find for the specific cases that make up this work. Whenever specific data could not be found, it was approximated from a variety of sources. The overall basis for this work is the OPSD website [23].

We firstly require a number of time periods, alongside their length in hours and the forecasted demand from each time period (in MW). An hourly time series for a 1-day interval for load forecast data was obtained as reported on the ENTSO-E transparency platform [7] for January 2, 2015. The data is available only for the entirety of Germany, but as per [15], we approximate the energy consumption for Baden-Württemberg based on its relative GDP. Namely, as of 2022, Germany had a non-price-adjusted GDP of 3,601,750,000,000 €, while Baden-Württemberg accounted for 538,948,000,000 € [33]. This translates to a share of around 15%, so this factor was used on each time series data point. Given this data, the time period of 1 hour and the number of periods representing 1 day were determined, with the results shown in Table 1.

We further require information on the generators, specifically the number of types and amount per type. For each type, we require the minimum and maximum load, start-up costs, generation costs, and base costs. For this information, we used information published by the German Bundesnetzagentur [19] as of October 1, 2020. After filtering and preprocessing the data, the database consisted of 83 generating units, which we summarize in Table 2.

The sources in Table 2 are displayed by generator category as well as parameter of interest. For each category of generator, we present the type of turbine (if present, steam, open-cycle gas turbine (OCGT), or combined-cycle gas turbine (CCGT)) and the use of combined heat and power (CHP), alongside the amount of such generators included in the data. For conversion purposes, we used a conversion rate of 0.93 €/USD.

Table 1 Resulting values for the load forecast (in MW) for Baden-Württemberg

0–1 h	**1–2 h**	**2–3 h**	**3–4 h**	**4–5 h**	**5-6 h**
5792.52	5659.20	5660.39	5824.54	6060.67	6494.16
6–7 h	**7–8 h**	**8–9 h**	**9–10 h**	**10–11 h**	**11–12 h**
7338.10	7712.19	7950.56	8147.03	8332.43	8323.75
12–13 h	**13–14 h**	**14–15 h**	**15–16 h**	**16–17 h**	**17–18 h**
8035.10	7977.79	7971.81	8153.76	8564.06	8569.30
18–19 h	**19–20 h**	**20–21 h**	**21–22 h**	**22–23 h**	**23–24 h**
8317.31	7748.70	7152.26	6879.02	6654.87	6335.70

Table 2 Researched parameters for each category of generator

Category	Turbine	CHP	Amount	Start-up	Depreciation	Minload	LCOE	Fuel cost	Source
Natural gas	Steam	Yes	14	2.8	57	38	81.45	25	[13, 18]
Natural gas	CCGT	No	1	2.8	60	45	104.25	25	[13, 18]
Natural gas	CCGT	Yes	1	2.8	60	45	81.45	25	[13, 18]
Natural gas	OCGT	No	4	0.1	24	20	87.42	25	[13, 34]
Oil	Steam	No	3	2.8	57	38	104.25	83	[13]
Oil	Steam	Yes	2	2.8	57	38	81.45	83	[13]
Oil	CCGT	No	2	2.8	60	45	104.25	83	[13]
Oil	OCGT	No	2	0.1	24	20	87.42	83	[13, 34]
Biomass and biogas	Steam	No	1	2.8	57	38	165.2	31.7	[13, 18]
Biomass and biogas	Steam	Yes	8	2.8	57	38	120.6	31.7	[13, 18]
Hard coal	Steam	–	14	5.9	49	38	155.35	10	[13, 18]
Hard coal (>500 MW)	Steam	–	3	2.7	105	38	155.35	10	[13, 18]
Hydro (run of river)	–	–	12	0	7.29	10	63.24	0	[6, 25, 37]
Hydro (pumped)	–	–	7	0	7.29	10	66.03	0	[6, 25, 37]
Nuclear	Steam	–	1	16.7	50	50	63.24	4.99	[13, 22]
Waste	Steam	–	8	2.8	57	38	58	50	[13, 17]

Start-up refers to the start-up fuel requirement for a cold start in $\mathrm{MWh}_{th}/\mathrm{MW}$. *Depreciation* refers to the start-up depreciation costs in €/MW incurred by higher stress and fatigue during non-optimal operation. *Minload* refers to the minimum percentage of rated power at which a generator operator will run in order to operate profitably and maximize lifetime.

LCOE is the levelized cost of electricity (€) that provides a value on the cost per MwH of a power plant in its lifetime, including investment expenditure, fixed and variable operating costs, interest rates, and plant lifetime. For this, we used a discount rate of 3.2% for biomass and biogas, 6.2% for hard coal, 5.2% for gas and oil [18], 7% for hydro and nuclear, and 6% for all plants with open-cycle gas turbines [25, 34]. Finally, *Fuel cost* refers to the cost to burn fuel in $\mathrm{MWh}_{th}/\mathrm{MW}$.

3.3 Implemented Stochastic SCUC Model

For the presented model, we analysed the usage of 83 different generators over 24 time periods (i.e., a uniform hourly resolution within 1 day). The full input data required for the model and the acquisition of the data are discussed in detail in Sect. 3.2.

The general approach is broken down into two main parts: firstly, we establish a deterministic base model for the SCUC derived from [39]; secondly, we model the uncertainties using scenario aggregation and add them to the baseline deterministic model. To model the security constraint, we apply a safety margin of 10%, meaning the generators require a reserve equal to 10% of anticipated demand while committed.

Sets

- $\mathcal{G}$—the set of generator units $g \in \mathcal{G}$ ($|G| = 83$);
- $\mathcal{P}$—the set of time periods $p \in \mathcal{P}$ ($|P| = 24$).

Variables

- $\mathrm{n}_{g,p} \in \{0, 1\}$—the number of generators of type $g \in \mathcal{G}$ that are committed during period $p \in \mathcal{P}$;
- $\mathrm{s}_{g,p} \in \{0, 1\}$—the number of generators of type $g \in \mathcal{G}$ that start during period $p \in \mathcal{P}$;
- $\mathrm{o}_{g,p} \in \mathbb{R}$—the total power generated by generators of type $g \in \mathcal{G}$ turning time period $p \in \mathcal{P}$.

Parameters (see Table 2)

- E_g—the levelized cost of electricity of generator $g \in \mathcal{G}$;
- $\underline{L}_g$ ($\overline{L}_g$)—the minimum (maximum) percentage of rated power at which a generator $g \in \mathcal{G}$ will run in order to operate profitably;
- $\underline{M}_g$—the minimum cost incurred when a generator is on (calculated as the floor value of the product of the LCOE and minimum load of a generator, converted into Euros);
- C_g—the cost of starting up generator $g \in \mathcal{G}$ and is calculated as

$$\text{start-up}_g \cdot \text{fuelcost}_g + \text{depreciation}_g$$

  (with these values taken from Table 2);
- D_p—the system demand at time period $p \in \mathcal{P}$.

3.3.1 Deterministic Model

We present the deterministic model below:

1. The objective function is defined with the goal of cost minimization. The three terms calculate the fixed costs, variable costs per unit of electricity generated, and start-up costs:

$$\min \sum_{g\in\mathcal{G}} \sum_{p\in\mathcal{P}} \underline{M}_g \cdot n_{g,p} + E_g \cdot (o_{g,p} - \underline{L}_g) + C_g \cdot s_{g,p} \tag{1}$$

2. To calculate the output of a generator during a given time period, we first restrict its output to fall within its predefined limits:

$$0 \le o_g - \underline{L}_g \le (\overline{L}_g - \underline{L}_g) \cdot n_{g,p} \quad \forall g \in \mathcal{G};\ p \in \mathcal{P} \tag{2}$$

3. Next, we must ensure that demand is satisfied for all time periods:

$$\sum_{g\in\mathcal{G}} o_{g,p} \ge D_p \quad \forall p \in \mathcal{P} \tag{3}$$

4. The following constraint ensures consistency with the start-up and running of generators:

$$n_{g,p} \le n_{g,p-1} + s_{g,p} \quad \forall g \in \mathcal{G};\ p \in \mathcal{P} \tag{4}$$

5. Finally, we consider the security constraint, which ensures that we have a 10% reserve regarding the demand:

$$\sum_{g\in\mathcal{G}} \overline{L}_g \cdot n_{g,p} \ge 1.1 \cdot D_p \quad \forall g \in \mathcal{G};\ p \in \mathcal{P} \tag{5}$$

We note that if we had multiple generators of each type, we would relax the binary variable $\mathrm{n}_{g,p}$ to be an integer and include the constraint

$$\mathrm{n}_{g,p} \leq \mathrm{N}_g \quad \forall g \in \mathcal{G};\ p \in \mathcal{P}$$

where N_g refers to the number of generators of type $g \in \mathcal{G}$ (we would also update the remaining constraints as necessary).

As a benchmark for comparison with the stochastic formulation, we ran formulations (1)–(5) using actual and forecasted load data from the OPSD dataset [23].

3.3.2 Scenario Generation

The UC problem is known to be complex, and for real-life applications, utility companies must solve the UC problem in advance. This relies on advanced predictions for the demand which naturally comes with some prediction error. Therefore, for the stochastic model, we consider the demand at each time period as an uncertain parameter.

To generate scenarios, we utilize the demand prediction error using a Markov chain Monte Carlo simulation adapted from [26]. With such scenarios, the deterministic formulations (1)–(5) is adapted to fit the stochastic demand (see, for example, [40]). Note that the scenario generation is independent of the approach and requires only forecasted and real data for the same period of time. For this purpose, we take the full 1-hour time series data from the OPSD website [23], consisting of 50,352 valid 1-hour time frames.

For each time frame, the error was calculated as the difference between the actual and forecasted data. Each error was categorized into 1 of 384 intervals of size 50 MW, each representing a state $m \in M$ (where $|M| = 384$). From this, we created an $m \times m$ successor matrix A, where entry A_{ij} corresponds to the frequency that state S_j is a successor of state S_i (for $i, j \in M$). To generate state-transition probabilities, we normalized the matrix such that the rows of A sum to 1.

The scenarios are generated as follows: for the chosen test day, we chose the first state based on its first time step. Based on the matrix A, we then probabilistically selected successor states for each of the 24-hour time periods. We multiplied each error by 50 (i.e., back to its original error category) and added this to the forecast data. We repeated this process to generate 100 scenarios $\mathbb{S}^{(k)}$, where $k \in \{1, \ldots, 100\}$.

3.3.3 Stochastic Model

To solve the stochastic formulation, we used a sample average approximation (SAA) approach to reformulate the stochastic problem as a deterministic problem. That is, we consider each of the scenarios as a deterministic subproblem. Let

$$f(x_p^s, u_p^s)$$

refer to formulation (1)–(5) for a given scenario $S \in \mathbb{S}$ at a given time period $p \in \mathcal{P}$, with state variables x_p^s and demand u_p^s. Then, the stochastic model is as follows:

$$\min \frac{1}{|\mathbb{S}|} \sum_{s \in \mathbb{S}} \sum_{p \in \mathcal{P}} f(x_p^s, u_p^s). \tag{6}$$

Overall, we consider a two-step process: in the first step, we generate the desired scenarios $\mathbb{S}$ independently (relating to the potential realizations of power demand) based on the real and forecasted data and the transition probabilities between different states. In the second step, we solve the deterministic problem (6) to provide the solution to the stochastic model.

We note that the value of the decision variables is taken from including all possible constraints into the model, thus guaranteeing the feasibility of constraints (2) and (4). However, at a given time period, the solution to the stochastic model (6) may be infeasible with regard to the realized demand (i.e. constraints (3) and (5)). Through the implemented SAA approach, the model seeks to mitigate the risk of this occurring while minimizing cost. Once the demand at a given time period is realized, we can adjust the solution to account for any infeasibility (or excess demand). However, we do not include this calculation in this work and instead seek to assess the realism and predictive quality of the scenario generation approach.

The scenario generation that we implement relies on the realization of the first time period (note that this is not necessarily required if we have fully forecasted data). The stochastic model can in fact be updated at each time step after each realization of the real power demand, providing that the successor matrix A is also updated to reflect the new transition probabilities from the current state.

We note that implementing a robust solution approach of assuming the worst-case demand (i.e., the maximum possible) in each period would lead to a fully feasible solution without the requirement for recourse actions but would also lead to much higher costs.

4 Results and Discussion

In this section, we analyse the presented models in three configurations, namely, formulations (1)–(5) with historical data, formulations (1)–(5) with forecasted data, and the stochastic approach with scenario generation presented in formulation (6). The models were implemented in Python and solved using Gurobi with standard settings.

The full results regarding which generators are switched on and off at which time periods are presented in Figs. 1, 2, 3, 4, and 5. The figures show heatmaps for the three models (white is committed, black is off) over the chosen time period.

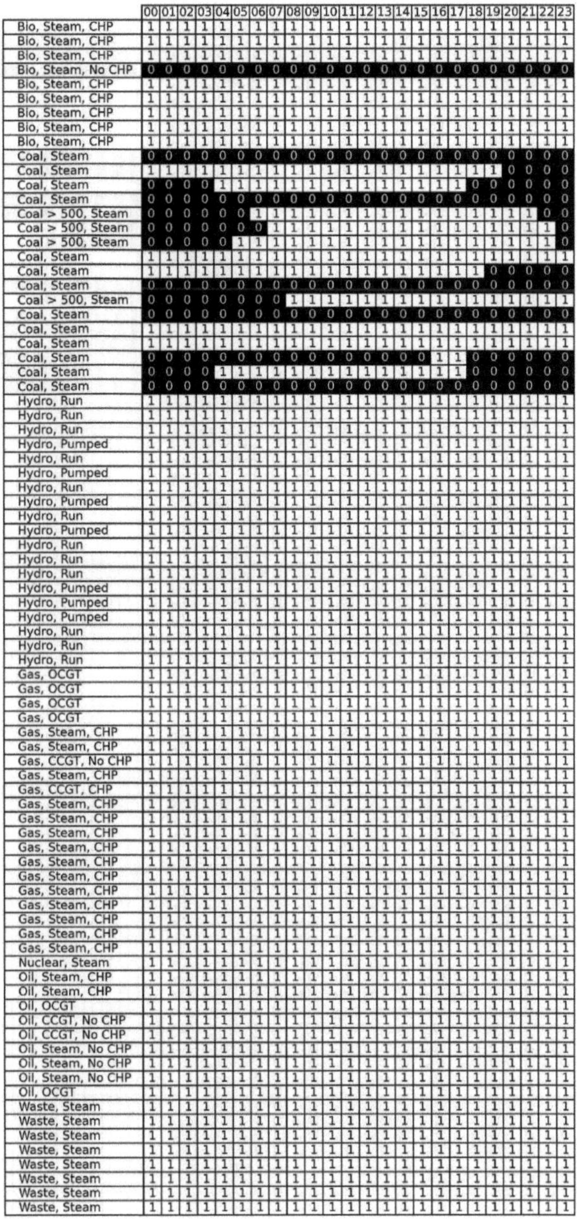

Fig. 1 Heatmap for historical load data

4.1 *Outcome*

For both deterministic scenarios, hard coal power plants were the only ones to start up and shutdown within the 24-hour time frame; aside from these and one non-CHP biogas and biomass plant, all others ran continuously. Regarding the stochastic model, the biogas and biomass plant was committed during peak loads, while a gas

Fig. 2 Heatmap for forecasted load data

	00	01	02	03	04	05	06	07	08	09	10	11	12	13	14	15	16	17	18	19	20	21	22	23
Bio, Steam, CHP	1	1	1	1	1	1	1	1	1	1	1	1	1	1	1	1	1	1	1	1	1	1	1	1
Bio, Steam, CHP	1	1	1	1	1	1	1	1	1	1	1	1	1	1	1	1	1	1	1	1	1	1	1	1
Bio, Steam, CHP	1	1	1	1	1	1	1	1	1	1	1	1	1	1	1	1	1	1	1	1	1	1	1	1
Bio, Steam, No CHP	0	0	0	0	0	0	0	0	0	0	0	0	0	0	0	0	0	0	0	0	0	0	0	0
Bio, Steam, CHP	1	1	1	1	1	1	1	1	1	1	1	1	1	1	1	1	1	1	1	1	1	1	1	1
Bio, Steam, CHP	1	1	1	1	1	1	1	1	1	1	1	1	1	1	1	1	1	1	1	1	1	1	1	1
Bio, Steam, CHP	1	1	1	1	1	1	1	1	1	1	1	1	1	1	1	1	1	1	1	1	1	1	1	1
Bio, Steam, CHP	1	1	1	1	1	1	1	1	1	1	1	1	1	1	1	1	1	1	1	1	1	1	1	1
Bio, Steam, CHP	1	1	1	1	1	1	1	1	1	1	1	1	1	1	1	1	1	1	1	1	1	1	1	1
Coal, Steam	0	0	0	0	0	0	0	0	0	0	0	0	0	0	0	0	0	0	0	0	0	0	0	0
Coal, Steam	0	0	0	0	0	0	0	0	0	0	0	0	0	0	0	0	0	0	0	0	0	0	0	0
Coal, Steam	1	1	1	1	1	1	1	1	1	1	1	1	1	1	1	1	1	1	1	0	0	0	0	0
Coal, Steam	0	0	0	0	0	0	0	0	0	0	0	0	0	0	0	0	0	0	0	0	0	0	0	0
Coal > 500, Steam	0	0	0	0	0	0	0	0	1	1	1	1	1	1	1	1	1	1	1	1	1	1	0	0
Coal > 500, Steam	0	0	0	0	1	1	1	1	1	1	1	1	1	1	1	1	1	1	1	0	0	0	0	0
Coal > 500, Steam	0	0	0	0	0	1	1	1	1	1	1	1	1	1	1	1	1	1	1	1	1	1	1	1
Coal, Steam	1	1	1	1	1	1	1	1	1	1	1	1	1	1	1	1	1	1	0	0	0	0	0	0
Coal, Steam	1	1	1	1	1	1	1	1	1	1	1	1	1	1	1	1	1	1	1	1	0	0	0	0
Coal, Steam	0	0	0	0	0	0	0	0	0	0	0	0	0	0	0	0	0	0	0	0	0	0	0	0
Coal > 500, Steam	0	0	0	0	0	0	1	1	1	1	1	1	1	1	1	1	1	1	1	1	1	1	1	1
Coal, Steam	0	0	0	0	0	0	0	0	0	0	0	0	0	0	0	0	0	0	0	0	0	0	0	0
Coal, Steam	0	0	0	0	0	0	0	0	0	0	0	0	0	0	0	0	0	0	0	0	0	0	0	0
Coal, Steam	0	0	0	0	0	1	1	1	1	1	1	1	1	1	1	1	1	1	1	1	1	1	1	0
Coal, Steam	0	0	0	0	0	0	0	0	0	0	0	0	0	0	0	0	0	0	0	0	0	0	0	0
Coal, Steam	0	0	0	0	0	0	0	0	0	0	0	0	0	0	0	0	0	0	0	0	0	0	0	0
Coal, Steam	0	0	0	0	0	0	0	0	0	0	0	0	0	0	0	0	0	0	0	0	0	0	0	0
Hydro, Run	1	1	1	1	1	1	1	1	1	1	1	1	1	1	1	1	1	1	1	1	1	1	1	1
Hydro, Run	1	1	1	1	1	1	1	1	1	1	1	1	1	1	1	1	1	1	1	1	1	1	1	1
Hydro, Run	1	1	1	1	1	1	1	1	1	1	1	1	1	1	1	1	1	1	1	1	1	1	1	1
Hydro, Pumped	1	1	1	1	1	1	1	1	1	1	1	1	1	1	1	1	1	1	1	1	1	1	1	1
Hydro, Run	1	1	1	1	1	1	1	1	1	1	1	1	1	1	1	1	1	1	1	1	1	1	1	1
Hydro, Pumped	1	1	1	1	1	1	1	1	1	1	1	1	1	1	1	1	1	1	1	1	1	1	1	1
Hydro, Run	1	1	1	1	1	1	1	1	1	1	1	1	1	1	1	1	1	1	1	1	1	1	1	1
Hydro, Pumped	1	1	1	1	1	1	1	1	1	1	1	1	1	1	1	1	1	1	1	1	1	1	1	1
Hydro, Run	1	1	1	1	1	1	1	1	1	1	1	1	1	1	1	1	1	1	1	1	1	1	1	1
Hydro, Pumped	1	1	1	1	1	1	1	1	1	1	1	1	1	1	1	1	1	1	1	1	1	1	1	1
Hydro, Run	1	1	1	1	1	1	1	1	1	1	1	1	1	1	1	1	1	1	1	1	1	1	1	1
Hydro, Run	1	1	1	1	1	1	1	1	1	1	1	1	1	1	1	1	1	1	1	1	1	1	1	1
Hydro, Run	1	1	1	1	1	1	1	1	1	1	1	1	1	1	1	1	1	1	1	1	1	1	1	1
Hydro, Pumped	1	1	1	1	1	1	1	1	1	1	1	1	1	1	1	1	1	1	1	1	1	1	1	1
Hydro, Pumped	1	1	1	1	1	1	1	1	1	1	1	1	1	1	1	1	1	1	1	1	1	1	1	1
Hydro, Pumped	1	1	1	1	1	1	1	1	1	1	1	1	1	1	1	1	1	1	1	1	1	1	1	1
Hydro, Run	1	1	1	1	1	1	1	1	1	1	1	1	1	1	1	1	1	1	1	1	1	1	1	1
Hydro, Run	1	1	1	1	1	1	1	1	1	1	1	1	1	1	1	1	1	1	1	1	1	1	1	1
Hydro, Run	1	1	1	1	1	1	1	1	1	1	1	1	1	1	1	1	1	1	1	1	1	1	1	1
Gas, OCGT	1	1	1	1	1	1	1	1	1	1	1	1	1	1	1	1	1	1	1	1	1	1	1	1
Gas, OCGT	1	1	1	1	1	1	1	1	1	1	1	1	1	1	1	1	1	1	1	1	1	1	1	1
Gas, OCGT	1	1	1	1	1	1	1	1	1	1	1	1	1	1	1	1	1	1	1	1	1	1	1	1
Gas, OCGT	1	1	1	1	1	1	1	1	1	1	1	1	1	1	1	1	1	1	1	1	1	1	1	1
Gas, Steam, CHP	1	1	1	1	1	1	1	1	1	1	1	1	1	1	1	1	1	1	1	1	1	1	1	1
Gas, Steam, CHP	1	1	1	1	1	1	1	1	1	1	1	1	1	1	1	1	1	1	1	1	1	1	1	1
Gas, CCGT, No CHP	1	1	1	1	1	1	1	1	1	1	1	1	1	1	1	1	1	1	1	1	1	1	1	1
Gas, Steam, CHP	1	1	1	1	1	1	1	1	1	1	1	1	1	1	1	1	1	1	1	1	1	1	1	1
Gas, CCGT, CHP	1	1	1	1	1	1	1	1	1	1	1	1	1	1	1	1	1	1	1	1	1	1	1	1
Gas, Steam, CHP	1	1	1	1	1	1	1	1	1	1	1	1	1	1	1	1	1	1	1	1	1	1	1	1
Gas, Steam, CHP	1	1	1	1	1	1	1	1	1	1	1	1	1	1	1	1	1	1	1	1	1	1	1	1
Gas, Steam, CHP	1	1	1	1	1	1	1	1	1	1	1	1	1	1	1	1	1	1	1	1	1	1	1	1
Gas, Steam, CHP	1	1	1	1	1	1	1	1	1	1	1	1	1	1	1	1	1	1	1	1	1	1	1	1
Gas, Steam, CHP	1	1	1	1	1	1	1	1	1	1	1	1	1	1	1	1	1	1	1	1	1	1	1	1
Gas, Steam, CHP	1	1	1	1	1	1	1	1	1	1	1	1	1	1	1	1	1	1	1	1	1	1	1	1
Gas, Steam, CHP	1	1	1	1	1	1	1	1	1	1	1	1	1	1	1	1	1	1	1	1	1	1	1	1
Gas, Steam, CHP	1	1	1	1	1	1	1	1	1	1	1	1	1	1	1	1	1	1	1	1	1	1	1	1
Gas, Steam, CHP	1	1	1	1	1	1	1	1	1	1	1	1	1	1	1	1	1	1	1	1	1	1	1	1
Gas, Steam, CHP	1	1	1	1	1	1	1	1	1	1	1	1	1	1	1	1	1	1	1	1	1	1	1	1
Gas, Steam, CHP	1	1	1	1	1	1	1	1	1	1	1	1	1	1	1	1	1	1	1	1	1	1	1	1
Nuclear, Steam	1	1	1	1	1	1	1	1	1	1	1	1	1	1	1	1	1	1	1	1	1	1	1	1
Oil, Steam, CHP	1	1	1	1	1	1	1	1	1	1	1	1	1	1	1	1	1	1	1	1	1	1	1	1
Oil, Steam, CHP	1	1	1	1	1	1	1	1	1	1	1	1	1	1	1	1	1	1	1	1	1	1	1	1
Oil, OCGT	1	1	1	1	1	1	1	1	1	1	1	1	1	1	1	1	1	1	1	1	1	1	1	1
Oil, CCGT, No CHP	1	1	1	1	1	1	1	1	1	1	1	1	1	1	1	1	1	1	1	1	1	1	1	1
Oil, CCGT, No CHP	1	1	1	1	1	1	1	1	1	1	1	1	1	1	1	1	1	1	1	1	1	1	1	1
Oil, Steam, No CHP	1	1	1	1	1	1	1	1	1	1	1	1	1	1	1	1	1	1	1	1	1	1	1	1
Oil, Steam, No CHP	1	1	1	1	1	1	1	1	1	1	1	1	1	1	1	1	1	1	1	1	1	1	1	1
Oil, Steam, No CHP	1	1	1	1	1	1	1	1	1	1	1	1	1	1	1	1	1	1	1	1	1	1	1	1
Oil, OCGT	1	1	1	1	1	1	1	1	1	1	1	1	1	1	1	1	1	1	1	1	1	1	1	1
Waste, Steam	1	1	1	1	1	1	1	1	1	1	1	1	1	1	1	1	1	1	1	1	1	1	1	1
Waste, Steam	1	1	1	1	1	1	1	1	1	1	1	1	1	1	1	1	1	1	1	1	1	1	1	1
Waste, Steam	1	1	1	1	1	1	1	1	1	1	1	1	1	1	1	1	1	1	1	1	1	1	1	1
Waste, Steam	1	1	1	1	1	1	1	1	1	1	1	1	1	1	1	1	1	1	1	1	1	1	1	1
Waste, Steam	1	1	1	1	1	1	1	1	1	1	1	1	1	1	1	1	1	1	1	1	1	1	1	1
Waste, Steam	1	1	1	1	1	1	1	1	1	1	1	1	1	1	1	1	1	1	1	1	1	1	1	1
Waste, Steam	1	1	1	1	1	1	1	1	1	1	1	1	1	1	1	1	1	1	1	1	1	1	1	1
Waste, Steam	1	1	1	1	1	1	1	1	1	1	1	1	1	1	1	1	1	1	1	1	1	1	1	1

turbine gas plant and a steam turbine oil-fired plant were switched on after two time steps (i.e., after non-peak load hours).

In all three cases, the hard coal power plants provided the most flexibility. We note that the presented models only consider price and not start-up times, which may discount the start-up and shutdown of coal power plants in more realistic scenarios,

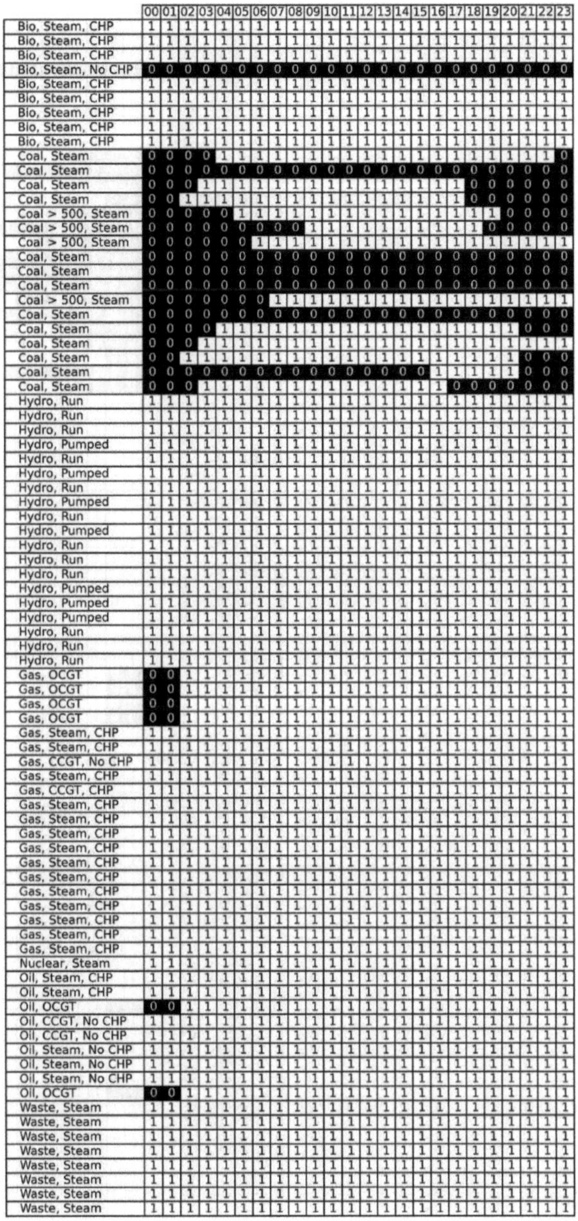

Fig. 3 Heatmap for the stochastic program

as well as during recourse actions after the demand is realized. These plants have high running costs due to carbon costs from CO_2 emissions that are considered in the LCOE, and therefore the model seeks to reduce the amount of periods for which they are active.

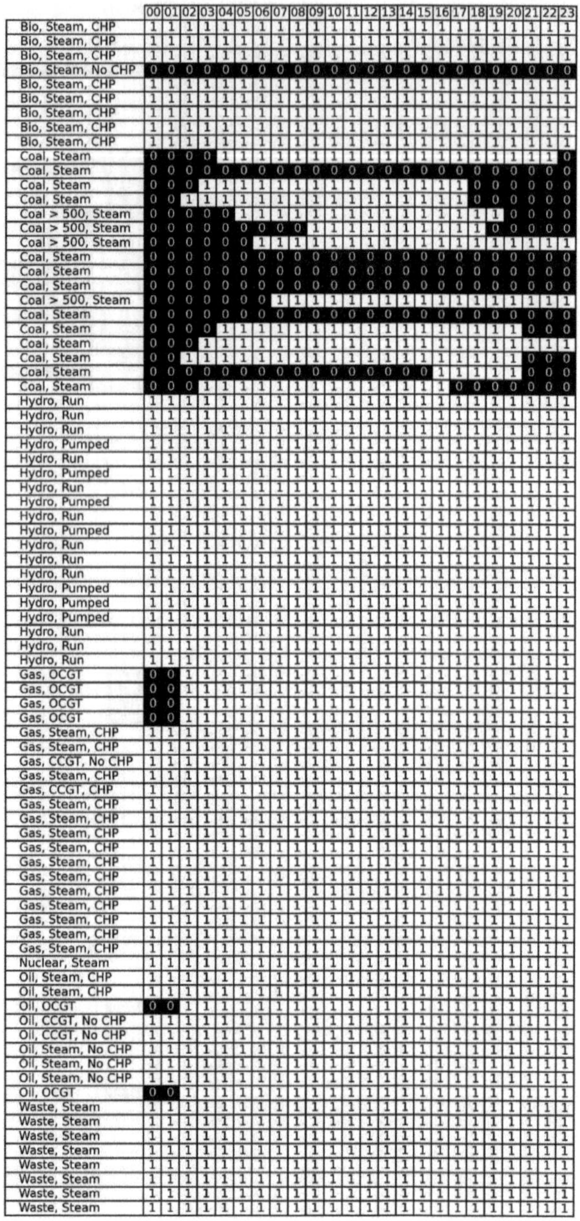

Fig. 4 Heatmap for the maximum demand

4.2 *Objective*

Regarding the model objectives, the stochastic model performed the best in terms of costs as it allowed more generators to be switched off—its total cost was

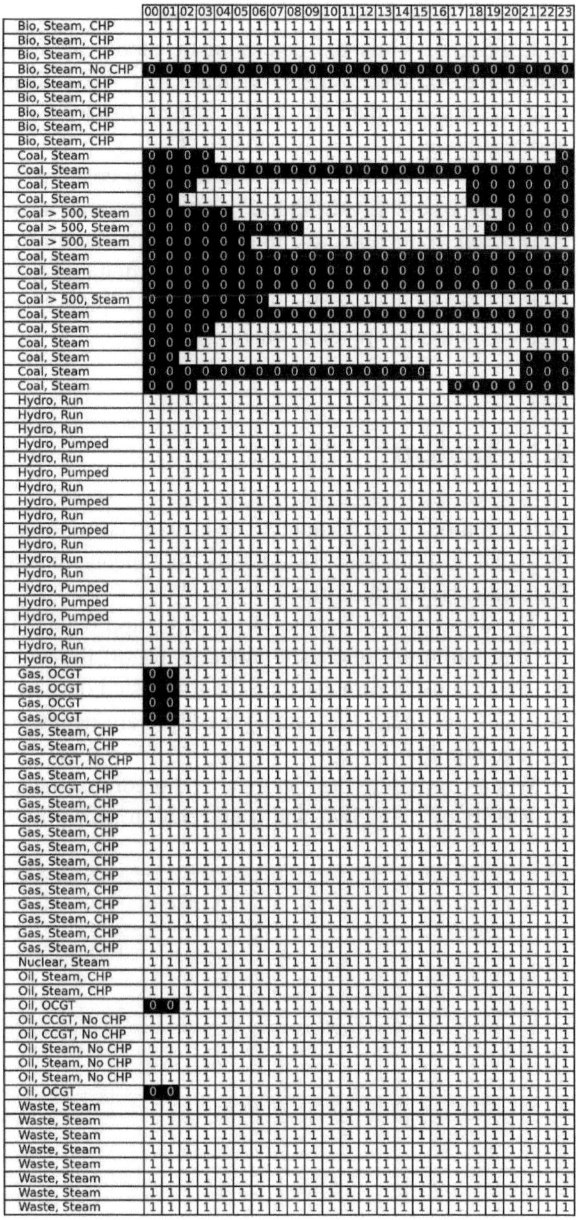

Fig. 5 Heatmap for the average demand

15,904,102€. We further experimented with different settings, taking the maximum and average of the demand in the generated scenarios. The total costs were, respectively, 18,214,273€ and 15,899,731€, suggesting that safeguarding against the worst-case scenario can be costly.

The deterministic model with predicted data had an objective value of 16,821,389€, while the deterministic model with historical data had the highest value—18,505,832€. This suggests that the forecasts and generated scenarios tend to underestimate the loads; in fact, the load is 1.78 times more likely to be underestimated than overestimated, leading to 2.4 times more underestimation of MW values. This would also lead to larger costs from corrective action in recourse decisions where higher-than-expected demand is realized; however, this is not considered in the objective function calculation for the stochastic model.

Nevertheless, multiplying the average demand by the average LCOE provides a realistic expectation of the overall price—17,468,256€. This suggests that the three predicted models are somewhat realistic.

4.3 Analysis

For each model, we computed the average commitment profiles (i.e., load profiles), the mean absolute error (MAE), the root mean squared error (RMSE), and pairwise Pearson correlation scores.

The average load profile of the three approaches is presented in Fig. 6. We can see that while the stochastic model is far off the historical data initially, it follows the trend more accurately than the forecasted data as time progresses. We note that this does not mean the same generators are necessarily employed, only the same average number.

When comparing the historical and predicted data, we see an MAE of 0.045181, an RMSE of 0.212558, and a correlation score of 0.808853. On the other hand,

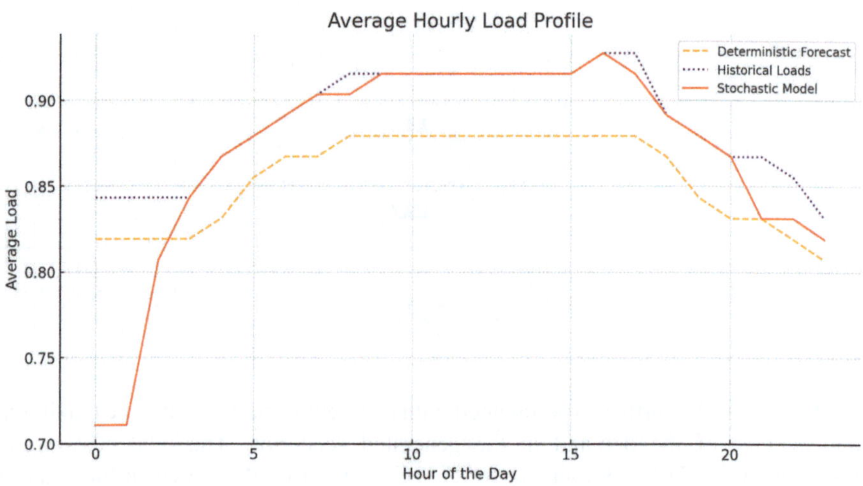

Fig. 6 Average load per hour of the day

comparing the stochastic data with the historical data yields an MAE of 0.089859, an RMSE of 0.299766, and a correlation score of 0.580297. These results suggest that the forecasted model is more accurate than the stochastic model at predictions.

5 Conclusions

This paper highlights the significance of incorporating uncertainty into SCUC models to effectively manage the complexities of modern power systems, especially in real-life applications. Specifically, we implemented a scenario-based version of the SCUC problem using realistic data from power plants based in Baden-Württemberg, Germany.

While this stochastic approach yielded good results on the average load profiles, the statistical metrics for comparison with historical data still favoured a deterministic forecast approach. Nevertheless, the Monte Carlo simulation approach was effective for scenario generation, offering a robust probabilistic framework for modelling uncertainties, leading to realistic results.

Future work should address the limiting assumptions in the model and data, such as including transmission line requirements within the security constraint. Further advancements can consider implementing state-of-the-art scenario-generation approaches for time-series data, as well as a full two-stage stochastic programming approach with recourse actions.

References

1. Abdou, I., Tkiouat, M.: Unit commitment problem in electrical power system: A literature review. Int. J. Electr. Comput. Eng. (2088-8708) **8**(3), 1357–1372 (2018)
2. Arroyo, J.M., Conejo, A.J.: Optimal response of a thermal unit to an electricity spot market. IEEE Trans. Power Syst. **15**(3), 1098–1104 (2000)
3. Benders, J.F.: Partitioning procedures for solving mixed-variables programming problems. Numer. Math. **4**, 238–252 (1962)
4. Bouffard, F., Galiana, F.D.: Stochastic security for operations planning with significant wind power generation. In: 2008 IEEE Power and Energy Society General Meeting-Conversion and Delivery of Electrical Energy in the 21st Century, pp. 1–11. IEEE, Piscataway (2008)
5. Codato, G., Fischetti, M.: Combinatorial Benders' cuts for mixed-integer linear programming. Oper. Res. **54**(4), 756–766 (2006)
6. Eggen, A.O., Belsnes, M.: Operation related maintenance and reinvestment costs for hydropower scheduling. Energy Syst. 1–21 (2023). https://doi.org/10.1007/s12667-023-00589-w
7. ENTSO-E Transparency Platform. https://transparency.entsoe.eu/. Accessed 05 Jul 2024
8. Frank, S., Rebennack, S.: An introduction to optimal power flow: Theory, formulation, and examples. IIE Trans. **48**(12), 1172–1197 (2016)
9. Fu, Y., Shahidehpour, M., Li, Z.: Security-constrained unit commitment with AC constraints. IEEE Trans. Power Syst. **20**(2), 1001–1013 (2005)

10. Fu, Y., Shahidehpour, M., Li, Z.: AC contingency dispatch based on security-constrained unit commitment. IEEE Trans. Power Syst. **21**(2), 897–908 (2006)
11. Füllner, C., Rebennack, S.: Stochastic dual dynamic programming and its variants. Preprint. Karlsruhe Institute of Technology (2021)
12. Füllner, C., Rebennack, S.: Non-convex nested Benders decomposition. Math. Program. **196**(1), 987–1024 (2022)
13. Gambardella, C.: On the economics of real-time pricing in low-carbon electricity markets. PhD thesis (2020)
14. Geoffrion, A.M.: Generalized Benders decomposition. J. Optim. Theory Appl. **10**, 237–260 (1972)
15. Huang, B.-N., Hwang, M.J., Yang, C.W.: Causal relationship between energy consumption and GDP growth revisited: a dynamic panel data approach. Ecol. Econ. **67**(1), 41–54 (2008)
16. Huang, Y., Pardalos, P.M., Zheng, Q.P.: Electrical Power Unit Commitment: Deterministic and Two-Stage Stochastic Programming Models and Algorithms. Springer, Berlin (2017)
17. Jahresabschluss MVV Energie AG. https://www.mvv.de/fileadmin/user_upload/Investoren/de/geschaeftsjahr_2021/Hauptversammlung/Jahresabschluss_MVV_Energie_AG_GJ_2021.pdf. Accessed 05 Jul 2024
18. Kost, C., Mayer, J.N., Thomsen, J., Hartmann, N., Senkpiel, C., Philipps, S., Nold, S., Lude, S., Saad, N., Schlegl, T.: Levelized Cost of Electricity-Renewable Energy Technologies. Fraunhofer ISE: Study (2013)
19. Kraftwerksliste. https://www.bundesnetzagentur.de/DE/Fachthemen/ElektrizitaetundGas/Versorgungssicherheit/Erzeugungskapazitaeten/Kraftwerksliste/start.html. Accessed 05 Jul 2024
20. Montero, L., Bello, A., Reneses, J.: A review on the unit commitment problem: Approaches, techniques, and resolution methods. Energies **15**(4), 1296 (2022)
21. Nasrolahpour, E., Ghasemi, H.: A stochastic security constrained unit commitment model for reconfigurable networks with high wind power penetration. Electr. Power Syst. Res. **121**, 341–350 (2015)
22. Nuclear costs in context. https://www.nei.org/resources/reports-briefs/nuclear-costs-in-context. Accessed 05 Jul 2024
23. Open Power System Data. https://open-power-system-data.org/. Accessed 05 Jul 2024
24. Padhy, N.P.: Unit commitment-a bibliographical survey. IEEE Trans. Power Syst. **19**(2), 1196–1205 (2004)
25. Projected costs of generating electricity 2020. https://www.iea.org/reports/projected-costs-of-generating-electricity-2020. Accessed 05 Jul 2024
26. Rachunok, B., Staid, A., Watson, J.-P., Woodruff, D.L., Yang, D.: Stochastic unit commitment performance considering Monte Carlo wind power scenarios. In: 2018 IEEE International Conference on Probabilistic Methods Applied to Power Systems (PMAPS), pp. 1–6. IEEE, Piscataway (2018)
27. Rebennack, S., Pardalos, P.M., VF, M., Iliadis, N.A.: Handbook of Power Systems I. Springer, Berlin (2010)
28. Römisch, W., Vigerske, S.: Recent progress in two-stage mixed-integer stochastic programming with applications to power production planning. Handbook of Power Systems I, pp. 177–208 (2010)
29. Ruzic, S., Rajakovic, N.: A new approach for solving extended unit commitment problem. IEEE Trans. Power Syst. **6**(1), 269–277 (1991)
30. Shahidehpour, M., Yamin, H., Li, Z.: Market Operations in Electric Power Systems: Forecasting, Scheduling, and Risk Management. John Wiley & Sons, Hoboken (2002)
31. Shahidehpour, M., Fu, Y., Wiedman, T.: Impact of natural gas infrastructure on electric power systems. Proc. IEEE **93**(5), 1042–1056 (2005)
32. Sinske, M., Rebennack, S.: Benders Decomposition, pp. 1–5. Springer International Publishing, Cham (2020)

33. Statistische Ämter des Bundes und der Länder | Gemeinsames Statistikportal. https://www.statistikportal.de/de/ugrdl/ergebnisse/wirtschaft-und-bevoelkerung/bipbws. Accessed 05 Jul 2024
34. Timilsina, G.R.: Demystifying the Costs of Electricity Generation Technologies. The World Bank, Washington (2020)
35. Tuohy, A., Meibom, P., Denny, E., O'Malley, M.: Unit commitment for systems with significant wind penetration. IEEE Trans. Power Syst. **24**(2), 592–601 (2009)
36. Van Dinter, J., Rebennack, S., Kallrath, J., Denholm, P., Newman, A.: The unit commitment model with concave emissions costs: a hybrid Benders decomposition with nonconvex master problems. Ann. Oper. Res. **210**, 361–386 (2013)
37. Wagner, H.-J., Mathur, J., Wagner, H.-J., Mathur, J.: Hydraulic turbines: types and operational aspects. Intro. Hydro Energy Syst. Basics Technol. Oper. 71–93 (2011)
38. Wang, J., Shahidehpour, M., Li, Z.: Security-constrained unit commitment with volatile wind power generation. IEEE Trans. Power Syst. **23**(3), 1319–1327 (2008)
39. Williams, H.P.: Model Building in Mathematical Programming. John Wiley & Sons, Hoboken (2013)
40. Wu, L., Shahidehpour, M., Li, T.: Stochastic security-constrained unit commitment. IEEE Trans. Power Syst. **22**(2), 800–811 (2007)
41. Zou, J., Ahmed, S., Sun, X.A.: Stochastic dual dynamic integer programming. Math. Program. **175**, 461–502 (2019)

New Centrality Indices in Network Analysis

Fuad Aleskerov and Daniil Tkachev

1 Introduction

Network analysis can be used to model banking operations, transport flows between cities, trade relations among countries, etc. [1–5]. Vertices in networks correspond to the objects; edges show information about connections.

Network analysis methods are used to find solutions to various problems of interaction among vertices in the network. Such tasks include searching for communities in networks (see [6, 7]), determining network capacity [8, 9], finding the shortest paths between vertices in the network [10, 11], optimization problems [12], etc. One of the main tasks of network analysis is identifying the most important vertices in the network under consideration. The importance of vertices depends on topology of the network, weights of edges, parameters of vertices, and group influence of vertices to a vertex.

The measure of vertices importance in the network is determined by centrality indices [13–16]. There are a huge number of different centrality indices. We present below some popular centrality indices.

The paper is organized as follows. In Sect. 3, Bundle and Pivotal centrality indices are considered. In Sect. 4, properties of the Bundle and Pivotal indices are discussed. In Sects. 5 and 6, minimal Bundle and weighted Bundle indices are introduced. Indirect influence between nodes is discussed in Sect. 7. A comparison of the results of the Bundle, Pivotal, and minimal Bundle centrality calculation is

---

F. Aleskerov
HSE University, Moscow, Russia

Institute of Control Sciences of Russian Academy of Sciences, Moscow, Russia

D. Tkachev (✉)
HSE University, Moscow, Russia

B. Goldengorin (ed.), *Theory, Algorithms, and Experiments in Applied Optimization*, Springer Optimization and Its Applications 226,
https://doi.org/10.1007/978-3-031-91357-0_2

considered in Sect. 8. Testing of the new centrality indices on rare earth compound trade model is presented in Sect. 9.

2 Classic Centrality Indices

2.1 Degree Centrality

One of the most popular centrality indices is Degree centrality [15]. Depending on the type of considered network degree centrality, there are different versions of degree centrality, weighted degree centrality, in/out-degree centrality, and weighted in/out-degree centrality.

Degree centrality is determined for undirected unweighted graph G with the set of vertices V and the set of edges E. The degree centrality for each vertex i is defined as

$$\text{Degree}(i) = |A_i|,$$

where A_i is the set of adjacent vertices to i.

Weighted degree centrality is determined for undirected weighted graph G with the set of vertices V and the set of edges W with weights w_{ij}. The degree centrality for each vertex i is defined as

$$\text{Degree}^w(i) = \sum_{j \in A_i} w_{ij},$$

where w_{ij} is the weight of the edge between vertices i and j.

In/out-degree centrality are determined for directed unweighted graph G with the set of vertices V and the set of edges E.

In-degree centrality for each vertex i is defined as

$$\text{In} - \text{degree}(i) = \left|\overleftarrow{N_i}\right|,$$

where $\overleftarrow{N_i}$ is the set of vertices j such that there exists the edge (j, i).

Out-degree centrality for each vertex i is defined as

$$\text{Out} - \text{degree}(i) = \left|\overrightarrow{N_i}\right|,$$

where $\overrightarrow{N_i}$—set of vertices j such that there exists the edge (i,j).

Weighted in/out-degree centrality are determined for directed weighted graph G with the set of vertices V and the set of edges W with weights w_{ij}.

Weighted in-degree and out-degree centrality for each vertex i are defined as

$$\text{In} - \text{degree}^{w}(i) = \sum\nolimits_{j \in \overleftarrow{N}_i} w_{ij},$$

$$\text{Out} - \text{degree}^{w}(i) = \sum\nolimits_{j \in \overrightarrow{N}_i} w_{ji}.$$

2.2 Closeness Centrality

Closeness centrality takes into account the distance between considered vertex i and other vertices [18]. The less the distance between considered vertex i and other vertices the higher the centrality value for vertex i.

Basically, closeness centrality is defined for undirected unweighted graph G with the set of vertices V and the set of edges E. However, it is modified for directed and weighted graph. Closeness centrality for each vertex i is defined as

$$\text{Closeness}(i) = \sum\nolimits_{j \in V} \frac{1}{d_{ij}},$$

where d_{ij} is the length of the shortest path between i and j.

2.3 Betweenness Centrality

Betweenness centrality takes into account the frequency of occurrence of the vertex i on the shortest paths between any vertices j, k [17]. The higher the frequency of occurrence of the vertex i the higher the centrality value of i.

Basically, betweenness centrality is defined for undirected unweighted graph G with the set of vertices V and the set of edges E. However, it is modified for directed and weighted graph. Betweenness centrality for each vertex i is defined as

$$\text{Betweenness}(i) = \sum\nolimits_{j,k \in V} \frac{\sigma_{jk}(i)}{\sigma_{jk}},$$

where $\sigma_{jk}(i)$ is the number of shortest paths from j to k through i, σ_{jk} is the number of shortest paths from j to k.

Considered centrality indices are used in many works to find importance of vertices in the network. Due to their prevalence, these indices are called "classic centrality indices." However, these indices have some limitations and disadvantages. One of the main disadvantages of classic centrality indices is that they do not take into account parameters of vertices and possibility of group influence [14].

Bundle and Pivotal centrality indices are introduced by Aleskerov and Yakuba [14]. These centrality indices determine the importance of vertices in networks, taking into account group influence and vertices parameters. These Bundle and Pivotal indices do not require specific network transformations unlike SRIC and LRIC, introduced by Aleskerov et al. [19].

Bundle and Pivotal centrality indices have been used in network models for different problems [1, 2, 20, 21]. However, the Bundle and Pivotal indices are constructed in such a way that they do not take into account directly the weights of edges. It determines our task to introduce a new centrality index.

3 Bundle and Pivotal Centrality Indices

Directed weighted graph G is a triple (V, E, W), where V is the set of vertices, E is the set of edges $\{i, j\}$, $i, j \in V$, $|V| = n$, W is the function which each edge $\{i, j\}$ matches a real number w_{ij}, i.e., $W : E \to \mathbb{R}$. For example, $V = \{A, B, C,\}$, $W = \{w_{AC} = 100, w_{BC} = 50, w_{CD} = 200\}$.

In Fig. 1, the corresponding directed weighted graph is presented.

In Aleskerov and Yakuba [14], Bundle and Pivotal centrality indices were introduced.

3.1 *Bundle Index (BI)*

It is assumed that the maximum number of vertices which can simultaneously influence a vertex is k. For each vertex i, it is defined a critical group S_i,

$$S_i \subseteq V \backslash \{i\}, |S_i| \leq k, \forall j \in S_i, w_{ji} \neq 0, \sum_{j \in S_i} w_{ji} \geq q_i$$

In other words, critical group S_i is the set of vertices which can simultaneously influence i. The value q_i can have different meanings depending on the model under

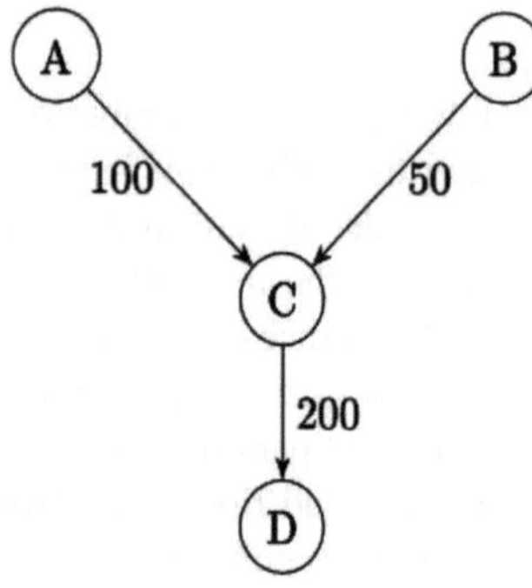

Fig. 1 Directed weighted graph

study and generally represents the parameter describing the vertex. The value k is introduced only for computational problems (often the value $k = 5$ for real data).

For each vertex i and the set of vertices S, it is defined the value $\mathrm{BI}_i(S)$

$$\mathrm{BI}_i(S) = \begin{cases} 1, \text{if } S \text{ is critical group for } i \\ 0, \text{otherwise} \end{cases}$$

The Bundle index BI(i) for each vertex i is defined as

$$\mathrm{BI}(i) = \sum\nolimits_{S \in 2^V \setminus \{\varnothing\}} \mathrm{BI}_i(S)$$

Let us consider the network in Fig. 1. Assume that the maximum number of vertices which can simultaneously influence a vertex is 2 and quota for each vertex is 100, i.e., $k = 2, q_i = 100$ for each i.

The vertex C has two critical groups—$\{A, B\}$ and $\{A\}$. The vertex D has one critical group—$\{C\}$. The values of Bundle index are calculated as

$$\mathrm{BI}(A)=0, \mathrm{BI}(B)=0, \mathrm{BI}(C)=\mathrm{BI_C}(\{A\})+\mathrm{BI_C}(\{A, B\})=2, \mathrm{BI}(D)=\mathrm{BI}_D(\{C\})=1.$$

3.2 *Pivotal Index (PI)*

The vertex j is called pivotal for a vertex i in the critical group S if

$$\sum\nolimits_{k \in S} w_{ki} \geq q_i, \ \sum\nolimits_{k \in S \setminus \{j\}} w_{ki} < q_i.$$

In other words, vertex j is pivotal for vertex i in the critical group S if the sum of edges' weights from the vertices in S without j is less than the quota.

Pivotal index PI(i) for each vertex i is defined as

$$\mathrm{PI}(i) = \sum\nolimits_{S \in 2^V \setminus \{\varnothing\}} |S| \cdot \mathrm{PI}_i(S),$$

where $\mathrm{PI}_i(S)$ is the number of pivotal vertices for i in the critical group S.

Consider the network in Fig. 1. The vertex A is pivotal for the critical group $\{A, B\}$ and the critical group $\{A\}$ for C. The vertex C is pivotal for the critical group $\{C\}$ for D. Thus, the values of Pivotal index PI(i) for vertices are

$$\mathrm{PI}(A) = 0, \mathrm{PI}(B) = 0,$$

$$\mathrm{PI}(C) = |\{A\}| \cdot \mathrm{PI}_C(\{A\}) + |\{A, B\}| \cdot \mathrm{PI}_C(\{A, B\}) = 3,$$

$$\mathrm{BI}(D) = |\{C\}| \cdot \mathrm{PI}_D(\{C\}) = 1.$$

4 Properties of the Bundle and Pivotal Indices

Consider some models where Bundle and Pivotal indices are used to identify most important vertices.

Case 1. Bank-Borrowers Network

In Aleskerov and Yakuba [14], centrality indices Bundle and Pivotal are used to identify banks' vulnerability to bankruptcy from borrowers in the bank-borrowers network. Consider an example of a bank-borrowers network.

Let us assume that the agent A borrows \$60 from bank C; the agent B borrows \$40 from bank C and \$100 from bank D. If borrowers do not return to C or to D \$100 in sum, bank C (or D) goes bankrupt. It means that banks C and D have equal quotas q, $q_C = q_D = 100$. It is presented in Fig. 2.

It is clear to see that for bank D if the debt from B is not received it is enough to go bankrupt. But bank C go bankrupt only if both agents (A and B) simultaneously do not return the debt. Thus, we can assume that bank D is more vulnerable to the bankruptcy from borrowers than bank C.

Nevertheless, the values of the centrality indices Bundle and Pivotal show the opposite case. The values of Bundle and Pivotal indices are the following

$$\mathrm{BI}(A) = \mathrm{BI}(B) = 0, \mathrm{BI}(C) = \mathrm{BI}(D) = 1,$$

$$\mathrm{PI}(A) = \mathrm{PI}(B) = 0, \mathrm{PI}(C) = 4, \mathrm{PI}(D) = 1.$$

Bank C and bank D are equally vulnerable to the bankruptcy from borrowers by the Bundle centrality index. Bank C is more vulnerable to the bankruptcy from borrowers than bank D by the Pivotal centrality index.

The reason of this is that the Bundle and Pivotal centrality indices linearly depend on the number of critical groups of considered vertex. Moreover, for all vertices i

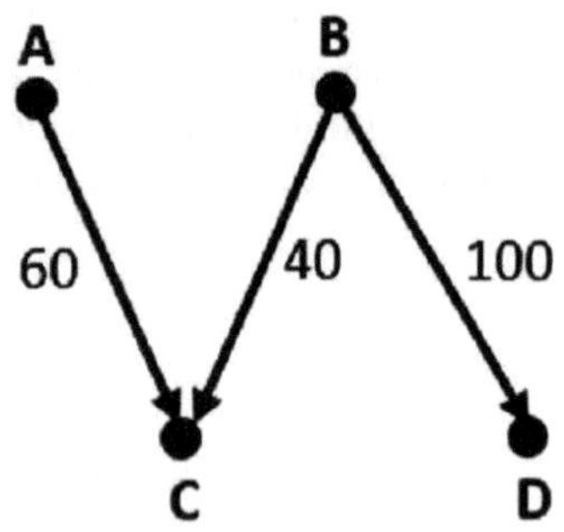

Fig. 2 Bank-borrowers network

and their critical groups S, the set $S \cup \{j\}$ is the critical group for i for all vertices j. To solve this, we introduce Minimal Bundle centrality index.

Case 2. Journals Citation Network
In Aleskerov et al. [2], the Bundle and Pivotal centrality indices are used to identify most cited economic journals of Russia in the economic journals of Russia citation network. Consider an example of journals citation network.

Let us assume that journal C has been cited by journal A (100 times), B (8 times) (see Fig. 3).

The values of quotas are equal to 10% of the citation sum as in Aleskerov et al. [2]. Thus, $q_A = q_B = 0$, $q_C = 10.8$, the values of the Bundle and Pivotal indices for journals are the following

$$\mathrm{BI}(A) = \mathrm{BI}(B) = 0, \mathrm{BI}(C) = 2,$$

$$\mathrm{PI}(A) = \mathrm{PI}(B) = 0, \mathrm{PI}(C) = 3.$$

Now assume that the citation of journal C increased by A and B to 950 and 50, respectively (see Fig. 4).

Then, $q_A = q_B = 0$, $q_C = 100$, the values of Bundle and Pivotal indices for journals are the following

$$\mathrm{BI}(A) = \mathrm{BI}(B) = 0, \mathrm{BI}(C) = 2,$$

$$\mathrm{PI}(A) = \mathrm{PI}(B) = 0, \mathrm{PI}(C) = 3.$$

Despite the fact that the number of citations of the journal has increased significantly, the values of the Bundle and Pivotal indices have not been changed.

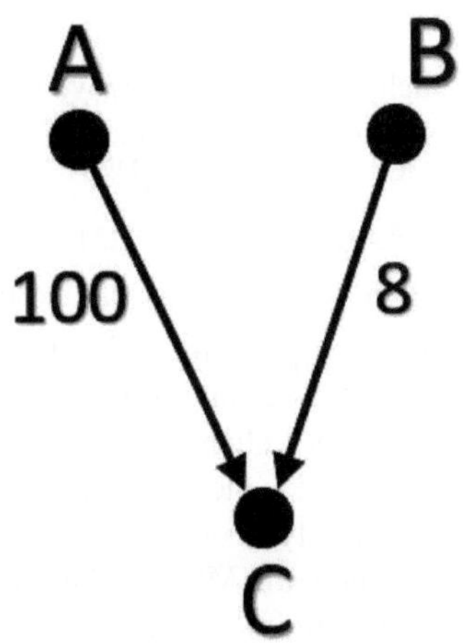

Fig. 3 Journal citation network-1

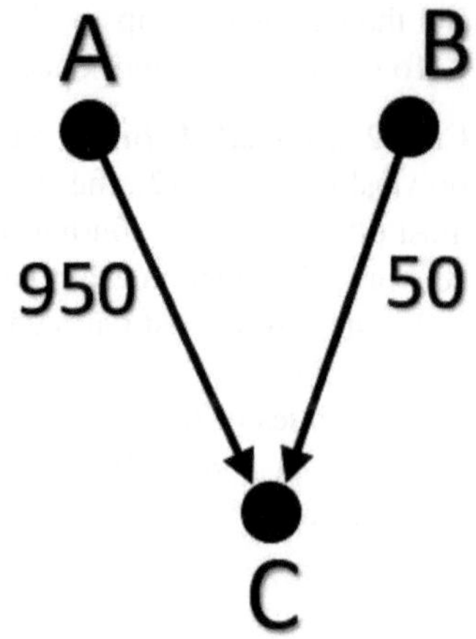

Fig. 4 Journal citation network-2

The reason of this is that the Bundle and Pivotal indices do not directly take into account weights of the edges. To solve these, we introduce new centrality indices.

5 Minimal Bundle Centrality Index mBI(i)

Consider the critical group S_i for i, and let all vertices in S_i be pivotal,

$$S_i \subseteq V \backslash \{i\}, |S_i| \leq k, \forall j \in S_i, w_{ji} \neq 0,$$

$$\forall l \in S_i \sum_{j \in S_i} w_{ji} \geq q_i, \sum_{j \in S_i \{l\}} w_{ji} < q_i.$$

Such critical groups S_i we call minimal critical groups and denote as S_i^{MC}. Let us assume that the vertex i in the considered network has d minimal critical groups $S_i^{\mathrm{MC}} - S_i^{\mathrm{MC}_1}, \ldots, S_i^{\mathrm{MC}_d}$.

Assume that $S_i^{\mathrm{MC}_d}$, $1 \leq r \leq d$, has m vertices—$j_1, \ldots, j_m$. Each vertex j_p, $1 \leq p \leq m$, in $S_i^{\mathrm{MC}_d}$ can be in two positions. The first position is that j_p reset the weight $w_{j_p i}$ by the value such that the sum of full reset of vertices from $S_i^{\mathrm{MC}_d}$ is higher than q_i taking into account other weights. The second position is that vertex j_p does not change the weight $w_{j_p i}$. There are 2^m variations of all cases with m vertices. It is clearly seen that the sum of full reset of vertices connection is higher than q_i only in one case, when all vertices in S_i^{MC} are in the first position.

Denote the MCG(i) the set of all minimal critical groups for the vertex i.

The minimal Bundle index mBI(i) for i is calculated as

$$\mathrm{mBI}(i) = \sum\nolimits_{S_i^{\mathrm{MC}} \in \mathrm{MCG}(i)} \frac{1}{2^{|S_i^{\mathrm{MC}}|}}.$$

The normalized minimal Bundle index for i NmBI(i) can be calculated as

$$\mathrm{NmBI}(i) = \frac{\mathrm{mBI}(i)}{\sum_{j \in V} \mathrm{mBI}(j)}.$$

For example, let us calculate the values of the minimal Bundle index for the network in Fig. 1. The values of minimal Bundle index for A and B are the following MBI(A) = MBI(B) = 0. The vertex C has one minimal critical groups—$\{A\}$. The vertex D has one critical group—$\{D\}$. The values of minimal Bundle index for C, D are calculated as

$$\mathrm{mBI}(C) = \mathrm{mBI}(D) = \frac{1}{2^1} = \frac{1}{2}.$$

Therefore, the vertices C and D are both equal by the minimal Bundle index.

6 The Weighted Centrality Indices $\mathbf{wBI^1}(i)$ and $\mathbf{wBI^2}(i)$

In this section, we introduce weighted Bundle centrality indices $\mathrm{wBI}^1(i)$ and $\mathrm{wBI}^2(i)$.

The Weighted Bundle Centrality Index $\mathbf{wBI^1}(i)$
For each vertex i and its critical group of vertices S_i, it is defined the value $\mathrm{wBI}_i^1(S)$

$$\mathrm{wBI}_i^1(S) = \begin{cases} \frac{\sum_{k \in S} w_{ki}}{\sum_{j \in V} w_{ji}}, \text{ if } S \text{ is critical group for } i \\ 0, otherwise. \end{cases}$$

where q_i is the value of quota for i.

The weighted Bundle index $\mathrm{wBI}^1(i)$ for each vertex i is defined as

$$\mathrm{wBI}^1(i) = \sum\nolimits_{S \in 2^V \setminus \{\varnothing\}} \mathrm{wBI}_i^1(S)$$

The Weighted Bundle Centrality Index $\mathbf{wBI^2}(i)$
Weighted Bundle centrality index $\mathrm{wBI}^2(i)$ calculation is similar to the $\mathrm{wBI}^1(i)$ calculation. The main difference between $\mathrm{wBI}^1(i)$ and $\mathrm{wBI}^2(i)$ is that the $\mathrm{wBI}^2(i)$ takes into account the weights of all edges in the network.

For each vertex i and its critical group of vertices S_i, it is defined the value $\mathrm{wBI}_i^2(S_i)$

$$\text{wBI}_i^2(S) = \begin{cases} \frac{\sum_{k\in S} w_{ki}}{\sum_{j,l\in V} w_{jl}}, \text{ if } S \text{ is critical group for } i \\ 0, otherwise \end{cases}$$

The weighted Bundle index $\text{wBI}^2(i)$ for each vertex i is defined as

$$\text{wBI}^2(i) = \sum\nolimits_{S\in 2^V\setminus\{\varnothing\}} \text{wBI}_i^2(S).$$

Example. Let us evaluate the values of the weighted Bundle indices $\text{wBI}^1(i)$ and $\text{wBI}^2(i)$ for the network in Fig. 1. The vertex C has two critical groups—$\{A, B\}$ and $\{A\}$. The vertex D has one critical group—$\{C\}$. The values of the weighted Bundle index $\text{wBI}^1(i)$ are

$$\text{wBI}^1(A) = \text{wBI}^1(B) = 0,$$

$$\text{wBI}^1(C) = \text{wBI}_C^1(\{A\}) + \text{wBI}_C^1(\{A, B\}) = \frac{100 + 150}{150} \approx 1.67,$$

$$\text{wBI}^1(D) = \text{wBI}_D^1(\{C\}) = \frac{200}{200} = 1,$$

Therefore, the vertex C is more important by the weighted Bundle index $\text{wBI}^1(i)$.

The values of the weighted Bundle index $\text{wBI}^2(i)$ are

$$\text{wBI}^2(A) = \text{wBI}^2(B) = 0,$$

$$\text{wBI}^2(C) = \text{wBI}_C^2(\{A\}) + \text{wBI}_C^2(\{A, B\}) = \frac{100 + 150}{350} \approx 0.71,$$

$$\text{wBI}^2(D) = \text{wBI}_D^2(\{C\}) = \frac{200}{350} = 0.57.$$

7 Indirect Influence Between Nodes

In Aleskerov and Yakuba [14], the indirect influence between nodes construction is considered in the following way.

To construct the indirect influence from j to i of the length d, the values are evaluated

$$P_{jk_{1,1}\ldots k_{1,d-1}i} = \min\left(w_{jk_{1,1}}^0, \ldots, w_{k_{1,d-1}i}^0\right),$$

$$P_{jk_{2,1}\ldots k_{2,d-1}i} = \min\left(w^0_{jk_{2,1}}, \ldots, w^0_{k_{2,d-1}i}\right),$$

$$\ldots$$

$$P_{jk_{t,1}\ldots k_{t,d-1}i} = \min\left(w^0_{jk_{t,1}}, \ldots, w^0_{k_{t,d-1}i}\right),$$

$$w^{d-1}_{ji} = \max\left(P_{jk_{1,1}\ldots k_{1,d-1}i}, P_{jk_{2,1}\ldots k_{2,d-1}i} \ldots, P_{jk_{2,1}\ldots k_{2,d-1}i}\right),$$

where $\min\left(w^0_{jk_{n,1}}, \ldots, w^0_{k_{n,d-1}i}\right)$—the minimum values of edges weights in the path of the length d form j to i, $\{k_{1,1}, \ldots, k_{1,d}\}$—the set of nodes in the path under consideration.

It is constructed the directed weighted graph $G^d(V, W^d)$, V—set of vertices, W^{d-1}—the set of edges $\{i, j\}, i, j \in V$ with weights w^d_{ji}. The graph G^d represents an indirect influence of the length d. The minimal Bundle and weighted Bundle centrality indices calculation in the G^d takes into account indirect influence of nodes.

Example. Let us evaluate the G^2, $d = 2$ for the graph in Fig. 1 and calculate minimal Bundle and weighted Bundle indices values (see Fig. 5).

$$P_{ACD} = \min(100, 200) = 100,\ P_{BCD} = \min(50, 200) = 50,$$

$$w^2_{AD} = 100,\ w^2_{BD} = 50.$$

The minimal Bundle and weighted Bundle centrality indices values for the graph in Fig. 5 are shown in Table 1.

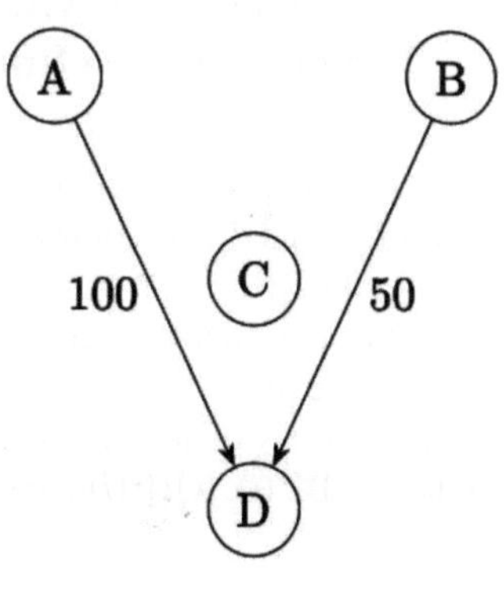

Fig. 5 Indirect influence of nodes

Table 1 Indirect influence

	A	B	C	D
minimal Bundle (indirect)	0	0	0	$\frac{1}{2}$
weighted Bundle $\mathrm{wBI}^1(i)$ (indirect)	0	0	0	$\frac{5}{3}$
weighted Bundle $\mathrm{wBI}^2(i)$ (indirect)	0	0	0	$\frac{5}{3}$

Table 2 Centrality indices comparison—1

	Figure 2				Figure 3		
	A	B	C	D	A	B	C
Minimal Bundle	0	0	$\frac{1}{4}$	$\frac{1}{2}$	0	0	$\frac{1}{2}$
Weighted Bundle $\mathrm{wBI}^1(i)$	0	0	1	1	0	0	$\frac{208}{108}$
Weighted Bundle $\mathrm{wBI}^2(i)$	0	0	$\frac{1}{2}$	$\frac{1}{2}$	0	0	$\frac{208}{108}$
Bundle	0	0	1	1	0	0	2
Pivotal	0	0	2	1	0	0	3

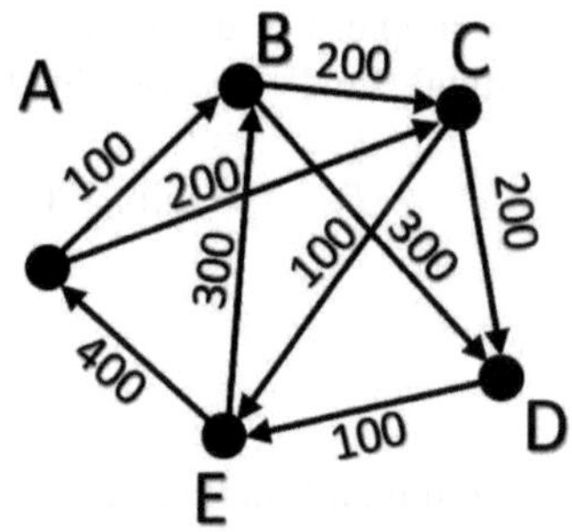

Fig. 6 Example of network

8 Comparing the Minimal and Weighted Bundle Centrality Indices with the Bundle and Pivotal Centrality Indices

Let us evaluate the values of the minimal and weighted Bundle indices for networks in Figs. 2 and 3 and compare values with the Bundle and Pivotal centrality values (see Table 2).

Let us consider the network in Fig. 6.

Assume that $k = 5$, values of quotas q_i are equal to 300.

The values of the Bundle, Pivotal, minimal Bundle, and the weighted Bundle indices $\mathrm{wBI}^1(i)$, $\mathrm{wBI}^2(i)$ are shown in Table 3.

Table 3 Centrality indices values

	A	*B*	*C*	*D*	*E*
Bundle	1	2	1	2	0
Pivotal	1	3	4	3	0
mBI	0.5	0.5	0.25	0.5	0
$wBI^1(i)$	1	1.75	1	1.6	0
$wBI^2(i)$	0.21	0.37	0.21	0.42	0

9 Rare Earth Compounds Trade Models

We construct the network of the rare earth compound trade and identify the most vulnerable countries to import using minimal Bundle index and weighted Bundle centrality indices and compare the results with the values of the Bundle and Pivotal centrality. Similar models of oil trade [21] and export/import operations for main crops can be found in Aleskerov et al. [1].

We consider data on the export/import of rare earth elements, countries' population, and countries' GDP per capita for 2005–2020. Data were downloaded from the World Bank website[1] and the UN Comtrade databases[2], which are maintained by the UN Statistics Division.

The network of the rare earth compounds trade model is presented as a weighted directed graph, where the vertices indicate countries and the direction and weight of the edge indicate the direction and the value of the export per capita of rare earth compounds from one country to another, expressed in US dollars.

In the rare earth compounds export-import model, the critical group will be the set of exporting countries the cessation of exports of which could lead to a shortage of rare earth compounds in the country. Quota q_i for country i is calculated according to the following principle.

First, the average rare earth compound imports M over the last 5 years are calculated. Then the coefficient K of change of real GDP compared to the previous year is calculated. If the coefficient K is <1 (falling GDP), the quota q_i is equal to the product of M and K. If the real GDP has risen ($K > 1$), the quota q_i remains equal to M.

For example, assuming that the values of country i import for 2017–2022 are 114, 126, 120, 130, 150 tons per capita, respectively. Then, the average value of import by previous 5 years is 128 tons. Values of GDP in country i for 2022 and 2023 are \$1580 and \$1450, respectively. Then, the coefficient of change of real GDP $K \approx 0.92 < 1$, $q_i = 0.92 \bullet 128 = 117.76$ tons per capita.

The import dependence of the rare earth compound countries is identified by minimal Bundle, weighted Bundle, and Bundle and Pivotal centrality indices. The

[1] UNComtrade Database [Electronic source].—URL: https://comtrade.un.org/data/ (Date of the application 04.11.2024)

[2] World Bank. World Bank Open Data [Electronic source]—URL: https://data.worldbank.org/indicator/NY.GDP.MKTP.CD (Date of the application 04.11.2024)

Table 4 TOP-10 most rare earth compound import depended countries in 2020

	Bundle (BI)	Pivotal (PI)	mBI	$wBI^1(i)$	$wBI^2(i)$
1	Japan	Japan	Japan	Japan	Japan
2	India	India	India	India	India
3	Switzerland	Switzerland	Germany	Germany	USA
4	Germany	Germany	Switzerland	Switzerland	Republic of Korea
5	USA	USA	USA	USA	Germany
6	Turkey	Turkey	Republic of Korea	Republic of Korea	Switzerland
7	Republic of Korea	Republic of Korea	Russia	Russia	Vietnam
8	Russia	Russia	Turkey	Turkey	Russia
9	Poland	Poland	Poland	Poland	Thailand
10	Spain	Spain	Spain	Spain	Taiwan

TOP-10 most rare earth compound import depended countries in 2020 are presented in Table 4.

10 Conclusion

We have considered shortages of the Bundle and Pivotal centrality indices use and introduce the minimal Bundle and the weighted Bundle centrality indices. New centrality indices have been tested for the model of rare earth compounds based on real data. For this case, the values of new centrality indices were compared with the values of the Bundle and Pivotal centrality indices.

Acknowledgments This article is an output of a research project "Study of the models and methods of decision-making under conditions of deep uncertainty. Extension and testing of developed methods" implemented as part of the Basic Research Program at the National Research University Higher School of Economics (HSE University).

References

1. Aleskerov, F.T., Dutta, S., Egorov, D., Tkachev, D.: Networks under deep uncertainty concerning food security. J. New Econ. Assoc. **64**(3), 12–29 (2024)
2. Aleskerov, F., Kazachinskaya, A., Karabekyan, D., Semina, A., Yakuba, V.: Economic journals of Russia, their characteristics and network analysis. J. New Econ. Assoc. Index. **2**, 170–182 (2021) (in Russian)
3. Kalyagin, V.A., Pardalos, P.M., Rassias, T.M. (eds.): Network Models in Economics and Finance. Springer Optimization and Its Applications, vol. 100, pp. 1–310 (2014)
4. Pardalos, P.M., Zamaraev, V.: The impact of social networks on sports. In: Social Networks and the Economics of Sports, pp. 1–8. Springer (2014)

5. Vogiatzis, C., Pardalos, P.M.: Evacuation modeling and betweenness centrality. In: Dynamics of Disasters—Key Concepts, Models, Algorithms, and Insights, vol. 185, pp. 345–359. Springer (2016)
6. Chwe, M.S.Y.: Communication and coordination in social networks. Rev. Econ. Stud. **67**(1), 1–16 (2000)
7. Roberts, S.G., Dunbar, R.I.: Communication in social networks: Effects of kinship, network size, and emotional closeness. Pers Relatsh. **18**(3), 439–452 (2011)
8. Persico, V., Marchetta, P., Botta, A., Pescapè, A.: Measuring network throughput in the cloud: The case of Amazon EC2. Comput. Netw. **93**, 408–422 (2015)
9. Wongpiromsarn, T., Uthaicharoenpong, T., Wang, Y., Frazzoli, E., Wang, D.: Distributed traffic signal control for maximum network throughput. In: Conference Record - IEEE Conference on Intelligent Transportation Systems, pp. 588–595. IEEE (2012)
10. Mitchell, J.S.: Geometric shortest paths and network optimization. In: Handbook of Computational Geometry, vol. 334, pp. 633–702. Elsevier Science (2000)
11. Xu, M.H., Liu, Y.Q., Huang, Q.L., Zhang, Y.X., Luan, G.F.: An improved Dijkstra's shortest path algorithm for sparse network. Appl. Math. Comput. **185**(1), 247–254 (2007)
12. Thai, M.T., Pardalos, P.M. (eds.): Handbook of Optimization in Complex Networks: Theory and Applications, vol. 57, pp. 1–546. Springer Science & Business Media (2012)
13. Aleskerov, F., Shvydun, S., Meshcheryakova, N.: New Centrality Measures in Networks: How to Take into Account the Parameters of the Nodes and Group Influence of Nodes to Nodes, pp. 1–114. CRC Press, USA (2021)
14. Aleskerov, F., Yakuba, V.: Matrix-Vector Approach to Construct Generalized Centrality Indices in Networks. Working Paper WP7 Series, pp. 1–21. Higher School of Economics Publ. House (2020)
15. Barabási, A.L.: Network science. Philos. Trans. R. Soc. A Math. Phys. Eng. Sci. **371**, 1–318 (2013)
16. Newman, M.E.: The structure and function of complex networks. SIAM Rev. **45**(2), 167–256 (2003)
17. Freeman, L.C.: A set of measures of centrality based on betweenness. Sociometry. **40**, 35–41 (1977)
18. Bavelas, A.: Communication patterns in task-oriented groups. J. Acoust. Soc. Am. **22**(6), 725–730 (1950)
19. Aleskerov, F., Meshcheryakova, N., Shvydun, S.: Centrality Measures in Networks Based on Nodes Attributes, Long-range Interactions and Group Influence. Working Paper WP7 Series, pp. 1–44. Higher School of Economics Publ. House (2016)
20. Aleskerov, F., Dutta, S., Egorov, D., Tkachev, D.: Networks under Deep Uncertainty. Procedia Comput. Sci. **214**, 1285–1292 (2022)
21. Aleskerov, F., Seregin, M., Tkachev, D.: The network analysis of oil trade under deep uncertainty. Procedia Comput. Sci. **221**, 1021–1028 (2023)

On Variants of PCGs: A Survey of Current Results and Open Problems

Tiziana Calamoneri, Angelo Monti, and Blerina Sinaimeri

1 Introduction

The pairwise compatibility graphs (PCGs) were originally introduced in the context of phylogenetics to address a problem related to the reconstruction of evolutionary trees [36]. These graphs have received considerable interest in the last years (see, for example, [14, 53] and the references therein) mainly due to the fact that PCGs provide a versatile mathematical tool to model diverse relationships, extending beyond biology to more general graph problems [44].

Interesting subclasses of PCGs have been explored by imposing stricter constraints either on the tree structure or on the interval conditions. Some well-known examples include *double-threshold* graphs (see, for example, [23, 34, 35, 38, 52, 60]) in the former case and *leaf powers* (see, for example, [50, 56], and [40]), introduced prior to PCGs, in the latter, both of which have attracted considerable interest.

Moreover, several generalizations of PCGs have been introduced recently. Among these, *multi-interval PCGs* [1, 20, 46, 51] extend the definition by associating a set of disjoint intervals, rather than a single interval, to determine edge inclusion. Specifically, an edge exists between two vertices if their distance in the tree falls within one of the specified intervals. Other extensions of the PCG class include *k-OR-PCGs* and *k-AND-PCGs* [20], which represent classes of graphs that can be expressed as the union and intersection, respectively, of k PCGs.

T. Calamoneri (✉) · A. Monti
Sapienza University of Rome, Rome, Italy
e-mail: calamoneri@di.uniroma1.it; monti@di.uniroma1.it

B. Sinaimeri
Luiss University, Rome, Italy
e-mail: bsinaimeri@luiss.it

B. Goldengorin (ed.), *Theory, Algorithms, and Experiments in Applied Optimization*, Springer Optimization and Its Applications 226,
https://doi.org/10.1007/978-3-031-91357-0_3

This paper aims to provide a comprehensive overview of the various PCG variants, reviewing the existing results and highlighting key open problems. We focus on the main challenges and suggest potential directions for future research.

In more details, in Sect. 2, we recall only a few fundamental properties of PCGs, referring the interesting reader to the surveys on PCGs [14, 53]. In Sect. 3, we explore several subclasses of PCGs formed by imposing constraints on the tree or interval structure. In Sect. 4, we discuss generalizations of PCGs, either by increasing the number of intervals or allowing multiple trees. Lastly, in Sect. 5, we focus on a specific restriction of these generalizations, namely, star-k-PCGs. In this paper, we have chosen not to provide formal definitions of the numerous referenced graph concepts to maintain the flow of the exposition. For detailed definitions, the reader is referred to [7].

2 Pairwise Compatibility Graphs

A graph is a PCG if there exists an edge-weighted tree such that each leaf of the tree is a vertex of the graph, and an edge $\{u, v\}$ is present in the graph if and only if the weight of the path connecting leaves u and v in the tree lies within a given interval. Throughout this paper, whenever we refer to an edge-weighted tree, we will always mean that the edge weights are positive. Given an edge-weighted tree T, calling $d_T(u, v)$ the sum of the weights of the edges on the unique path in T between u and v, we now provide the formal definition of a PCG.

Definition 1 ([36]) Given an edge-weighted tree T and an interval I of nonnegative real numbers, the *pairwise compatibility graph* (PCG) associated with T and I, denoted as $PCG(T, I)$, is a graph $G = (V, E)$ whose vertex set V coincides with $Leaves(T)$ and $e = \{u, v\}$, with $u \neq v$, belongs to E if and only if $d_T(u, v)$ lies within the interval I.

A graph G is a PCG if there exists the pair (T, I) such that $G = PCG(T, I)$.

The tree T is said to be *witness tree* of G. In Fig. 1a, an example of PCG is depicted; Fig. 1b, c show two witness trees with respect to interval $[5, 7]$.

Graphs That Are in PCG Many graph classes have been proven to belong to PCG, such as cliques, trees, forests, and their complement [33], disjoint union of cycles and single-chord cycles [62], cacti and triangle-free outerplanar 3-graphs [57], complete k-partite graphs [61], tree power graphs [62], as well as interval graphs [6], Dilworth k graphs [12], two-dimensional grids [29], and threshold graphs and some of their superclasses [15].

Note that the PCG class is not closed under complementing. Indeed, in Fig. 6a, a graph G that is not a PCG is shown; however, its complement, consisting of two disjoint cycles, each with $four$ vertices, is a PCG. Other studies [17, 61] have investigated the effects of several common graph operations on the closure properties of PCGs.

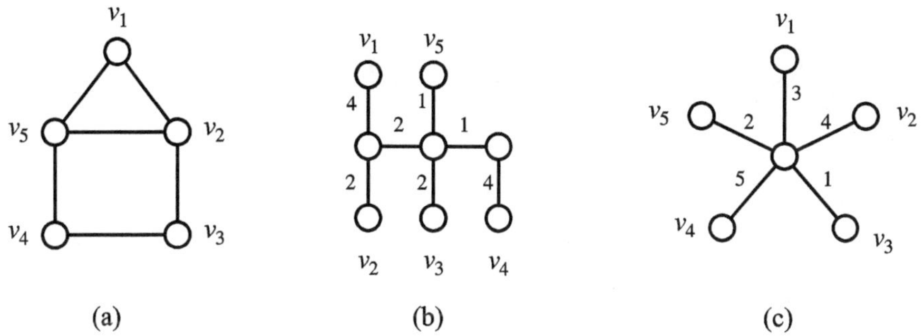

Fig. 1 (**a**) Example of a graph G that is a PCG, (**b**) and (**c**) two witness trees (a caterpillar and a star) for G, both with interval $[5, 7]$

We conclude this paragraph by highlighting that several attempts have been made to model the problem of determining whether a graph is a PCG [2, 3] or a PCG of a specific tree topology [18] using integer linear programming formulations. This approach has allowed to identify graphs with a small number of nodes as PCGs and discover small graphs that are not PCGs.

Graphs That Are Not in PCG When PCGs were introduced, it was conjectured that all graphs are PCGs [36]. This conjecture was disproven by Yanhaona et al. [62], who presented a graph with 15 vertices that is not a PCG. Later, Durocher et al. [25] proved the existence of a graph with *eight* vertices that is not a PCG, establishing it as the smallest known non-PCG, as all graphs with at most *seven* vertices are PCGs [16]. In fact, there are exactly 7 non-PCGs with 8 vertices [3], all depicted in Fig. 2, and 1494 minimal non-PCGs with 9 vertices, where a minimal non-PCG is a graph that is not a PCG, but each of its induced subgraph is [2]. Identifying the shared properties of these graphs could represent a progress toward a better understanding of PCGs. This naturally leads to the central open problem in the field:

Problem 1 Find a characterization of PCGs.

Trying to tackle this problem, since then, many other graphs have been proven to be non-PCGs. Mehnaz and Rahman [45] provided a subclass of bipartite graphs that are not PCGs; in [4], it is proved that wheels with at least *nine* vertices and squares of cycles of at least *eight* vertices are non-PCGs; disk graphs have been proven not to be in PCG in [53]. For what concerns planar graphs, the first example of a planar graph that is not a PCG was provided in [25] and consists of 16 vertices. However, the square of a cycle with eight vertices provides a smaller example of a planar non-PCG and is, in fact, the smallest known instance (see graph G_2 in Fig. 2).

Problem 2 Find subclasses of planar graphs or other well-studied graph families that are not included in the PCG class.

Lastly, if a graph H is not a PCG, then any graph containing H as an induced subgraph is not a PCG.

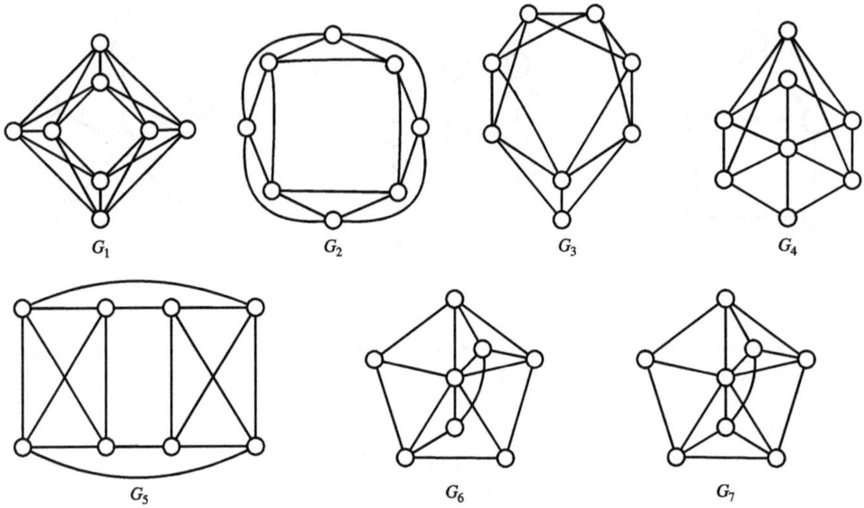

Fig. 2 The seven graphs with eight vertices that are not PCGs

Despite significant research, the following problem remains unsolved and is considered as the most critical open question in this field:

Problem 3 Determine the computational complexity of deciding whether a given graph G is PCG.

Working in this direction, a more general version of the problem has been proven NP-hard in [25]. Namely, it asks whether, given a graph $G = (V, E)$, a set $S \subseteq \bar{E}$ (where $\bar{E}$ is the set of non-edges of G), and an integer k, there exists a PCG $G' = PCG(T, I = [d_{min}, d_{max}])$ which contains G as a subgraph but does not contain any edge of S and ensures that at least k edges of S have distances greater than $d_{\max}$ between their corresponding leaves.

Additionally, Hossain et al. in [33] provided a necessary condition for a graph to be a PCG. Specifically, they showed that if a graph G contains two disjoint induced cycles (each either a chordless cycle with at least four vertices or any cycle with at least five vertices), then G is not a PCG. Furthermore, they show that given a graph G, if its complement does not contain a cycle, then G is a PCG.

They also identified four graph classes that lie between these conditions and gave examples of two graphs that belong to two of these classes but are not PCGs. Examples of graphs lying in the other two classes and not-PCGs are provided in [29].

3 Restrictions of PCGs

It is clear that, for a PCG $G = \text{PCG}(T, I)$, both the interval I and the tree T are crucial in defining G. Imposing constraints on either the interval I or on the witness tree T leads to important subclasses of PCGs. Specifically, the literature has explored limiting I to certain ranges, restricting T to particular structures such as stars or caterpillars, or applying conditions on the edge weights of T.

In the following subsections, we first examine the classes of LPGs, exact LPGs, and mLPGs that arise from natural constraints on the interval I. Next, we explore PCGs constructed using specific tree structures—namely, stars and caterpillars. Finally, we describe a simplified model where the tree is unweighted, meaning each edge has a weight of 1.

3.1 Restrictions on the Interval

Note that some natural restrictions that we can impose on the interval I are as follows: ignore the lower bound, ignore the upper bound, or consider I as containing only one element. In the following, we explore the subclasses of PCGs that arise from them.

3.1.1 LPGs

A graph is referred to as a *leaf power graph* if it is a pairwise compatibility graph for some edge-weighted tree, where only an upper bound is imposed on the distances between leaves. Formally:

Definition 2 ([50]) A graph $G = PCG(T, I)$ is a k-leaf power graph (k-LPG or, more simply, LPG) if $I = (0, k]$ for some positive integer k.

Thus, a graph $G = (V, E)$ is an LPG(T, I) if $e = \{u, v\}$, with $u \neq v$, belongs to E if and only if the distance $d_T(u, v) \leq k$. Notice that G is an induced subgraph of the graph power T^k induced by the leaves of T, which is the reason for the chosen name. Figure 3a shows an example of a 7-LPG, and its corresponding witness tree is depicted in Fig. 3c.

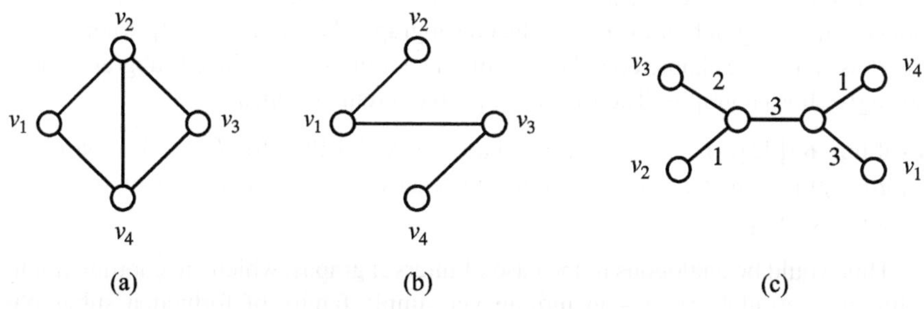

Fig. 3 (**a**) An LPG graph G_1, (**b**) an mLPG graph G_2, (**c**) a witness tree T showing $G_1 = LPG(T, (0, 7])$, and $G_2 = mLPG(T, [6, +\infty))$

This class of graphs was introduced before PCGs and has been widely studied in the literature. Interval graphs and the larger class of rooted directed path graphs are subclasses of LPGs [6, 10]. Moreover, LPGs are strongly chordal, meaning they are sun-free chordal graphs [27], but the converse does not hold. In fact, a graph is an LPG if and only if it is a fixed tolerance NeST graph [10, 31], and such graphs are a proper subclass of strongly chordal graphs [5]. For an excellent survey on LPGs, see [56].

With the goal of recognizing leaf power graphs, researchers have studied characterizations for specific fixed values of k. More specifically, 2-LPGs consist of collections of disjoint cliques, making them easy to recognize. The 3-LPGs are precisely the chordal graphs that are free of bulls, darts, and gems [8, 54] and can thus be recognized in linear time. For $k = 4$, a linear-time algorithm was presented in [9]. Chang and Ko [22] developed a linear-time recognition algorithm for 5-LPGs, and Ducoffe [24] demonstrated that 6-LPGs can be recognized in polynomial time. It has been proven that recognizing k-leaf powers is fixed-parameter tractable when parameterized by k and the degeneracy of the input graph [26]. Recently, Lafond [41] presented a polynomial-time algorithm for recognizing k-LPGs for any constant k that runs in $O(n^{f(k)})$ time, where f is a function that depends only on k (but grows at the rate of a power tower function). Note that the high dependency on the parameter k makes this algorithm unsuitable for practical use.

Problem 4 Determine the computational complexity of the k-LPG recognition problem for not constant values of k.

In particular, the following problem remains open:

Problem 5 ([41]) Is recognizing k-LPG fixed-parameter tractable in k? That is, can it be done in time $O(f(k)n^c)$ for some function f and some constant c?

Finally, characterizations using forbidden induced subgraphs have been considered. These are important because they provide practical methods for recognizing graph classes by checking for a finite set of minimal forbidden subgraphs. Although polynomial-time algorithms exist for recognizing k-LPGs when k is fixed, they are often impractical; thus, finding such a characterization would greatly aid their recognition. For $k \leq 4$, it is known that there exists a finite set F_k of graphs such that the class of k-LPG is characterized as the set of strongly chordal graphs that do not contain any graph in F_k as an induced subgraph. According to [42], when $k \geq 5$, k-LPGs cannot be characterized by excluding a finite set of induced subgraphs from strongly chordal graphs. The following problem remains open:

Problem 6 ([42]) Is it possible to characterize k-LPGs for $k \geq 5$ as strongly chordal graphs that avoid an infinite but easily describable family of forbidden induced subgraphs?

This would be analogous to the case of interval graphs, which are chordal graphs without asteroidal triples—an infinite yet simple family of forbidden subgraphs [43].

3.1.2 mLPGs

Analogous to leaf power graphs, which impose only an upper bound on distances between leaves, a graph is called a *min leaf power graph* (mLPG) if it is a pairwise compatibility graph derived from an edge-weighted tree with only a lower bound on the distances between leaves. More formally:

Definition 3 ([17]) A graph $G = PCG(T, I)$ is a *min leaf power graph* (k-mLPG or, more simply, mLPG) if $I = [k, +\infty)$ for some positive integer k.

Thus, a graph $G = (V, E)$ is an mLPG(T, I) if $e = \{u, v\}$, with $u \neq v$, belongs to E if and only if the distance $d_T(u, v) \geq k$. Figure 3b shows an example of a 7-mLPG, and its corresponding witness tree is depicted in Fig. 1c.

It is worth mentioning that the mLPG class has been defined as the complement of the LPG class. Indeed, a graph G is in LPG if and only if its complement is in mLPG [17]. This is the reason why their name evokes the power of graphs, although these graphs are not of this type.

It is worth noting that the two classes, LPG and mLPG, are incomparable as neither is entirely contained within the other. Indeed, the class of split matchings belongs to mLPG\LPG, while the class of split antimatchings belongs to LPG\mLPG [14, 15].

LPG ∩ mLPG Particular attention in the literature has been given to the characterization of the intersection of the mLPG and LPG classes. It is clear that a self-complemented class that is included either in LPG or in mLPG is included also in LPG ∩ mLPG. For example, threshold graphs [17], split permutation graphs and more in general, Dilworth 2 graphs [13] are in LPG ∩ mLPG. Nevertheless, the characterization of this class of graphs remains open.

Problem 7 Give a necessary and sufficient condition for a graph to be in the class LPG ∩ mLPG.

3.1.3 Exact LPGs

In the literature, there has also been interest in PCGs where the interval I consists of a single integer value. More formally:

Definition 4 ([11]) A graph $G = PCG(T, I)$ is an *exact k-leaf power graph* (exact k-LPG) if $I = [k, k]$ for some positive integer k.

Thus, a graph $G = (V, E)$ is an exact k-LPG if $e = \{u, v\}$, with $u \neq v$, belongs to E if and only if the distance $d_T(u, v) = k$. In Fig. 4, an exact 6-LPG is shown. Note that exact LPGs are not a subset of LPGs; in fact, there exist graphs that belong to exact LPG but not to LPG.

Brandstädt et al. in [11] provided characterizations of exact 3-LPGs and exact 4-LPGs, which imply that these graphs can be recognized in linear time. Additionally, they established several properties of exact 5-LPGs.

Problem 8 ([11]) Characterize exact k-leaf powers for $k \geq 5$.

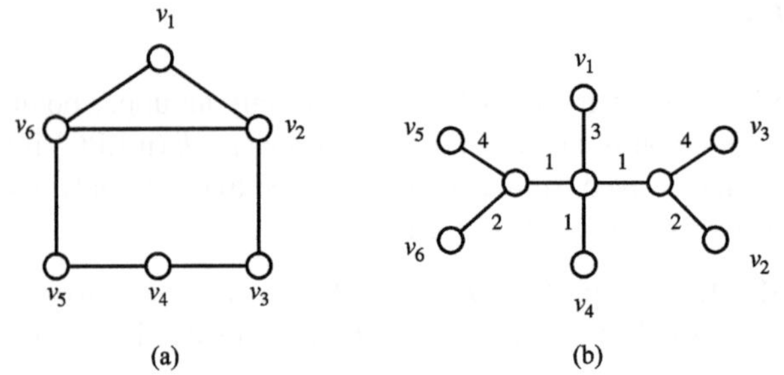

Fig. 4 (**a**) An exact 6-LPG, (**b**) the corresponding witness tree

3.2 Restrictions on the Edge-Weighted Tree

We can impose restrictions on the tree by specifying its topology—such as limiting it to stars or caterpillars—or by applying specific conditions to the edge weights. In particular, the former restriction, which focuses on the topology of the tree, has led to the study of graph classes that are intrinsically important.

3.2.1 PCGs of Stars

A *star* is a tree that consists of a single internal vertex (and $n - -1$ leaves). Stars are one of the most basic types of trees, and therefore it is natural to explore the case where the witness tree in the PCG is restricted to be a star. This leads to the notion of a *star-PCG* graph.

Definition 5 ([15]) A graph $G = PCG(T, I)$ is a *star-PCG* if the witness tree T is a star.

Figure 1a shows an example of a star-PCG, and its corresponding witness star is depicted in Fig. 1c.

Several classes of graphs are proven to be PCGs of stars, for instance, paths, caterpillars, square grids [49], and threshold graphs (that, more precisely, are in star-LPG ∩ star-mLPG) [13].

In [46], the smallest graph that is not a star-PCG is shown; it has exactly $five$ vertices, and there are only $four$ graphs with $five$ vertices that are not star-PCG (see Fig. 5).

Recognition and Characterization of Star-PCGs Although stars are a relatively simple topology, the recognition of star-PCGs is not an easy problem. It was only in [60] that Xiao and Nagamochi gave a complete characterization of the star-PCGs, showing that vertices of star-PCGs admit a special ordering. This characterization enabled them to develop an algorithm for recognizing star-PCGs with a time complexity of $O(n^3m)$, where n and m denote the number of vertices

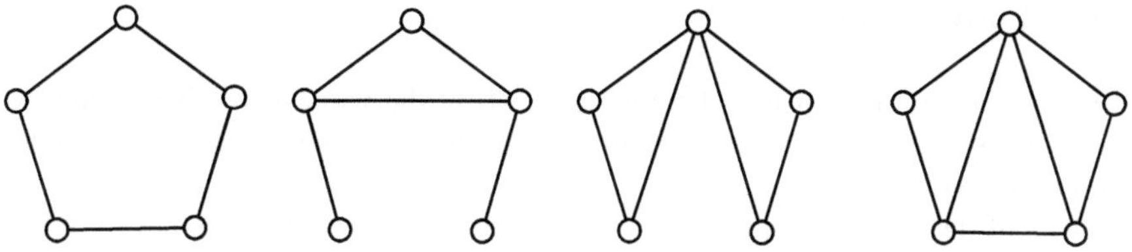

Fig. 5 The smallest graphs that are not star-PCGs

and edges in the graph, respectively. Next, Kobayashi et al. [38] introduced a new characterization of star-PCGs that relates them to bipartite permutation graphs; using it, they were able to improve the time complexity for recognizing star-PCGs proposing a linear-time recognition algorithm.

It is worth noting that star-PCGs are also known as *double-threshold graphs*. The following problem remains open:

Problem 9 ([38]) Find a characterization of star-PCGs through a family of forbidden induced subgraphs.

3.2.2 PCGs of Caterpillars

A caterpillar is a tree in which all the vertices are within distance 1 of a central path. The following is the definition of a caterpillar PCG.

Definition 6 ([15]) A graph $G = PCG(T, I)$ is a *caterpillar PCG* graph if the tree T is a caterpillar.

Figure 1a shows an example of a caterpillar PCG, and its corresponding caterpillar witness tree is depicted in Fig. 1b.

The simplicity of the structure is the reason why many researchers found caterpillars as witness trees of several graph classes, such as, for example, cycles, ladder graphs [57, 62], interval graphs [6], threshold signed graphs [13], threshold tolerance graphs [19], squared grids [29], circular graphs [62]. On the other hand, it can be proven that certain graphs, although PCGs, cannot have a caterpillar as a witness tree; examples are the wheels with a number of vertices included between 7 and 11 [18].

Problem 10 ([29]) Determine the computational complexity of deciding whether a given graph is a caterpillar PCG.

Problem 11 Find a characterization of caterpillar PCGs through a family of forbidden induced subgraphs.

3.2.3 Unweighted PCGs

It is quite natural to consider the constraint where the witness tree is unweighted, that is, the distance between two leaves is determined by the number of edges along the unique path connecting them. It is important to note that if we allow internal vertices of degree two in the witness tree, then this constraint does not introduce any additional restriction. Indeed, as shown in [17], for PCGs, it is always possible to restrict the interval and edge weights to nonnegative integers. Consequently, any witness tree can be converted into a tree with integer edge-weights, and further into an unweighted tree, by subdividing each edge of weight w into a path of $w - 1$ vertices. However, if we impose the condition that no internal vertices have degree two, the constraint of unweighted trees becomes meaningful. Moreover, this constraint is particularly relevant in phylogenetic trees, where internal vertices of degree two do not occur.

Brandstädt and Hundt characterize LPGs of unweighted caterpillars [6] and prove that the class of LPGs of unit weight caterpillars is equivalent to unit interval graphs. Calamoneri et al. [18] give some properties of the PCGs of caterpillars. The following problem remains open:

Problem 12 Find a characterization of PCGs of unweighted witness caterpillars.

4 Generalizations of PCGs

The definition of PCGs is based on a single tree and a single interval. However, many real-world evolutionary relationships are too complex to be fully captured by such a simple framework, and indeed, many graphs have been proven not to be PCGs. To address these limitations, three generalizations have been proposed: k-interval-PCGs extend PCGs by using a single tree with k intervals, while k-OR-PCGs and k-AND-PCGs generalize the framework by employing k trees, each with its own interval. They offer greater flexibility in modeling the complexities of evolutionary history, especially in contexts where evolutionary relationships are inferred from multiple data sources [28].

It is also worth noting that one might naturally consider an intermediate generalization using k trees and a single interval. However, as shown in [20], this approach does not yield any new classes distinct from k-OR-PCGs and k-AND-PCGs.

4.1 Multi-interval PCGs

The class of k-interval-PCGs was first introduced in [1], and they are often simply noted as k-PCGs.

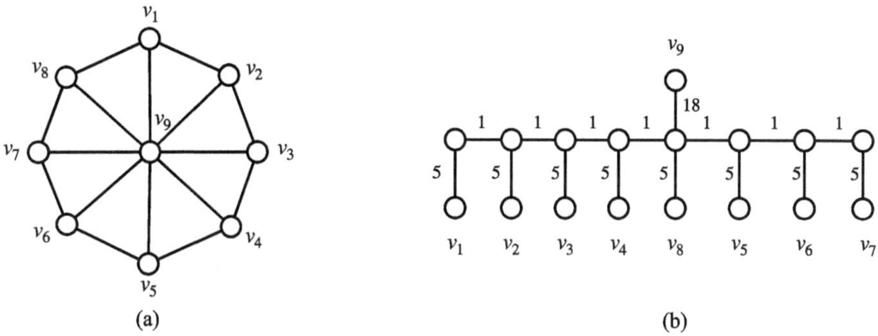

Fig. 6 (**a**) The wheel graph on nine vertices, which has been proven not to belong to the class of PCGs [4], (**b**) a witness tree, which, along with the two intervals $I_1 = [11, 11]$ and $I_2 = [19, +\infty)$, proves that the graph is a 2-interval-PCG [1]

Definition 7 ([1]) Given an edge-weighted tree T and $k \geq 1$ disjoint intervals $I_1, \ldots, I_k$ of nonnegative real numbers, the k-interval-PCG associated with T and $I_1, \ldots, I_k$, denoted as $k\text{-}PCG(T, I_1, \ldots, I_k)$, is a graph $G = (V, E)$ whose vertex set V coincides with $Leaves(T)$ and $e = \{u, v\}$ belongs to E if and only if $d_T(u, v) \in I_i$ for some i, $1 \leq i \leq k$.

A graph G is a k-interval-PCG if there exists a tuple $(T, I_1, \ldots, I_k)$ such that $G = k\text{-}PCG(T, I_1, \ldots, I_k)$.

In Fig. 6, we show a wheel as an example of 2-interval-PCG. Some classes of graphs, such as grids, have been proven to be in 2-interval-PCG [51].

Graphs That Are k-Interval-PCGs Every graph with m edges is an m-interval-PCG [1]. This result has been improved in [20] where it shown that every graph with m edges and n vertices is a k-interval-PCG with $k = \min\{m, \lceil \frac{n^2}{4} - \frac{n}{4} + \frac{1}{2} \rceil\}$. Therefore, we introduce the concept of the *multi-interval number* of a graph:

Definition 8 Given a graph G, we define *the multi-interval number*, $\rho(G)$, to be the smallest positive integer k, such that G is a k-interval-PCG.

Analogous to the PCG case, a fortiori a polynomial-time algorithm for recognizing k-interval-PCGs is not known. Hence, the following problem arises:

Problem 13 Given a graph G and an integer k, determine the exact complexity of deciding whether G is in k-interval-PCG. In particular, if k is part of the input, is it NP-hard to decide the membership to the class?

Graphs That Are Not k-Interval-PCGs In [1], the question of whether a constant number of intervals is sufficient for every graph to be a k-interval-PCG was left open. This question is later answered negatively in [20], where it is shown that for any $k \geq 2$ there exists a bipartite graph that requires at least k intervals and thus is not a $(k-1)$-interval-PCG. This means that there is no constant k for which the k-interval-PCG class contains all the graphs. Thus, it holds

$$PCG = 1\text{-interval-PCG} \subset 2\text{-interval-PCG} \subset \dots$$
$$\subset (k-1)\text{-interval-PCG} \subset k\text{-interval-PCG} \subset \dots.$$

Ahmed and Rahman [1] showed that wheel graphs with at least *nine* vertices, which are proven to be not PCGs, are instead 2-interval PCGs. Recently, in [21], it is shown that all graphs with at most eight vertices are 2-interval-PCGs, by providing a witness tree and two intervals for all the eight-vertex graphs that are not PCGs. One step ahead has been done in [30], where it is proved that also all graphs with at most *ten* vertices are 2-interval-PCGs. Actually, the only graph that is known not to be in 2-interval-PCG has 135 vertices [20]. This leads to the following natural question:

Problem 14 What is the smallest value of n such that there exists an n vertex graph G for which $\rho(G) > 2$?

If G is a k-interval-PCG, then its complement $\overline{G}$ is a $(k+1)$-interval-PCG [20]. It remains open the following problem:

Problem 15 Determine whether the class k-interval-PCG, with $k \geq 2$, is closed with respect to the complement.

The Multi-interval Number of Graph Classes For any graph G that is a PCG, it trivially holds that $\rho(G) = 1$. Let W_n be the wheel graph with n vertices. From the results in [1] and [4], it follows that $\rho(W_n) = 2$ for $n \geq 9$. Additionally, in [1], a restricted class of series-parallel graphs, known as SQQ series-parallel graphs, are proven to be in 2-interval-PCG, implying that $\rho(G) \leq 2$ for these graphs. The following problem remains open:

Problem 16 ([53]) Given a series-parallel graph G, determine the value $\rho(G)$.

Two-dimensional grids are PCGs [29]; thus, $\rho(G) = 1$ for these graphs. It is not known the value of ρ for grids of higher dimension, except for a subclass of three-dimensional grids [51]:

Problem 17 ([49]) Given a three-dimensional grid G, determine the value $\rho(G)$.

Problem 18 It would be interesting to explore the possible relation between the multi-interval number of a graph and some of its parameters, for example, the tree-width.

4.2 *OR-PCGs and AND-PCGs*

We now consider two generalizations of PCGs that permit the use of different trees, namely, OR-PCG s and AND-PCG s [20]. A graph is a k-OR-PCG (k-AND-PCG) if it can be represented as the union (respectively, intersection) of k PCGs. The concept of applying OR and AND operations to graph classes, as explored in [32], has been used to study the complexity of recognizing k-OR cographs, with the objective of characterizing orthology graphs derived from more than one phylogenetic tree.

4.2.1 OR-PCGs

We start providing the formal definition of k-OR-PCG.

Definition 9 ([20]) Let $T_1, \ldots, T_k$ be $k \geq 1$ edge-weighted trees, and let $Leaves(T_1) = \ldots = Leaves(T_k) = L$; let $I_1, \ldots, I_k$ be k (not necessarily disjoint) intervals of nonnegative real numbers; the k-OR-PCG associated with $T_1, \ldots, T_k$ and $I_1, \ldots, I_k$, denoted as k-OR-PCG$(T_1, \ldots, T_k, I_1, \ldots, I_k)$, is a graph $G = (V, E)$ whose vertex set V coincides with L and $e = \{u, v\}$ belongs to E if and only if there exists an i, $1 \leq i \leq k$, such that $d_{T_i}(u, v) \in I_i$.

A graph G is a k-OR-PCG if there exists a tuple $(T_1, \ldots, T_k, I_1, \ldots, I_k)$ such that $G =$ k-OR-PCG$(T_1, \ldots, T_k, I_1, \ldots, I_k)$.

In Fig. 7, we provide an example of a 2-OR-PCG. Any k-interval-PCG also belongs to k-OR-PCG, assuming the edge-weighted trees are identical. Therefore, for any k, k-interval-PCG $\subseteq$ k-OR-PCG. Notice that a k-OR-PCG G is not necessarily a k-interval-PCG, as G may not be represented as the union of k PCGs that use the same edge-weighted witness tree. Therefore, the following problem remains open:

Problem 19 For any $k \geq 2$, determine whether k-interval-PCG $\subsetneq$ k-OR-PCG.

The OR Number of Specific Graph Classes As already established in Sect. 4.1, every graph is a k-OR-PCG for some positive integer k. Thus, we introduce the following concept of *OR number* of a graph:

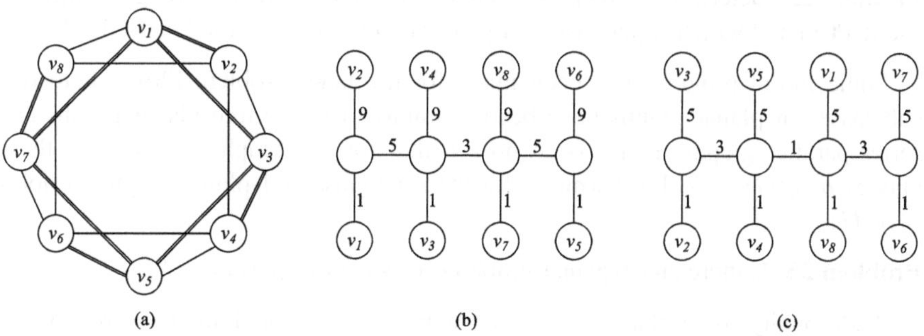

Fig. 7 **(a)** A graph G which is not a PCG [4]. The two graphs on the same set of vertices G_1 (induced by the double-lined edges) and G_2 (induced by the continuous-lined edges) are in PCG; we provide **(b)** a tree T_1 and an interval $I_1 = [6, 10]$ such that $G_1 =$ PCG(T_1, I_1) and **(c)** a tree T_2 and an interval $I_2 = [5, 6]$ such that $G_2 =$ PCG(T_2, I_2). It clearly holds that $G =$ 2-OR-PCG(T_1, T_2, I_1, I_2). (Figure taken from [20])

Definition 10 Given a graph G, we define its *OR number*, $\rho_{OR}(G)$, to be the smallest positive integer k, such that G is a k-OR-PCG.

Notice that for any graph G, it holds $1 \leq \rho_{OR}(G) \leq \rho(G)$.

In [20], using Ramsey theory, it is proved that for any integer $k > 0$, there exist bipartite graphs that are not k-OR-PCGs. Hence, the following open problem is of interest:

Problem 20 ([20]) For any integer $k > 2$, determine the smallest value of n such that there exists an n vertex graph G for which $\rho_{OR}(G) > k$.

Several results have been established regarding specific graph classes and their OR numbers. Clearly, we focus only on stating results for graph classes whose membership in PCG is not known because the OR number is 1 for PCGs.

An alternative definition of k-OR-PCG is that a graph $G = (V, E)$ is a k-OR-PCG$(T_1, \ldots, T_k, I_1, \ldots, I_k)$ if and only if there exist k graphs $G_1, \ldots, G_k$ on the same vertex set V such that for all i with $1 \leq i \leq k$ we have $G_i = PCG(T_i, I_i)$ and $E(G) = \bigcup_{i=1}^{k} E(G_i)$. Consequently, in [20], the extensive literature on graph edge coverings (e.g., see the survey paper [37]) has been exploited to obtain the following results for OR-PCGs. For every connected graph G with maximum degree 3, $\rho_{OR}(G) \leq 2$. Furthermore, for every regular graph with even degree Δ, it holds $\rho_{OR}(G) \leq \frac{\Delta}{2}$. For bipartite regular graphs with odd degree Δ, we have that $\rho_{OR}(G) \leq \lceil \frac{\Delta}{2} \rceil$. Additionally, $\rho_{OR}(G) \leq a$ for every graph G with arboricity a (we recall that a graph G has *arboricity a* if a is the minimum number of spanning forests needed to cover all the edges of G.)

For every planar graph G, it holds $\rho_{OR}(G) \leq 3$. Moreover, in [4, 25], it is shown that there exist planar graphs G for which $\rho_{OR}(G) = 2$. This leads to the following two problems:

Problem 21 ([20]) Given a planar graph G, determine $\rho_{OR}(G)$.

Problem 22 Determine which subclasses of planar graphs are contained in 2-OR-PCG and which superclasses of planar graphs are contained in 4-OR-PCG.

Some initial progress has been made for the latter problem [20], where two subclasses of planar graphs have been examined: triangle-free planar graphs and series-parallel graphs. It has been proven that for any graph G in one of these classes $\rho_{OR}(G) \leq 2$. Furthermore, for the superclass of 1-planar graphs, it holds $\rho_{OR}(G) \leq 4$.

Problem 23 Is there any 1-planar graph G for which $\rho_{OR}(G) = 4$?

Additionally, outerplanar graphs have been investigated in the same paper, showing that $\rho_{OR}(G) \leq 2$ for every outerplanar graph G.

Analogous to the previous classes, the following open problem holds:

Problem 24 Given a graph G and an integer k, determine the exact complexity of deciding whether G is in k-OR-PCG. In particular, if k is part of the input, is it NP-hard to decide membership in the class?

4.2.2 AND-PCGs

First, we give the formal definition of AND-PCG.

Definition 11 ([20]) Let $T_1, \ldots, T_k$ be $k \geq 1$ edge-weighted trees and $Leaves(T_1) = \ldots = Leaves(T_k) = L$; let $I_1, \ldots, I_k$ be k (not necessarily disjoint) intervals of nonnegative real numbers; the k-AND-PCG associated with $T_1, \ldots, T_k$ and $I_1, \ldots, I_k$, denoted as k-AND-PCG$(T_1, \ldots, T_k, I_1, \ldots, I_k)$, is a graph $G = (V, E)$ whose vertex set V coincides with L and $e = \{u, v\}$ belongs to E if and only if for all i, $1 \leq i \leq k$, $d_{T_i}(u, v) \in I_i$. A graph G is a k-AND-PCG if there exists a tuple $(T_1, \ldots, T_k, I_1, \ldots, I_k)$ such that $G = k$-AND-PCG$(T_1, \ldots, T_k, I_1, \ldots, I_k)$.

In Fig. 8, we provide an example of a graph which is a 3-AND-PCG.

Observe that PCG = 1-AND-PCG and that PCG $\subseteq$ 2-AND-PCG. The graph depicted in Fig. 8a shows that PCG $\subsetneq$ 3-AND-PCG. However, the following problem remains open:

Problem 25 Determine whether, for any $k \geq 1$, it holds

$$k\text{-AND-PCG} \subsetneq (k+1)\text{-AND-PCG}.$$

Graphs That Are Not k-AND-PCGs Let H be the graph with 24 vertices consisting of 3 disjoint copies of $K_{4,4}$. It was proved in [20] that its complement $\overline{H}$ is not a 2-AND-PCG.

Problem 26 What is the smallest value of n such that there exists a graph with n vertices that is not in 2-AND-PCG?

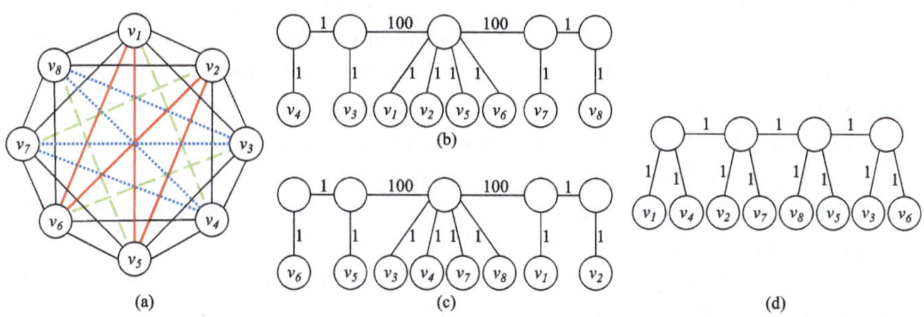

Fig. 8 **(a)** A graph G (considering only the black and continuous-lined edges) which is not a PCG [4] and three sets of edges: the set E_1 of blue and dotted edges, the set E_2 of red and double-lined edges, and the set E_3 of green and long dashed edges. The three graphs $G_1 = K_8 \setminus E_1$, $G_2 = K_8 \setminus E_2$, and $G_3 = K_8 \setminus E_3$ are in PCG; we provide the three intervals $I_1 = I_2 = [2, 103]$, $I_3 = [3, 5]$, and **(b)** a tree T_1 such that $G_1 = \text{PCG}(T_1, I_1)$, **(c)** a tree T_2 such that $G_2 = \text{PCG}(T_2, I_2)$, **(d)** a tree T_3 such that $G_3 = \text{PCG}(T_3, I_3)$. It clearly holds that $G =$ 3-AND-PCG$(T_1, T_2, T_3, I_1, I_2, I_3)$ (Figure taken from [20])

Moreover, it was shown in [20], using Ramsey theory, that for any integer $k > 0$ there exist bipartite graphs that are not k-AND-PCG. This naturally leads to the following general open problem:

Problem 27 ([20]) For any integer $k > 2$, what is the smallest value of n such that there exists an n vertex graph G for which $\rho_{AND}(G) > k$?

The AND Number of Specific Graph Classes In [20], it is shown that every graph is a k-AND-PCG and, thus, similar to OR-PCG, it is possible to define the AND number of a graph.

Definition 12 ([20]) Given a graph G, we define *the AND number*, $\rho_{AND}(G)$, to be the smallest positive integer k, such that G is a k-OR-PCG.

A graph G is called a *maximal non-PCG* if G is not PCG, but the addition of any edge produces a PCG. Note that such graphs do exist: for any graph G that is not a PCG, either G is already a maximal non-PCG, in which case we are done, or there is an edge that can be added while keeping the graph outside the class of PCGs. By iterating this process and observing that the complete graph is a PCG, we eventually arrive at a maximal non-PCG. In [20], it is proved that if G is a maximal non-PCG, then $\rho_{AND}(G) = 2$.

The *intersection dimension* of a graph G with respect to a class of graphs $\mathcal{C}$ is the minimum k such that G is the intersection of some k graphs belonging to $\mathcal{C}$ [39]. An equivalent definition of a k-AND-PCG is that a graph $G = (V, E)$ is an k-AND-PCG$(T_1, \ldots, T_k, I_1, \ldots, I_k)$ if and only if there exist k graphs $G_1, \ldots, G_k$ on the same vertex set V such that for all i, $1 \leq i \leq k$, we have $G_i = \text{PCG}(T_i, I_i)$ and $E(G) = \bigcap_{i=1}^{k} E(G_i)$. In other words, a graph G is a k-AND-PCG if and only if it can be represented as the intersection of k PCGs.

This alternative definition enables the exploration of the literature on the intersection dimension of graphs to establish their membership in the k-AND-PCG class. In particular, a graph G is a k-AND-PCG if and only if its intersection dimension with respect to the PCG class is at most k.

Among the different specializations of intersection dimension, the most well-known is the *boxicity* (i.e., the intersection dimension with respect to the class of interval graphs) [55]. Since interval graphs are known to be in PCG [6], it holds that a graph with boxicity b is a b-AND-PCG.

It is known that every planar graph has a boxicity of at most *three* [59] and every outerplanar graph has a boxicity of at most *two* [58]. Therefore, for every planar graph G, $2 \leq \rho_{AND}(G) \leq 3$, and for every outerplanar graph G, $1 \leq \rho_{AND}(G) \leq 3$.

Problem 28 ([20]) Determine the value of $\rho_{AND}(G)$ for any planar (outerplanar) graph G.

Related to the operation of graph complement, it is proved in [20] that if a graph G is a k-AND-PCG, its complement graph $\overline{G}$ is a $2k$-OR-PCG.

Moreover, while it is known that if G is a PCG, then its complement $\overline{G}$ is a 2-interval-PCG and therefore a 2-OR-PCG, the analogous related to the 2-AND-PCG class remains an open problem:

Problem 29 ([20]) If G is a PCG, determine whether $\overline{G}$ is a 2-AND-PCG.

The PCG class is not closed with respect to the complement, and it is not known whether the same holds for AND-PCG s:

Problem 30 ([20]) Determine whether the class AND-PCG is closed with respect to the complement.

Some results in this direction were presented in [20], where it was shown that if a graph G is a k-AND-PCG, its complement $\overline{G}$ is an $2k$-OR-PCG.

Problem 31 ([20]) If G is a k-OR-PCG, what can we say about the value of k' for which $\overline{G}$ is a k'-AND-PCG?

Analogous to the previous classes, the following problem remains open:

Problem 32 Given a graph G and an integer k, determine the exact complexity of deciding whether G is in k-AND-PCG. In particular, if k is part of the input, is it NP-hard to decide membership?

We conclude this section by highlighting that, as for PCGs, also in the context of 2-interval-PCGs the problem has been formulated as integer linear programming [21, 30]. While no such formulation currently exists for AND-PCGs and OR-PCGs, developing one might be useful in identifying small graphs that do not belong to these classes, thus contributing to their characterization.

5 Restrictions of Generalizations: Star-k-PCGs

It is natural to explore the case where the underlying tree in the multi-interval PCGs is restricted to a star. This leads to the notion of a *multi-interval-star-PCG* graph.

Definition 13 ([46]) A graph $G = k\text{-PCG}(T, I_1, \ldots, I_k)$ is a *star-k-PCG* if the tree T is a star.

In Fig. 9, an example of star 2-PCG is depicted.

Obviously, the class of star-1-PCGs coincides with star-PCG. In [46], using an ILP formulation, it is shown that the smallest graph that is not a 2-star-PCG has exactly *seven* vertices and that there are only *three* graphs with *seven* vertices that are not 2-star-PCG; see Fig. 10c.

In [38, 49], it is shown that the class of star-k-PCGs is equivalent to the class of $2k$-threshold graphs, a concept that has attracted significant interest in the research community since its introduction in [34], as demonstrated by the studies in [23, 34, 35, 52]. The connection between star-PCGs and double-threshold graphs has led

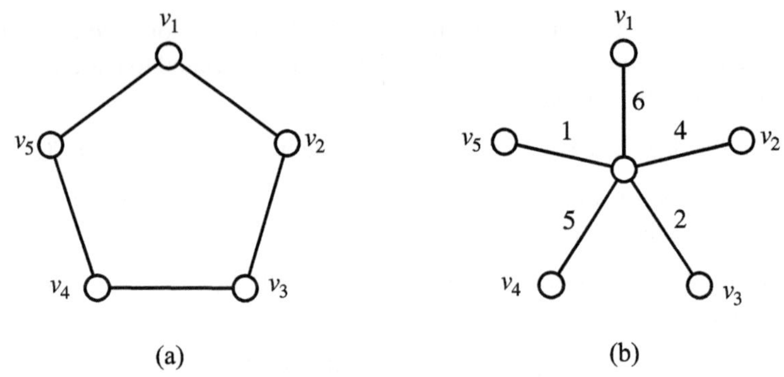

Fig. 9 (**a**) A graph G that is not star-1-PCG; (**b**) the witness star showing that G is star-2-PCG with $I_1 = [6, 7]$ and $I_2 = [10, 10]$

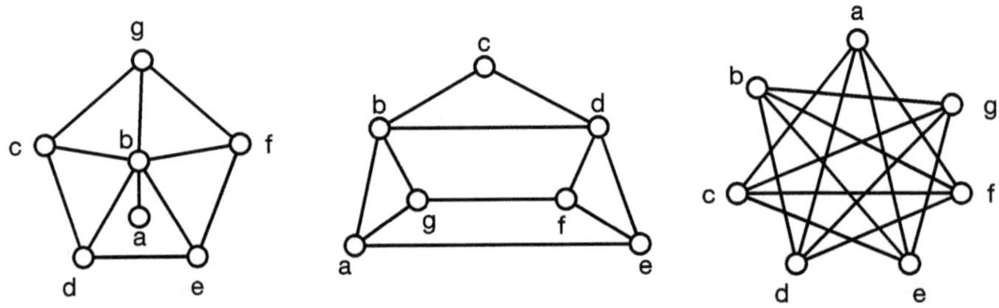

Fig. 10 The only three graphs on seven vertices that are not star-2-PCGs (Figure taken from [49])

to the development of a polynomial-time algorithm for identifying graphs that are star-1-PCGs [38, 60]. Nevertheless, the problem remains open for $k > 1$:

Problem 33 For any constant $k > 1$, determine whether a given graph is a star-k-PCG in polynomial time.

Every graph G is a star-k-PCG for some positive integer k [1]. Thus, it is natural to consider the *star number* of a graph G, denoted as $\gamma(G)$, that is the smallest positive integer k such that G is a star-k-PCG [48, 49].

In [46], it is shown that $\gamma(C_n) = 2$ for every cycle C_n with $n \geq 5$ and that $\gamma(G_{n_1,n_2}) = 2$ for the two-dimensional grid graphs G_{n_1,n_2}, with $\min\{n_1, n_2\} \geq 2$. Moreover, there are four-dimensional grids that are not star-2-PCG.

In [47], the effects on the star number of certain graph operations are examined. These results can be applied to determine the star numbers of caterpillars and lobster graphs (*i.e.*, graphs such that when we delete their leaves we obtain a caterpillar) and, more generally, improve the existing bounds on the star number of acyclic graphs:

Problem 34 Given an acyclic graph G, determine its star number.

Finally, as a generalization of star-PCGs was done for star-k-PCGs, it would be interesting to study the analogous generalization of caterpillar PCGs to caterpillar-k-PCGs.

Problem 35 Explore the class of caterpillar-k-PCGs.

References

1. Ahmed, S., Rahman, Md.S.: Multi-interval pairwise compatibility graphs. In: International Conference on Theory and Applications of Models of Computation, pp. 71–84. Springer, Berlin (2017)
2. Azam, N.A.: On Enumeration of Tree-Like Graphs and Pairwise Compatibility Graphs. PhD Thesis, Kyoto University (2021)
3. Azam, N.A., Shurbevski, A., Nagamochi, H.: A method for enumerating pairwise compatibility graphs with a given number of vertices. Discr. Appl. Math. **303**, 171–185 (2021)
4. Baiocchi, P., Calamoneri, T., Monti, A., Petreschi, R.: Some classes of graphs that are not PCGs. Theor. Comput. Sci. **791**, 62–75 (2019)
5. Bibelnieks, E., Dearing, P.M.: Neighborhood subtree tolerance graphs. Discr. Appl. Math. **43**(1), 13–26 (1993)
6. Brandstädt, A., Hundt, C.: Ptolemaic graphs and interval graphs are leaf powers. In: LATIN 2008: Theoretical Informatics: 8th Latin American Symposium, Búzios, April 7–11, 2008. Proceedings 8, pp. 479–491. Springer, Berlin (2008)
7. Brandstädt, A., Le, V.B., Spinrad, J.P.: Graph Classes: A Survey. Society for Industrial and Applied Mathematics, Philadelphia (1999)
8. Brandstädt, A., et al.: Structure and linear time recognition of 3-leaf powers. Inf. Proces. Lett. **98**(4), 133–138 (2006)
9. Brandstädt, A., Le, V.B., Sritharan, R.: Structure and linear-time recognition of 4-leaf powers. ACM Trans. Algorithms **5**(1), 1–22 (2008)
10. Brandstädt, A., Hundt, C., Mancini, F., Wagner, P.: Rooted directed path graphs are leaf powers. Discrete Math. **310**(4), 897–910 (2010)
11. Brandstädt, A., Le, V.B., Rautenbach, D.: Exact leaf powers. Theor. Comput. Sci. **411**(31–33), 2968–2977 (2010)
12. Calamoneri, T., Petreschi, R.: On pairwise compatibility graphs having Dilworth number k. Theor. Comput. Sci. **547**, 82–89 (2014)
13. Calamoneri, T., Petreschi, R.: On pairwise compatibility graphs having Dilworth number two. Theor. Comput. Sci. **524**, 34–40 (2014)
14. Calamoneri, T., Sinaimeri, B.: Pairwise compatibility graphs: a survey. SIAM Rev. **58**(3), 445–460 (2016)
15. Calamoneri, T., Petreschi, R., Sinaimeri, B.: On the pairwise compatibility property of some superclasses of threshold graphs. Discr. Math. Algorithms Appl. **5**(02), 1360002 (2013)
16. Calamoneri, T., Frascaria, D., Sinaimeri, B.: All graphs with at most seven vertices are pairwise compatibility graphs. Comput. J. **56**(7), 882–886 (2013)
17. Calamoneri, T., Montefusco, E., Petreschi, R., Sinaimeri, B.: Exploring pairwise compatibility graphs. Theor. Comput. Sci **468**, 23–36 (2013)
18. Calamoneri, T., Frangioni, A., Sinaimeri, B.: Pairwise compatibility graphs of caterpillars. Comput. J. **57**(11), 1616–1623 (2014)
19. Calamoneri, T., Sinaimeri, B., Gastaldello, M.: On pairwise compatibility of some graph (super) classes. Preprint. arXiv:1504.06454 (2015)
20. Calamoneri, T., Lafond, M., Monti, A., Sinaimeri, B.: On generalizations of pairwise compatibility graphs. Discr. Math. Theor. Comput. Sci. 26(Graph Theory) (2024)

21. Calamoneri, T., Monti, A., Petroni, F.: All graphs with at most 8 nodes are 2-interval-PCGs. Fundam. Inform. **193**(1), 1–9 (2024)
22. Chang, M.-S., Ko, M.-T.: The 3-steiner root problem. In: International Workshop on Graph-Theoretic Concepts in Computer Science, pp. 109–120. Springer, Berlin (2007)
23. Chen, G., Hao, Y.: Multithreshold multipartite graphs. J. Graph Theory **100**(4), 727–732 (2022)
24. Ducoffe, G.: The 4-steiner root problem. In: International Workshop on Graph-Theoretic Concepts in Computer Science, pp. 14–26. Springer, Berlin (2019)
25. Durocher, S., Mondal, D., Rahman, Md.S.: On graphs that are not pcgs. Theor. Comput. Sci. **571**, 78–87 (2015)
26. Eppstein, D., Havvaei, E.: Parameterized leaf power recognition via embedding into graph products. Algorithmica **82**(8), 2337–2359 (2020)
27. Farber, M.: Characterizations of strongly chordal graphs. Discr. Math. **43**(2–3), 173–189 (1983)
28. Glover, N., Dessimoz, C., Ebersberger, I., Forslund, S.K., Gabaldón, T., Huerta-Cepas, J., Martin, M.-J., Muffato, M., Patricio, M., Pereira, C., da Silva, A.S., Wang, Y., Sonnhammer, E., Thomas, P.D.: Advances and applications in the quest for orthologs. Mol. Biol. Evol. **36**(10), 2157–2164 (2019)
29. Hakim, S.A., Papan, B.B., Rahman, Md.S.: New results on pairwise compatibility graphs. Inf. Proces. Lett. **178**, 106284 (2022)
30. Hayat, S., Azam, N.A.: A method to generate multi-interval pairwise compatibility graphs. ArXiv:2410.10525 (2024)
31. Hayward, R.B., Kearney, P.E., Malton, A.: Nest graphs. Discr. Appl. Math. **121**(1–3), 139–153 (2002)
32. Hellmuth, M., Wieseke, N.: On tree representations of relations and graphs: symbolic ultrametrics and cograph edge decompositions. J. Comb. Optim. **36**(2), 591–616 (2018)
33. Hossain, Md.I., Salma, S.A., Rahman, Md.S., Mondal, D.: A necessary condition and a sufficient condition for pairwise compatibility graphs. J. Graph Algorithms Appl. **21**(3), 341–352 (2017)
34. Jamison, R.E., Sprague, A.P.: Multithreshold graphs. J. Graph Theory **94**(4), 518–530 (2020)
35. Jamison, R.E., Sprague, A.P.: Double-threshold permutation graphs. J. Algebraic Comb. **56**(1), 23–41 (2022)
36. Kearney, P., Munro, J.I., Phillips, D.: Efficient generation of uniform samples from phylogenetic trees. In: International Workshop on Algorithms in Bioinformatics, pp. 177–189. Springer, Berlin (2003)
37. Knauer, K., Ueckerdt, T.: Three ways to cover a graph. Discr. Math. **339**(2), 745–758 (2016)
38. Kobayashi, Y., Okamoto, Y., Otachi, Y., Uno, Y.: Linear-time recognition of double-threshold graphs. Algorithmica **84**(4), 1163–1181 (2022)
39. Kratochvil, J., Tuza, Z.: Intersection dimensions of graph classes. Graphs Comb. **10**(2), 159–168 (1994)
40. Lafond, M.: Recognizing k-leaf powers in polynomial time, for constant k. In: Proceedings of the 2022 Annual ACM-SIAM Symposium on Discrete Algorithms (SODA), pp. 1384–1410. SIAM, Philadelphia (2022)
41. Lafond, M.: Recognizing k-leaf powers in polynomial time, for constant k. ACM Trans. Algorithms **19**(4), 1–35 (2023)
42. la Tour, M.D., Lafond, M., Ndiaye, N., Vetta, A.: k-leaf powers cannot be characterized by a finite set of forbidden induced subgraphs for $k \geq 5$ (2024). https://arxiv.org/abs/2407.02412
43. Lekkeikerker, C., Boland, J.: Representation of a finite graph by a set of intervals on the real line. Fundam. Math. **51**(1), 45–64 (1962)
44. Long, Y., Stadler, P.F.: Exact-2-relation graphs. Discr. Appl. Math. **285**, 212–226 (2020)
45. Mehnaz, S., Rahman, Md.S.: Pairwise compatibility graphs revisited. In: 2013 International Conference on Informatics, Electronics and Vision (ICIEV), pp. 1–6. IEEE, Piscataway (2013)
46. Monti, A., Sinaimeri, B.: On star-multi-interval pairwise compatibility graphs. In: Lin, C.-C., Lin, B.M.T., Liotta, G. (eds.). WALCOM: Algorithms and Computation, pp. 267–278. Springer Nature, Cham (2023)

47. Monti, A., Sinaimeri, B.: Effects of graph operations on star pairwise compatibility graphs (2024). https://arxiv.org/abs/2410.16023
48. Monti, A., Sinaimeri, B., et al.: On graphs that are not star-k-pcgs (short paper). In: CEUR Workshop Proceedings, vol. 3587, pp. 92–97 (2023)
49. Monti, A., Sinaimeri, B.: On star-k-PCGs: Exploring class boundaries for small k values (2024). https://arxiv.org/abs/2209.11860
50. Nishimura, N., Ragde, P., Thilikos, D.M.: On graph powers for leaf-labeled trees. J. Algorithms **42**(1), 69–108 (2002)
51. Papan, B.B., Pranto, P.B., Rahman, Md.S.: On 2-interval pairwise compatibility properties of two classes of grid graphs. Comput. J. **66**(5), 1256–1267 (2023)
52. Puleo, G.J.: Some results on multithreshold graphs. Graphs Comb. **36**(3), 913–919 (2020)
53. Rahman, Md.S., Ahmed, S.: A survey on pairwise compatibility graphs. AKCE Int. J. Graphs Comb. **17**(3), 788–795 (2020)
54. Rautenbach, D.: Some remarks about leaf roots. Discr. Math. **306**(13), 1456–1461 (2006)
55. Roberts, F.S.: On the boxicity and cubicity of a graph. In: Tutte, W.T. (ed.) Recent Progress in Combinatorics, pp. 301–310, x. Academic Press, New York (1969)
56. Rosenke, C., Le, V.B., Brandstädt, A.: Leaf powers. In: Encyclopedia of Mathematics and its Applications, pp. 168–188. Cambridge University Press, Cambridge (2021)
57. Salma, S.A., Rahman, Md.S., Hossain, Md.I.: Triangle-free outerplanar 3-graphs are pairwise compatibility graphs. J. Graph Algorithms Appl. **17**(2), 81–102 (2013)
58. Scheinerman, E.R.: Intersection Classes and Multiple Intersection Parameters of Graphs. Princeton University, Princeton (1984)
59. Thomassen, C.: Interval representations of planar graphs. J. Comb. Theory Ser. B **40**(1), 9–20 (1986)
60. Xiao, M., Nagamochi, H.: Characterizing star-PCGs. Algorithmica **82**(10), 3066–3090 (2020)
61. Xiao, M., Nagamochi, H.: Some reduction operations to pairwise compatibility graphs. Inf. Proces. Lett. **153**, 105875 (2020)
62. Yanhaona, M.N., Bayzid, Md.S., Rahman, Md.S.: Discovering pairwise compatibility graphs. Discr. Math. Algorithms Appl. **2**(04), 607–623 (2010)

Simulation Experiments to Assess the Effectiveness of Randomization in Security Screening at Sports and Entertainment Venues

Dennis Egan, Kevin Mc Inerney, Paul Kantor, Christie Nelson, and Fred Roberts

1 Introduction

Our nation's sports and entertainment venues (stadiums, arenas, etc.) host millions of patrons annually, form the basis for a multi-billion-dollar industry, and present an inviting target for terrorists. Security screening at sports and entertainment venues has developed to be more and more sophisticated under rapidly shifting threat and security response environments. When it is not possible to inspect everyone, or when it is just a good idea to confuse potential bad actors, introducing randomization into security screening is a potentially very useful method. Security directors could greatly benefit from methodologies and tools that can help them make informed decisions on the introduction of new or modified security protocols and processes such as various randomized screening methods during a planning stage that takes place long before game-day execution. Simulating security processes through simulation software is one such way to test security scenarios before deploying them, in order to gain an understanding of the possible benefits/effectiveness and costs such as patron dissatisfaction or extra manpower or introduction of security vulnerabilities. Simulation software can also help security directors to weigh pros and cons of various potential changes both against various options and current practices. These points are illustrated using ideas for how to randomize the patron screening process in order to make it more difficult for bad actors to overcome security procedures. Simulation is one method that can be used to start to understand how possible randomization initiatives might work for individual venues. Specifically, we develop a baseline security simulation and three variants with new security procedures, and run simulations to compare the results of the

D. Egan · K. Mc Inerney · P. Kantor · C. Nelson · F. Roberts (✉)
Rutgers University, New Brunswick, NJ, USA
e-mail: froberts@dimacs.rutgers.edu

B. Goldengorin (ed.), *Theory, Algorithms, and Experiments in Applied Optimization*, Springer Optimization and Its Applications 226,
https://doi.org/10.1007/978-3-031-91357-0_4

different procedures on various metrics such as the time after event start when all patrons enter the stadium and the size of the largest crowd waiting to be inspected.

The rest of this paper is organized as follows. Section 2 explores the concept of randomization as applied to security screening, and Sect. 3 surveys selected literature on the use of simulation in security screening at stadiums and large entertainment venues. In Sect. 4, we discuss the simulation design used in our experiments. Specifically, Sect. 4.1 lays out the process of inspection that is simulated, and all its components, and other subsections describe the Stadium Simulator we built and used (with details left to an Appendix), the need for multiple runs, and the performance measures we adopted to assess the results of the simulations. Section 5 describes the experiments and their results, and Sect. 6 discusses the results and the basic conclusions.

2 Randomization

One important concept that we explore in this paper is randomization. Randomization aims at making it more complicated/confusing/expensive for adversaries, which acts as a deterrent. There are also other goals for randomization: monitoring operational integrity (e.g., by randomly rechecking credentials of employees); stimulating the capability or alertness of security personnel (e.g., through use of red-teams); and achieving intermediate levels of security when threat intelligence and/or budget considerations do not recommend 100% application (e.g., when inspecting some fraction of persons or covering part of a venue with cameras is better than not doing anything).

It is well recognized that no security initiative is perfect. Randomization may allow you to do more at less cost, but of course may also "catch" less. Professional sports leagues all seek 100% inspections. However, it should be noted that any added screening, particularly when applied randomly, raises the adversary's costs. Conceivably, the deterrent effect (if it could be measured) might make randomization preferred to 100% checking. In any case, when a process is too expensive to do 100% of the time, randomization can still reduce threats and increase security; it is a low-cost way to introduce a higher level of security.

Randomized secondary screening recognizes that primary screening is not 100% effective; it "fills the gap," increases the probability of detection (as any added security initiative does), leads to increased deterrence, and makes patrons feel more secure. Adding a randomized secondary check can improve security in other ways. For instance, when wanding is done for randomly selected patrons who have "passed" a walk-through metal detector (WTMD) test, this provides valuable information on the "detection rate" of the WTMD plus its operation.

There are advantages to being unpredictable. Randomization "makes the bad guys work harder;" it gives them pause for thought. Randomization diminishes the effectiveness of surveillance by the adversary. The goal is to defeat a sophisticated surveillance team. Randomization keeps those with intent to do harm off balance.

Also, randomization serves as a deterrent: If procedures are seen to be uncertain, unpredictable, adversaries might alter their calculation of the likelihood of success or failure. Deterrence is especially effective when it is known that a random security process is being implemented, but the exact protocol or randomization scheme is not visible.

Randomization is already in place in other security settings than stadiums. The TSA, US Coast Guard, airports, Federal Air Marshals Service, light rail systems, and other organizations have all deployed randomized designs to improve physical security. For example, TSA's Pre-Check and CBP's Global Entry program pre-vet enrolled travelers, but some are still randomly chosen for more screening. TSA also uses a WTMD random alarm capability at checkpoints for secondary inspection. Some of the random secondary screening involves use of explosive swabs (and sometimes this also occurs at passenger gates). TSA also randomizes the use of passenger screening K-9s (including vapor wake dogs) and randomized placement of behavioral detection officers. They use a random approach to decide whether to do a specific security practice on a given day; or use a random approach to choose which prepared plan to use on a given day. See Stewart and Mueller [1] for more information.

As a second example, the NJ Transit Police Department has done random baggage screening: At fixed baggage check points run by the NJTPD Emergency Service Unit they have screened baggage with simple randomization: every *n*th person depending on volume; however, the station and location at which the baggage check point is located are determined by intelligence analysis or risk analysis.

Some efforts at randomization in security involve the use of sophisticated tools of game theory based on adversary-defender games where the adversary takes advantage of some knowledge of the defender's strategy. This idea has been pioneered by the CREATE US Department of Homeland Security Center of Excellence at the University of Southern California, led by Professor Milind Tambe. The work on game theory and security has led to a wide range of actual deployed applications: ARMOR (Assistant for Randomized Monitoring Over Routes) at LAX airport for scheduling checkpoints and K-9 patrols [2]; IRIS (Intelligent Randomization in Scheduling) [3] for the US Federal Air Marshal Service to deploy air marshals on US air carriers; GUARDS (Game-theoretic Unpredictable and Randomly Deployed Security) [4] developed for the TSA for randomizing security activities to protect airport infrastructure; PROTECT (Port Resilience Operational/Tactical Enforcement to Combat Terrorism) [5] was for the US Coast Guard to schedule randomized patrols within ports and deploying escort boats for the US Coast Guard to protect ferries; and TRUSTS (Tactical Randomization for Urban Security in Transit Systems) [6] for the Los Angeles Sheriff's Department to schedule multi-operation patrolling (fare evasion, counter-terrorism and crime) on LA area metro trains. DARMS (Dynamic Aviation Risk Management Solution) [7] has been developed by the TSA with help from CREATE to give both passengers and flights risk scores and determine the level of inspection of a passenger and which flight to assign them based on risk scoring. ARMOR-FISH [8] was developed for the Coast Guard to help prevent illegal, unreported, and unregulated fishing.

The same tools have been used to assign randomized patrols to catch poachers in wildlife preserves. ARMOR was the first application of these methods. It provided the police at LAX Airport an hour-by-hour schedule for where to set up checkpoints and K-9 patrols. In early studies, the benefits of ARMOR were quantified in terms of contraband found and reduced overtime pay.

One proposed method that has not yet (to our knowledge) been explored in practice is to use a random mixture of security levels across a collection of networked screening devices, such as the metal detectors at a large stadium. This approach [9, 10] would permit a venue to find the "sweet spot" that minimizes patron dissatisfaction, while increasing security to a level that falls **between** the fixed settings available on standard walk-through detectors.

3 Simulation for Security Screening: Selected Prior Work

Our research was influenced by earlier simulation work involving sports stadiums and other large venues. Hall, et al. [11] give an overview of many aspects of sport stadium security and safety, including ingress and egress.

Most of the stadium simulation work described in the literature has been concerned with modeling evacuations of such venues. For example, Kar and Zale [12] present an evacuation model using macro-simulation for University of Southern Mississippi's football stadium, Gianchini, et al. [13] present a simulation of evacuation of Scott Stadium at University of Virginia, and Mahmudzadeh et al. [14] present a simulation of evacuation of Naghshe Jahan Stadium in the city of Isfahan, Iran. Klupfel and Meyer-Konig [15] use a cellular automaton approach to simulate evacuations from large and complex structures. Kirchner, et al. [16] present a multi-agent-based framework for simulating human and social behavior during emergency evacuation, Pan, et al. [17] also take behavior into account in simulating evacuation, and so do Chin, et al. [18, 19]. Rodriguez, et al. [20] study how agent egress is affected by the environment in both real-world and large-scale virtual environments.

The literature on simulations for ingress inspections has primarily been devoted to airports. For example, Edmunds [21] uses simulation to estimate traveler wait times and queue lengths resulting from the United States Visitor and Immigrant Status Indicator Technology Program (US-VISIT) in terms of traveler wait times and queue lengths, using simulation. Holguin-Veras, et al. [22] develop a discrete event simulation of security screening operations at an idealized airport, based on composite data from airports in Atlanta and Tampa. This work takes into account the need for secondary inspections. Pendergraft, et al. [23] apply discrete event simulation modeling to understand the operational dynamics of passenger security screening at Baltimore-Washington International Airport.

Ricks, et al. [24] develop a patron screening model that addresses the speed of screening using different procedures: walk-through metal detectors, wandings, and pat-downs, and creates a real-time simulation of queue formation using the physical arrangement of gates at an NFL stadium. This allows for analysis based on different patron screening procedures.

Chang, et al. [25] use a series of parallel simulations in the Unity3D simulation framework to determine, based on a given threat detection threshold, the number of sensors of various modalities required, and where to place them.

4 The Simulation Design

4.1 Process Simulation for Stadium Security Screening

In very general terms, simulation describes the method of building an analytical model of a process of interest such as security screening at a sports and entertainment venue. It involves breaking the process down into individual steps that we refer to as "components" (e.g., a bag check) and decision points (e.g., to ask a patron who alarmed at the WTMD to go to secondary screening, or allow them to continue on to ticketing if they did not alarm). Each of these components takes various amounts of time that can be estimated by collecting field data combined with subject matter expert input. Creating such a simulation to model for a complex process such as patron screening involves selection of values such as time spent at a component of the process (e.g., wanding) and the probability of failure ("alarming" for a prohibited item or being selected for a random process) for all the different components and at each of the many different decision points. Choice of these parameter values can have subtleties. For example, secondary wanding in the winter takes longer than in the summer. There are also entities or agents (here we can think of an agent as a single patron) that travel through the process from start to finish. In patron screening, when an agent "finishes" the process is the point in time when the patron enters the venue. Some patrons may exit the system at earlier points to return items to their cars, or because they have invalid tickets, etc. The goal is to re-create the system close to how it exists within the real world.

When such a model is prepared, it should first be tested to make sure it replicates the process as it currently operates. This may include talking to experts who know how the venue looks at various points of time during the process in question, doing time analysis through security video, and collecting field data at live events. Once the model is validated, options for changes or enhancements can be made and it is possible to test the impact of the changes based on certain measurable outcomes. The configuration and related parameter values as set within the system represent the "baseline" against which optional configurations can be evaluated. The options for changes can be further structured as business case scenarios that model implementation-feasible options where different designs, inputs, and outcome measures are clearly defined. Using simulations, scenarios can be tested and compared to the baseline condition to determine possible impacts before real-time implementation.

The process of screening patrons entering sports and entertainment venues can be modeled using simulation. The process consists of a system of interactions and decisions. Bags that patrons want to bring with them into the venue can be checked for size restrictions or whether or not they meet a materials standard (e.g., must be clear). Bags can additionally be screened for prohibited contents. To further address contraband issues, venues use a variety of screening methods such as hand-held metal detectors or wands, walk-through metal detectors (WTMDs), physical pat-downs (gender appropriate), K-9s, or other methods. Ticket checking through collection or scanning, although not strictly a security measure, can also be thought of as part of this process, as it is usually the last part of the full patron screening process before someone can gain entry to a venue for an event. (One might use the ticket taking step to enhance security, e.g., if scanning a ticket provides background information about the ticket holder that can be utilized for enhanced screening.)

When a simulation is used to describe and evaluate security measures or processes, a baseline is defined and alternative security scenarios can be modeled and compared for costs and benefits to the baseline. For example, the baseline could consist of a set number of security lanes with a progressive flow from bag checks to WTMDs to wanding to ticket collectors. The level of security for each process can be defined by detection level (a WTMD might detect 80% of the contraband that passes through it, for example) and service time (it may take 2 s to check a bag for contents or 7 s to wand a patron).

The arrival rate of patrons is equally as important as the configuration of the system screening processes. It is obvious that the flow of patrons and what happens to them in the security process are dependent upon the types and times of the security processes as well as how many people are trying to use the system. A system where arrivals come at a faster rate than the time to clear the services is one where back-ups and queues occur. Growing queues affect patron satisfaction and lead to security vulnerability issues. Thus, the baseline system and any modified system must reflect the anticipated rate of patron arrivals. It is important to be able to estimate the flow of patron arrivals. How many patrons are arriving per some timeframe? Do they arrive at the same rate as the event start approaches or are there arrival peaks? Is there more than one arrival peak? How many arrive in groups? What is a typical group size?

A key goal of simulation modeling of the baseline case is to develop it in a way that it represents as closely as possible the real-world situation and circumstances. It is therefore extremely important to be able to know or estimate the parameters that define the system processes and participants. For example, how long does it take on average to process a patron through a WTMD? How often does wanding result in a false alarm? If a patron fails the WTMD test, how long does the secondary screening take? What percentage of patrons causes the WTMD to alarm?

Ultimately, the effectiveness of a security initiative can be measured by its ability to detect contraband. How contraband is defined may vary from venue to venue (guns? knives? brass knuckles?) A simulation model requires an estimate of the number (percentage) of patrons carrying contraband. This can be based on experience with contraband detection (though people might stop carrying contraband as a result of the deterrent effect of security initiatives, with a resulting decrease in contraband detected). So, ultimately, the percentage of patrons carrying contraband has to be guessed, and several different estimates of this percentage should be experimented with.

The assignment of correct or best estimate values to the parameters that define the system forms the basis of model fidelity. Model fidelity can be checked by comparing measures obtained during simulation runs to expected data. Venue security directors often know how many patrons are in the venue at different times prior to or at event start, and this is one way to validate a model by comparing these numbers to simulation outputs. The model can also be validated using other venue security director data or experience, as well as through ticket scanning data or video analysis.

The value of simulation we emphasize is to assess the impact of proposed new or amended security practices compared to the baseline. For example, a security director wants to know if increasing the detection rate of the WTMDs will increase the time it takes the last patron in the line to enter the venue after the start of an event—we call this the ***queue clearance time***. To change detection rate, the baseline is changed in some way. In this case, that may simply be an increase in detection rate at the WTMDs, or it may be to add an extra screening component such as randomly determined assignment to secondary screening. Outcome measures such as average waiting time or queue clearance time can be calculated. As an example, if the last patron clears security under the baseline process 5 min after event start, we may see a simulation result based on a new security initiative that shows that the clearance time has now increased to 12 min. With this information, combined with estimates of increased contraband detection, the security director can then evaluate the trade-off of increased security and decreased patron satisfaction.

4.2 Simulation Software Used

Simulation software allows the user to perform computerized "runs" under both a baseline and a modification to the baseline, and then to compare results. Such software allows the user to set system configurations, set parameter values, and establish patron arrival patterns. Trial and error testing can be used to evaluate how well the baseline parameters replicate the real-world conditions. Desired output measures can be selected so that the security initiative under consideration can be compared to the various scenario ideas.

The ***Stadium Simulator*** we built and used represents a "gate," which is thought of as a set of "lanes." Each lane includes a series of possible checks or operations. Reflecting operational observations, the simulator also allows for a "queue" just before each operation, to represent the people who have to wait to go through the next operation. The simulator needs to know some key information about the "arriving people." The simulator can represent people as arriving singly or in groups (in the work we describe, the relative probabilities are set as single control parameters, using a Poisson distribution). The simulator can also represent two patterns of arrival to account for an early wave (perhaps due to a promotional giveaway) and a later wave arriving closer to the start of the event. In our work, the rate of arrival (of the groups) is modeled by an exponential probability distribution (which is, somewhat confusingly, known as "Poisson arrival"). The simulator also sets the probability that a simulated patron will actually be carrying something that ought to be detected—"contraband." We model only one form of contraband. When the patrons arrive "at the gate," the simulator randomly routes them equally to any of the lanes.

The simulator also needs information about the behavior or characteristics of each of the operations that process the patrons. In our studies, these are called "outer bag size check" (check of bags at an outer perimeter), "inner bag contents check," "walk-through metal detector" (the WTMDs can have different security settings), "secondary screening" (typically wanding, and after a person sets off an alarm at the initial screening or is randomly selected for further screening), and "ticket check." Not all of these operations need to be activated in a given baseline or experimental modification. What is the behavior of these operations? Each operation has a "detection rate," which represents the fraction of the contraband that it will actually detect. This number must be determined by observation and is really a combination of the effectiveness of any security devices, and the alertness of the operators, who may miss alarm signals from time to time. No detector is perfect, and so we also model the false alarm rate—the chance that the system gives an alarm when there is no contraband. Finally, and most importantly, each operation takes time, and this time is usually variable. This is represented by a probability distribution, typically being uniform between a shortest time and a longest time. For more information about the Stadium Simulator we used, see the Appendix.

4.3 Multiple Runs

We developed a large number of randomization protocols for screening patrons to experiment with through our Stadium Simulator, then limited the number to those we felt would provide the most useful feedback and reflected reasonably practical randomization options.

The simulation models the arrival of thousands of patrons, over periods of the order of an hour. Because of the randomness in arrival times, and in the operations of screening them, one should run the simulator multiple times, to see how much random variation there is and to get an overview of what might be happening. Then, the results of the runs for the baseline can be compared to the runs for the experimental change. The simulation generates a huge amount of data. The venue security director will need to decide what information will be most helpful. Does he or she want to see the result of each run? Or the average value of the outcomes (e.g., average time spent in security) on each baseline run vs. on each experimental run? Or the "worst case" (longest time spent in security) on each baseline run vs. on each experimental run?

4.4 Performance Measures

In the following discussion, we concentrate on four representative measures that describe the effectiveness and possible costs/outcomes of a screening security initiative. The first measure relates to patron satisfaction. It is the average time a patron spends in the entire security process including waiting in line, and going through each component of screening. The next is queue clearance time, the number of minutes between event start time and the time when the last patron who arrives *before the event starts* completes the last of the screening operations. These are examples of "patron satisfaction" indicators.

Another measure is more related to security, by reducing the size of the "soft target." This measure is the greatest size attained during the course of the security screening of the total set of people not yet cleared, across all the lanes, and all the operations in each lane. This number gives a measure of maximum vulnerable group size, for a group that could be attacked by a heavy truck attack, an active shooter, a suicide bomber, etc. The fourth measure we use in the discussion is the contraband detection rate, or its complement, the contraband "miss" rate. A potential security benefit of adding randomized security protocols is to improve the rate at which contraband is detected.

5 The Experiments

The purpose here is to demonstrate the value of simulation by showing the process of baseline development for a sample case and comparison testing for three experimental scenarios with identified output measures. These experimental scenarios are based on ideas about security options already used or (in one case) developed by our team. Each involves a type of randomization.

The stories that emerge from these scenarios are the true message of how this approach can provide insight into complex and dynamic security processes and become one of many useful information sources and decision support tools for venue security directors.

The following examples represent one possible baseline and three experimental scenarios. The selection of parameter values used in the examples is purely illustrative, though we based them on our own observations at a variety of events. Some of these values represent significant levels of guesswork. The baseline development also went through various trial and error adjustments to arrive at system values that seem in-line with our direct observations of actual game-day patron screening. Parameter values and the results provided as part of the scenario comparisons are only to illustrate simulation and its potential to identify system change impacts.

5.1 The Baseline Simulator Parameters

The baseline parameter set represents the current state to which experiments of other configurations can be compared. Table 1 outlines the parameters used by the Stadium Simulator and their baseline values.

Table 1 Stadium simulation software parameters with baseline values

Parameter	Brief description	Baseline values
Patron arrivals/arrival rate	Patrons arrive at a stadium at a certain rate. This rate has a distribution, or in other words, there is some fluctuation in how many come at a specific moment. The software can additionally account for more than one peak arrival (e.g., an early peak for a giveaway and a later peak for pre-kickoff (pre-event start) rush)	A uniform distribution was selected with four patrons arriving per second
Event start time and simulation run time	The software allows the user to set the number of hours/minutes before event start when the security processes begin. The simulation run time sets the amount of time after event start that the simulation will run to allow all the patrons to clear the system	Event starts 60 min after simulation is initiated; the simulation run time was set for 1 h and 30 min
Group A & B Lanes—Pre-screening bag size checks; parameters of baseline and experiment differ for different lanes	This sets the total number of security lanes across two groups for a pre-screening process such as checking the size of the patron's bag for compliance	Total of ten lanes; for the baseline, all ten lanes have identical screening processes and parameter settings. In experiments, we sometimes put half of the ten lanes into each group, and in others, we put nine in Group A and one in Group B
Bag size check	For the bag size check pre-screening, the user can set the percentage of bags that will fail compliance, the time it takes to complete the process and the detection rate of successfully identifying an oversized bag. A false positive rate of incorrectly identifying the bag size is an option	% w/oversized bags—10% Screening time—1 s Detection rate—45% False positives—0 Random alarm rate—0
Optional: Bag size check followed by random alarm to signal additional screening	At the bag check size screening location, security could implement a random selection process for a secondary screening. If selected, parameter fields can set the rates related to a secondary screening process	No random screening process is selected for the baseline

(continued)

Table 1 (continued)

Parameter	Brief description	Baseline values
Group A and B security screening lanes; parameters of baseline and experiment differ for different lanes	This sets the total number of security lanes across two groups for the screening processes such as bag contents or WTMDs. Thus it allows for different configurations across various number of lanes	Total of ten lanes; for the baseline, all lanes have identical screening processes and parameter settings. In experiments, we sometimes put half of the lanes into each group, and in others we put nine in Group A and one in Group B
Bag contents check	These parameters define the bag contents check as one of the screening options in the security process	% patrons w/contraband—Varies among 1%, 10%, and 25%. Screening time—2 s Detection rate—45% False positive rate—0%
WTMDs	These parameters define the WTMDs as one of the screening options in the security process	% patrons w/contraband—Varies among 1%, 10%, and 25% (this is set at the bag contents level and remains through the remainder of the process) Screening time—3 s Detection rate—80% False positives—0 Random alarm rate—0
Optional: WTMDs—Random alarm/screening	At the WTMD location, security could implement a random selection process for a secondary screening. If selected, the patron will be sent to secondary screening as if he/she was detected with contraband as part of the walk-through screening process	No random screening process is selected for the baseline.
Secondary screening (if patron failed the WTMD screening)	If a patron fails the WTMD, they are sent to a secondary screening, usually hand-held metal detector wanding. This is different than the random screening processes that patrons could be sent to regardless of whether they alarm	Screening time—7 s Detection rate—95% False positives—0
Ticket takers	Although not a formal security screening, the ticket taking, scanning procedure is within the full process for patrons to gain entry into the event and therefore is part of the processing time	Screening time—2 s Failure rate—0 (Failure rate is the percentage of tickets that do not scan properly)

5.2 Experiment 1: Increased WTMD Security Level

The following experiments are illustrations. They were developed from ideas and discussions with sports and entertainment venue partners. The experimental conditions were operationalized by changing a subset of the baseline parameters.

***Experiment Number 1*—Increase the security level of some of the WTMDs in use**

Motivation: The WTMDs in use have this option built in. This is a concept that is consistent with implementing a random process for the purpose of interfering with the adversary's ability to make a good calculation of likelihood of success. If an adversary has knowledge that perhaps one or more WTMDs is operating at a higher security level but does not know which one(s) these are, it may deter them from attempting to bring in metallic contraband such as knives or guns.

The simulation output measures include various measures of the impact of the approach: possible increased contraband detection as well as the possible negative impacts on security and patron satisfaction such as potentially longer times in the security process, possibly larger crowds awaiting security screening, and potentially longer waits for the last arrivals to get into the venue following the start of an event.

To be clear, the Stadium Simulator is not attempting to measure or predict the impact on deterrence. The scenario is based on the idea that a terrorist is a rational actor who is evaluating information and uncertainty in their decision to attack or not. The value of the simulation is in examining the process and the resulting measurable impacts.

The experimental condition was operationalized by changing the security setting level on a single WTMD, i.e., patrons in one of the ten lanes were exposed to the higher setting. We set the detection rate to 95% (as opposed to the baseline rate of 80%) and added 1 s to the service time at that WTMD (increasing the throughput time from 3 to 4 s).[1] So, all patrons in the lane with the higher WTMD security setting experienced an extra 1 s delay, and more of them experienced secondary screening (an extra 7 s for wanding) than in the baseline due to the higher detection rate. We also experimented with different values for the proportion of patrons actually carrying contraband. The percentage of patrons carrying contraband was set either at 10% or 1%.

The results for our four outcome measures are shown in Table 2. In this table and those following, we show averages of each of the four measures computed across 20 simulated runs.

[1] Note that we do not have a good estimate of the service time for processing patrons at the higher security level setting. We think it should take longer on average because the WTMD alarm will sound more frequently, leading to extra time to divest small metallic items, etc. We experimented with increasing the service time up to 8 s and found the expected impacts. For example, the extra 4 s service time resulted in an increase of over 200 people in the maximum number of patrons in security.

Table 2 Experiment number 1 outcomes

	Average time in security (minutes)	Queue clearance time (minutes after kickoff (event start))	Maximum number of patrons in the security process	Contraband detection rate (percentage)
Baseline (10% with contraband)	2.55	6.61	940	87.1%
Baseline (1% with contraband)	2.54	6.60	941	86.3%
Experimental (10% with contraband)	3.16	15.40	1086	87.5%
Experimental (1% with contraband)	3.22	15.70	1087	87.1%

The results show that just changing one of ten lanes to a higher security setting had a small effect on average time in security: an increase of $3.16 - 2.55 = 0.61$ min, i.e., approximately 40 s. The change had a large effect on queue clearance time, an increase of $15.40 - 6.61$ min or approximately 9 min. It increased the maximum number of patrons in the security process by $1086 - 940$ or approximately 145 people. On the other hand, the improvement in overall detection rate was very small. Furthermore, changing the percentage of patrons with contraband from 10% to 1% made essentially no difference in these metrics. If we further analyze what happens in the lane with the WTMD set to the higher security setting, since the entire increase in the average waiting time was due to that lane, we can estimate that patrons in that lane spent an average of 9.0 min waiting (compared to patrons in every other lane waiting an average of 2.54 min). Clearly the queue for that lane built up to much larger numbers. Since the entire increase in the maximum number of patrons in the security process is due to the lane with the higher security setting, we can estimate that the average maximum queue length for the nine lanes at the lower setting is 94 patrons, while the lane with the higher setting has an average maximum length of 240 patrons.

We doubt that any security director would implement this approach based on the trade-off among increasing waiting times, queue clearance time, and maximum number of patrons awaiting security against only a tiny increase in overall detection rate. On the basis of this one experiment, we are not by any means concluding that the strategy of setting the security level on one or more WTMDs higher is a bad idea. The conclusion depends heavily on the parameters we used in reaching this conclusion. We just use this example to illustrate the point that such experimentation before rolling out a new security initiative is a good idea.

We wondered why there was such a big increase in queue clearance time, and this led us to see that our simulation model was doing something unrealistic; it was not allowing patrons to switch from a longer line to a shorter one, but instead assumed that the random assignment to a line was permanent. In this way, the experimentation taught us something about what is and what is not realistically possible. We would certainly find pushback from patrons if we forced them to stay in the line we randomly chose for them and that line moved so much more slowly than others. Running and analyzing a simulation can teach you a lot about your assumptions.

5.3 *Experiment 2: Random Additional Screening Following Bag Size Check*

***Experiment Number 2*—Add random additional screening following the bag size check; at the periphery where bag sizes are checked, a random selection of patrons who pass that check will be sent to an additional screening process (e.g., explosive swab test, use of K-9s, wanding, etc.)**

Motivation: This additional random security at the periphery of a venue would be a highly visible way to attempt to detect non-metallic threats (e.g., explosives) before they could be brought closer to the venue and more dense crowds. This protocol would have both a deterrent effect and an improved detection effect.

This experimental condition was operationalized by activating a random process in a single lane following the bag size check. The process randomly selected some of the patrons in that lane for additional screening. We varied the percentage of people randomly selected for the additional screening, and we set the service time for the additional screening at 7s. This time is well within our observations for wanding times. Other kinds of screening, e.g., explosive swab testing, could take longer. For all conditions, we set the patron contraband rate at 10%. The results are shown in Table 3.

Here we see essentially no effect when only 20% of the patrons from a single lane are selected for additional screening. All four metrics are very similar to the baseline. We figured that since the additional screening was affecting only 2% of all patrons (i.e., 20% of the patrons in one of ten lanes), its effect may be too small.

Table 3 Experiment number 2 outcomes

	Average time in security (minutes)	Queue clearance time (minutes after kickoff (event start))	Maximum number of patrons in the security process	Contraband detection rate (percentage)
Baseline (10% with contraband)	2.55	6.61	940	87.1%
Experimental (20% random selection)	2.49	6.38	934	86.7%
Experimental (75% random selection)	3.20	31.83	949	86.5%

As a result, we raised the random selection to 75% of the patrons in that one lane being sent for additional screening, for a total of 7.5% of all patrons getting the extra screening. This did make a difference in three of the four metrics. Notably, it increased the queue clearance time by more than 25 min. This means that at event start time, the last patron standing in the lane for which 75% of the patrons were sent to additional screening would still have to wait over a half hour to get into the game, compared to the other queues that cleared in about six and a half minutes.

Given these results, a security director clearly would not opt for this way to randomly select patrons for additional screening. Moreover, patrons would not passively stay in the line for a half hour while patrons in all the other lines were getting into the game much faster. This again points to the unrealistic nature of not allowing people to switch lanes.

Suppose a security director still wanted to try additional screening. These results suggest that the director might try multiple secondary screeners, i.e., processing people for secondary screening in parallel, if the resources were available. The results also suggest that a distributed approach to extra security screening could be tried, i.e., providing extra screening on multiple lanes so as to prevent the very large buildup of a single queue.

5.4 Experiment 3: Random Alarms on Some WTMDs

***Experiment Number 3*—Add random alarming on some WTMDs; an alarm will go-off randomly for some proportion of patrons who pass through the WTMD without it detecting anything. Those patrons will be sent for secondary screening, e.g., wanding**

Motivation: This is of course already used by TSA, stadiums, and others. In this case, the security benefit is two-fold. As a randomized process, it adds uncertainty to the adversary's risk evaluation. This uncertainty comes from irregularities in observations that the adversary may be making as they formulate their attack strategy. The second benefit may be the effect on staff alertness if the WTMD alarms more frequently.

This experiment was operationalized as follows. The baseline was set for 25% of patrons carrying contraband. As can be seen by comparing the following table to the 10% and 1% contraband conditions shown in previous tables, the 25% baseline condition essentially did not change any of the four measures we have considered. To begin to get a sense of what might happen if security were distributed beyond a single lane, we set the WTMDs on half of the lanes (5) to randomly alarm for 25% of the patrons for whom no metal was detected, while having no random security inspection selection on the remaining lanes. Those patrons who randomly alarmed were sent to a secondary screening process for which service time was 7 s (typical of wanding times). As a result, this experiment provides extra security screening for approximately 12.5% of all patrons (25% in five of the ten lanes). The results are shown in Table 4.

Table 4 Experiment number 3 outcomes

	Average time in security (minutes)	Queue clearance time (minutes after kickoff (event start))	Maximum number of patrons in the security process	Contraband detection rate (percentage)
Baseline (25% with contraband)	2.53	6.60	939	86.7%
Experimental (25% random selection on half the WTMDs)	2.78	8.81	1008	89.3%

The results show differences in all four measures. The average time in security increases, but only by about 15 s overall. The average time for the lanes with the randomly alarming WTMDs is 3.03 min compared 2.53 min for the lanes without random alarms. However, the 30-s difference would be barely noticeable. The queue clearance times increase by about 2.2 min. Since the max queue clearance times almost certainly come from those lanes with the extra security screening, this means that at the time the event starts the last person standing in those lanes will have to wait about 2.2 min longer than the last person standing in lanes without the extra security. However, everyone gets in <10 min after event start. The maximum number of patrons in the security process increases from 939 to 1008 indicating a greater vulnerability built up due to the extra screening. Finally, there is a small improvement in the contraband detection rate from 86.7% to 89.3%. Another way to view this number is that in the baseline, 13.3% of all contraband goes undetected, while in the experimental scenario that drops to 10.7% of all contraband going undetected. In the lanes with the additional screening, we can estimate that 91.9% of all contraband is detected (8.1% undetected) in those lanes.

Of the experimental scenarios evaluated here, this "distributed approach" may be the most promising—based on the measured effects resulting from the simulation. This experimental scenario provides extra security screening to 12.5% of all patrons compared to 10% (Experiment Number 1) and 7.5% (Experiment Number 2) with a better patron satisfaction/increased security trade-off. There are fairly small increases in average waiting times and queue clearance times, along with a small improvement in contraband detection. A venue security director would have to weigh these trade-offs, while also taking into account the larger maximum number of people in security due to the experimental scenario. The director would undoubtedly want to try other combinations, experiment with the patron contraband rate, and determine whether distributing randomly selected additional security to different points involves too many extra resources. It is important to repeat that the better performance of this third scenario vs. the other two may just be a reflection of the parameters selected (and the unrealistic modeling of lane changing behavior). The main point of presenting these examples is to illustrate how one might use simulation to test effectiveness of a potential new initiative.

6 Discussion

The situation at sports and entertainment venues introduces two important complexities. First, at least at the largest venues, the operators can close or open lanes, to make sure that the crowd is not too large, and delays are not too long, an option that may not be possible in all security screening settings. Indeed, a simulation can help the security director plan for times to open or close new lanes (and also use this to understand implications for needed manpower). Second, the patrons are not assigned to lanes randomly or due to some algorithm. They get to choose. But they do not always make a simple choice, such as choosing the shortest line, because the crowd may be so large that they cannot even see the shortest line.

The fact that the Stadium Simulator we used does not include patron choice explains the large increase in clearance time when just one lane has its detection rate increased. We have looked at the graphic display of the simulation visualization (an additional feature of the simulator not shown here), a few minutes before clearance, and have seen that the simulator has dozens of simulated patrons waiting in one lane, while the other lanes are almost empty.

The Stadium Simulator we used is limited in its approach to analyzing patron contraband. We certainly recommend expanding the notion of "contraband." Currently we have defined only one kind of contraband, in addition to the size of a patron's bag which is detected and dealt with separately. "Contraband" items could be broken down into at least the following categories: (a) metallic and threatening; (b) metallic and non-threatening; and (c) explosives. Each category of contraband could be included in the simulation along with appropriate kinds of detectors and impacts. For example, both threatening and non-threatening metallic items can set off a WTMD alarm, causing delays, etc., but these would be handled differently by security staff. At this point, most sports and entertainment venue security cannot detect explosives, except perhaps by deploying specially trained K-9s. Given the prevalence of explosives involved in terror attacks, more emphasis on detecting explosives (e.g., through explosive swabs, and backpack X-rays, in addition to more K-9s) may be called for in the future, and could be included in future security simulations. Contraband could also be expanded to include (d) non-metallic and non-threatening items (e.g., water bottles found in a bag search).

Acknowledgements This material is based upon work supported by the U.S. Department of Homeland Security under Grant Award Number 22STESE00001-04-00 from DHS Office of University Programs via Northeastern University, U.S. Department of Homeland Security under Grant Award Number 2009-ST-061-CC10002 from DHS Office of University Programs via Rutgers University, and Task Order HSHQDC-17-J-00302—BOA HSHQDC-16-A-B0005 to Rutgers University. FSR also acknowledges support from NSF grant 2122119 to Rutgers University.

Disclaimer The views and conclusions contained in this document are those of the authors and should not be interpreted as necessarily representing the official policies, either expressed or implied, of the U.S. Department of Homeland Security, the DHS Office of SAFETY Act Implementation, the National Science Foundation, or any other institution or organization.

Dedication We are pleased to dedicate this paper to our friend and colleague Panos Pardalos on the occasion of his 70th birthday.

Appendix: The Stadium Simulator

The Stadium Simulator runs on Java, using the JavaFX framework to produce the UI; the jfreechart library to create graphs and data visualizations; and the JUNG library to visualize the layout of the various security components. The Stadium Simulator employs a message-passing approach to simulating patrons moving through security. Components are connected in a directed acyclic graph (DAG), which is defined by an XML file tailored to the specific situation being simulated. Using an arbitrary DAG as the underlying connective infrastructure allows for immense flexibility in how the system is setup. An included UI simplifies the creation of the XML file for common security setups.

The system runs in discrete time-steps, each representing one-tenth of a second. Using the underlying DAG, each component is aware of all components up-stream of it in the security setup.

There are two kinds of messages:

1. Pull messages: Sent by a down-stream component to one or more up-stream components, this message indicates that the component is ready to receive a specific number of patrons.
2. Data message: Sent in response to a pull message, this message contains a list of patrons that are moving to the component that sent the pull message.

At each time-step, empty components will send a pull message to up-stream components (those closer to the parking lot) requesting patrons. If there are available patrons, the up-stream component will respond with a data message with those patrons. If not, the pull message will be ignored for this time-step and the empty component will re-send the message next time-step.

Components of the system fall into four categories:

1. Arrivals: These simulate the arrival of patrons into the security setup by generating new patron objects. The arrival can be simulated using a supplied probability distribution, such as a Poisson distribution. The simulator can also represent two patterns of arrival to account for an early wave (perhaps due to a promotional giveaway) and a later wave arriving closer to the start of the event. Each patron is marked with its time of arrival. This can also mark patrons as having contraband or being part of the red team, which is used by screener components.
2. Departures: These simulate a patron leaving the security setup, either by entering the stadium or by being sent back to the parking lot. These components can record how long each individual patron was in the security setup. They can also register if any contraband/red team members reach them, either marking them as having been successfully sent back to the parking lot or registering a system failure if such a patron reaches the stadium.
3. Screeners: Any number of different security screening systems including WTMDs, pat-downs, wanding, bag checks, etc. can be used. The patron's delay is simulated using a provided probability distribution. A second probability

distribution determines the detection rate. Patrons with detected contraband or who are recognized as part of the red team are redirected to a different output, often a departure component representing a patron being sent back to the parking lot.

4. Lines: A simple queue of patron objects waiting for some other component, often a screener.

References

1. Stewart, M., Mueller, J.: Risk-based passenger screening: risk and economic assessment of TSA PreCheck increased security at reduced cost? NOVA. The University of Newcastle's Digital Repository (2016)
2. Pita, J., Jain, M., Western, C., Paruchuri, P., Marecki, J., Tambe, M., Ordonez, F., Kraus, S.: Deployed ARMOR protection: the Application of a Game Theoretic Model for Security at the Los Angeles International Airport. In: Proceedings of the Seventh International Joint Conference on Autonomous Agents and Multiagent Systems, AAMAS08, May 2008. ACM, New York (2008)
3. Tsai, J., Rathi, S., Kiekintveld, C., Tambe, M., Ordonez, F.: IRIS: a tool for strategic security allocation in transportation networks. In: Proceedings of the International Conference on Autonomous Agents and Multiagent Systems, AAMAS, May 2009
4. Pita, J., Kiekintveld, C., Tambe, M., Steigerwald, E., Cullen, S.: GUARDS - innovative application of game theory for national airport security. In: Proceedings of the International Joint Conference on Artificial Intelligence (IJCAI), July 2011
5. An, B., Shieh, E., Yang, R., Tambe, M., Baldwin, C., DiRenzo, J., Maule, B., Meyer, G.: PROTECT – a deployed game theoretic system for strategic security allocation for the United States coast guard. AI Mag. **33**, 96–110 (2012)
6. Jiang, A., Yin, Z., Tambe, M., Kiekintveld, C., Sullivan, J.P., Sandholm, T., Leyton-Brown, K.: TRUSTS: scheduling randomized patrols for fare inspection in transit systems using game theory. AI Mag. **33**, 59–72 (2012)
7. Brown, M., Sinha, A., Schlenker A.: One size does not fit all: a game-theoretic approach for dynamically and effectively screening for threat. In: Proceedings of the Thirtieth AAAI Conference on Artificial Intelligence, February 2016
8. Fang, F., Stone, P., Tambe, M.: When security games go green: designing defender strategies to prevent poaching and illegal fishing. In: IJCAI'15 Proceedings of the 24th International Conference on Artificial Intelligence, pp. 2589–2595 (2015)
9. Dick, A.: A method and system for dynamically optimizing the security of persons gathered in bounded spaces. Web page. https://techfinder.rutgers.edu/tech/A_Method_and_System_for_Dynamically_Optimizing_the_Security_of_Persons_Gathered_in_Bounded_Spaces (n.d.). Accessed 1 Nov 2024
10. Kantor, P.B., Egan, D.E., Menkov, V.: Method and system for dynamically improving the performance of security screening. U.S. Patent Application US16/494,138 (2018)
11. Hall, S.A., Cooper, W.E., Marciani, L., McGee, J.A.: Security Management for Sports and Special Events. Human Kinetics (2012)
12. Kar, B., Zale, J.J.: A GIS-based football stadium evacuation model. Southeast. Geogr. **52**, 70–89 (2012)
13. Gianchini, P., Gounsoulin, J.M., Hart, K.W., et al.: Risk-informed assessment of Scott Stadium evacuation through agent-based simulation. In: Proc. Systems and Information Engineering Design Symposium (SIEDS), pp. 163–168. IEEE (2010)
14. Mahmudzadeh, A., Ghorbani, M., Hakimelahi, A.: Providing an emergency evacuation model for the stadium. Transp. Res. Procedia. **48**, 620–631 (2020)

15. Klupfel, H., Meyer-Konig, T.: Simulation of the evacuation of a football stadium using the CA model PedGo. In: Traffic and Granular Flow '03, pp. 423–428. Springer (2005)
16. Kirchner, A., Klupfel, H., Nishinari, A., et al.: Simulation of competitive egress behaviour. Physica A. **324**, 689–697 (2003)
17. Pan, X., Han, C.S., Dauber, K., Law, K.H.: A multi-agent based framework for the simulation of human and social behaviors during emergency evacuations. AI Soc. **1**, 113–132 (2007)
18. Chin, K., Young, T., Chorlton, B., Aucoin, D., Gales, J.: Crowd behaviour in Canadian football stadia — part 1: data collection. Can. J. Civ. Eng. **49**, 1254–1262 (2022). https://doi.org/10.1139/cjce-2021-0425
19. Chin, K., Young, T., Chorlton, B., Aucoin, D., Gales, J.: Crowd behaviour in Canadian football stadia —part 2: modelling. Can. J. Civ. Eng. **49**, 1263–1271 (2022). https://doi.org/10.1139/cjce-2021-0426
20. Rodriguez, S., Giese, A., Amato, N., et al.: Environmental effect on egress simulation. In: Proc. of the 5th Intern. Conf. on Motion in Games (MIG), 2012. Lecture Notes in Computer Science. Springer (2012)
21. Edmunds, T.: Simulation analysis of inspections of international travelers at Los Angeles International Airport for US-VISIT. Lawrence Livermore National Laboratory, UCRL-TR-202734 (2004, March)
22. Holguin-Veras, J., Xu, N., Bhat, C.: An assessment of the impacts of inspection times on the airline industry's market share after September 11th. J. Air Transp. Manag. **23**, 17–24 (2012)
23. Pendergraft, D.R., Robertson, C.V., Shrader, S.: Simulation of an airport passenger security system. In: Proceedings 2004 Winter Simulation Conference (2004)
24. Ricks, B.C., Nakamura, B., Almaz, A., DeMarco, R., Hui, C., Kantor, P., Matlin, A., Nelson, C., Powell, H., Roberts, F., Thompson, B.: Modeling the impact of patron screening at an NFL stadium. In: Guan, Y., Liao, H. (eds.) Proceedings of 2014 IIE Industrial and Systems Engineering Research Conference (ISERC2014) (2014)
25. Chang, K.C., Lane, N., Uhmeyer, A., Jones, M., Hubbell, M., Reuther, A., Robert Seater, R.: Simulation approach to sensor placement using unity 3D. In: 2018 IEEE High Performance extreme Computing Conference (HPEC), pp. 1–6. https://api.semanticscholar.org/CorpusID:54440874 (2018)

Sentiment and Conflict Prediction in Urban Development: Data-Driven Approach

Nailia Gabdrakhmanova and Maria Pilgun

1 Introduction

Real-time data analysis is in demand in various fields, such as in the field of medical laboratory diagnostics [1]; for observations of coupled biological, chemical, and physical processes in the ocean from the macro to micro scale [2]; in real-time intelligent systems [3]; and for high-speed 3D railroad tie deflection mapping in real time using an array of air coupled non-contact transducers [4], etc.

The objectives of the study were to develop and test an algorithm that includes the integration of neural network and mathematical models for analyzing digital content to identify semantic accents that characterize the dissatisfaction of residents, as well as to assess the positioning features of the project in the media space, segments of the greatest informational attention, social tension among residents of the metropolis during the urban planning project implementation, as well as to predict the development of the situation.

In particular, the study was aimed at analyzing the dynamics of information activity on digital resources to track changes in the moods of active residents of the city and certain districts of the big city involved in the discussion of the construction of the second ring metro line (RL2), to identify key content topics that initiate users' involvement in the discussion of the project, and to timely forecast emerging

N. Gabdrakhmanova (✉)
Peoples Friendship University of Russia, Moscow, Russia
e-mail: gabdrakhmanova-nt@rudn.ru

M. Pilgun
Department of General and Comparative-Historical Linguistics, Lomonosov Moscow State University, Moscow, Russia

Research Institute of Prospective Directions and Technologies, Russian State Social University, Moscow, Russia

B. Goldengorin (ed.), *Theory, Algorithms, and Experiments in Applied Optimization*, Springer Optimization and Its Applications 226,
https://doi.org/10.1007/978-3-031-91357-0_5

and/or developing conflict situations. The analysis was conducted during the active construction phase, which included the construction of three new stations of the second ring subway line (RL2), named here station #1, station #2, and station #3, as well as the reconstruction of station #4.

2 Materials and Methods

The material for the study was the verbal content generated by users on digital platforms, dedicated to the implementation of the RL2 project, as well as digital footprints. The data were collected between October 1, 2020 and June 10, 2021. Quantitative characteristics of data:

Number of tokens—62,657,289;
Number of messages—7063;
Maximum number of messages per day—614;
Number of active authors—933;
Activity (posts per author)—7,57;
Number of sources—213.

Microblogs, messengers, social networks, and video platforms lead in audience reach (Fig. 1).

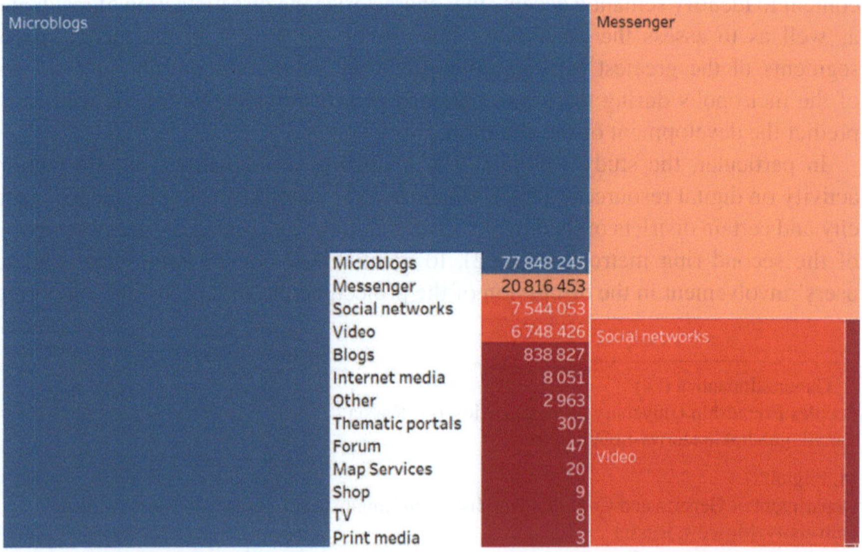

Fig. 1 Digital sources by audience reach

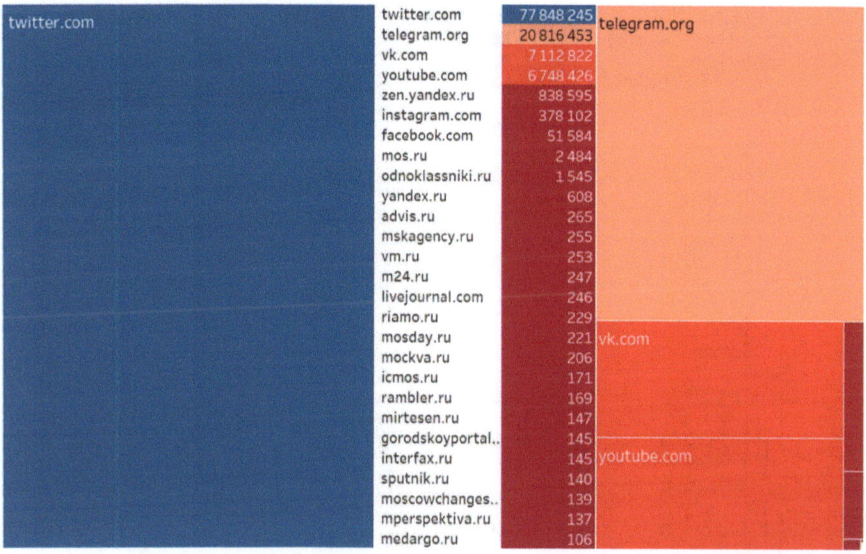

Fig. 2 Types of Sources by Audience Reach

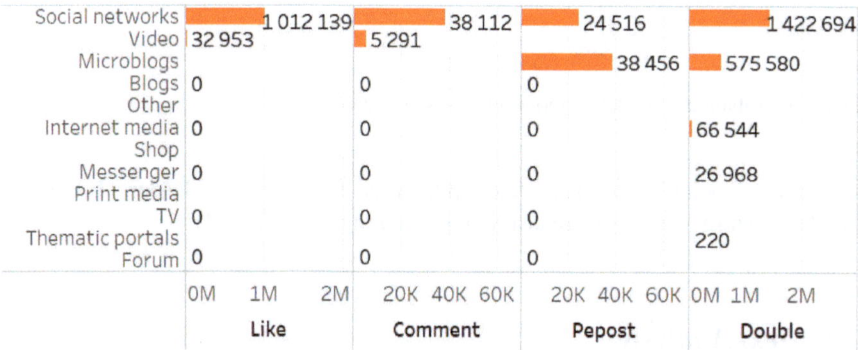

Fig. 3 Distribution of digital footprints of actors by source type

Twitter, Telegram, VKontakte, and YouTube rank among the top sources by audience reach (Fig. 2).

To generate content related to the project's implementation, actors preferred using social networks, video hosting platforms, and microblogs (Fig. 3).

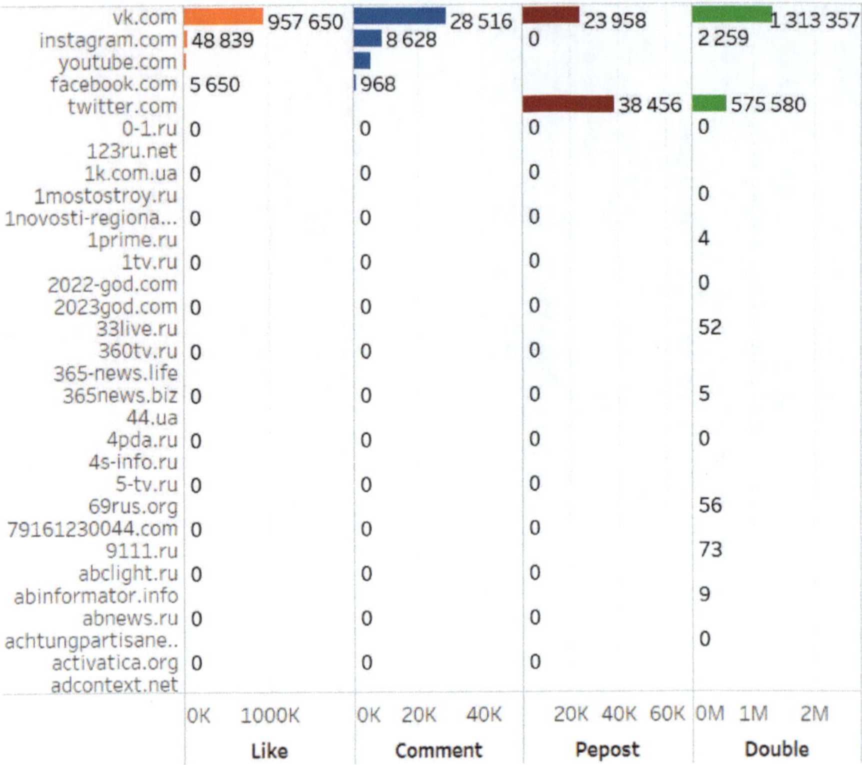

Fig. 4 Distribution of digital footprints of actors by sources

Among social networks, VKontakte was the absolute leader. Instagram, YouTube, and Facebook were also popular among users (Fig. 4).

2.1 *Text Analysis*

The article presents a methodology for determining the perception of a certain situation by actors, which is implemented based on the results of the analysis of the content generated by users and their digital footprints. The study involved an interdis-ciplinary approach. To interpret the data, neural network text analysis, analysis of lexical associations using the TextAnalyst 2.3 technology, a detailed description of which is presented in [5, 6]; content analysis [7–9] using the AutoMap service; and senti-ment analysis using the Eureka Engine sentiment module were performed. For visual analytics, the Tableau platform was used.

The design of the text analysis is presented in the flowchart:

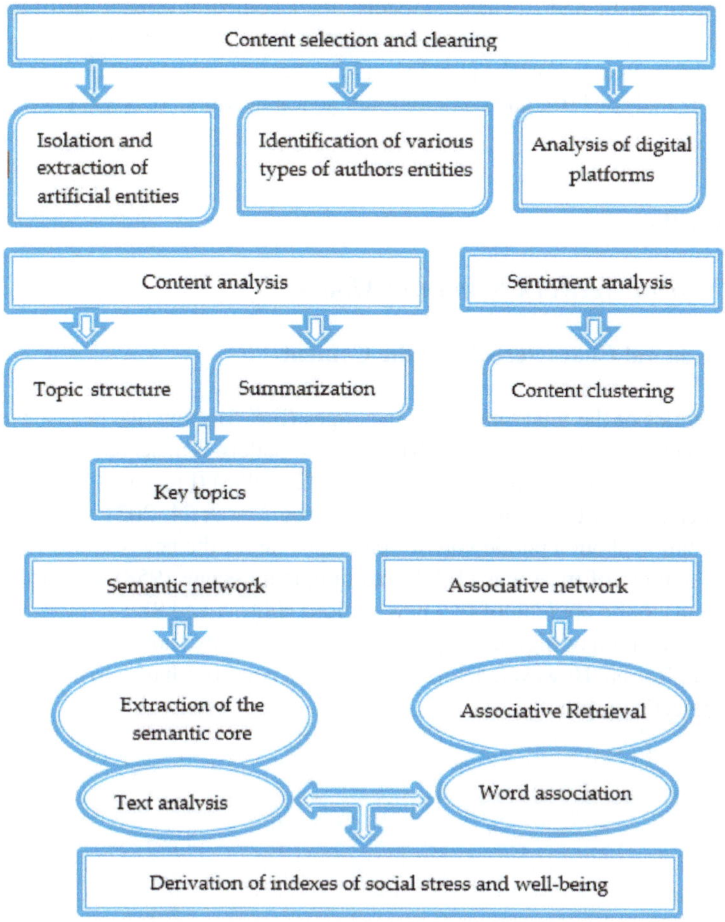

2.2 *Approaches to Mathematical Modeling*

It follows from the objective formulated above that the mathematical model shall be dynamic. Dynamic mathematical models make it possible to analyze the situation and make predictions several steps ahead. In this article, the authors have developed several different types of mathematical models to study the potential of a conflict situation based on digital data. At the first stage, the problem was formalized, and a time series was constructed based on pre-processed digital data. Currently, quite a lot of methods have been developed for the analysis of time series [10–13]. The article presents the results of developing models according to observational data using deterministic and stochastic models; and a comparative analysis of models in terms of the prediction accuracy was conducted. When developing models, ordinary differential equations (ODEs) [14, 15], stochastic differential equations (SDE) [15], regression analysis [16], and convolutional neural networks (CNN) [17] were used.

3 Results

3.1 Neural Network Semantic Model

3.1.1 General Characteristics of the Content

The dynamics of the total number of messages (Fig. 5) show two peaks: March 24, 2021 (2241) and April 1, 2021 (1699). A similar pattern is observed in the dynamics of unique messages, with a peak on April 1, 2021 (958) (Fig. 6).

The peak in the total number of messages on March 24, 2021 (2241) is driven by information about a public vote to select a name for the new station. The peak in the total number of messages (1699) and unique messages (958) on April 1, 2021, corresponds to the opening of two new stations (Figs. 7 and 8).

The geolocation of digital footprints of actors logically shows a predominance of users from Russia. However, actors from various countries have also shown interest in the project's progress.

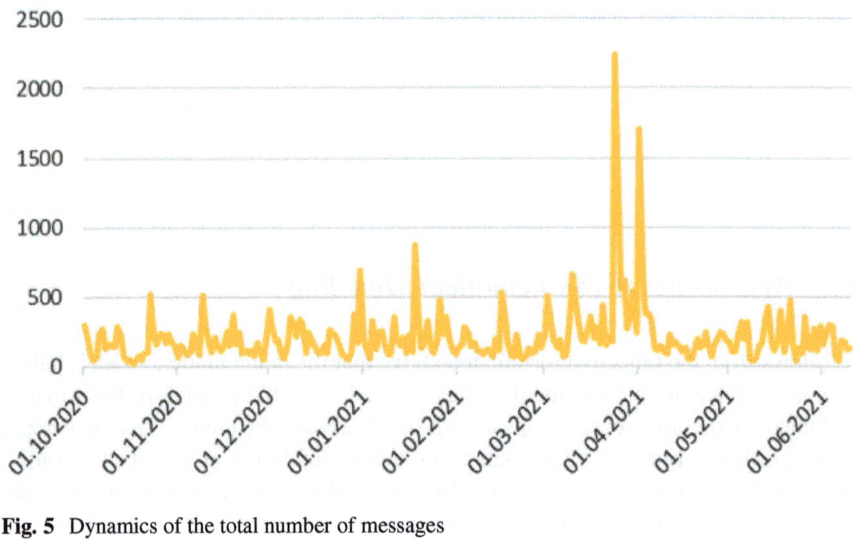

Fig. 5 Dynamics of the total number of messages

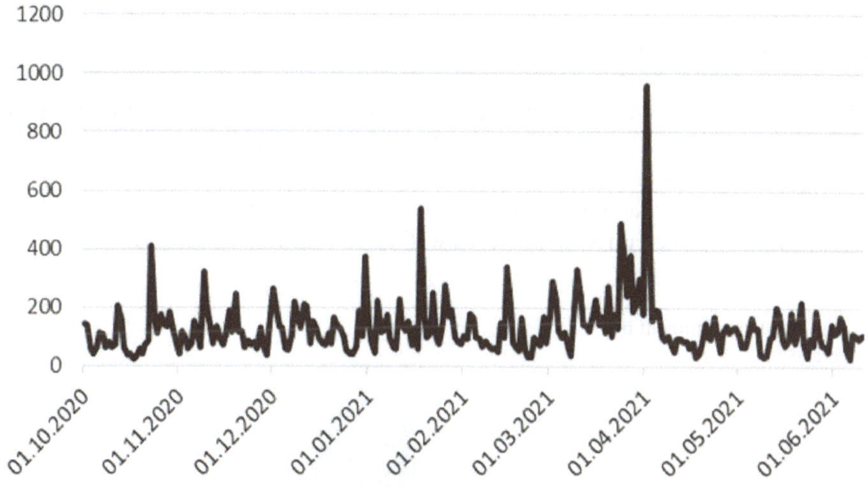

Fig. 6 Dynamics of the number of unique messages

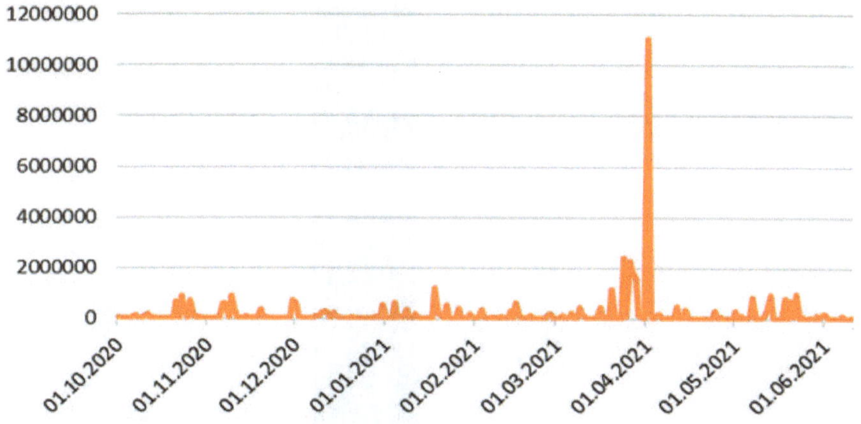

Fig. 7 Dynamics of the number of views

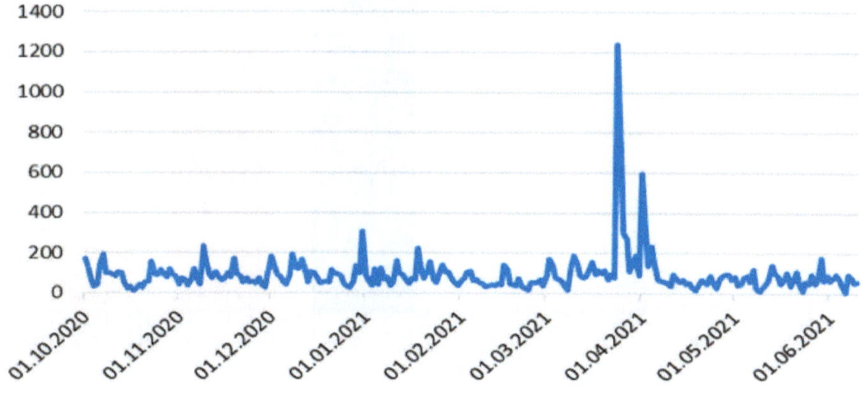

Fig. 8 Dynamics of author activity

3.1.2 Sentiment of the Content

The sentiment feature of the vast majority of messages and digital footprints in the context of references to the project is neutral (Figs. 9 and 10).

Negative and neutral posts were generated on the following platforms (in descending order): VKontakte, Telegram, and Twitter. In contrast, positive messages were primarily posted on VKontakte, as well as on mos.ru and Telegram.

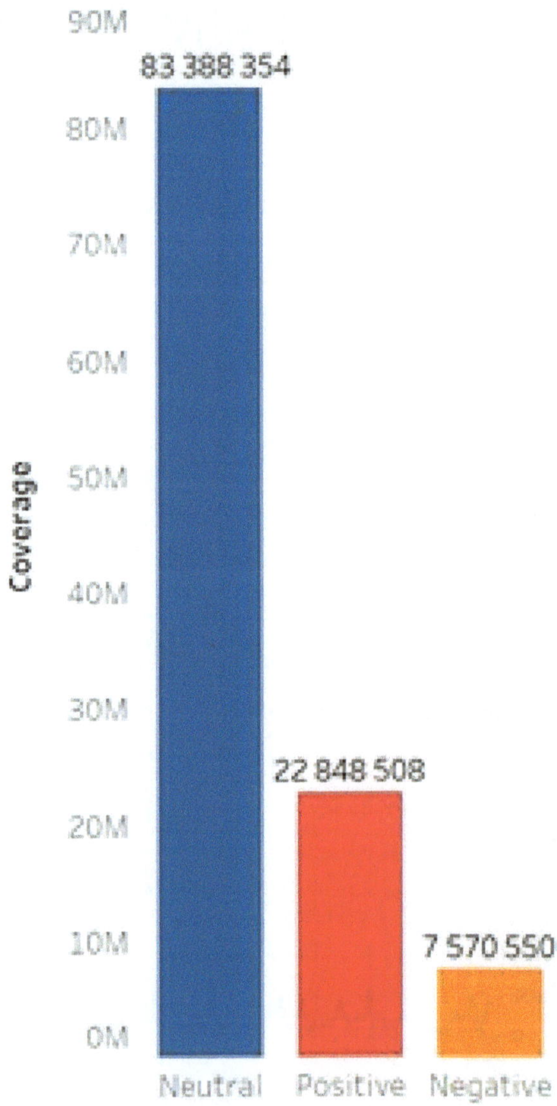

Fig. 9 Content sentiment by coverage

Fig. 10 Users' digital footprints sentiment

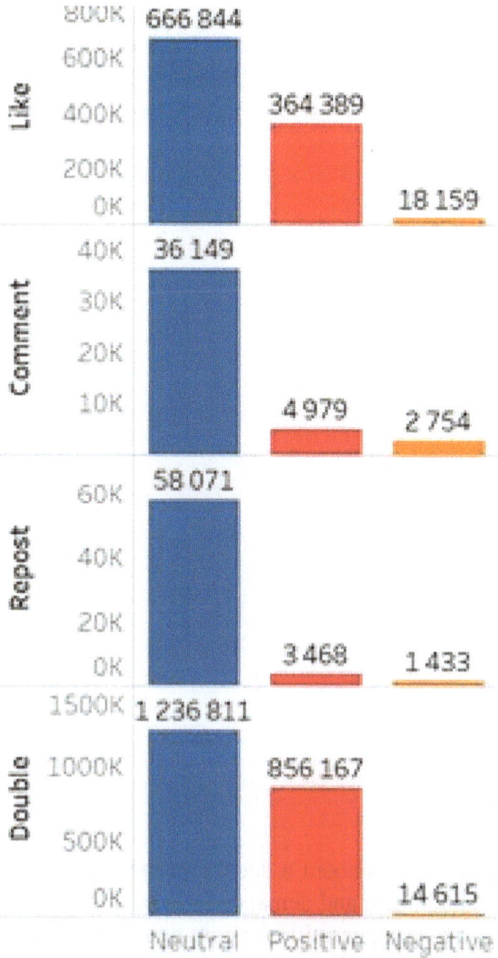

Data analysis revealed a significant increase in relevant content in 2021 compared to 2020 (Fig. 11). Of course, it is important to consider that most of the content is generated by biased media. However, it is worth noting that the number of negative messages increased by 2.1 times, neutral messages by 2.2 times, and positive messages by 3.1 times (Fig. 12).

The top sources by sentiment are presented in Figs. 13, 14, and 15.

Key negative semantic accents presented in the user-generated content

- Complaints against designers and builders for urban planning mistakes that do not consider residents' needs.
- Citizens perceive the location of new stations as poorly planned.

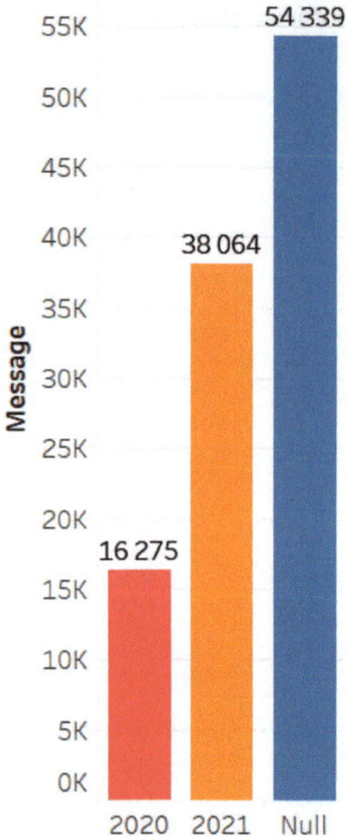

Fig. 11 Dynamics of the total number of messages

- Concerns about a decline in the quality of life, potential damage to residential buildings, and demands for insurance contracts for housing, reduced utility fees, and compensation for communal payments.
- Residents disapprove of the placement of station exits and the removal or relocation of pedestrian crossings.
- Increased noise levels in residential buildings during construction, especially from detour roads passing under windows.
- Construction work being carried out at night.
- Narrowing of sidewalks and relocation of pedestrian zones directly under residential windows.
- Risk of technological disasters during construction.
- Worsening traffic conditions in the area due to road closures during construction.
- Destruction of green spaces in the city.
- Concerns that the new lines will overcrowd the transport system, making it difficult for residents of older districts to board.
- Changes to established public transport routes due to construction.

Fig. 12 Dynamics of messages by sentiment

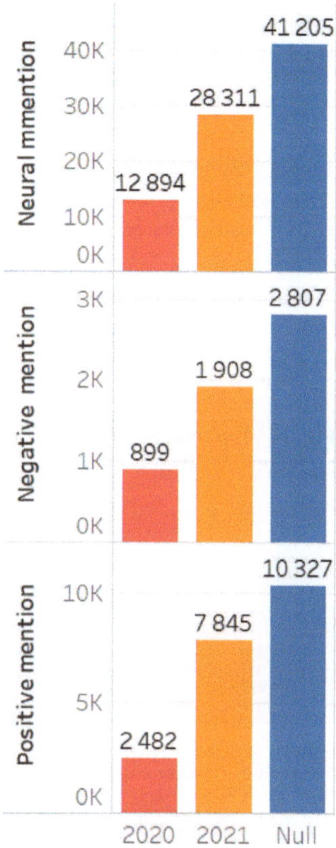

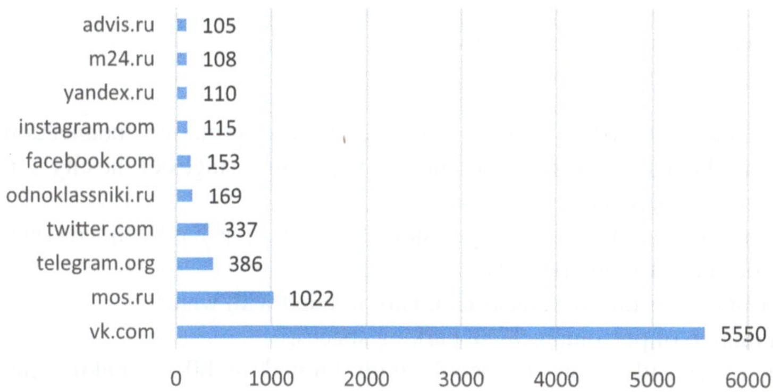

Fig. 13 Top sources with positive sentiment

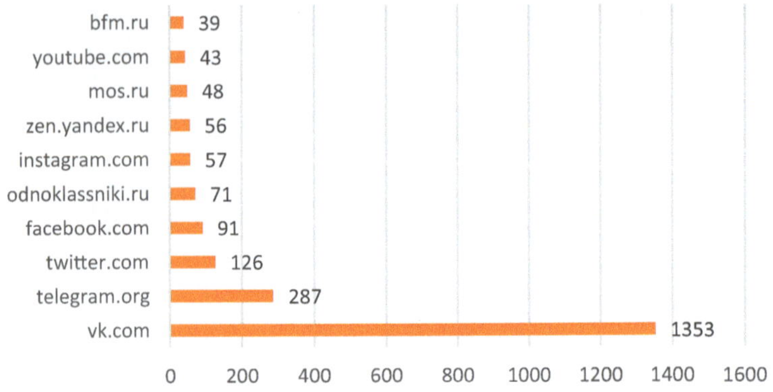

Fig. 14 Top sources with negative sentiment

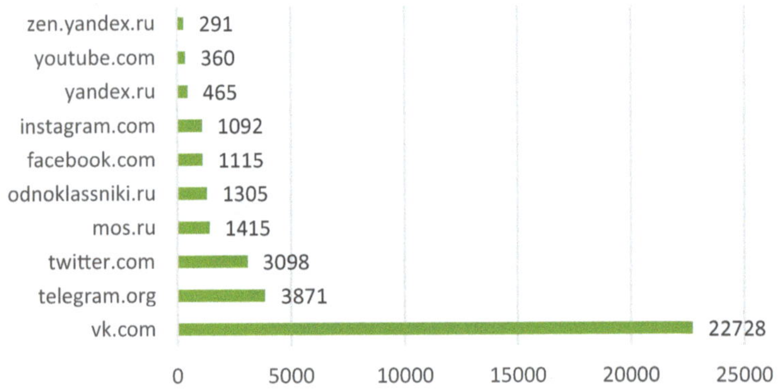

Fig. 15 Top sources with neutral sentiment

- Complicated transfer systems, ineffective transportation connections, and decreased planning efficiency compared to previous stages of the city's transport system development.
- Citizens believe that the quality of new station designs has declined compared to previous construction periods.
- Excavation pits dug for construction are flooded with water.
- Elimination of personal vehicle parking spaces.
- Lack of consideration for people with limited mobility, making the urban environment less accessible.
- Negative aspects of construction aggravate existing conflicts between residents and developers, particularly dis-satisfaction with frequent resurfacing of pedestrian pathways.

Table 1 Social stress and social well-being indices

Facility name	Social stress index	Social well-being indices
RL2	5.39	12.75

3.1.3 Level of Social Tension

The result of the analysis of the consolidated database showed a low level of social stress of 5.39 and a moderate value of the social well-being index of 12.75 (Table 1).

Analysis of the data showed a low level of social tension caused by the RL2 construction. In terms of the number of negative messages, the station #2 is slightly ahead of other stations; the minimum number of negative reactions is observed for the station #1. In the positive cluster, the station #4 is the leader; the minimum number of positive messages is related to the station #1. In the neutral cluster, the station #3 takes the first place; the minimum number of neutral messages is related to the construction of the station #4.

3.2 Mathematical Models

The problem under consideration belongs to the class of problems that are difficult to formalize. Based on the above analysis of the situation according to digital data, the following formalization of the problem was adopted. The following groups have been introduced: P is a group of actors with a positive attitude toward the project; pos(t) is the number of comments of the group P at the moment t, N is a group of actors with a negative attitude toward the project; neg(t) is the number of comments of the group N at the moment t; U is a group of actors with a neutral attitude toward the project; u(t) is the number of comments of the group U at the moment t. In view of the problem under consideration, we assumed that the percentage of all three groups of the part society that is not involved in social networks is the same as for the part of the society that is involved. Due to the fact that we are primarily interested in the dynamics of the number of actors with a positive and negative attitude, and given that the activity of posts varies depending on the day of the week, holidays, etc. the following normalization is adopted. Two new variables have been introduced: $p(t) = \text{pos}(t)/u(t)$, $n(t) = \text{neg}(t)/u(t)$. In Fig. 16, the dynamics of the initial data $p(t)$ and $n(t)$ is presented.

3.2.1 General Formulation

Time series $\{p(t)\}$, $\{n(t)\}$ are given, where $t = 1,.., N$, which characterize the level of positive and negative attitude toward the project, respectively. The sampling step is constant. It is necessary to build dynamic models for predicting and analyzing the emergence of a conflict situation according to these data.

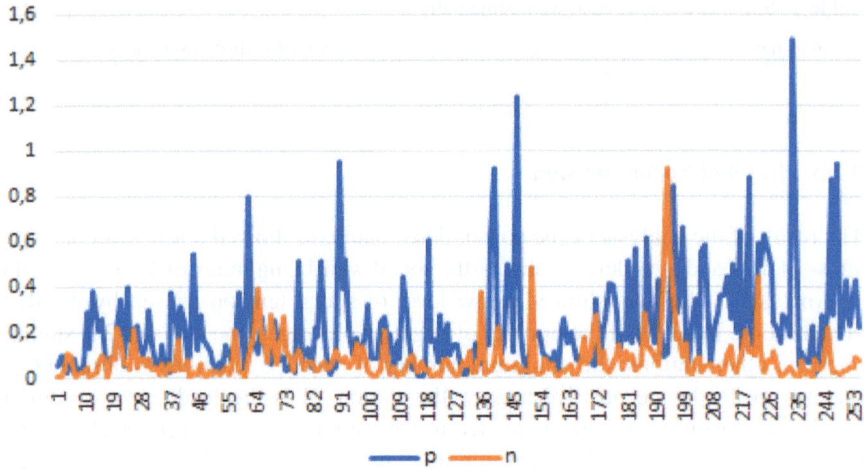

Fig. 16 The dynamics of the initial data $p(t)$ and $n(t)$ is presented

3.2.2 Mathematical Models with Regression Analysis

The best known and widely used models for time series analysis are various auto-regressive models. Trends of the time series $\{p(t)\}$, $\{n(t)\}$ are built using regression analysis methods. As a result of estimating the model parameters, the following equations are obtained:

$$p(t) = 0,1 + 0,001\ t;\ n(t) = 0.06 + 0.0001\ t$$

The obtained estimates of the trend coefficients preceding t indicate that the series $\{p(t)\}$ grows faster than the series $\{n(t)\}$. Mathematical models based on ordinary differential equations (ODEs). To analyze the behavior of a dynamic system, it is important to have continuous mathematical models. The most suitable continuous models are differential equations. In this section, we present the results of building mathematical models based on ODEs. When choosing the general form of the differential equation, the problem under consideration can be represented as a competition problem: two-species struggle in populations, an arms race, military operations, etc. [10]. We have looked at various ODE models for describing competition. After comparing the results of the built models, systems of autonomous differential equations were chosen to describe the processes. The chosen equation is as follows:

$$\begin{cases} \frac{dp}{dt} = \alpha_1 \bullet n + \beta_1 \bullet p + \gamma_1 \\ \frac{dn}{dt} = \alpha_2 \bullet n + \beta_2 \bullet p + \gamma_2 \end{cases} \quad (1)$$

Model parameter estimates (1) were found using the multiple regression method and the Nelder–Mead method. At the first step, using regression analysis, we found coefficient estimates using the regression analysis method for each Eq. (1) separately; then this solution was adjusted using the Nelder–Mead method for the system. The process of coefficient approximation resulted in the following values:

$$\alpha_1 = 0.07;\ \beta_1 = 0.8;\ \gamma_1 = -0.18;\ \alpha_2 = -0.05;\ \beta_2 = 0.5;\ \gamma_2 = -0.03.$$

For ODE models, it is important that the estimates of parameters α and α_2 are within the following range:

$$0 < \alpha_1 < 1;\ 0 < \alpha_2 < 1.$$

The model obtained (1) was studied in the nOp plane in order to determine the qualitative behavior of the $p(t)$, $n(t)$ functions in time. Equation (1) has the following equilibrium position:

$$\begin{cases} \frac{dp}{dt} = 0 \\ \frac{dn}{dt} = 0 \end{cases}$$

The equilibrium values of n and p are obviously found from the following conditions:

$$\begin{cases} \alpha_1 \bullet n + \beta_1 \bullet p + \gamma_1 = 0 \\ \alpha_2 \bullet n + \beta_2 \bullet p + \gamma_2 = 0 \end{cases}$$

The following values of equilibrium states are calculated: $p_0 = 0.288$, $n_0 = 0.08$.

The values of the equilibrium states match the estimates of the mathematical expectation of the $p(t)$, $n(t)$ time series. We used the result obtained when choosing a stochastic differential equation (SDE) model. This result can also indicate the adequacy of the built model. Analyzing the estimates of the coefficients for the system of differential Eq. (1), it can be noted that the series $\{p(t)\}$ grows faster than the series $\{n(t)\}$.

3.2.3 Mathematical Models Based on Stochastic Differential Equations (SDEs)

To describe the evolution of systems with interacting elements, there are two approaches: the construction of deterministic or stochastic models. Stochastic calculus is a powerful tool for describing many processes. A stochastic differential equation (SDE) is a differential equation in which one or more terms represent a stochastic process. The SDE solution is also a stochastic process. Unlike deterministic ones, stochastic models make it possible to take into account the probabilistic nature of birth-and-death processes, as well as the effects of the external environment, which cause random fluctuations in the model parameters. The most commonly used example of an SDE is an equation with a white noise term. For each time series, a mathematical model was built in the form of the Ornstein–Uhlenbeck process:

$$dp = \alpha_1 \bullet (\beta_1 - p(t)) + \sigma_1 \bullet \delta W_1 \tag{2}$$

$$dn = \alpha_2 \bullet (\beta_2 - n(t)) + \sigma_2 \bullet \delta W_2 \tag{3}$$

where W is the Wiener process. To estimate the parameters of models (2) and (3), the Vasicek model construction algorithm was used [18]. As a result of building the model, the following parameters were obtained:

$$\alpha_1 = 0.79;\ \beta_1 = 0.23;\ \sigma_1 = 0.21;\ \alpha_2 = 0.46;\ \beta_2 = 0.08;\ \sigma_2 = 0.08.$$

For SDE models, for an equilibrium to exist, the following condition must be met:

$$\beta_1 \bullet \beta_2 > \alpha_1 \bullet \alpha_2.$$

This condition is met for the built models.

3.2.4 Mathematical Model with CNNs

Traditionally designed for 2D image data, CNNs can be used to model univariate and multivariate time series prediction problems. In our problem, the data are considered in the form of a multivariate time series. The input of the CNN was data with an interval of 2. The choice of the size of the number of input time steps has an important impact on how much data will be used for training. An estimate of the size of the number of input time steps was obtained using estimates of time series autocorrelations. When training the neural network, the input of the CNN was the vector ([$x(k$-2),$y(k$-2)], ([$x(k$-1),$y(k$-1)])), and the vector [$x(k)$,$y(k)$] was taken as a response vector. The process was implemented in the Google Colaboratory environment (Google Research product used for code writing); the models were implemented and trained using machine learning libraries such as tensorflow, keras, sklearn. Additionally, the mathematical libraries Numpy, Pandas and Matplotlib were used for calculations and plotting. When training the CNN, the ADAMAXoptimizer was used, and ReLU was used as an activation function. The following training results were obtained: Epoch 98/100: 164/164- 0 s- loss: 0.0140- 405 ms/epoch- 2 ms/step; Epoch 99/100: 164/164- 0 s- loss: 0.0141- 404 ms/epoch- 2 ms/step; Epoch 100/100: 164/164- 0 s- loss: 0.0139- 396 ms/epoch- 2 ms/step; train score: 0.18 RMSE; test score: 0.30 RMSE. Comparative analysis of the models obtained. To assess the quality of the mathematical models built, we used the average retro prediction error for ten steps ahead for the time series $\{p(t)\}$ and $\{n(t)\}$. The estimate of the mean error is found by the formula:

Table 2 Mean retro prediction error for all models

Model	$p(t)$	$n(t)$
AR	0.04	0.009
ODE	0.1624	0.016
SDE	0.084	0.025
CNN	0.025	0.003

$$E = \frac{1}{n}\left(\sum_{i=1}^{n}\left(x_i - x_i^m\right)\right)^2$$

where x_i is the value of the time series at the *i-th* point of the retro prediction; $(x_i)^m$ is the model value of the time series at the *i-th* point of the retro prediction. Table 2 shows the calculated values of the mean prediction error for all models.

It follows from the table that the models built using convolutional neural networks yield the smallest prediction error. However, the built analytical models are important for the analysis of the situation. Such models are models built on the basis of ODEs and SDEs. Comparing the average retro prediction errors of the ODE and SDE models, we can conclude that it is important to take into account the stochastic nature of the processes.

4 Conclusions

The main (key) tool in the study undertaken is neural network models and differential equations. At the first stage, neural networks were used for text analysis and data filtering. At the next stage, on the basis of the obtained solutions, the problem was formalized and mathematically formulated. It is shown that it is possible to analyze and monitor the situation based on the forecast of time series using CNNs. As a result of solving the problem, the following specific conclusions were made, regarding this situation.

The study showed that the integration of neural network and mathematical models for the analysis of digital content made it possible to identify real points of dissatisfaction and positioning of the project in the media space, segments of the greatest informational attention, and social tension among residents of the big city and its districts around the RL2 construction place, as well as to predict the development of the situation. The level of social stress and digital aggression determined in this study showed how sensitive residents of the big city and residents of construction areas are to the information activity and involvement of network users in the discussion of the RL2 construction and ongoing construction work on the sector and also made it possible to draw conclusions about the level of social tension and approval of the current situation around the object of study. Certain potential risks of conflict with residents may arise during the construction of the station #2. Meanwhile, data analysis and calculation of social stress and social well-being indices made it possible to predict the absence of strong opposition to the construction on the part of the residents of the district. The course of events confirmed the correctness of the results obtained.

References

1. Gressner, A.M., Arndt, T. (eds.): Lexikon der Medizinischen Laboratoriumsdiagnostik. Direct Analysis in Real Time. Springer Reference Medizin. Springer, Berlin, Heidelberg (2019). https://doi.org/10.1007/978-3-662-48986-4-311065
2. Venkatesan R., Tandon, A., D'Asaro, E., Atmanand, M.A. (eds.): Observing the Oceans in Real Time, 323 p. Springer, Cham (2018)
3. Mizera-Pietraszko, J., Pichappan, P., Mohamed, L. (eds.): Lecture Notes in Real-Time Intelligent Systems, 526 p. Springer, Cham (2019
4. Datta, D., Hosseinzadeh, A.Z., Cui, R., di Scalea, F.L.: High-Speed 3D Railroad Tie Deflection Mapping in Real-Time Using an Array of Air-Coupled Non-contact Transducers, vol. 254. Springer, Cham (2023)
5. Kharlamov, A., Pilgun, M. (eds.): Neuroinformatics and Semantic Representations: Theory and Applications, 317 p. Cambridge Scholars Publishing, Newcastle upon Tyne (2020)
6. Pilgun, M., Raskhodchikov, A.N., Antonova, K.O.: Effects of Covid-19 on multilingual communication. Front. Psychol. **12** (2022)
7. Krippendorff, K.: Content Analysis. An Introduction to Its Methodology, 3rd edn, 441 p. SAGE Publications, Los Angeles (2012)
8. Mayring, P.: Qualitative content analysis. In: A Companion to Qualitative Research, pp. 159–176. SAGE (2004)
9. White, M.D., Marsh, E.: Content analysis: a flexible methodology. Libr. Trends. **55**(1), 22–45 (2006)
10. Time series analysis. In: Schintler, L.A., McNeely, C.L. (eds.) Encyclopedia of Big Data. Springer, Cham. (2022)
11. Nokeri, T.C.: Time-series analysis. In: Data Science Revealed. Apress, Berkeley, CA (2021). https://doi.org/10.1007/978-1-4842-6870-4-3
12. Onder, I., Wei, W.: Time series analysis. In: Egger, R. (ed.) Applied Data Science in Tourism. Tourism on the Verge. Springer, Cham (2022). https://doi.org/10.1007/978-3-030-88389-8-22
13. Gabdrakhmanova, N., Fedin, V., Matsuta, B.: The modeling of forecasting new situations in the dynamics of the economic system on the example of several financial indicators. Procedia Comput. Sci. **186**, 512 (2020)
14. Arnold, V.: Geometric Methods in the Theory of Ordinary Differential Equations. MTSNMO, Moscow (2012)
15. Samarsky, A., Mikhailov, A.: Principles of Mathematical Modeling: Ideas, Methods, Examples. CRC Press, London (2001)
16. Aivazyan, S.: Methods of Econometrics. Magistr, INFRA-M (2010)
17. Scholle, F.: Deep Learning in Python. SPb, Peter (2021)
18. Shiryaev, A.: Fundamentals of Stochastic Financial Mathematics. Fasis, Moscow (1998)

Flexible Pooling Pattern Design with Integer Programming

Uwe Gotzes, Annika Buchholz, Josef Kallrath, Niels Lindner, and Thorsten Koch

1 Introduction

The COVID-19 pandemic has highlighted the urgent need to conduct large numbers of polymerase chain reaction (PCR) tests rapidly and cost-effectively. A *sample* is a collected swab of a tested individual and is not to be confused with a random selection from a population in the statistical sense. As testing samples is time-consuming, *pooling* of samples is a well-established method to increase laboratory testing capacity, particularly in the detection of diseases such as COVID-19 [18, 20]. The idea is to combine several samples into a pool, so that a negative result for the pool implies a negative result for all its samples.

In order to reduce the probability of false positives, it is crucial to detect the cases when not all of the samples in a pool are positive. To solve this group testing problem, which dates back to syphilis detection [6], two approaches are common: Traditional pooling strategies rely on adaptive testing, where each test depends on

U. Gotzes
Universität Duisburg-Essen, Essen, Germany
e-mail: uwe.gotzes@uni-due.de

A. Buchholz · N. Lindner (✉)
Zuse Institute Berlin, Berlin, Germany
e-mail: buchholz@zib.de; lindner@zib.de

J. Kallrath
University of Florida, Gainesville, FL, USA
e-mail: jkallrath@ufl.edu

T. Koch
Technische Universität Berlin, Berlin, Germany

Zuse Institute Berlin, Berlin, Germany
e-mail: koch@zib.de

B. Goldengorin (ed.), *Theory, Algorithms, and Experiments in Applied Optimization*, Springer Optimization and Its Applications 226,
https://doi.org/10.1007/978-3-031-91357-0_6

the results of previous ones, requiring a laboratory feedback loop and at least two PCR runs to confirm positive cases. This process can be time-consuming and delay critical decisions. For a comprehensive discussion on adaptive testing methods, see [7]. More recent works employ, for example, Bayesian inference [14] and game-theoretic approaches [3].

We instead pursue a nonadaptive testing scheme following Täufer [17], where each sample is distributed in a prescribed manner over several pools, yielding reliable results after a single round of testing. By carefully designing the pooling patterns and leveraging knowledge about the prevalence of the disease, he provides explicit error bounds on the number of false positives, which scale favorably with pool size and sample multiplicity. This approach allows for significantly streamlined PCR testing and reductions in detection times for large-scale testing scenarios. A thorough information-theoretic view with focus on algorithms for nonadaptive group testing is given in [1].

For the schemes that we will consider, the following parameters are relevant: A number of samples N should be assigned to pools in such a way that each pool contains material from at most n samples (*pool size*), and conversely, each sample should be distributed among k pools (*multiplicity*).

However, in [17], the pool design is constrained to specific parameters. The number of samples N is assumed to be the square of the pool size n for multiplicity $k = 3$, and for multiplicities $3 < k \leq n + 1$, the pool size is additionally required to be a prime number. According to [17], in practice, one will want to work with much lower multiplicities than $n + 1$, since a high multiplicity would require many tests and defeat the purpose of pooling. From a practical perspective, it seems reasonable to generate large pools by a sequence of unions of equally diluted pools. This leads to pool sizes that are powers of 2, which are not prime numbers except for the case $n = 2$ itself. Täufer, relying on close connections between the parameters, suggests adding negative dummy samples to achieve the prerequisites. With our method, we can however refrain from adding dummy samples and directly solve the models with the desired parameters.

In this paper, we describe and solve integer programming and (pseudo-)Boolean satisfiability models to flexibly compute pool design patterns. The models allow users to freely select parameters such as the number N of samples (swabs) to be tested, the maximum pool size n, and the number of pools each swab appears in (multiplicity k). The solution ensures that any two swabs co-occur in a predefined maximum number of different pools, which is critical for accurate identification of positive samples. Swabs are allocated to pools (wells) on a microplate, with the option to use a 24-well microplate, a 96-well microplate, or any other size. Given the initially unknown or uncertain rate of positive cases, a positive swab is likely to be identified without needing a second round of testing, provided the parameters are chosen appropriately.

Our primary contribution is providing a practical and constructive perspective on designing flexible pooling patterns without the restrictive parameters required in previous works. Our approach allows for arbitrary choices of sample numbers, pool sizes, and multiplicities, making it highly adaptable to real-world applications.

Moreover, we present a method to quickly compute optimal pooling strategies to minimize the number of tests necessary, based on mathematical optimization techniques. This method could be integrated into automated testing systems, further enhancing its practical utility.

The mathematical framework presented in this paper not only does apply to PCR testing but applies to a multitude of areas, including, e.g., drug design [11], cybersecurity [19], deep learning [12], environmental modeling [15], wastewater testing for viral loads [4], and food production [10]. In quality testing of products, for instance, samples from different production lines can be pooled and tested for contamination, faulty components, or adherence to standards [16].

We illustrate the relation between the parameters (N, n, k) and the false-positive rate in Sect. 2. Our mathematical model is described in Sect. 3, along with several formulations as integer programming and satisfiability problems. An algorithm to extend solutions to larger instances is given in Sect. 4. We discuss our computational results in Sect. 5, before concluding the paper in Sect. 6.

2 Choosing Parameters

In this section, we discuss background on how to select the parameters (N, n, k) as described above.

In our approach, as in [17], we assume perfect performance of the PCR test and that the probability ρ of an arbitrary sample being positive is small compared to the inverse pool size $1/n$. An individual is considered positive if all the pools containing their sample return positive results. An item is deemed negative if it appears in at least one negative pool. Multipooling ensures that all infected individuals are correctly identified, as all pools containing their samples will test positive.

Assume that $0 < \varepsilon_{\text{fp}} < 1$ represents the acceptable proportion of false positives among all positive results. From [17], we know that if $\rho \leq \frac{1}{2}$, then for any multipooling strategy with pool size n and multiplicity k, the probability of a positive test being a false positive does not exceed ε_{fp} if

$$k \geq \frac{\ln\left(\dfrac{\varepsilon_{\text{fp}}}{1-\varepsilon_{\text{fp}}}\right) + \ln\left(\dfrac{\rho}{1-\rho}\right)}{\ln\big(1-(1-\rho)^{n-1}\big)}. \tag{1}$$

As noted in [17], the number of tests required in a multipool strategy is Nk/n, representing an improvement over individual testing by a factor of n/k. A key insight is that the lower bound on k in Inequality (1) scales favorably with larger pool sizes n. Specifically, k grows logarithmically with both the pool size n and $1/\varepsilon_{\text{fp}}$, as discussed in [17].

To illustrate this, we compute the required multiplicity k for different values of prevalence ρ and pool sizes n, keeping $\varepsilon_{\text{fp}} = 0.01$. The results are shown in Table 1.

Table 1 Computed lower bounds for k from inequality (1) for different ρ and n, with $\varepsilon_{fp} = 0.01$

n	$\rho = 0.1$	$\rho = 0.01$	$\rho = 0.001$
2	2.95	2.00	1.67
4	5.20	2.61	1.98
8	10.44	3.42	2.32
16	29.46	4.67	2.73
32	174.62	6.97	3.30
64	5181.52	12.14	4.11

For a low prevalence, the multiplicity required k grows slowly with increasing pool size n, which is favorable for practical applications.

Equation 1 can be converted to ε_{fp} which can be useful if ρ, n, and k are given:

$$\varepsilon_{fp} \leq \frac{\left(1-(1-\rho)^{n-1}\right)^k \cdot (1-\rho)}{\rho + \left(1-(1-\rho)^{n-1}\right)^k \cdot (1-\rho)} \tag{2}$$

We also want to mention the formula for the portion of samples falsely declared positive among all samples declared positive in a multipool strategy [17]:

$$\mathbb{P}(\text{sample } s \text{ negative} \mid \text{all pools } p \text{ containing } s \text{ positive})$$
$$= \frac{(1-\rho)\left(1-(1-\rho)^{n-1}\right)^k}{\left(\rho + (1-\rho)\left(1-(1-\rho)^{n-1}\right)^k\right)}. \tag{3}$$

In our setting, Eq. (3) becomes a $\leq$-inequality, because our pools may contain less than n elements.

Yelin et al. [20] tested RNA extracts of positive samples for the presence of SARS-CoV-2, alone or in pools of different sizes (2-, 4-, 8-, 16-, 32-, and 64-sample pools) with negative samples. They found that a single positive sample can be detected in pools of up to 32 samples using standard kits and protocols, with an estimated false-negative rate of 10%. Detection of positive samples diluted in even up to 64 samples may also be attainable, although this may require additional amplification cycles. Due to our assumption of perfect test performance, we will not consider pool sizes larger than 32.

3 Mathematical Models

The question of which combinations (N, n, k) allow for a multipool design appears to be a complex combinatorial problem, as indicated by Täufer [17]. In this section, we describe the building blocks of our model, including the sets, parameters, and variables used, as well as the objective function and constraints.

Table 2 Parameters of the model

Symbol	Description
N	Total number of swabs
M	Total number of wells that can accommodate (pooled) swab material
n	Maximum number of swabs per pool
$\overline{p}$	Maximum number of pools that contain the same two swabs
k	Number of pools that contain material from a specific swab

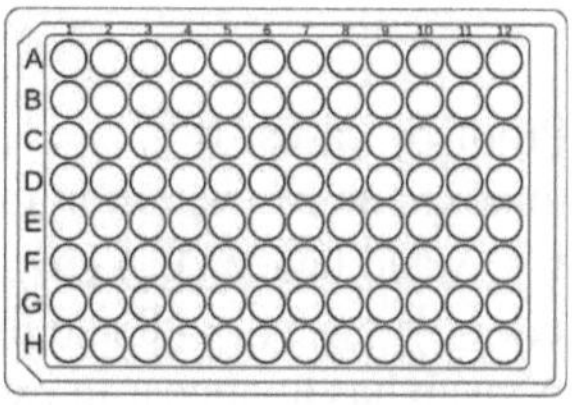

Fig. 1 Schematic of 96-well microplate

3.1 Parameters, Sets, and Variables

The parameters of our model are summarized in Table 2.

Practically, $\overline{p}$ is always chosen as one, because larger values create unnecessary and unwanted correlations between different pools. In addition, computing the probability of false positives becomes difficult for values $\overline{p} \neq 1$. Also, the choice of k according to Inequality (1) is no longer valid in this case.

We introduce the set of swabs $S := \{1, \dots, N\}$ and the set of wells (pools) on a microplate $P := \{1, \dots, M\}$. For example, for a microplate as depicted in Fig. 1, $P = \{1, \dots, 96\}$. For microplates of other sizes, P needs to be adjusted accordingly. Each well can accommodate one pool of swabs for testing.

We now introduce binary variables $x_{sp} \in \{0, 1\}$ for all $s \in S$ and $p \in P$ and $z_p \in \{0, 1\}$ for all $p \in P$. $x_{sp} = 1$ if and only if the material of swab s is present in pool p. Similarly, $z_p = 1$ if and only if well p is used to host a pool.

3.2 Binary Quadratically Constrained Program

We want to minimize the number of pools needed:

$$\min \sum_{p \in P} z_p \tag{4}$$

For a swab to be in a pool, the pool must be "open":

$$x_{sp} \leq z_p \quad \text{for all } s \in S, p \in P \tag{5}$$

At most n swabs are allowed in each pool:

$$\sum_{s \in S} x_{sp} \leq n \quad \text{for all } p \in P \tag{6}$$

We can aggregate (5) and (6) to

$$\sum_{s \in S} x_{sp} \leq n \cdot z_p \quad \text{for all } p \in P. \tag{7}$$

Proof by cases: If $z_p = 1$, then Inequalities (6) and (7) become identical, and (5) imposes no restriction in this case. $z_p = 0$ in Inequality (5) implies $x_{sp} = 0$ for all $s \in S$ and hence $\sum_{s \in S} x_{sp} \leq 0 = n \cdot z_p$. On the other hand, $z_p = 0$ in (7) implies $x_{sp} \leq \sum_{s \in S} x_{sp} = 0$ and therefore $x_{sp} \leq z_p$ for all $s \in S$.
A swab has to appear in exactly k pools:

$$\sum_{p \in P} x_{sp} = k \quad \text{for all } s \in S \tag{8}$$

Any two swabs co-occur in at most $\overline{p}$ pools:

$$\sum_{p \in P} x_{ip} \cdot x_{jp} \leq \overline{p} \quad \text{for all } (i, j) \in S \times S \text{ with } i < j \tag{9}$$

The solution of this optimization problem forms an *incidence structure*. If we additionally require the sum in constraint (6) and constraint (7), respectively, to be greater than 1 and set $\overline{p} = 1$ in constraint (9), the solution corresponds to a *partial linear space*. With further regularity conditions—specifically equality in (6)—the solution, if one exists, is a so-called *tactical configuration* and constitutes a *multipool* according to the definition in [17]. In this case, the optimal value is Nk/n, whereas in the more general non-multipool case, the number of required tests (pools) will be larger. The lower bound for k given in Inequality (1) is valid regardless of the choice of N and n and whether equality or inequality constraints are used. A pooling matrix $A \in \{0, 1\}^{M \times N}$ has an entry $a_{ij} = 1$ iff sample n is in pool m. In nonadaptive testing, a pooling matrix for a multipool is called a $(k-1)$-disjunct matrix, and it is known that it can identify up to $(k-1)$ positive samples [13]. We and [17] are interested in scenarios where the number of infected samples can exceed the multiplicity k.

As we assume perfect PCR performance, our goal of pooling is to reduce the number of tests and avoid a second round of testing. From this perspective, scenarios where $k \geq n$ (see Table 1) are not practical, as in this case $Nk/n \geq N$, meaning the number of tests will not be reduced. In practice, PCR tests have error rates, and pooling can still be beneficial even if it does not reduce the number of tests compared to individual testing. Combinatorial designs, particularly balanced incomplete block designs (BIBDs), offer advantages in mitigating the impact of these errors through redundancy. In our approach, at most $\overline{p}$ pools contain any two distinct samples, and not all pools need to have the same size. In BIBDs, all blocks (pools) contain the same number of elements (samples), and each pair of distinct elements appears together in the same number of blocks. The number of blocks is always greater than or equal to the number of elements, but BIBDs provide a built-in level of redundancy that helps identify inconsistent results, thereby improving overall robustness in error-prone testing environments (cf. [5]).

For $\overline{p} = 1$, constraint (9) can equivalently be formulated from a "pool perspective" rather than a "swab perspective":

$$\sum_{s \in S} x_{sp} \cdot x_{sq} \leq \overline{p} \quad \text{for all } (p, q) \in P \times P \text{ with } p < q \tag{10}$$

These are fewer constraints than (9) since the number of samples is typically larger than the number of pools. Inequality (9) limits the number of pools in which pairs of swabs appear together. For any pair (i, j), the number of pools where both swabs i and j are assigned together must not exceed $\overline{p}$. It controls swab co-occurrence. Inequality (10) limits the number of swabs shared between pairs of pools. Specifically, for any pair of pools (p, q), the number of swabs that are in both pools p and q must not exceed $\overline{p}$. It controls pool overlap. With $\overline{p} = 1$, both conditions restrict overlaps to at most one occurrence. In (9) any two swabs can be together in at most one pool. In (10), any two pools can share at most one common swab. More formally, we can show $\neg$ (9) $\Leftrightarrow \neg$ (10). We start with $\neg$ (10) $\Rightarrow \neg$ (9) and assume $\sum_{s \in S} x_{sp} \cdot x_{sq} > 1$ for one pair $(p_1, p_2) \in P \times P$. Then there are $s_1, s_2 \in S$ with $x_{s_1 p_1} = x_{s_1 p_2} = x_{s_2 p_1} = x_{s_2 p_2} = 1$, so that $\sum_{p \in P} x_{s_1 p} \cdot x_{s_2 p} \geq x_{s_1 p_1} \cdot x_{s_2 p_1} + x_{s_1 p_2} \cdot x_{s_2 p_2} = 2$. The other inclusion follows in the same manner. For $\overline{p} = 0$, Inequality (9) prevents any two swabs from being together in a pool, effectively forcing pools to contain at most one element. Inequality (10) requires that pools are completely disjoint in this case. When $\overline{p} > 1$, Inequality (9) allows swabs to co-occur in up to $\overline{p}$ pools, focusing on limiting how often swabs are together in the same pool. Inequality (10) allows pools to share up to $\overline{p}$ swabs, focusing on limiting pool overlap. Look at $\overline{p} = 2$, $S = \{1, 2\}$, and $P = \{1, 2, 3\}$ and all $x_{sp} = 1$. In this case, $\sum_{s \in S} x_{sp} \cdot x_{sq} = 2$ but $\sum_{p \in P} x_{ip} \cdot x_{jp} = 3$. Thus, except for $\overline{p} = 1$, the conditions impose different restrictions on the assignment of swabs to pools and are not equivalent. From our perspective, the constraints are equivalent since $\overline{p} = 1$ in practice.

The integer quadratic model with the objective function (4) and a choice of constraints according to

$$\Big(\big((5) \wedge (6)\big) \vee (7)\Big) \wedge \Big((8) \wedge \big((9) \vee (10)\big)\Big)$$

can be transformed into an integer linear model (BLP) by applying the standard technique to linearize products of binary variables in constraints like (9) or (10):

$$y = x_1 \cdot x_2 \Leftrightarrow (y \le x_1 \wedge y \le x_2 \wedge x_1 + x_2 \le 1 + y) \text{ for } x_1, x_2, y \in \{0, 1\} \quad (11)$$

By adding cuts of the form $\sum x^*_{sp} \le c - 1$ to the model—where x^*_{sp} are variables from a previous optimal solution that were set to 1 and c is their count—different solutions for the same parameters can easily be obtained. This technique allows us to explore alternative optimal solutions and enhances the flexibility of our approach.

3.2.1 Symmetry Breaking

Empirically, it is of computational benefit to include the constraints

$$z_p \ge z_{p+1} \quad \text{for } p \in \{1, \ldots, M-1\} \quad (12)$$

that demand that pools with smaller indices are opened at first. The model is valid without these constraints, but they rule out a large number of essentially equivalent solutions.

3.2.2 Maximizing the Number of Swabs for a Given Microplate Size

The model can easily be adapted to maximize the number of swabs that can fit in M wells for given M, n, and k. In this case, we need variables $u_s \in \{0, 1\}, s \in S$, which indicate whether a given swab can be accommodated. Then, we add u_s to (8)

$$\sum_{p \in P} x_{sp} = k \cdot u_s \quad \text{for all } s \in S \quad (13)$$

and use it together with constraints (6) and (9) to $\max \sum_{s \in S} u_s$.

3.3 *Quadratic Unconstrained Binary Optimization*

In addition to the binary linear programming formulation, we also explored the problem using Quadratic Unconstrained Binary Optimization (QUBO) to enhance flexibility and accuracy. Furthermore, QUBOs are expected to be solved quickly by future quantum computers, especially quantum annealers. Many heuristics and metaheuristics (such as greedy algorithms, local search, or genetic algorithms) perform well on QUBO problems. However, due to the quadratic constraints, we need to linearize the model, which leads to a large number of variables. Furthermore, the QUBO formulation requires the introduction of binary slack variables as described in Table 3.

The objective function has been adjusted to yield a quadratic objective without constraints. The penalty $\mathcal{P}$ is a scalar that must be selected to be sufficiently large to ensure that all constraints of the original model are satisfied:

Table 3 Binary slack variables of the QUBO model

Symbol	Domain
$\sigma 1_{pn}$	$(p, n) \in P \times N_1, \quad N_1 = \{0, \dots, n\}$
$\sigma 2_{sp}$	$(s, p) \in S \times P$
$\sigma 3_{ijn}$	$(i, j, n) \in S \times S \times N_3, \quad N_3 = \{0, 1\}$
$\sigma 4_{pij}$	$(p, i, j) \in P \times S \times S$
$\sigma 5_{pij}$	$(p, i, j) \in P \times S \times S$
$\sigma 6_{pij}$	$(p, i, j) \in P \times S \times S$
y_{pij}	$(p, i, j) \in P \times S \times S$

$$
\begin{aligned}
\min \sum_{p \in P} z_p + \mathcal{P} \Bigg(& \sum_{p \in P} \Big(\sum_{s \in S} x_{sp} + \sum_{n \in N_1} \sigma 1_{pn} - n \cdot z_p \Big)^2 \\
& + \sum_{s \in S} \Big(k + \sum_{p \in P} \sigma 2_{sp} - \sum_{p \in P} x_{sp} \Big)^2 \\
& + \sum_{i<j}^{N} \Big(\sum_{p \in P} y_{pij} + \sum_{n \in N_3} \sigma 3_{ijn} - 1 \Big)^2 \\
& + \sum_{i<j}^{N} \Big(\sum_{p \in P} y_{pij} + \sigma 4_{pij} - x_{ip} \Big)^2 \\
& + \sum_{i<j}^{N} \Big(\sum_{p \in P} y_{pij} + \sigma 5_{pij} - x_{jp} \Big)^2 \\
& + \sum_{i<j}^{N} \Big(\sum_{p \in P} x_{ip} + x_{jp} + \sigma 6_{pij} - 1 - y_{pij} \Big)^2 \Bigg)
\end{aligned} \tag{14}
$$

This problem is convex, which suits solvers based on relaxing the binary variables.

3.4 Pseudo-Boolean and Satisfiability Models

Our binary quadratic program can naturally be interpreted in terms of pseudo-Boolean optimization. Some dedicated pseudo-Boolean solvers can deal with both linear and quadratic constraints. Otherwise, as shown above, the nonlinear constraint (10) may be linearized.

In particular, we can rewrite our model as a satisfiability problem, just asking whether a solution exists for a particular setting of (M, N, n, k). This allows us to substantially simplify our model, as we can eliminate the z_p variables. The objective function (4) is no longer needed and (7) is replaced by (6).

This approach fits very well with modeling as a pseudo-Boolean problem. Furthermore, we can now use another approach to solve our problem via reformulation as a Boolean satisfiability (SAT) problem, transforming the constraints into conjunctive normal form. One way to achieve this is the technique of sequential weight counters [9].

For example, for the constraints (10), this results in the following set of clauses, introducing N new variables c_s for each pair $(p, q) \in P \times P$:

$$\neg c_{s-1} \vee c_s \qquad \text{for } s \in \{2, \dots, N\}, \tag{15}$$

$$\neg x_{sp} \vee \neg x_{sq} \vee c_s \qquad \text{for } s \in \{1, \dots, N\}, \tag{16}$$

$$\neg x_{sp} \vee \neg x_{sq} \vee \neg c_{s-1} \qquad \text{for } s \in \{2, \dots, N\} \tag{17}$$

The correctness of this transformation can be seen as follows: Let s be the smallest number such that $x_{sp} = x_{sq} = 1$. Then $c_s = 1$ by (16), and hence $c_t = 1$ for all $t \in \{s, \dots, N\}$ by (15). This makes it impossible to find a $t \in \{s+1, \dots, N\}$ with $x_{tp} = x_{tq} = 1$, because then $c_{t-1} = 0$ by (17).

As SAT problems are decision problems, these reformulations will not minimize but decide whether there is a feasible solution with a given number M of pools. Applying the sequential weight counter procedures yields the following clauses with Nn additional variables d_{si} for $s \in \{1, \dots, N\}$ and $i \in \{0, \dots, n-1\}$, for each $p \in P$:

$$\neg d_{s-1\,i} \vee d_{si} \qquad \text{for } s \in \{2, \dots, N\}, i \in \{0, \dots, n-1\}, \tag{18}$$

$$\neg x_{sp} \vee \neg d_{s-1\,i-1} \vee d_{si} \qquad \text{for } s \in \{2, \dots, N\}, i \in \{1, \dots, n-1\}, \tag{19}$$

$$\neg x_{sp} \vee d_{s\,0} \qquad \text{for } s \in \{1, \dots, N\}, \tag{20}$$

$$\neg x_{sp} \vee \neg d_{s-1\,n-1} \qquad \text{for } s \in \{2, \dots, N\} \tag{21}$$

The rationale is that d_{si} indicates whether more than i of the first s swabs have been selected. The constraints (8) can be relaxed to

$$\sum_{p \in P} x_{sp} \geq k \qquad \text{for all } s \in S, \tag{22}$$

which is in turn equivalent to

$$\sum_{p \in P} (1 - x_{sp}) \leq M - k \qquad \text{for all } s \in S. \tag{23}$$

Its SAT reformulation by means of sequential weight counters is analogous to the one of (6), replacing $\neg x_{sp}$ with x_{sp} and requiring $M(M-k)$ additional variables.

3.5 Solver Overview

We now have multiple possibilities to solve our problem: minimizing the number of required pools for a given (N, n, k) or maximizing the possible number of swabs for a given (M, n, k) by a

- Binary linear program (BLP)
- Quadratic unconstrained binary optimization (QUBO) programming problem
- Binary quadratically constrained program (BQCP)
- Pseudo-Boolean optimization problem (PBO)

or given preset (M, N, n, k) check for a feasible solution by a

- Binary linear program (BLP)
- Binary quadratically constrained program (BQCP)
- Pseudo-Boolean satisfaction problem (PBS)
- Boolean satisfiability problem (SAT)

4 Extensions for Larger Problems

Building on an existing (N, n, k)-solution (or feasible point), we can derive two key components for solving an $(N + 1, n, k)$-instance:

1. An effective bound, which is simply the optimal value (or the best lower bound) of the (N, n, k)-instance
2. A strong starting solution for the $(N + 1, n, k)$-instance by applying Algorithm 1 to the (N, n, k)-solution

This method allows us to iteratively address larger instances effectively.

Algorithm 1: Sample Insertion Algorithm – generate $(N + 1, n, k)$-solution

```
Input: Existing solution for (N, n, k)-instance; N ≫ n > k
Output: Solution for (N + 1, n, k)-instance
Insert sample N + 1 into as many pools as possible (up to k) that have less than n samples;
Let ℓ be the number of such pools (possibly zero);
if ℓ = k then
    Solution for (N + 1, n, k)-instance found;
else
    m ← k − ℓ;
    Remove one sample from each of any m pools that do not accommodate sample
      N + 1;
    Ensure that the removed samples are distinct across the selected pools;
    Insert sample N + 1 into these m pools that now have space after the removal;
    Form a new pool from the removed m samples and sample N + 1;
    Solution for (N + 1, n, k)-instance found;
end
```

5 Computational Results

Our computations have been carried out on an Intel Xeon Gold 6132 machine with 14 cores, 28 threads, a 2.6–3.7-GHz clock, and 384 GB RAM. We use Gurobi 11 [8] to solve the BQCP formulation, which is internally treated as a BLP, and also for the PBS model. The SAT models are handled with CaDiCaL [2]. Table 4 lists some results obtained with the different approaches. Columns n_{min} and n_{max} refer to the minimum and maximum pool sizes utilized in the respective solution. Notably, BQCP is able to solve many instances to optimality and, by its nature, gives lower and upper bounds on the optimal number M of pools. We use this bound range to invoke the feasibility-based models SAT and PBS that require a fixed value of M. For PBS, we also include as a lower bound the highest value of M for which the model has been proven to be infeasible (plus one). We omit such lower bounds for SAT, as no unsatisfiability proof could be obtained within the wall time limit of 30 minutes and the bound range provided by BQCP.

Table 4 Computational results: input parameters (N, n, k)—BQCP, lower (LB) and upper (UB) bounds for the optimal number M of pools, total running time in seconds, time to UB solution, minimum (n_{min}) and maximum (n_{max}) pool size for the UB solution; SAT, best solution found, running time; PBS, best lower bound (1+ highest infeasible M), best upper bound, running times

Input			BQCP						SAT		PBS			
N	n	k	LB	UB	time[s]	best[s]	n_{min}	n_{max}	UB	time[s]	LB	time [s]	UB	time[s]
24	4	3	18	18	1	1	4	4	18	1	18	1	18	1
24	6	4	16	19	1800	545	4	6	20	103	17	94	19	119
24	8	5	20	–	1800	–	–	–	24	1274	22	310	27	24
32	4	3	24	24	1	1	4	4	24	2	24	1	24	1
32	6	4	22	22	203	203	5	6	22	11	22	1	22	515
32	8	5	21	35	1800	1608	2	6	29	342	26	991	31	6
48	4	3	36	36	9	9	4	4	36	5	36	2	36	2
48	6	4	32	32	304	304	1	6	32	18	32	1	32	9
48	8	5	30	–	1800	–	–	–	37	66	31	237	37	1562
64	4	3	48	48	32	32	4	4	48	11	48	8	48	5
64	6	4	43	43	182	182	1	6	43	2	43	3	43	39
64	8	4	32	34	1800	1055	5	8	33	23	32	2	32	510
64	8	5	40	–	1800	–	–	–	43	360	40	5	45	381
64	16	4	16	46	1800	1208	1	9	31	1129	25	408	31	759
80	10	3	24	24	796	796	10	10	25	6	24	1	24	14
96	4	3	72	72	116	116	4	4	72	51	72	47	72	24
96	6	4	64	64	841	841	–	–	64	119	64	36	64	164
96	8	5	60	–	1800	–	–	–	60	734	60	57	62	944
320	10	3	96	96	874	874	1	7	96	1391	96	943	96	197

We conclude from Table 4 that the hardness of the instances correlates heavily with the multiplicity k: The BQCP could not be solved to optimality for any of the five instances with $k = 5$, and for four of them, no solution was found at all, whereas SAT performed particularly well in this case. On the other hand, all instances with $k = 3$ were solved to optimality, both with BQCP and PBS.

In general, SAT and PBS perform well in finding feasible solutions, with no clear winner. They are often faster than BQCP, but of course with the drawback that the number of pools needs to be specified as an input to SAT and BQCP.

For $\rho = 0.01$, a $(64, 8, 2)$-multipool will have 31.4% false positives among all positives. Theorem 2 in [17] allows for the construction of a $(64, 8, 3)$-multipool which reduces the probability of a positive being false positive to 3.01%. Täufer [17] points out that switching to $k = 4$ to achieve similar false-positive rates when increasing the pool size is not straightforward. This is where our method comes into play. We configure our model with $(64, 8, 4)$ and get the pooling matrix with $M = 32$ pools shown in Fig. 2 as a result. In the visualization, a square represents a value of 1, while all other entries are 0. The probability of a positive being false positive is 0.21% then. If we switch over to $(64, 16, 4)$, this probability would be less than or equal to 3.65%. A pooling matrix for these parameters is shown in Fig. 3. It can be seen that the potential of at most 16 swabs per pool is by far not utilized. At most nine swabs occur in one pool together (row 7, for instance) to achieve the multiplicity of four while minimizing the number of pools needed. For this instance, $M = 31$ tests are needed and the lower bound is 25.

A popular choice is $n = 10$, $k = 3$ and $\overline{p} = 1$. With these parameters, for example, 320 swabs can be allocated to a 96-well microplate, and 1280 swabs to a 384-well microplate. We can easily derive a corresponding pooling matrix for the larger microplate after rearrangement using the solution variables x_{sp}. Figure 4 shows a 320×96 matrix, indicating that 96 pool tests are performed on 320 individuals, with each individual's swab appearing in 3 pools. If the prevalence ρ is not larger than 0.003 and $\varepsilon_{\text{fp}} \leq 0.01$ is accepted, one could use the derived pooling matrix for programming an automated pipetting robot (Fig. 5).

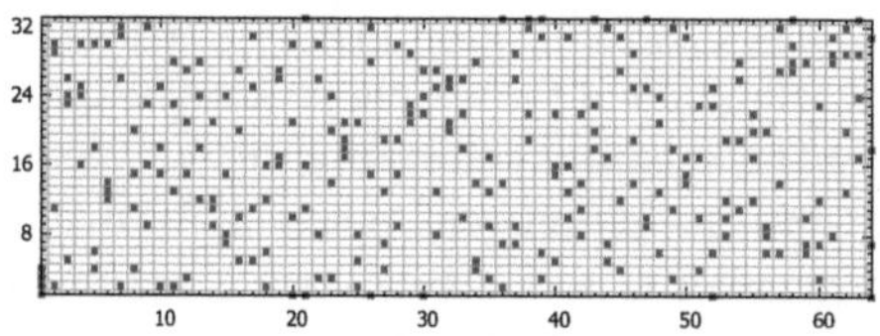

Fig. 2 Pooling matrix for 32 tests on 64 individuals: multiplicity 4, pool size 8

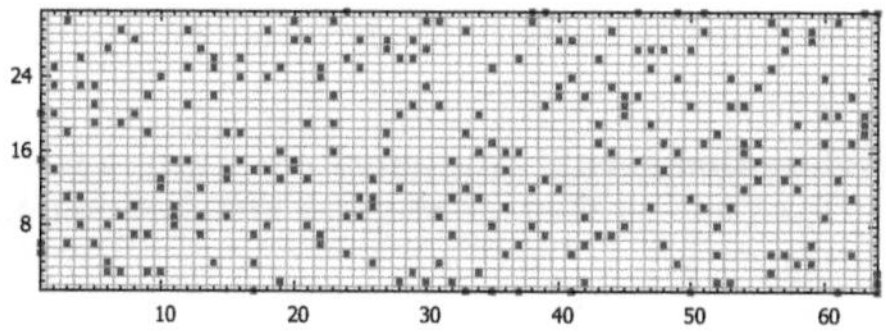

Fig. 3 Pooling matrix for 31 tests on 64 individuals: multiplicity 4, pool size 16

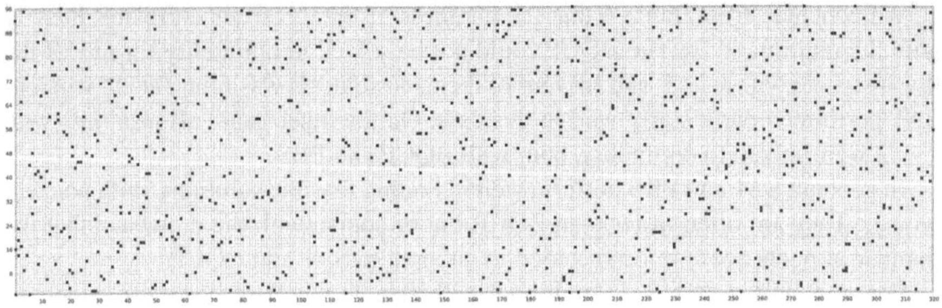

Fig. 4 Pooling matrix for 96 tests on 320 individuals: multiplicity 3, pool size 10

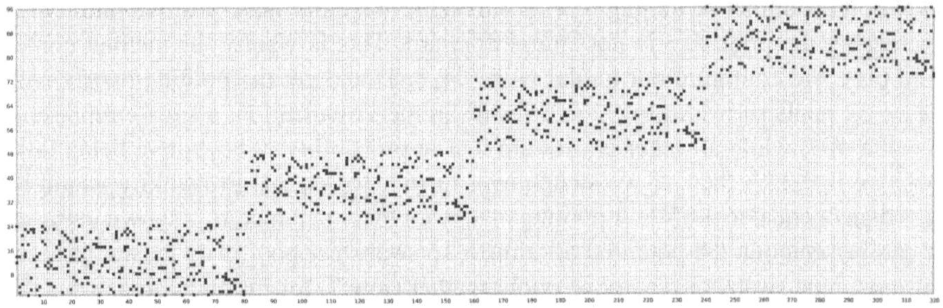

Fig. 5 Block structured pooling matrix for 96 tests on 320 individuals by rearranging a solution for 24 tests on 80 individuals: multiplicity 3, pool size 10

6 Conclusion and Outlook

The integer programming and satisfiability models that have been formulated permit the computation of efficient pool designs. Our methods allow a flexible and unrestricted handling of the parameters, i.e., the number of collected samples, pool size, and multiplicity, to minimize the number of required pools. We demonstrate that the computation of nonadaptive pooling strategies is practically feasible by means of optimization methods. This results in an increase in testing capacity, an acceleration of the testing process, and a reduction in overall costs.

The seamless integration of our method into automated testing systems offers significant commercial potential for manufacturing and industrial quality control in a range of sectors, whenever it is reasonable to combine samples in a pool.

We believe that further improvements are possible. A dedicated system with tailored starting heuristics and automatic switching between optimization and satisfiability models could decrease the time to a practical solution further.

Acknowledgments This work has been conducted in the Research Campus MODAL funded by the German Federal Ministry of Education and Research (BMBF) (Fund Nos. 05M14ZAM, 05M20ZBM).

References

1. Aldridge, M., Johnson, O., Scarlett, J.: Group testing: an information theory perspective. Found. Trends Commun. Inf. Theory **15**(3–4), 196–392 (2019)
2. Biere, A., Faller, T., Fazekas, K., Fleury, M., Froleyks, N., Pollitt, F.: CaDiCaL 2.0. In: Gurfinkel, A., Ganesh, V. (eds.) Computer Aided Verification - 36th International Conference, CAV 2024, Montreal, QC, July 24–27, 2024, Proceedings, Part I, vol. 14681. Lecture Notes in Computer Science, pp. 133–152. Springer, Berlin (2024)
3. Chionas, G., Kowalski, D.R., Krysta, P.: Combinatorial group testing with selfish agents. In: Proceedings of the 37th International Conference on Neural Information Processing Systems, NIPS '23, Red Hook. Curran Associates Inc., Newry (2023)
4. Cleary, B., Hay, J.A., Blumenstiel, B., Harden, M., Cipicchio, M., Bezney, J., Simonton, B., Hong, D., Senghore, M., Sesay, A.K., Gabriel, S., Regev, A., Mina, M.J.: Using viral load and epidemic dynamics to optimize pooled testing in resource-constrained settings. Sci. Transl. Medi. **13**(589), eabf1568 (2021)
5. Colbourn, C.J. (ed.) CRC Handbook of Combinatorial Designs, 1st edn. CRC Press, Boca Raton (1996)
6. Dorfman, R.: The Detection of Defective Members of Large Populations. Ann. Math. Stat. **14**(4), 436–440 (1943)
7. Du, D., Hwang, F.: Combinatorial Group Testing and Applications. World Scientific Publishing, Singapore (2000)
8. Gurobi Optimization, LLC. Gurobi Optimizer Reference Manual (2024)
9. Hölldobler, S., Manthey, N., Steinke, P.: A Compact Encoding of Pseudo-Boolean Constraints into SAT. In: Glimm, B., Krüger, A. (eds.) KI 2012: Advances in Artificial Intelligence, pp. 107–118. Springer, Berlin (2012)
10. Jagadeesan, B., Gerner-Smidt, P., Allard, M.W., Leuillet, S., Winkler, A., Xiao, Y., Chaffron, S., Van Der Vossen, J., Tang, S., Katase, M., McClure, P., Kimura, B., Ching Chai, L., Chapman, J., Grant, K.: The use of next generation sequencing for improving food safety: translation into practice. Food Microbiol. **79**, 96–115 (2019)
11. Kainkaryam, R., Woolf, P.: Pooling in high-throughput drug screening. Curr. Opin. Drug Discov. Devel. **12**, 339–50 (2009)
12. Liang, W., Zou, J.: Neural group testing to accelerate deep learning. In: 2021 IEEE International Symposium on Information Theory (ISIT), pp. 958–963 (2021)
13. Mazumdar, A.: On almost disjunct matrices for group testing. In: Chao, K.-M., Hsu, T.s., Lee, D.-T. (eds.) Algorithms and Computation, pp. 649–658. Springer, Berlin (2012)
14. Sakata, A.: Active pooling design in group testing based on Bayesian posterior prediction. Phys. Rev. E **103**, 022110 (2021)
15. Schaeffer, D.J., Kerster, H.W., Janardan, K.G.: Monitoring toxics by group testing. Environ. Manag. **6**, 467–469 (1982)
16. Sobel, M., Groll, P.A.: Group testing to eliminate efficiently all defectives in a binomial sample. Bell Syst. Techn. J. **38**, 1179–1252 (1959)
17. Täufer, M.: Rapid, large-scale, and effective detection of COVID-19 via non-adaptive testing. J. Theor. Biol. **506**, 110450 (2020)
18. Verdun, C.M., Fuchs, T., Harar, P., Elbrächter, D., Fischer, D.S., Berner, J., Grohs, P., Theis, F.J., Krahmer, F.: Group testing for SARS-CoV-2 allows for up to 10-fold efficiency increase across realistic scenarios and testing strategies. Front. Public Health **9**, 583377 (2021)
19. Xuan, Y., Shin, I., Thai, M.T., Znati, T.: Detecting application denial-of-service attacks: a group-testing-based approach. IEEE. Trans. Parallel Distrib. Syst. **21**(8), 1203–1216 (2010)
20. Yelin, I., Aharony, N., Tamar, E.S., Argoetti, A., Messer, E., Berenbaum, D., Shafran, E., Kuzli, A., Gandali, N., Shkedi, O., et al. Evaluation of COVID-19 RT-qPCR test in multi sample pools. Clin. Infect. Dis. **71**(16), 2073–2078 (2020)

Formulations of the Currency Portfolio Optimization Problem and Sensitivity Analysis of a Currency Basket

Hongxuan X. Huang

1 Introduction

In the market economy, the concept of asset is used to describe various resources with certain economic values and/or expectations. An asset usually provides some benefits for its owner on a short-term or long-term basis. According to their similar characteristics and behaviors, assets in the marketplace are often classified into four main classes such as equities, fixed incomes, cash and cash equivalents, and alternative investments. Currently, most investment assets include both tangible and intangible instruments that investors buy and sell during a certain period for the purposes of restoring their wealth or generating additional income, which involves real estate, commodities, futures, other financial derivatives, and even cryptocurrencies [4]. All of asset classes that a person or an institution possesses can be summarized as the *wealth* and represented by amounts of various currencies in different currency zone, which is referred to as a *currency portfolio*.

The concept of the *currency portfolio* is similar to that of a *portfolio* as a grouping of financial assets including stocks, bonds, and cash equivalents as well as their fund counterparts [10]. Usually, the process of selecting a portfolio is divided into two stages: the first stage estimates the future performance of available securities, and the second stage makes a decision about the choice of a portfolio, which is based on the rule of expected returns–variance of returns [10]. H.M. Markowitz had discussed the method of normative portfolio analysis (i.e., mean-variance portfolio

This paper is dedicated to Professor Panos M. Pardalos on the occasion of his 70th birthday.

H. X. Huang (✉)
Department of Industrial Engineering, Tsinghua University, Beijing, P.R. China
e-mail: hxhuang@tsinghua.edu.cn

B. Goldengorin (ed.), *Theory, Algorithms, and Experiments in Applied Optimization*, Springer Optimization and Its Applications 226,
https://doi.org/10.1007/978-3-031-91357-0_7

analysis) up to 1990, which describes a norm of behavior that investors should pursue for constructing a portfolio rather than predicting their actual behaviors [12]. An introduction to the theory of portfolio selection could be found in Chapter 3 in the book edited by F.J. Fabozzi and H.M. Markowitz [3]. In addition, the ideas introduced in the paper of the *Utility of Wealth* by H.M. Markowitz had been developed in the prospect theory, a framework proposed by D. Kahneman and A. Tversky, which explains the difference between present wealth and customary wealth by the coding of gains and losses relative to a reference point [9, 11]. It is worth to note that a unique currency or an abstract utility is assumed implicitly to measure the wealth and future performance of various assets in the financial market [3, 11]. This paper will present an approach to formulate the currency portfolio optimization problem and a virtual standard currency (VSC) to measure the value of assets in different currency zones. The VSC concept is concerned with the virtual (standard) exchange rate (VSER) vectors, whose existence is proved in Ref. [7]. In fact, the concept of a currency portfolio can be regarded as an extension of a classical portfolio in order to deal with the impact of foreign exchange rates. Any pair of currencies in the currency basket is assumed to be freely convertible.

It is well known that each country has its own sovereign currency circulating domestically and exchanging from/into foreign currencies during trading with other countries. The value of trade assets fluctuates with foreign exchange rates along the time, which leads to the amount of wealth in financial market up and down with the import and export of goods and services. Usually, the amount of wealth is measured by a combination of different currencies, which consists of a currency portfolio. For example, a multinational corporation may operate its business across different currency areas and makes its revenue or profit in a highly impressionable state. An instance comes from McDonald's Corporation that has 33, 000 subsidiaries being located around the world and operating across 119 countries. According to the US and international accounting standards, McDonald's Corporation as a multinational firm is required to prepare a financial statement in each period by converting each division's net profit into a single reporting currency (e.g., the US dollar) in which its main business is conducted. The firm suffered a net income 4 percent *decrease* in the second quarter of 2012 as a strong dollar took a bite out of profits. Total revenue for the third quarter was \$6.92 billion, up slightly from \$6.91 billion a year ago. When stripping out the impact of exchange rates, the company's revenue *rose* 5% [2]. Since each subsidiary operates locally and uses a specific local currency to serve customers, it is worth to evaluate objectively the value of a multinational parent corporation under the consideration of local currencies and exchange rates.

Another instance is that the exchange rates between the US dollar (USD) and the Chinese Renminbi Yuan (CNY or RMB) changed significantly in 2015. The Bloomberg-compiled data showed that there were combined foreign exchange losses of 48.7 billion CNY (\$7.5 billion) for 980 listed Chinese companies, almost 13 times the amount in 2014. In particular, the airline sector was hurt most with a combined foreign exchange losses of 17.9 billion CNY for 2015, compared with 951.7 million CNY a year earlier [14]. In addition, the impact of Renminbi Yuan devaluation included that profits at those firms slumped 11% to 789.2 billion CNY in 2015. For McDonald's Corporation mentioned above, it got fewer book dollars if euros or other local currencies were converted into dollars since unfavorable currency exchange rates made the dollar strong against other currencies [2]. Since a currency portfolio is a typical kind of wealth denoting various assets in local markets by the corresponding local currencies, it is natural to replace a tool measuring a person's wealth of an asset portfolio by another one measuring the value of a currency portfolio in the global financial market. Therefore, there exists such a fundamental question as follows:

How to evaluate and optimize currency portfolios under the fluctuation of exchange rates?

This is referred to as the *currency portfolio optimization*(CPO) problem.

A *foreign exchange rate*, also known as a *forex rate* or *currency quotation* B/C, is the price of one currency B in terms of another currency C. Although people can evaluate the value of a currency portfolio in terms of a specific currency chosen in advance, the choice of a representative currency and elements in a currency portfolio(referred as the *currency basket*) is a perceptual matter under the fluctuation of exchange rates along the time. In fact, the current measurement of wealth is more *subjective* than *objective* for dealing with the influence of uncertainty on exchange rates. For example, suppose that the bid and ask prices of 1 euro are \$1.11463 and \$1.11787, respectively, in terms of the US dollar in a local foreign currency market, which is published by such a market maker as the Bank of China at Beijing time 4:58:11 a.m. on September 21, 2024. The indirect quotation of the exchange rate EUR/USD is \$1 = € 1/1.11787 = € 0.89456 (five decimal places are kept here). If an investor has both euro € 100 and the US dollar \$100, denoted by a currency portfolio (€ 100, \$100), then the real value of his/her currency portfolio is still undetermined objectively. When it is measured in terms of the US dollar, then the amount of the investor's wealth is € 100 + \$100 = \$211.46, compared to the amount of his/her wealth in terms of the currency *euro*, € 100 + \$100 = € 189.46.

Let us define two different kinds of currency portfolios (EUR, USD) as follows:

$$(100 + \alpha/1.11463 - \beta, 100 - \alpha + 1.11463\beta), \forall\alpha, \beta \in [0, 100] \tag{1}$$

$$(100 + \alpha/1.11787 - \beta, 100 - \alpha + 1.11787\beta), \forall\alpha, \beta \in [0, 100] \tag{2}$$

By using the direct quotation of the forex rate EUR/USD, it is easy to check that the first type of portfolios given by (1) has the same value \$211.46 as that of (€ 100, \$100) in terms of the US dollar but with different value in terms of euro. Similarly, by using the indirect quotation of the forex rate EUR/USD, the second type of portfolios given by (2) has the same value € 189.46 as that of (€ 100, \$100) in terms of euro but with different value in terms of the US dollar.

In contrary, given a freely convertible currency basket, the concept of *virtual standard currency* (VSC) can be used for approximating foreign exchange rates at a certain time with respect to a market maker [7]. For example, given the direct and indirect quotations of the forex rate EUR/USD as above, the virtual standard exchange rate (VSER) is obtained by applying the rank-one approximation to the forex matrix

$$A = \begin{pmatrix} 1 & 1.11463 \\ 0.89456 & 1 \end{pmatrix},$$

which is defined by

$$u^* = \begin{pmatrix} 1.05606 \\ 0.94623 \end{pmatrix}, \quad v^* = \begin{pmatrix} 0.94623 \\ 1.05606 \end{pmatrix}.$$

Therefore, if the parameter u_i^* is regarded as the value of the currency $i \in \{EUR, USD\}$, then the value of the currency portfolio (€ 100, $100) is

$$100 * u_1^* + 100 * u_2^* = 100 * 1.05606 + 100 * 0.94623 \approx 200.23$$

in terms of the VSC measurement. Similarly, each currency portfolio (1) and (2) has a corresponding value in terms of the VSC measurement, given by

$$\begin{aligned}
&(100 + \alpha/1.11463 - \beta) * 1.05606 + (100 - \alpha + 1.11463\beta) * 0.94623 \\
&= 200.23 + 0.00122\alpha - 0.00136\beta
\end{aligned}$$

and

$$\begin{aligned}
&(100 + \alpha/1.11787 - \beta) * 1.05606 + (100 - \alpha + 1.11787\beta) * 0.94623 \\
&= 200.23 - 0.00152\alpha + 0.00170\beta,
\end{aligned}$$

respectively. Each portfolio has a particular VSC value dependent on the parameters α and β. Furthermore, for the currency portfolio optimization (CPO) problem associated with the currency basket $\{EUR, USD\}$, it can be formulated by two kinds of variables, in which one describes different amounts of currencies in portfolios and the other indicates the amount of each currency exchanged into another [6]. It has been proved that the rank-one approximation (ROA) approach also has an intrinsic robust property, which is capable of evaluating objectively and optimizing efficiently currency portfolios over time.

The remaining part of the paper is arranged as follows: Firstly, the currency portfolio optimization problem is formulated together with its equivalent network model, which is a variant of the maximum amplifying flow problem in Sect. 2. Secondly, the solution techniques for the reduction of decision variables are discussed in Sect. 3. One reduced formulation is based on the new concept of the max-product operation and its property of the associative law, which is firstly proposed in the paper and also implies a procedure to solve the network flow model of the CPO problem. Another reduced formulation is based on the rank-one approximation (ROA) approach, which is referred to as the ROA model and keeps only one kind of variables describing the amount of each currency in portfolios. In addition, the sensitivity analysis of a currency basket is discussed in Sect. 4, which estimates the ROA solution to a forex rate matrix from its augmented or reduced forex rate matrix. In other words, the effect of a currency increasing and decreasing on the VSC measurement can be estimated during solving the ROA problem with respect to a current basket of currencies. In Sect. 5, numerical experiments are performed over a sequence of forex rate matrices, and the performance of several methods is verified for the ROA approach, various formulations of the CPO problem, and the sensitive analysis of a currency basket. Finally, some concluding remarks are given in Sect. 6.

2 Formulations of the Currency Portfolio Optimization

In this section, some assumptions and notations are presented firstly in order to formulate the currency portfolio optimization problem. Then a basic model is developed for the CPO problem, which is a variant of the linear programming problem. Some equivalent or approximate models are also discussed later together with a new concept of the max-product operation and the rank-one approximation (ROA) method for approximating foreign exchange rates.

2.1 Assumptions and Notations

There are four assumptions for various formulations of the currency portfolio optimization problem as follows:

A1: Every real currency in a portfolio can be freely exchanged into others in the market at any stage during working time periods. Without loss of generality, suppose that there are n currencies exchanged mutually and freely in the currency market and denoted by the set $S = \{1, 2, \cdots, n\}$.

A2: The financial market is perfect without transaction costs for any pair of currency exchanges except for a difference (or spread) between the bid and ask prices of one unit of a foreign currency F in terms of the domestic currency D. The direct quotation of the forex rate F/D between F and D is expressed by $F\$1 = D\$\lambda$, where λ is the bid price of one unit of the currency F in terms of the currency D. In the meantime, the ask price of one unit of currency F, denoted by $F\$1 = D\$(\lambda + \delta)$, usually is greater than its bid price, where $\delta > 0$. The indirect quotation of the forex rate F/D is given by $D\$1 = F\$\frac{1}{\lambda+\delta}$.

A3: The foreign exchange rates are associated with a market maker and given or forecasted in a certain forward period, that is, the certainty assumption holds in order to explore the essential properties with the currency portfolio optimization problem.

A4: By simplicity, any interest rate is not considered for every currency during the maximization of currency portfolios.

Suppose all forex rates over time are grouped by a sequence of matrices $A^k, k = 1, 2, \cdots, N$, where N is the number of time periods and k is the index of the period k. The element a_{ij}^k in the matrix A^k is the bid price of one unit of the currency i in terms of the currency j, i.e., the direct quotation of the forex rate i/j. It can be verified that the ask price of the currency i is equal to $1/a_{ji}^k$ in terms of the currency j. In addition, at the period k, an initial currency portfolio, which is denoted by x^{k-1}, will be transformed into another currency portfolio x^k at the end of the same period. Furthermore, let the decision variable y_{ij}^k denote the amount of the currency i exchanged into the currency j during the period k, where $i, j = 1, 2, \cdots, n$ and $k = 1, 2, \cdots, N$.

2.2 The Basic Model for the CPO Problem

In the financial market, there are many market makers providing the business of foreign currency exchanges. Suppose that a sequence of forex rate matrices $\{A^k\}$ is estimated or forecasted by an investor at the beginning of time periods $k = 1, 2, \cdots, N$, which may be based on information about the currency market and a set of specific market makers. For a basket of currencies $S = \{1, 2, \cdots, n\}$, the investor's goal is to maximize the amount of a particular currency $m \in S$ at the end of period N, where $m \in S$ is selected in advance.

Let both vectors x^{k-1} and x^k denote currency portfolios at the beginning and the ending of time period k, respectively, together with decision variables y_{ij}^k defining the amount of currency i exchanged into another currency j at the same period k. The relationship among the decision variables $\{y_{ij}^k\}$ and currency portfolios x^{k-1}, x^k can be expressed by the following linear equations,

$$\sum_{j=1}^{n} y_{ij}^k = x_i^{k-1}, \forall i = 1, 2, \cdots, n, \tag{3}$$

$$\sum_{i=1}^{n} a_{ij}^k y_{ij}^k = x_j^k, \forall j = 1, 2, \cdots, n, \tag{4}$$

where $k = 1, 2, \cdots, N$. Since the investor's goal implies an objective function defined by x_m^N, which is an element of the currency portfolio x^N at the end of time period N, the basic model for the currency portfolio maximization problem is formulated by

$$\max \ x_m^N, \tag{5}$$

$$s.t. \ x^{k-1} = Y^k e, \tag{6}$$

$$x^k = \left(A^k \circ Y^k\right)^T e, \tag{7}$$

$$x_i^k \geq 0, y_{ij}^k \geq 0, \forall i, j = 1, 2, \cdots, n, k = 1, 2, \cdots, N, \tag{8}$$

where the vector $e = (1, 1, \cdots, 1)^T \in \Re^n$, x^0 is an initial currency portfolio, A^k is a cross-currency rate matrix at the period k associated with the market maker M, x^k describes a currency portfolio at the end of period k, $Y^k = (y_{ij}^k) \in \Re^{n \times n}$ is a matrix consisting of amount of currency exchanges, and the operator $\circ$ indicates the Hadamard product of two matrices. Note that the Hadamard product $A^k \circ Y^k$ of two matrices A^k and Y^k is also known as the Schur product or the entrywise product, that is, $\left(A^k \circ Y^k\right)_{ij} = a_{ij}^k y_{ij}^k$ for all $i, j = 1, 2, \cdots, n, k = 1, 2, \cdots, N$.

The basic model (5) ∼(8) is a variant of the linear programming problem. It has nN decision variables x_i^k and n^2N decision variables y_{ij}^k together with $2nN$ linear equality constraints. If the high-frequency data about forex rates are used for the currency portfolio optimization, the number of the time periods N will be so huge that there is a large requirements for the memory space and computer time. This motivates looking for decision variable reduction techniques for the basic model (5) ∼(8) or finding its equivalent formulation of the CPO problem. Note that the objective function in the basic model is related to the currency m, which is chosen in advance and seems to be lack of generality. Actually, there exists a generality for an optimal allocation of currencies in portfolios through the rank-one approximation approach, which is based on the concepts of the virtual standard currency and the virtual standard exchange rates [6].

2.3 Equivalent Network Formulation for the CPO Problem

For the optimization of currency portfolios with respect to a market maker M and a sequence of forex rate matrices A^k, $k = 1, 2, \cdots, N$, let us construct a maximum amplifying flow network and see also Fig. 1 as follows:

- Two currency portfolios $x^s = (x_1^s, x_2^s, \cdots, x_n^s)^T$ and $x^t = (x_1^t, x_2^t, \cdots, x_n^t)^T$ correspond to a source node s at the beginning of time period 0 and a sink node t at the ending of time period $N + 1$, respectively.
- There are $N + 2$ stages, denoted by $0, 1, \cdots, N, N + 1$, corresponding to the source node s, the sink node t, and all of N time periods, $k = 1, 2, \cdots, N$.
- Each node $c_{i,k}$ in the middle of network denotes the amount of currency $i \in S$ in a currency portfolio at the end of time period $k = 0, 1, 2, \cdots, N$, where $S = \{1, 2, \cdots, n\}$ is the basket of currencies. In Fig. 1, the total number of nodes is $n(N + 1) + 2$ including the source node s and the sink node t, where n is the number of currencies.

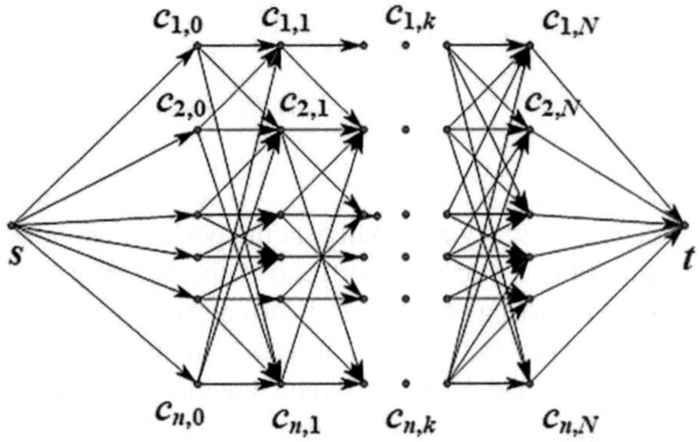

Fig. 1 Illustration for the network of currency flows over time periods $k = 1, 2, \cdots, N$

- There are three types of arcs in Fig. 1. The first type of arc is from the source node s to a currency node at the beginning of the period 1, whose weight is set to a real number 1. The second type of arc is from a currency node at the end of the period N to the sink node t, whose weight is also set to a real number 1. The third type of arc is from the currency i at the stage $k-1$ to the currency j at the stage k, whose weight is set to a_{ij}^k, a component in the forex rate matrix A^k, where $k = 1, 2, \cdots, N$.
- Without loss of generality, given a specific pair of currencies c and m, assume that the source node s corresponds to a portfolio with a single currency c at the beginning of the stage 0, while the sink node t corresponds to the resulting portfolio with a single currency m at the end of the stage $N+1$.

A typical formulation of the currency portfolio optimization problem is defined by the maximization of an amplifying flow from the source node s associated with the currency c to the sink node t with respect to the currency m, that is,

$$\max_{i_1,i_2,\cdots,i_{N-1}} \rho_{cm}^N = \prod_{k=1}^{N} a_{i_{k-1}i_k}^k, \tag{9}$$

where $i_0 = c, i_N = m, i_k \in \{1, \cdots, n\}, k = 1, 2, \cdots, N-1$.

In general, starting from a currency portfolio $x^s = x^0$ at the beginning of the first period, a flow will be amplified through a series of currencies $i_1, i_2, \cdots, i_{N-1}$ to a certain amount of wealth measured in the currency m by the following combinatorial optimization problem:

$$\max_{i_1,i_2,\cdots,i_{N-1}} x_m^N = \sum_{c=1}^{n} x_c^0 \rho_{cm}^N \tag{10}$$

The combinatorial optimization model (10) is equivalent to the basic model (5) ~(8) since all network flows satisfy the linear superposition property. The capacity of every arc is unlimited, but the amount of each currency flowing through an arc could be adjusted by the weight of the corresponding arc. For the convenience of simplicity, the adjustment of a currency flow is described by such a term as the *amplifying flow*, and the model (10) is referred to as the *maximum amplifying flow problem*, which is a variant of the maximum flow problem.

Since the initial currency portfolio $x^s = x^0$ is chosen arbitrarily, it is better that the performance of the currency portfolio optimization is evaluated by the ratio of the target amount x_m^N of the currency m divided by the reference amount $x_m^1 = \sum_{c=1}^n x_c^0 \rho_{cm}^1$, which is also measured in the same currency m and defined by

$$\rho_m^N = \frac{x_m^N}{x_m^1} = \frac{\sum_{c=1}^n x_c^0 a_{cm}^1 \frac{\rho_{cm}^N}{a_{cm}^1}}{\sum_{c=1}^n x_c^0 a_{cm}^1}, \tag{11}$$

where $\rho_{cm}^1 = a_{cm}^1$. In particular, it is easy to see that the maximum of the ratio ρ_m^N in (11) with respect to all currency portfolios x^0 is equal to $\max_{c \in S}\{\rho_{cm}^N / a_{cm}^1\}$.

For the convenience of simplicity, let us define a *standard ratio* for the currency portfolio optimization in terms of the currency m by

$$\rho_m^N = \max_{c \in S}\{\rho_{cm}^N / a_{cm}^1\}$$

which is the optimal value of the following optimization problem,

$$\rho_m^N = \max_{\substack{x^0 \neq 0,\, x^0 \geq 0 \\ i_1, i_2, \cdots, i_{N-1}}} \left\{ \frac{\sum_{c=1}^n x_c^0 \rho_{cm}^N / a_{cm}^1}{\sum_{c=1}^n x_c^0} \right\}, \tag{12}$$

where ρ_{cm}^N is given by (9).

3 Solution Techniques for Solving the CPO Problem

In this section, a new approach is proposed to analyze the problem of currency portfolio optimization, which is based on the concept of *max-product operation* similar to the inner product in an n-dimensional Euclidean space. In addition, another approach such as the rank-one approximation(ROA) method is also discussed for approximating forex rates, which is related to the concepts of the *virtual standard currency*(VSC) and the *virtual standard exchange rates*(VSER). It is worth to note that there exists a close relationship between the max-product operation and the ROA approach for analyzing the CPO problem.

3.1 Definition of the Max-Product Operation

The *max-product operation* in this paper is defined by a special product between two nonnegative vectors or between a nonnegative vector and a nonnegative matrix, which is also extended to analyze the relationship between two nonnegative matrices.

Definition 1 Given two nonnegative vectors a and b in the orthant $\Re^n_+ = \{x = (x_1, x_2, \cdots, x_n)^T \in \Re^n \mid x_i \geq 0, i = 1, 2, \cdots, n\}$, the *max-product operation between a and b* is defined by

$$a^T \otimes b = \max_i \{a_i b_i\}. \tag{13}$$

Note that the max-product operation between two vectors a and b is commutative, that is, $a^T \otimes b = b^T \otimes a$.

Definition 2 Given a nonnegative vector $a \in \Re^n_+$ and a nonnegative matrix $B \in \Re^{n\times n}_+$, the *max-product operation between a and B* is defined by a row vector as follows,

$$\begin{aligned} a^T \otimes B &= (a^T \otimes B_{\cdot 1}, a^T \otimes B_{\cdot 2}, \cdots, a^T \otimes B_{\cdot n}) \\ &= (\max_i\{a_i b_{i1}\}, \max_i\{a_i b_{i2}\}, \cdots, \max_i\{a_i b_{in}\}), \end{aligned} \tag{14}$$

where $B_{\cdot j}$ is the j-th column vector in the matrix B. Similarly, the *max-product operation between B and a* is defined by a column vector as follows,

$$\begin{aligned} B \otimes a &= (B_{1\cdot} \otimes a, B_{2\cdot} \otimes a, \cdots, B_{n\cdot} \otimes a)^T \\ &= (\max_j\{a_j b_{1j}\}, \max_j\{a_j b_{2j}\}, \cdots, \max_j\{a_j b_{nj}\})^T, \end{aligned} \tag{15}$$

where $B_{i\cdot}$ is the i-th row vector in the matrix B.

According to Definition 2, it is easy to verify that $B \otimes a = (a^T \otimes B^T)^T$. In addition, the max-product operation can be easily extended to such a scenario as the max operation between two nonnegative matrices. Here, a similar definition is given below about the max operation of two matrices.

Definition 3 Given two nonnegative matrices $A, B \in \Re^{n\times n}_+$, the *max-product operation between A and B* is defined by such a matrix $C = (c_{ij}) \in \Re^{n\times n}_+$ as follows,

$$C = (c_{ij}) = \begin{pmatrix} A_{1\cdot} \otimes B \\ A_{2\cdot} \otimes B \\ \cdots \\ A_{n\cdot} \otimes B \end{pmatrix} \triangleq A \otimes B, \tag{16}$$

where $c_{ij} = A_{i\cdot} \otimes B_{\cdot j}$, $A_{i\cdot}$ is the i-th row vector in the matrix A and $B_{\cdot j}$ is the j-th column vector in the matrix B.

Based on Definition 2, it is easy to check that the property $a^T \otimes B = \left(B^T \otimes a\right)^T$ holds and $\forall A, B \in \Re_+^{n\times n}$, $A \otimes B = \left(B^T \otimes A^T\right)^T$. Hence, we have an equivalent representation of $A \otimes B$ with respect to two nonnegative matrices A and B, denoted by

$$A \otimes B = (A \otimes B_{\cdot 1}, A \otimes B_{\cdot 2}, \cdots, A \otimes B_{\cdot n}), \tag{17}$$

where $B_{\cdot j}$ is the j-th column vector in the matrix B.

3.2 The Max-Product Operation's Associative Property

In this subsection, it is shown that the max-product operation is associative for a sequence of nonnegative matrices. For three nonnegative matrices A, B, and C, the associative property is that there is no difference between $(A \otimes B) \otimes C$ and $A \otimes (B \otimes C)$ (see Proposition 1). In such a case, without loss of generality, the max-product operations among matrices A, B, and C are simply denoted by $A \otimes B \otimes C$.

Proposition 1 *Suppose that $A, B, C \in \Re_+^{n\times n}$ are three nonnegative matrices. Then the following equality*

$$(A \otimes B) \otimes C = A \otimes (B \otimes C) \tag{18}$$

holds, where the operation $\otimes$ is defined in Definition 3.

Proof According to the Definition 3, the i-th row in the matrix $(A \otimes B) \otimes C$ is given by

$$(A_{i\cdot} \otimes B) \otimes C,$$

in which the l-th component is denoted by

$$\max_j\{\max_k\{a_{ik}b_{kj}\} \cdot c_{jl}\}.$$

Similarly, the l-th component in the i-th row of the matrix $A \otimes (B \otimes C)$ is given by

$$\max_k\{a_{ik} \cdot \max_j\{b_{kj}c_{jl}\}\}.$$

It is necessary and sufficient to show that

$$\max_k\{a_{ik} \cdot \max_j\{b_{kj}c_{jl}\}\} = \max_j\{\max_k\{a_{ik}b_{kj}\} \cdot c_{jl}\}, \forall i, l = 1, 2, \cdots, n. \tag{19}$$

Since the following inequalities

$$a_{ik}b_{kj} \leq \max_k\{a_{ik}b_{kj}\}, \forall i, j, k$$

hold and $c_{jl} \geq 0, \forall j, l$, the following inequalities

$$a_{ik}b_{kj}c_{jl} \leq \max_k\{a_{ik}b_{kj}\} \cdot c_{jl}, \forall i, j, k, l$$

also hold. This implies that both inequalities

$$a_{ik} \cdot \max_j\{b_{kj}c_{jl}\} \leq \max_j\{\max_k\{a_{ik}b_{kj}\} \cdot c_{jl}\}, \forall i, k, l$$

and

$$\max_k\{a_{ik} \cdot \max_j\{b_{kj}c_{jl}\}\} \leq \max_j\{\max_k\{a_{ik}b_{kj}\} \cdot c_{jl}\}, \forall i, l$$

hold for all of elements in nonnegative matrices A, B, and C.

Similarly, based on $a_{ik} \geq 0, \forall i, k$ and the property of the max function over a finite real number set, it could be proved that the inequality above also holds reversely, that is, we have

$$\max_k\{a_{ik} \cdot \max_j\{b_{kj}c_{jl}\}\} \geq \max_j\{\max_k\{a_{ik}b_{kj}\} \cdot c_{jl}\}, \forall i, l.$$

Hence, all of equalities in (19) hold. This implies that for any three nonnegative matrices $A, B, C \in \Re_+^{n\times n}$, the associative law holds with respect to the max-product operation. □

3.3 The Max-Product Solution to the CPO Problem

For the basic model (5) ∼(8), the currency portfolio x^k at the end of time period k satisfies the constraints (7), which is equivalently expressed by

$$\begin{aligned} x^k &= \left(A^k \circ Y^k\right)^T e \\ &= \begin{pmatrix} a_{11}y_{11} + a_{21}y_{21} + \cdots + a_{n1}y_{n1} \\ a_{12}y_{12} + a_{22}y_{22} + \cdots + a_{n2}y_{n2} \\ \cdots \\ a_{1n}y_{1n} + a_{2n}y_{2n} + \cdots + a_{nn}y_{nn} \end{pmatrix} \end{aligned}$$

$$= \begin{pmatrix} A_{\cdot 1}^T Y_{\cdot 1} \\ A_{\cdot 2}^T Y_{\cdot 2} \\ \cdots \\ A_{\cdot n}^T Y_{\cdot n} \end{pmatrix}, \tag{20}$$

where the superscript k for some elements is removed without misunderstandings and $A_{\cdot j}$ and $Y_{\cdot j}$ denote the j-th column vectors in matrices A^k and Y^k, respectively. Since the basic model (5) ∼(8) is a special case of the linear programming problem, which is also equivalent to a modified maximum flow problem, this indicates that the maximum amplifying flow problem (10) has an equivalent solution to the basic model and vice versa. In particular, if the currency portfolio x^k in (20) is optimal, then it is equivalently expressed by

$$x^k = \left(\left(A^k\right)^T \otimes Y^k\right) e, \tag{21}$$

where $\otimes$ denotes the max-product operation and all of components in the vector $e \in \Re^n$ is 1. It can be easily checked that for any forex rate matrix A and any currency portfolio x, we have

$$\left(A^T \otimes diag(x)\right) e = A^T x.$$

For the network formulation of the CPO problem illustrated in Fig. 1 and any pair of currencies c and m given in advance, the maximum of an amplifying flow from the source node s to the sink node t through currencies c and m is the element at the intersection of the c-th row and the m-th column in a compound matrix. The compound matrix involves the max-product operation and its associative law, which is defined by

$$\prod_{k=1}^{N} \otimes A^k := A^1 \otimes \cdots \otimes A^N. \tag{22}$$

Furthermore, for an initial currency portfolio $x^0 \in \Re^n_+$, the maximum amount of wealth in terms of the currency m is given by the m-th component in the vector x^N at the end of time period N. According to the recurrence of the equality (21), an amplifying currency portfolio is generated by

$$x^N = \left(\prod_{k=1}^{N} \otimes A^k\right)^T x^0,$$

which implies the optimal value of the basic model (5) ∼(8)

$$\bar{x}_m^N = \max x_m^N = (e^m)^T \left(\prod_{k=1}^{N} \otimes A^k \right)^T x^0, \tag{23}$$

where e^m is a unit vector in $\Re^n$ whose the m-th component is 1 and others are 0.

Proposition 2 *If the forex rate matrices $A^k, k = 1, 2, \cdots, N$ are given by $A^k = u^k \left(v^k\right)^T$, where $u^k > 0, v^k > 0, \forall k$, then the following conclusion holds for the compound matrix $\prod_{k=1}^{N} \otimes A^k$, that is,*

$$\prod_{k=1}^{N} \otimes A^k = \left\{ \prod_{k=1}^{N-1} \left(v^k\right)^T \otimes u^{k+1} \right\} u^1 \left(v^N\right)^T. \tag{24}$$

Proof Based on the assumption of $A^k = u^k \left(v^k\right)^T$ and the associative law of the max-product operations for the sequence of matrices $\{A^k\}$, we firstly analyze the property of the matrix $A^k \otimes A^{k+1}$.

It is easily verified that the following equality

$$\begin{aligned} A^k \otimes A^{k+1} &= \left\{u^k \left(v^k\right)^T\right\} \otimes \left\{u^{k+1} \left(v^{k+1}\right)^T\right\} \\ &= \left(\left(u^k \left(v^k\right)^T\right) \otimes u^{k+1} v_1^{k+1}, \cdots, \left(u^k \left(v^k\right)^T\right) \otimes u^{k+1} v_n^{k+1}\right) \\ &= \left(\left(v^k\right)^T \otimes u^{k+1}\right) \left(u^k v_1^{k+1}, \cdots, u^k v_n^{k+1}\right) \\ &= \left(\left(v^k\right)^T \otimes u^{k+1}\right) \left(u^k \left(v^{k+1}\right)^T\right) \end{aligned}$$

holds for $k = 1, 2, \cdots, N-1$, where the second equality uses the equivalent representation (17) of Definition 3 and the third equality comes from (15) in Definition 2.

Then, the mathematical induction is applied to all pairs of two matrices such as $A^k \otimes A^{k+1}, k = 1, 2, \cdots, N-1$, which implies the result in (24). □

The result in Proposition 2 is summary to that in Theorem 4.1 in Ref. [6]. The ROA approach is focused on approximating the solution to the CPO problem through approximating a sequence of forex rate matrices. In fact, in order to analyze the CPO problem, there exists a close relationship between the ROA approach and the max-product operation, which is presented in Corollary 1.

Corollary 1 *Given a finite sequence of cross-currency rate matrices $\{A^k\}$, suppose that there are two sequences of vectors $\{u^k\}$ and $\{v^k\}$ such that the matrix $(u^k)(v^k)^T$ is the robust rank-one approximation to A^k, $k = 1, 2, \cdots, N$. Then the rank-one approximation (ROA) model (25) ∼(27) can be constructed to approximate the basic model (5) ∼(8), which is given by*

$$\max \ x_m^N, \tag{25}$$

$$s.t. \ \ (u^k)^T x^{k-1} = e^T (V^k)^{-1} x^k, \tag{26}$$

$$x^k \in \Re^n_+, k = 1, 2, \cdots, N, \tag{27}$$

where x^0 *is an initial currency portfolio. In addition, if a sequence of intermediate variables* $\{t_k\}$ *is introduced by defining*

$$t_k = (u^k)^T x^{k-1}, k = 1, 2, \cdots, N, \tag{28}$$

then there exists an optimal sequence of currency portfolios $\{(x^*)^k\}$ *to the ROA model (25) ~(27) that satisfies*

$$(x^*)_i^k = \begin{cases} v^k_{p_k} t^*_k, & i = p_k, \\ 0, & i \neq p_k, \end{cases}, \tag{29}$$
$$i = 1, 2, \cdots, n, \quad k = 1, 2, \cdots, N,$$

where

$$\begin{aligned} t_1^* &= (u^1)^T x^0, \\ t_k^* &= (u^k)^T (x^*)^{k-1} \\ &= u^k_{p_{k-1}} (x^*)^{k-1}_{p_{k-1}}, \quad k = 2, \cdots, N, \\ p_k &= arg \max_i \{v_i^k u_i^{k+1}\}, \quad k = 1, 2, \cdots, N-1, \end{aligned}$$

and $p_N = m$*, where the currency* m *is specially selected in advance to evaluate the amount of wealth in currency portfolios.*

Note that the max-product operation in Proposition 2 coincides to that in Corollary 1 for defining the index p_k, that is,

$$\left(v^k\right)^T \otimes u^{k+1} = v^k_{p_k} u^{k+1}_{p_k}, \quad k = 1, 2, \cdots, N-1.$$

4 The Sensitivity Analysis of a Currency Basket

Let us denote the set of all positive vectors in the n-dimensional Euclidean space by $\Re^n_{++} = \{x \in \Re^n \mid x = (x_1, x_2, \cdots, x_n)^T, x_i > 0, i = 1, 2, \cdots, n\}$. Similarly, the set of all $n \times n$-dimensional matrices with positive components is denoted by $\Re^{n\times n}_{++}$. In this section, the sensitivity analysis of a currency basket is discussed with respect to a forex matrix $A \in \Re^{n\times n}_{++}$ and its *augmented forex matrix* defined by

$$B = \begin{pmatrix} A & s \\ b^T & 1 \end{pmatrix}, \tag{30}$$

where $s \in \Re^n_+$ is the bid price vector of currencies in terms of the augmented currency $n+1$ and $b \in \Re^n_+$ is the indirect quotation vector of the corresponding forex rates. Accordingly, the forex rate matrix A is referred to as a reduced forex matrix of the forex matrix B. For the convenience of simplicity, the forex matrices A and B correspond to the currency baskets $S = \{1, 2, \cdots, n\}$ and $\tilde{S} = S \cup \{n+1\}$, respectively. Note that the reciprocal of an element b_i in the indirect quotation vector b denotes the ask price of currency i and the inequality $s_i b_i < 1$ holds for $i = 1, 2, \cdots, n$ with respect to the currency basket $\tilde{S}$. The sensitivity analysis of a currency basket in this paper is regarded to as estimating the *virtual standard exchange rate* (VSER) vectors with respect to an increasing or decreasing currency basket from those associated with the original currency basket.

Based on the concept and properties of the *Virtual Standard Currency* (VSC) explored in Refs. [6, 7], the VSER vector (u^*, v^*) is the unique optimal solution to the following *rank-one approximation*(ROA) problem

$$\min_{u,v} \left\{ \|A - uv^T\|^2 \mid \|u\|_2 = \|v\|_2, u, v \in \Re^n_{++} \right\}, \tag{31}$$

where $A \in \Re^{n \times n}_{++}$ is a forex rate matrix with respect to a currency basket $S = \{1, 2, \cdots, n\}$ and a market maker, both vectors u and v are decision variables, $\| \cdot \|$ in the objective function of the ROA problem (31) denotes the Frobenius norm or the Euclidean norm (2-norm) in $\Re^{n \times n}$, and $\| \cdot \|_2$ in the equality constraint denotes the 2-norm of vectors in $\Re^n$.

Furthermore, for the ROA problem (31), both positive vectors $u^* = \sqrt{\sigma}\bar{u}$ and $v^* = \sqrt{\sigma}\bar{v}$ satisfy the following optimality conditions,

$$A\bar{v} = \sigma\bar{u}, \tag{32}$$

$$A^T\bar{u} = \sigma\bar{v}, \tag{33}$$

$$\|\bar{u}\|_2 = \|\bar{v}\|_2 = 1, \tag{34}$$

where σ is the largest singular value of the forex matrix A.

Note that for the augmented forex matrix B, there exists a corresponding ROA problem, similar to (31) with respect to B and its virtual standard exchange rates.

4.1 The Properties of the VSER Vectors with the Basket $\tilde{S}$

Suppose that λ is the largest singular value of the augmented forex matrix B and both vectors $w^* = \sqrt{\lambda}\bar{w}$ and $r^* = \sqrt{\lambda}\bar{r}$ are the virtual standard exchange rates with respect to B, that is,

$$B \approx w^*(r^*)^T = \lambda\bar{w}\bar{r}^T,$$

where $\bar{w} = (w^1, w_0)^T$ and $\bar{r} = (r^1, r_0)$ are unit vectors, $w^1, r^1 \in \Re^n_{++}$, $w_0, r_0 \in \Re_{++}$. Based on the optimality conditions of the ROA problem associated with the augmented forex matrix B, both unit vectors $\bar{w}$ and $\bar{r}$ satisfy the following system of equations,

$$B\bar{r} = \lambda\bar{w}, \tag{35}$$

$$B^T\bar{w} = \lambda\bar{r}, \tag{36}$$

which is equivalent to a system of equations involving variables w_0, r_0, w^1, r^1, and λ as follows,

$$A\left(r^1\right) = \lambda w^1 - r_0 s, \tag{37}$$

$$b^T r^1 + r_0 = \lambda w_0, \tag{38}$$

$$A^T\left(w^1\right) = \lambda r^1 - w_0 b, \tag{39}$$

$$s^T w^1 + w_0 = \lambda r_0, \tag{40}$$

where A is a reduced forex rate matrix of B.

In addition, some additional properties among w_0, r_0, w^1, and r^1 can be shown and presented below.

Proposition 3 *Suppose that A is a forex matrix and B is an augmented forex matrix given in (30). If λ is the largest singular value of B and both vectors $w^* = \sqrt{\lambda}\bar{w}$ and $r^* = \sqrt{\lambda}\bar{r}$ are the virtual standard exchange rates related to B, where $\bar{w} = (w^1, w_0)^T$ and $\bar{r} = (r^1, r_0)$ are unit vectors, then the following properties*

$$\lambda r_0^2 + \lambda w_0^2 + (w^1)^T A r^1 = \lambda + r_0 w_0, \tag{41}$$

$$s^T A r^1 + \|s\|_2^2 r_0 + \lambda w_0 = \lambda^2 r_0, \tag{42}$$

$$b^T A^T w^1 + \|b\|_2^2 w_0 + \lambda r_0 = \lambda^2 w_0 \tag{43}$$

hold with respect to the corresponding parameters and the forex matrix A, where $\|\cdot\|_2$ denotes the 2-norm of an n-dimensional vector.

Proof By multiplying the both sides of Eq. (37) from the left by $(w^1)^T$, we obtain

$$(w^1)^T A\left(r^1\right) = \lambda(w^1)^T w^1 - r_0(w^1)^T s.$$

By using the equality $(w^1)^T w^1 = 1 - w_0^2$ and (40), we have

$$(w^1)^T A\left(r^1\right) = \lambda\left(1 - w_0^2\right) - r_0\left(\lambda r_0 - w_0\right),$$

which is equivalent to (41).

Similarly, if both Eqs. (37) and (39) are multiplied by s^T and b^T from the left, respectively, the following both equations

$$s^T A\left(r^1\right) = \lambda s^T w^1 - r_0 s^T s$$

and

$$b^T A^T\left(w^1\right) = \lambda b^T r^1 - w_0 b^T b$$

could be obtained. Based on Eqs. (38) and (40), the items $b^T r^1$ and $s^T w^1$ in the expressions above could be substituted, and other results in this proposition are also concluded. □

4.2 The Estimation of the VSER Vectors with the Basket S

In this and next subsection, an approach is proposed to estimate the virtual standard exchange rate (VSER) vectors u^* and v^* with the currency basket S from its augmented currency basket $\tilde{S}$. Both vectors u^* and v^* correspond to the unique solution to the ROA problem (31) with respect to A. The corresponding unit vectors $\bar{u}$ and $\bar{v}$ satisfy the optimality conditions (32) ∼(33).

According to linear systems (37) and (39), it is easily verified that both vectors r^1 and w^1 satisfy the following system of equations

$$A^T A\left(r^1\right) = \lambda^2 r^1 - \left(\lambda w_0 b + r_0 A^T s\right) \tag{44}$$

and

$$AA^T\left(w^1\right) = \lambda^2 w^1 - \left(\lambda r_0 s + w_0 Ab\right), \tag{45}$$

respectively. Since B is a positive matrix and $w_0 > 0, r_0 > 0$, we have such an inequality as $\lambda > \sigma$, where λ and σ are the largest singular values of the matrices B and A [5]. Hence, both components r^1 and w^1 in the unit vectors $\bar{r} = (r^1, r_0)$ and $\bar{w} = (w^1, w_0)^T$ could be estimated by the following formulas

$$r^1 = \left(\lambda^2 I - A^T A\right)^{-1}\left(\lambda w_0 b + r_0 A^T s\right), \tag{46}$$

and

$$w^1 = \left(\lambda^2 I - AA^T\right)^{-1}\left(\lambda r_0 s + w_0 Ab\right), \tag{47}$$

respectively, where I denotes the identity matrix in $\Re^{n\times n}$. Note that both unit vectors $\bar{w}$ and $\bar{r}$ are induced from the virtual standard exchange rates $w^* = \sqrt{\lambda}\bar{w}$ and $r^* = \sqrt{\lambda}\bar{r}$, which is associated with the augmented forex matrix B.

Suppose that the forex matrix A has a singular value decomposition (SVD) as follows,

$$A = UDV^T, \tag{48}$$

where $U = \left(u^1, \cdots, u^n\right) \in \Re^{n\times n}$ and $V = \left(v^1, \cdots, v^n\right) \in \Re^{n\times n}$ are orthogonal matrices and $D = diag\{\sigma_1, \cdots, \sigma_n\}$ is a diagonal matrix generated with singular values of A. If the SVD expression (48) is substituted into (46) and (47), respectively, then both vectors r^1 and w^1 are equivalently described by

$$r^1 = \lambda w_0 V\left(\lambda^2 I - D^2\right)^{-1} V^T b + r_0 V\left(\lambda^2 I - D^2\right)^{-1} DU^T s \tag{49}$$

and

$$w^1 = \lambda r_0 U\left(\lambda^2 I - D^2\right)^{-1} U^T s + w_0 U\left(\lambda^2 I - D^2\right)^{-1} DV^T b, \tag{50}$$

respectively.

Similarly, the right-hand sides in both Eqs. (37) and (39) also have equivalent expressions as follows,

$$\lambda w^1 - r_0 s = r_0 U\left(\lambda^2 I - D^2\right)^{-1} D^2 U^T s + \lambda w_0 U\left(\lambda^2 I - D^2\right)^{-1} DV^T b, \tag{51}$$

and

$$\lambda r^1 - w_0 b = w_0 V\left(\lambda^2 I - D^2\right)^{-1} D^2 V^T b + \lambda r_0 V\left(\lambda^2 I - D^2\right)^{-1} DU^T s, \tag{52}$$

respectively.

In particular, if the forex matrix A is a rank-one matrix, that is,

$$A = \sigma_1 u^1 (v^1)^T,$$

then the singular value vectors of A could be constructed from those of the augmented forex matrix B. Such a possibility is presented in the following Proposition 4.

Proposition 4 *Suppose that A is a rank-one forex matrix and B is an augmented forex matrix given in (30). Letting two unit vectors $\bar{u}$ and $\bar{v}$ correspond to the virtual standard exchange rates (VSER) with respect to A and other unit vectors $\bar{w}$ and $\bar{r}$ similarly exist with respect to B, then the unit VSER vectors associated with the* *forex matrix A (or the currency basket S) can be estimated and described by the following expressions,*

$$\bar{u} = \frac{\lambda w^1 - r_0 s}{\sqrt{\lambda^2(1 - w_0^2) + r_0^2\|s\|_2^2 - 2r_0\lambda s^T w^1}}, \tag{53}$$

and

$$\bar{v} = \frac{\lambda r^1 - w_0 b}{\sqrt{\lambda^2(1 - r_0^2) + w_0^2\|b\|_2^2 - 2w_0\lambda b^T r^1}}, \tag{54}$$

where $\|\cdot\|_2$ *denotes the* 2*-norm and* λ *is the largest singular value of the augmented matrix B.*

Proof Since A is a rank-one forex matrix, suppose that $A = \sigma\bar{u}\bar{v}^T$, where σ is the largest singular value of A and both unit vectors $\bar{u}$ and $\bar{v}$ correspond to the virtual standard exchange rates (VSER) with respect to A. For the singular value decomposition $A = UDV^T$, the first column vector in the orthogonal matrices U and V satisfy $u^1 = \pm\bar{u}$ and $v^1 = \pm\bar{v}$, respectively [7]. In particular, the singular values of A satisfy $\sigma_1 = \sigma$ and $\sigma_i = 0, i = 2, 3, \cdots, n$.

Based on the system of Eqs. (51), it is obtained that

$$\lambda w^1 - r_0 s = \frac{r_0\sigma^2\left((u^1)^T s\right) + w_0\lambda\sigma\left((v^1)^T b\right)}{\lambda^2 - \sigma^2} u^1, \tag{55}$$

which indicates that the vector $\lambda w^1 - r_0 s$ is parallel to the unit vector $\bar{u}$ with respect to A.

Similarly, under the rank-one assumption of the matrix A, the system of Eqs. (52) is equivalent to

$$\lambda r^1 - w_0 b = \frac{w_0\sigma^2\left((v^1)^T b\right) + r_0\lambda\sigma\left((u^1)^T s\right)}{\lambda^2 - \sigma^2} v^1, \tag{56}$$

which is parallel to the unit vector $\bar{v}$ with respect to A.

In addition, both vectors $\bar{w} = (w^1, w_0)^T$ and $\bar{r} = (r^1, r_0)$ are unit; it is easily shown that

$$\|\lambda w^1 - r_0 s\|_2 = \sqrt{\lambda^2(1 - w_0^2) + r_0^2\|s\|_2^2 - 2r_0\lambda s^T w^1}$$

and

$$\|\lambda r^1 - w_0 b\|_2 = \sqrt{\lambda^2(1 - r_0^2) + w_0^2\|b\|_2^2 - 2w_0\lambda b^T r^1}.$$

Hence, for the rank-one forex matrix A, the estimators (53) and (54) hold for the unit vectors $\bar{u}$ and $\bar{v}$, respectively. □

4.3 Relationship Among Largest Singular Values

In this subsection, a proposition is firstly presented to estimate the largest singular value of the reduced forex rate matrix A from parameters with the augmented forex matrix B. Then another proposition is given to estimate the largest singular value of B based on parameters with A. These results describe the relationship among largest singular values and the corresponding VSER vectors.

Proposition 5 *Suppose that A is a rank-one forex matrix and B is an augmented forex matrix given in (30). Let us denote the largest singular values of A and B by σ and λ, respectively. If the VSER vectors of B are denoted by $w^* = \sqrt{\lambda}\bar{w}$ and $r^* = \sqrt{\lambda}\bar{r}$, where $\bar{w} = (w^1, w_0)^T$ and $\bar{r} = (r^1, r_0)^T$ are unit vectors, then there exists a relationship between σ and λ as follows,*

$$\sigma^2 = \lambda \frac{s^T A\left(\lambda r^1 - w_0 b\right)}{s^T A r^1 + r_0\left(\bar{u}^T s\right)^2} = \lambda \frac{s^T A A^T w^1}{s^T A r^1 + r_0\left(\bar{u}^T s\right)^2}, \tag{57}$$

$$\sigma^2 = \lambda \frac{b^T A^T\left(\lambda w^1 - r_0 s\right)}{b^T A^T w^1 + w_0\left(\bar{v}^T b\right)^2} = \lambda \frac{b^T A^T A r^1}{b^T A^T w^1 + w_0\left(\bar{v}^T b\right)^2}, \tag{58}$$

where the vectors b and s are from the augmented forex rate matrix B and described by (30) and the unit vectors $\bar{u}$ and $\bar{v}$ are estimated by (53) and (54), respectively.

Proof The second equalities in both conclusions (57) and (58) come from two properties (37) and (39), respectively. Therefore, it is sufficient to show that the first equalities in both conclusions (57) and (58) hold for the matrix A and parameters associated with B.

Based on the assumption that the forex matrix A is a rank-one matrix, it is known that the expression of $\lambda r^1 - w_0 b$ in (56) holds and the VSER unit vectors of A satisfy $\bar{v} = \pm v^1$ and $\bar{u} = \pm u^1$, where u^1 and v^1 are the left and right singular characteristic vectors of A. Multiplying both sides of (56) from the left with the matrix A and using the property (32), we have

$$A\left(\lambda r^1 - w_0 b\right) = \sigma^2 \frac{w_0 \sigma\left((v^1)^T b\right) + r_0 \lambda\left((u^1)^T s\right)}{\lambda^2 - \sigma^2} u^1. \tag{59}$$

In addition, by using the property (55), it can be verified that the following equality holds:

$$w^1 - \frac{r_0}{\lambda}s + \frac{r_0\left((u^1)^T s\right)}{\lambda}u^1 = \left\{\frac{r_0\sigma^2\left((u^1)^T s\right)}{\lambda(\lambda^2-\sigma^2)} + \frac{w_0\sigma\left((v^1)^T b\right)}{\lambda^2-\sigma^2} + \frac{r_0\left((u^1)^T s\right)}{\lambda}\right\}u^1$$
$$= \frac{w_0\sigma\left((v^1)^T b\right) + r_0\lambda\left((u^1)^T s\right)}{\lambda^2-\sigma^2}u^1 \tag{60}$$

After comparing the right-hand sides of (59) and (60), we obtain an equality as follows:

$$A\left(\lambda r^1 - w_0 b\right) = \frac{\sigma^2}{\lambda}\left(\lambda w^1 - r_0 s + r_0\left((u^1)^T s\right)u^1\right)$$

Furthermore, multiplying both sides of the equality above from the left by λs^T, we also get

$$\lambda s^T A\left(\lambda r^1 - w_0 b\right) = \sigma^2\left(\lambda s^T w^1 - r_0\|s\|_2^2 + r_0\left((u^1)^T s\right)^2\right). \tag{61}$$

Note that the equality (40) implies that $s^T w^1 = \lambda r_0 - w_0$. If $s^T w^1$ is substituted into the right-hand side of the Eq. (61), then the following equality

$$\lambda s^T A\left(\lambda r^1 - w_0 b\right) = \sigma^2\left(s^T A r^1 + r_0\left((u^1)^T s\right)^2\right)$$

is obtained by using the property (42), which is equivalent to the first equality in the conclusion (57).

Similarly, it can be proved that

$$A^T\left(\lambda w^1 - r_0 s\right) = \frac{\sigma^2}{\lambda}\left(\lambda r^1 - w_0 b + w_0\left((v^1)^T b\right)v^1\right).$$

If the above equality is multiplied by λb^T from the left and combines with the properties (38) and (43), then the following expression

$$\lambda b^T A^T\left(\lambda w^1 - r_0 s\right) = \sigma^2\left(b^T A^T w^1 + w_0\left((v^1)^T b\right)^2\right)$$

is obtained, which is equivalent to the first equality in the conclusion (58). □

Proposition 6 *Suppose that A is a rank-one forex matrix and B is an augmented forex matrix given in (30). Let us denote the largest singular values of matrices A and B by σ and λ, respectively. If the VSER vectors of the matrix A are denoted by $u^* = \sqrt{\sigma}\bar{u}$ and $v^* = \sqrt{\sigma}\bar{v}$, where $\bar{u}$ and $\bar{v}$ are unit vectors, then the value of λ can be estimated from σ by using the following equality,*

$$\lambda^2 = \frac{\sigma^2 + \|b\|_2^2 + \|s\|_2^2 + 1}{3} - \frac{\sqrt[3]{2}\zeta_1}{3\left(\zeta_0 + \sqrt{4\zeta_1^3 + \zeta_0^2}\right)^{1/3}} + \frac{\left(\zeta_0 + \sqrt{4\zeta_1^3 + \zeta_0^2}\right)^{1/3}}{3\sqrt[3]{2}}, \tag{62}$$

where

$$\zeta_0 = 3p_1 - p_2^2,$$
$$\zeta_1 = 27p_0 + 9p_2p_1 - 2p_2^3$$

and the parameters $p_0, p_1, and p_2$ are coefficients in a cubic equation (66) with one unknown variable, coming from parameters $\sigma, \bar{u}$, and $\bar{v}$ with respect to A and both vectors b and s in the augmented forex matrix B.

Proof Since A is a rank-one forex matrix, the conclusion (57) in Proposition 5 holds for the corresponding augmented forex matrix B, which implies

$$\left(\lambda^2 - \sigma^2\right) s^T A r^1 = r_0 \sigma^2 \left(\bar{u}^T s\right)^2 + w_0 \lambda s^T A b,$$

where $\bar{u}$ is the unit vector parallel to the VSER vector u^* of A. Based on the property (42) in Proposition 3, an equivalent formulation of $s^T A r^1$ is obtained and substituted into the above equation, which induces a property involving unknowns r_0, w_0, and λ as follows:

$$r_0 \left(\left(\lambda^2 - \sigma^2\right)\left(\lambda^2 - \|s\|_2^2\right) - \sigma^2 \left(\bar{u}^T s\right)^2\right) = w_0 \lambda \left(s^T A b + \lambda^2 - \sigma^2\right) \tag{63}$$

Similarly, based on the property (43) in Proposition 3 and the conclusion (58) in Proposition 5, another property is also obtained as follows,

$$w_0 \left(\left(\lambda^2 - \sigma^2\right)\left(\lambda^2 - \|b\|_2^2\right) - \sigma^2 \left(\bar{v}^T b\right)^2\right) = r_0 \lambda \left(s^T A b + \lambda^2 - \sigma^2\right), \tag{64}$$

which involves the same unknowns r_0, w_0, and λ.

Note that two parameters w_0 and r_0 are the last component in unit vectors $\bar{w} = (w^1, w_0)^T$ and $\bar{r} = (r^1, r_0)^T$, respectively. Since $w^* = \sqrt{\lambda}\bar{w}$ and $r^* = \sqrt{\lambda}\bar{r}$ are the VSER vectors of the augmented forex matrix B, both parameters r_0 and w_0 in (63) and (64) are positive. Therefore, there is a quartic equation with one unknown x as follows

$$\left(x^2 + \left(\sigma^2 - \|s\|_2^2\right)x - \sigma^2 \left(\bar{u}^T s\right)^2\right)\left(x^2 + \left(\sigma^2 - \|b\|_2^2\right)x - \sigma^2 \left(\bar{v}^T b\right)^2\right)$$
$$= \left(x + \sigma^2\right)\left(x + s^T A b\right)^2, \tag{65}$$

where $x = \lambda^2 - \sigma^2$ involving the largest singular values σ and λ.

In addition, based on the assumption of the rank-one forex matrix $A = \sigma \bar{u} \bar{v}^T$, it is easy to check that $s^T A b = \sigma \left(\bar{u}^T s\right) \left(\bar{v}^T b\right)$, which implies that the quartic equation (65) has a root zero. Since the forex matrix B is augmented from the forex matrix A, their largest singular values satisfy the inequality $\lambda > \sigma$, which implies that the root zero is extraneous. Therefore, the quartic equation (65) is equivalent to such a cubic equation as follows,

$$x^3 + p_2 x^2 + p_1 x = p_0, \tag{66}$$

with one unknown variable x except for a root zero, where

$$\begin{aligned}
p_0 &= \sigma^2 \left(\sigma^2 - \|b\|_2^2\right) \left(\bar{u}^T s\right)^2 + \sigma^2 \left(\sigma^2 - \|s\|_2^2\right) \left(\bar{v}^T b\right)^2 + \left(s^T A b\right)^2 + 2\sigma^2 \left(s^T A b\right), \\
p_1 &= \left(\sigma^2 - \|b\|_2^2\right) \left(\sigma^2 - \|s\|_2^2\right) - \sigma^2 \left(\left(\bar{u}^T s\right)^2 + \left(\bar{v}^T b\right)^2\right) - \sigma^2 - 2\left(s^T A b\right), \\
p_2 &= 2\sigma^2 - \|b\|_2^2 - \|s\|_2^2 - 1.
\end{aligned}$$

For the cubic equation (66) with one unknown variable, let us introduce a new unknown variable y such that $x = y - \frac{p_2}{3}$ and substitute the expression of x into (66); another equivalent cubic equation with unknown variable y is given by

$$y^3 + q_1 y = q_0, \tag{67}$$

where

$$\begin{aligned}
q_0 &= \frac{27p_0 + 9p_2 p_1 - 2p_2^3}{27}, \\
q_1 &= \frac{3p_1 - p_2^2}{3}.
\end{aligned}$$

By using the Cardano formula [8], a real root of the cubic equation (67) is found and defined by

$$y = \sqrt[3]{\frac{q_0}{2} + \sqrt{\left(\frac{q_0}{2}\right)^2 + \left(\frac{q_1}{3}\right)^3}} + \sqrt[3]{\frac{q_0}{2} - \sqrt{\left(\frac{q_0}{2}\right)^2 + \left(\frac{q_1}{3}\right)^3}}. \tag{68}$$

Hence, a real solution to the cubic equation (66) is constructed by

$$x = \lambda^2 - \sigma^2 = y - \frac{p_2}{3}.$$

Since the unknown variable x corresponds to both the largest singular values λ and σ associated with the forex matrices B and A, respectively, it is easily verified that the conclusion (62) in Proposition 6 holds for

$$\lambda^2 = \sigma^2 + x = \sigma^2 + y - \frac{p_2}{3},$$

where y is given in (68) and corresponds to parameters $\zeta_1 = 3q_1 = 3p_1 - p_2^2$ and $\zeta_0 = 27q_0 = 27p_0 + 9p_2p_1 - 2p_2^3$. □

Note that there is two equivalent expressions for ζ_0 and ζ_1 in Proposition 6 as follows,

$$\begin{aligned}\zeta_0 &= 3p_1 - p_2^2\\ &= 2 + 6\|b\|_2^2 + 6\|b\|_2^4 + 2\|b\|_2^6 + 6\|s\|_2^2 + 3\|b\|_2^2\|s\|_2^2 - 3\|b\|_2^4\|s\|_2^2 + 6\|s\|_2^4\\ &\quad -3\|b\|_2^2\|s\|_2^4 + 2\|s\|_2^6 + 18\left(s^T Ab\right) + 18\|b\|_2^2\left(s^T Ab\right) + 18\|s\|_2^2\left(s^T Ab\right)\\ &\quad +27\left(s^T Ab\right)^2 - 3\sigma^2 - 6\sigma^2\|b\|_2^2 - 3\sigma^2\|b\|_2^4 - 6\sigma^2\|s\|_2^2 + 12\sigma^2\|b\|_2^2\|s\|_2^2\\ &\quad -3\sigma^2\|s\|_2^4 + 18\sigma^2\left(s^T Ab\right) - 3\sigma^4 - 3\sigma^4\|b\|_2^2 - 3\sigma^4\|s\|_2^2 + 2\sigma^6 + 9\sigma^2\left(s^T\bar{u}\right)^2\\ &\quad -18\sigma^2\|b\|_2^2\left(s^T\bar{u}\right)^2 + 9\sigma^2\|s\|_2^2\left(s^T\bar{u}\right)^2 + 9\sigma^4\left(s^T\bar{u}\right)^2 + 9\sigma^2\left(b^T\bar{v}\right)^2\\ &\quad +9\sigma^2\|b\|_2^2\left(b^T\bar{v}\right)^2 - 18\sigma^2\|s\|_2^2\left(b^T\bar{v}\right)^2 + 9\sigma^4\left(b^T\bar{v}\right)^2,\end{aligned}$$

$$\begin{aligned}\zeta_1 &= 27p_0 + 9p_2p_1 - 2p_2^3\\ &= -1 - 2\|b\|_2^2 - \|b\|_2^4 - 2\|s\|_2^2 + \|b\|_2^2\|s\|_2^2 - \|s\|_2^4 - 6\left(s^T Ab\right) + \sigma^2\\ &\quad +\sigma^2\|b\|_2^2 + \sigma^2\|s\|_2^2 - \sigma^4 - 3\sigma^2\left(s^T\bar{u}\right)^2 - 3\sigma^2\left(b^T\bar{v}\right)^2,\end{aligned}$$

which is directly defined by parameters $\sigma, \bar{u}, \textit{and } \bar{v}$ with respect to the forex matrix A and vectors b, s in the augmented forex matrix B described by (30).

4.4 The Estimation of the VSER Vectors with the Basket $\tilde{S}$

An expression about $\lambda w^1 - r_0 s$, which is given by (51), holds for any forex matrix A and its augmented forex matrix B. If the forex matrix A is a rank-one matrix, then an equivalent expression is also given by (55). Based on (55), the vector w^1 has an equivalent expression as follows:

$$w^1 = r_0\left(\frac{1}{\lambda}s + \frac{\sigma^2\left(\left(\bar{u}^T s\right)\right)}{\lambda(\lambda^2 - \sigma^2)}\bar{u}\right) + w_0\frac{\sigma\left(\left(\bar{v}\right)^T b\right)}{\lambda^2 - \sigma^2}\bar{u} \tag{69}$$

After the largest singular value λ of the augmented matrix B is estimated, the linear equation (64) could be solved with an unknown variable r_0, which corresponds to such a function of r_0 dependent with w_0 as

$$r_0 = w_0\frac{\left(\left(\lambda^2 - \sigma^2\right)\left(\lambda^2 - \|b\|_2^2\right) - \sigma^2\left(\bar{v}^T b\right)^2\right)}{\lambda\left(s^T Ab + \lambda^2 - \sigma^2\right)}. \tag{70}$$

If the expression of r_0 such as (70) is substituted into (69), an expression of w^1 is also obtained as a function of w_0 as follows:

$$w^1 = w_0 \left\{ \frac{\left((\lambda^2-\sigma^2)(\lambda^2-\|b\|_2^2)-\sigma^2(\bar{v}^T b)^2\right)}{\lambda(s^T Ab+\lambda^2-\sigma^2)} \left(\frac{1}{\lambda}s + \frac{\sigma^2((\bar{u}^T s))}{\lambda(\lambda^2-\sigma^2)}\bar{u}\right) \right.$$
$$\left. + \frac{\sigma\left((\bar{v})^T b\right)}{\lambda^2-\sigma^2}\bar{u} \right\}.$$

Since $\bar{w} = (w^1, w_0)^T$ is a unit vector, a value of w_0 could be estimated by using the equality $(w^1)^T w^1 + w_0^2 = 1$.

Similarly, the estimation of the unit vector $\bar{r} = (r^1, r_0)^T$ can be obtained through representing r^1 as a function of r_0

$$r^1 = r_0 \left\{ \frac{\left((\lambda^2-\sigma^2)(\lambda^2-\|s\|_2^2)-\sigma^2(\bar{u}^T s)^2\right)}{\lambda(s^T Ab+\lambda^2-\sigma^2)} \left(\frac{1}{\lambda}b + \frac{\sigma^2((\bar{v}^T b))}{\lambda(\lambda^2-\sigma^2)}\bar{v}\right) \right.$$
$$\left. + \frac{\sigma\left((\bar{u})^T s\right)}{\lambda^2-\sigma^2}\bar{v} \right\}$$

and then computing the component r_0 from the property $(r^1)^T r^1 + r_0^2 = 1$.

The results above for estimating unit vectors $\bar{w} = (w^1, w_0)^T$ and $\bar{r} = (r^1, r_0)^T$ are summarized in the following proposition.

Proposition 7 *Suppose that A is a rank-one forex matrix and B is an augmented forex matrix given in (30). Let us denote the largest singular values of A and B by σ and λ, respectively, and the VSER vectors of B by $w^* = \sqrt{\lambda}\bar{w}$ and $r^* = \sqrt{\lambda}\bar{r}$, where $\bar{w} = (w^1, w_0)^T$ and $\bar{r} = (r^1, r_0)^T$ are unit vectors. If the square of the largest singular value λ with respect to the augmented matrix B, λ^2, is estimated in advance by (62) in Proposition 6, then all of components in both vectors $\bar{w}$ and $\bar{r}$ are estimated by*

$$w_0 = \frac{1}{\sqrt{\|\theta\|_2^2 + 1}}, \tag{71}$$

$$w^1 = \frac{1}{\sqrt{\|\theta\|_2^2 + 1}}\theta, \tag{72}$$

$$r_0 = \frac{1}{\sqrt{\|\gamma\|_2^2 + 1}}, \tag{73}$$

$$r^1 = \frac{1}{\sqrt{\|\gamma\|_2^2 + 1}}\gamma, \tag{74}$$

where both θ and γ are n-dimensional vectors given by

$$\theta = \frac{(\lambda^2-\sigma^2)(\lambda^2-\|b\|_2^2)-\sigma(b^T v^*)^2}{\lambda^2(s^T Ab+\lambda^2-\sigma^2)} \left(s + \frac{\sigma(s^T u^*)}{\lambda^2-\sigma^2}u^*\right) + \frac{(b^T v^*)}{\lambda^2-\sigma^2}u^*$$

and

$$\gamma = \frac{(\lambda^2-\sigma^2)(\lambda^2-\|s\|_2^2)-\sigma(s^T u^*)^2}{\lambda^2(s^T Ab+\lambda^2-\sigma^2)}\left(b + \frac{\sigma(b^T v^*)}{\lambda^2-\sigma^2}v^*\right) + \frac{(s^T u^*)}{\lambda^2-\sigma^2}v^*,$$

respectively, and u^* *and* v^* *are the VSER vectors of the forex matrix* A *such that* $\|u^*\|_2 = \|v^*\|_2$.

5 Numerical Verifications

In this section, numerical examples are presented to verify the performance of approaches for analyzing the currency portfolio optimization problem. All of numerical experiments were conducted in the Lenovo computer X220i with 2.30GHz Intel(R) Core(TM) i3-2350M CPU and 6.0GB memory.

5.1 *Source of Data About the Forex Rates*

Data about forex rates were collected from the internet website of the Bank of China (https://www.boc.cn/) through the user's account, which spread from September 2, 2024, to October 1, 2024, except for holidays and weekends. For example, a forex spot rate matrix in Table 1 is given by the Bank of China, which includes all of cross foreign exchange spot rates among nine currencies at Beijing Time 20:23 p.m. on October 1, 2024. There are 74 foreign exchange spot rate matrices, which induces time series in unequal time periods. The currency basket here consists of nine currencies corresponding to the US dollar (USD), Euro (EUR), Pound sterling (GBP), Australian dollar (AUD), Swiss franc (CHF), Canadian dollar (CAD), Singapore dollar (SGD), Hong Kong dollar (HKD), and Japanese yen (JPY). Each element a_{ij} in Table 1 denotes the amount of the currency j equivalent to one currency i for $i, j = 1, 2, \cdots, 9$. The index i corresponds to the i-th currency code in the sequence of *EUR*, *GBP*, *AUD*, *USD*, *CHF*, *CAD*, *SGD*, *HKD*, *JPY*.

Note that the approaches proposed in this article can also be applied to other currency baskets such as *majors*, *G10*, *Asia*, *EMEA*, *Latin America*, *Middle East*, and *precious metals*. For example, the basket of *G10* includes currencies corresponding to countries (or currency zones) in the *Group of Ten*, a famous organization consisting of 11 industrialized countries (Belgium, Canada, France, Germany, Italy, Japan, the Netherlands, Sweden, Switzerland, the United Kingdom, and the United States). These countries usually organize meetings for consulting each other and cooperating on economic, monetary, and financial matters [1]. The meaning of currency and precious metal codes can be found in *ISO 4217 – Currency Code Maintenance: Get the Correct Currency Code* [13].

Table 1 A cross foreign exchange rate matrix on October 1, 2024 (Bank of China)

Names	EUR	GBP	AUD	USD	CHF	CAD	SGD	HKD	JPY
EUR	1	0.82955	1.59949	1.10601	0.93593	1.49432	1.42205	8.59459	158.789
GBP	1.19937	1	1.92372	1.33005	1.12561	1.79728	1.71039	10.33579	190.984
AUD	0.62205	0.51747	1	0.68959	0.58351	0.93163	0.88667	5.35894	99.015
USD	0.90120	0.74966	1.44502	1	0.84559	1.34996	1.28475	7.76384	143.471
CHF	1.06304	0.88440	1.70460	1.17862	1	1.59334	1.51634	9.16463	169.361
CAD	0.66624	0.55419	1.06800	0.73855	0.62538	1	0.94956	5.73904	106.067
SGD	0.69985	0.58215	1.12212	0.77589	0.65692	1.04806	1	6.03015	111.432
HKD	0.11589	0.09640	0.18582	0.12847	0.10878	0.17358	0.16523	1	18.432
JPY	0.00627	0.00522	0.01005	0.00695	0.00589	0.00939	0.00894	0.05398	1

5.2 The Performance of the ROA Approach

In numerical experiments, a sequence of forex rate matrices $\{A^k\}$ is decomposed into two sequences of virtual standard exchange rate vectors $\{u^k\}$ and $\{v^k\}$ by using the rank-one approximation method. There are 74 records that include both the ask and bid prices for every pair of currencies in the basket:

$$S = \left\{ \begin{array}{ccccccccc} 1, & 2, & 3, & 4, & 5, & 6, & 7, & 8, & 9 \\ \Updownarrow & \Updownarrow & \Updownarrow & \Updownarrow & \Updownarrow & \Updownarrow & \Updownarrow & \Updownarrow & \Updownarrow \\ EUR, & GBP, & AUD, & USD, & CHF, & CAD, & SGD, & HKD, & JPY \end{array} \right\}$$

Each record corresponds to an instance of the ROA problem (31) with respect to a cross foreign exchange rate matrix A. For each instance associated with the index k of time periods, the cross-currency rate matrix is constructed from both ask and bid prices, denoted by A^k. The ROA approach is adopted to solve the ROA optimization problem (31), in which $A = A^k$, and find two column vectors u^k and v^k such that $\|u^k\|_2 = \|v^k\|_2$ and $\|A^k - u^k(v^k)^T\|$ is minimal over all of rank-one matrices in $\Re^{n\times n}$.

Let us denote the rank-one matrix $B^k = u^k(v^k)^T$ and the error matrix $E^k = A^k - B^k$ at the period k. For each pair of the corresponding elements in A^k and B^k, there exists an approximation error $e^k_{ij} = a^k_{ij} - u^k_i v^k_j (k = 1, 2, \cdots, 74,\ i, j = 1, 2, \cdots, 9)$. The uncertainty of all error matrices $E^k, k = 1, 2, \cdots, 74$ is described in Table 2.

For every error matrix $E^k = A^k - B^k$, the maximum and the minimum deviations among all of components are collected together with the minimum of differences between $a^k_{ij} * a^k_{ji}$ and 1 for all i, j, k. In addition, the objective function $\|A^k - B\|$ in (31) has the minimum value $\|A^k - B^k\|$ in the domain of the rank-one matrices, which is computed with respect to the Euclidean and Frobenius norm, respectively. The performance of the rank-one approximation is measured by five sequences of errors with the same length 74, denoted by $\min\{A^k \circ (A^k)^T - 1\}$, $\max\{E^k\}$,

Table 2 Statistical results of the extreme errors in $E^k = A^k - B^k$

Estimated values	Extreme errors in the error matrix E^k				
($N = 74$)	$\min\{A^k \circ (A^k)^T - 1\}$	$\max\{E^k\}$	$\min\{E^k\}$	$\|E^k\|_2$	$\|E^k\|_F$
Maximum	−0.00469	0.00340	−0.00118	0.00534	0.00698
Minimum	−0.00551	0.00228	−0.00177	0.00385	0.00585
Average	−0.00494	0.00239	−0.00137	0.00413	0.00613
Standard deviation	0.00021	0.00014	0.00011	0.00022	0.00021

$\min\{E^k\}$, $\|E^k\|_2$, and $\|E^k\|_F$, respectively. The fundamental statistical properties for each sequence are summarized by a column in Table 2. It is easily known that the average of each sequence of extreme errors has an order about 10^{-3} between the forex matrix A^k and its ROA counterpart B^k. The standard deviation for every sequence of extreme errors has an order about 10^{-4}. Since the spread of forex exchange rates, which usually is associated with the market maker (Bank of China) and the corresponding currency basket, has a similar order of approximation errors, it can be concluded that the forex matrix A^k is approximated well by the rank-one matrix $B^k = u^k(v^k)^T$, whose errors are even smaller than the uncertainty of the bid and ask prices of currencies.

5.3 *The Performance of Formulations for the CPO Problem*

In this subsection, four kinds of models for the currency portfolio optimization problem are compared with the same initial currency portfolio as

$$x^0 = (19.30, 17.39, 18.39, 29.59, 37.77, 63.43, 38.00, 46.39, 75.57)^T.$$

Every element in the vector x^0 comes from a random number in [0, 100] and is kept two decimal places here. More decimal places are used during the optimization of currency portfolios under the computational environment MATLAB. After the initial portfolio x^0 is generated, it is used for starting different models. The optimization methods are compared here and classified into the following groups of models:

M1 The basic model (5) ∼(8) is solved by calling the subroutine linprog.m in the MATLAB, a computational environment. Given a specific currency $m \in S$ in advance, the initial objective value at x^0 is computed by $x_m^1 = (A_{\cdot m}^1)^T x^0$ for the currency m, where $A_{\cdot m}^1$ denotes the m-th column of the forex rate matrix A^1 at the time period 1. The final value of the objective function in (5) is given by x_m^N, where $N = 74$. The performance of the basic model is measured by the ratio of x_m^N to x_m^1, denoted by $\rho_m^N = x_m^N / x_m^1$ for $m = 1, 2, \cdots, 9$.

M2 The equivalent model (23) is solved by computing the max-product value (22) of the cross forex exchange rate matrices $A^k, k = 1, 2, \cdots, 74$. For a specific currency $m \in S$, the initial objective value at x^0 is given by $x_m^1 = (A_{\cdot m}^1)^T x^0$, where $A_{\cdot m}^1$ denotes the m-th column of the forex rate matrix A^1 at the time period 1. The optimal value of the objective function is given by $\bar{x}_m^N$ in (23), where $N = 74$. The performance of the max-product model (23) is similarly measured by the ratio of $\bar{x}_m^N$ to x_m^1, denoted by $\bar{\rho}_m^N = \bar{x}_m^N / x_m^1$ for $m = 1, 2, \cdots, 9$.

M3 For the ROA model (25) ∼(27) in Corollary 1, it is also solved by calling the subroutine linprog.m in the computational environment MATLAB. Similarly, the initial objective value at x^0 is computed by $\hat{x}_m^1 = v_m^1 (u^1)^T x^0$ for the currency m specified in advance, where both vectors u^1 and v^1 are the unique solution to the ROA problem (31) for approximating the forex matrix $A = A^1$ at the time period 1. The optimal value of the objective function in the ROA model (25) ∼(27) is given by (24), that is,

$$\hat{x}_m^N = v_m^N \left\{ \prod_{k=1}^{N-1} \left(v^k\right)^T \otimes u^{k+1} \right\} \left(u^1 x^0\right),$$

where $N = 74$. The performance of the ROA model is measured by $\hat{\rho}_m^N = \hat{x}_m^N / \hat{x}_m^1$, $m = 1, 2, \cdots, 9$.

M4 This group of models combines the ROA model (25) ∼(27) in Corollary 1 with the sequence of forex rate matrices $\{A^k\}$, in which the optimal solution is given by the ROA model, but the value of objective function is calculated by using the practical forex rate matrices. Hence, the initial objective value at x^0 is computed by $x_m^1 = (A_{\cdot m}^1)^T x^0$ for the currency m chosen in advance. The optimal value of the objective function is similarly given by

$$\tilde{x}_m^N = (A_{\cdot m}^N)^T x^{N-1},$$

where $N = 74$. The optimal sequence of exchanging currencies is given by the ROA model (25) ∼(27), but the currency portfolios are computed by the sequence of $\{A^k\}$. The performance of the hybrid model is measured by $\tilde{\rho}_m^N = \tilde{x}_m^N / x_m^1$, $m = 1, 2, \cdots, 9$.

For the optimization models explained above, their numerical verification results are presented in Table 3, where the first group (group M1) have the same initial and final (optimal) values of the objective function as those of the group M2. In addition, the group M4 has the same initial value of the objective function as that of both groups M1 and M2. Based on the results about ρ_m^N and $\hat{\rho}_m^N$, the performance of the group M3 is similar to that of the first/second group (i.e., the group M1 and the group M2), in which their errors have an order about 10^{-3} or 10^{-4}. Based on comparisons among the values of ρ_m^N, $\hat{\rho}_m^N$, and $\tilde{\rho}_m^N$, the performance of the group M4 is worst among four currency portfolio optimization models, but their errors still have an order about 10^{-2}, which indicates the feasibility of the hybrid model. Numerical experiments validate that the ROA approach is a powerful tool to approximate forex rate matrices and the optimal value of the currency portfolio optimization problem.

Table 3 Performance of the currency portfolio optimization models

Currency	Initial values $(x_m^1, \hat{x}_m^1, x_m^1)$			Final values $(x_m^N, \hat{x}_m^N, \tilde{x}_m^N)$			Ratios $(\rho_m^N, \hat{\rho}_m^N, \tilde{\rho}_m^N)$		
Names m	M1 = M2	M3	M4	M1 = M2	M3	M4	M1 = M2	M3	M4
EUR	192.57	192.59	192.57	220.04	220.16	216.64	1.1427	1.1432	1.1250
GBP	162.13	162.17	162.13	183.10	183.16	180.11	1.1293	1.1294	1.1109
AUD	314.47	314.50	314.47	353.00	352.95	347.13	1.1225	1.1223	1.1038
USD	212.87	212.88	212.87	244.17	244.03	240.04	1.1470	1.1463	1.1277
CHF	180.89	180.87	180.89	206.46	206.61	203.26	1.1414	1.1423	1.1236
CAD	287.28	287.22	287.28	329.75	329.76	324.33	1.1478	1.1481	1.1290
SGD	278.03	278.02	278.03	313.76	313.84	308.73	1.1285	1.1289	1.1104
HKD	1659.47	1659.46	1659.47	1895.90	1896.35	1863.69	1.1425	1.1428	1.1231
JPY	31,100	31,100	31,100	35,034	35,041	34,528	1.1265	1.1267	1.1102

5.4 The Performance of the Sensitivity Analysis

In this subsection, the cross foreign exchange spot rate matrix in Table 1 is considered again, which is similar to an augmented forex rate matrix in (30) and denoted by $B \in \Re^{9\times 9}$. The upper left 8×8 matrix is a reduced forex rate matrix, denoted by A, which corresponds to the following currency basket:

$$S\backslash\{JPY\} = \{EUR, GBP, AUD, USD, CHF, CAD, SGD, HKD\}$$

In the meantime, two vectors s and b are given by

$$s = (158.789, 190.984, 99.015, 143.471, 169.361, 106.067, 111.432, 18.432)^T$$

and

$$b = (0.00627, 0.00522, 0.01005, 0.00695, 0.00589, 0.00939, 0.00894, 0.05398)^T,$$

corresponding to the last column and the last row in Table 1, respectively.

Suppose that the solution (u^*, v^*) to the ROA problem (31) is obtained in advance with respect to the currency basket $S\backslash\{JPY\}$, that is,

$$u^* = (1.95862, 2.35541, 1.22131, 1.76929, 2.08843, 1.30792, 1.37423, 0.22784)^T$$

and

$$v^* = (0.50945, 0.42384, 0.81672, 0.56468, 0.478096, 0.76307, 0.72622, 4.38810)^T.$$

Let us denote two unit vectors parallel to u^* and v^* by $\bar{u}$ and $\bar{v}$, respectively. Note that these vectors can be used to approximate the forex rate matrix A such that the minimal errors are given by

$$\|A - u^*(v^*)\|_2 = \|A - \sigma\bar{u}\bar{v}^T\|_2 = 0.00315$$

and

$$\|A - u^*(v^*)\|_F = \|A - \sigma\bar{u}\bar{v}^T\|_F = 0.00563.$$

Based on the sensitivity analysis method in Sect. 4, we could estimate the solution (w^*, r^*) to the ROA problem (31) with respect to the augmented forex matrix B.

Suppose that the largest singular value of the matrix A is known as 22.01875. For the largest singular value of the augmented forex matrix B, it is estimated by the expression (62) in Proposition 6 and equals to the following positive real number:

$$\lambda = 381.12038$$

In addition, by using Proposition 7 and λ estimated above, two intermediate vectors θ and γ are estimated and given by

$$\theta = (158.791, 190.986, 99.016, 143.473, 169.363, 106.068, 111.433, 18.432)^T$$

and

$$\gamma = (0.00628, 0.00523, 0.01007, 0.00696, 0.00590, 0.00941, 0.00896, 0.05412)^T,$$

respectively. Furthermore, through the formula (71) ∼(74), the sensitivity analysis method generates two unit vectors as follows,

$$\bar{w} = (0.41733, 0.50195, 0.26023, 0.37708, 0.44512, 0.27877, 0.29287, 0.04844, 0.00263)^T$$

and

$$\bar{r} = (0.00627, 0.00522, 0.01006, 0.00695, 0.00589, 0.00940, 0.00894, 0.05403, 0.99833)^T,$$

which corresponds to the VSER vectors $w^* = \sqrt{\lambda}\bar{w}$ and $r^* = \sqrt{\lambda}\bar{r}$ for approximating the augmented forex rate matrix B. It is verified that these vectors are also the optimal solution to the ROA problem (31) with the minimal errors

$$\|B - w^*(r^*)^T\|_2 = \|B - \lambda\bar{w}\bar{r}^T\|_2 = 0.00407$$

and

$$\|B - w^*(r^*)^T\|_F = \|B - \lambda\bar{w}\bar{r}^T\|_F = 0.00656.$$

Conversely, if the optimal solution to the ROA problem (31) with respect to the forex rate matrix B is known in advance, then the sensitivity analysis method in Sect. 4 also succeeds to estimate the largest singular value σ of the reduced forex matrix A and the corresponding VSER vectors u^* and v^* to approximate the matrix A.

6 Concluding Remarks

Several formulations of the currency portfolio optimization (CPO) problem are developed and compared in the paper, in which that the network formulation is equivalent to the basic model. The network formulation is based on the maximum amplifying flow framework and gives an optimal solution to the CPO problem through the max-product operations strategy. The concept of the max-product operation is firstly proposed here and some properties are also obtained including the associative law. In addition, the rank-one approximation (ROA) approach is presented to approximate cross Foreign eXchange (abbreviated by FX or forex) spot rate matrices with respect to a currency basket and a market maker. Based on the ROA approach, a new method for the sensitivity analysis of a currency basket is proposed, which estimates the largest singular value of an augmented or reduced forex matrix together with its unit left and right singular characteristic vectors. The ROA method gives another strategy to formulate the currency portfolio optimization problem, which is referred to as the ROA model. The optimal sequential solution to the ROA model is also integrated with original forex rate matrices, which corresponds to a feasible solution to the CPO problem. Through numerical experiments, the performances of these methods are verified, which includes the ROA approach, four types of models for the CPO problem, and the sensitive analysis of a currency basket.

Acknowledgments The research of the author was partially supported by the Deputy Department Head's Discipline Construction Fund for the project as *Data Driven Foreign Exchange Market Structures and Profit-loss Analysis for Trans-currency Zone's Supply Chain* in the Department of Industrial Engineering, Tsinghua University, Beijing, China. The author gratefully acknowledges Prof. X.B. Zhao, the deputy department head in 2021, for his valuable suggestions and comments on the project, as well as the anonymous for the constructive suggestions to improve the paper.

References

1. Bank for International Settlements: The Group of Ten. The Bank for International Settlements (BIS) (2016) Available via DIALOG. https://www.bis.org/list/g10publications/. Cited 21 Dec. 2016
2. Choi, C.: McDonald's Profit Slips on Stronger Dollar, Slowing Sales; Revenue Misses Forcasts. The Associates Press, New York (2012). Available via DIALOG. https://www.canadianbusiness.com/business-news/mcdonalds-profit-slip. Cited 5 Oct 2018
3. Fabozzi, F.J., Markowitz, H.M., Kolm, P.N., Gupta, F.: Portfolio selection. In: Fabozzi, F.J., Markowitz, H.M. (eds.) Theory and Practice of Investment Management, pp. 45–78. Wiley, Hobokin (2011)
4. Ganti, A.: What are asset classes? More than just stocks and bonds. In: Investopedia (Updated 12 Jun 2024). Dotdash Meredith, New York (2024). Available via DIALOG. https://www.investopedia.com/terms/a/assetclasses.asp. Cited 15 Sept 2024
5. Henryk, M.: Nonnegative Matrices. Wiley, New York (1988)
6. Huang, H.X., Zhang, Z.J.: An intrinsic robust rank-one-approximation approach for currency portfolio optimization. Quant. Finance Econ. **2**, 160–189 (2018). https://doi.org/10.3934/QFE.2018.1.160
7. Huang, H.X., Zhang, Z.J.: Virtual standard currency for approximating foreign exchange rates. Int. J. Electr. Commerce **23**, 33–62 (2019). https://doi.org/10.1080/10864415.2018.1512273
8. Katz, V.J.: A History of Mathematics: An Introduction. Pearson Education Asia Limited and Higher Education Press, Beijing (2004)

9. Kahneman, D., Tversky, A.: Prospect theory: an analysis of decision under risk. Econometrica **47**, 263–292 (1979)
10. Markowitz, H.M.: Portfolio selection. J. Finan. **7**, 77–91 (1952)
11. Markowitz, H.M.: The utility of wealth. J. Polit. Econ. **60**, 151–158 (1952)
12. Markowitz, H.M.: Normative portfolio analysis: past, present, and future. J. Econ. Bus. **42**, 99–103 (1990)
13. SIX Financial Information AG: Currency codes of the International Standard ISO 4217. The International Organization for Standardization (ISO) and its Swiss member SNV (Swiss Association for Standardization) (2024) Available via DIALOG. http://www.currency-iso.org/en/home/tables/table-a1.html. Cited 2 Oct. 2024
14. Tu, L.T., Wei, M.: China Forex Losses Jump 13-Fold as Investors Brace for More. Bloomberg (2016). Available via DIALOG. http://www.bloomberg.com/news/articles/2016-04-05/chinaforex-losses-jump-13-fold-to-7-5-billion-with-more-ahead. Cited 6 April 2016

On the Consistency and Large Deviations of the Method of Empirical Means in Stochastic Programming Problems

Pavel Knopov , Tatiana Ermolieva, and Evgeniya Kasitskaya

1 Introduction

The stochastic programming problem arises when one needs to make decisions in conditions of uncertainty and risk [1, 2]. The mean value of a factor of quality of control depending on a random parameter is optimized.

Indirect methods for solving stochastic programming problems involve approximating a stochastic problem with an approximate deterministic problem. One of the main indirect methods of stochastic programming is the so-called method of empirical means [3–7]. In this method, factors of quality of control are approximated by their empirical estimates. One of the main problems is estimation accuracy and convergence when the area of observations increases.

Large deviations theory (see, for example, [8–12]) is a part of probability theory that considers cases in which empiric estimates deviate from true values of parameters more than from "normal" values, i.e., more than a value that is effectively described by the central limit theorem. A more accurate calculation of the probability of such events demands a more accurate study of the integrals of exponential functionals.

This problem arises in many different contexts. Large deviation theory is applied in probability theory, mathematical statistics, operations research, informatics, statistical physics, financial mathematics, and other spheres.

P. Knopov (✉) · E. Kasitskaya
V. M. Glushkov Institute of Cybernetics, National Academy of Sciences of Ukraine, Kyiv, Ukraine
e-mail: knopovPS@nas.gov.ua; Kasitska@nas.gov.ua

T. Ermolieva
International Institute for Applied Systems Analysis, Laxenburg, Austria
e-mail: ermol@iiasa.ac.at

B. Goldengorin (ed.), *Theory, Algorithms, and Experiments in Applied Optimization*, Springer Optimization and Its Applications 226,
https://doi.org/10.1007/978-3-031-91357-0_8

Large deviation theory describes rare events. Nevertheless, the necessity of studying rare events is not in doubt because their occurrence can cause many different problems and demands much energy for their liquidation.

As written in many sources (for example, in [8, 9]), “theory” of large deviations is absent. In addition to basic definitions, which are standard, many methods and approaches exist that allow the analysis of such rare events. Often, identical results can be achieved in different manners. Common probabilistic estimates are transferred to partial situations that are under study.

2 The Method of Empirical Means for Discrete and Continuous Models with Dependent Observations

Let us consider stochastic programming problems where the empirical functions are constructed by observations of stationary random processes with a discrete or continuous parameter.

Let $(Y, L(Y))$ be some measurable space, where Y is a metric space, $L(Y)$ is a minimal σ-algebra on Y, and $\|\cdot\|$ is a norm set in Y. Let ξ_i, $i \in N$ be independent identically distributed observations of a random variable or a stationary in a strict sense ergodic random sequence, defined on a probability space $(\Omega, \Im, P)$ with values in measurable space $(Y, L(Y))$, where Ω is a space of elementary events, $\Im$ is a σ-algebra of elementary events on Ω, and P is a probability measure on $\Im$ such that $P(\Omega) = 1$. We assume that I is a closed subset in $\Re^l$, $l \geq 1$, possibly $I = \Re^l$, and that $f : I \times Y \to \Re$ is a nonnegative function satisfying the following conditions:

1. $f(\mathbf{u}, z)$, $\mathbf{u} \in I$, is continuous for all $z \in Y$;
2. for any $\mathbf{u} \in I$, the mapping $f(\mathbf{u}, z)$, $z \in Y$ is $L(Y)$ measurable.

The problem consists of finding the minimum point of the function and its minimal value (the stochastic optimization problem).

$$\min F(u) = \min Ef(u, \xi_1), \quad u \in I$$

This problem is approximated by the following problem: find the minimum points of the function

$$F_n(\mathbf{u}) = \frac{1}{n} \sum_{i=1}^{n} f(\mathbf{u}, \xi_i),$$

and its minimal value.

We present some examples of regression models, which are widely known to specialists in the field of theoretical and applied statistics.

1. $y_i = \sum\limits_{t=1}^{p} x_{it}\alpha_t^0 + \varepsilon_i, \quad i = 1, \ldots, n.$

   Here, ε_i and $i = 1, \ldots, n$ are independent or stationary dependent random variables, and $\mathbf{x}_i = \left\{x_{it}, t = \overline{1, p}\right\}$ and $i = 1, \ldots, n$ are independent identically distributed random vectors that are independent of ε_i and $i = 1, \ldots, n$.

   The vector $\boldsymbol{\alpha}^0 = \left(\alpha_1^0, \ldots, \alpha_p^0\right)$ is unknown and is estimated.
2. $y_i = g(\mathbf{x}(i), \boldsymbol{\alpha}^0) + \varepsilon_i, \mathbf{x}(i) \in R^p,$

   where the p-dimensional vectors $\mathbf{x}(i)$ and ε_i are mutually independent, and each of the sequences $\{\mathbf{x}(i)\}$ and $\{\varepsilon_i\}, i = 1, \ldots, n$, is the sequence of independent or stationary random vectors or variables.

Some cost functions characterizing the accuracy of the estimate are as follows:

1. $F_n(\alpha) = \frac{1}{n}\sum\limits_{i=1}^{n}\left[y_i - \sum\limits_{t=1}^{p} x_{it}\alpha_t\right]^2;$
2. $F_n(\alpha) = \frac{1}{n}\sum\limits_{i=1}^{n}\left[y_i - g\left(\mathbf{x}(i), \boldsymbol{\alpha}\right)\right]^2;$
3. $F_n(\alpha) = \frac{1}{n}\sum\limits_{i=1}^{n}\mid y_i - \sum\limits_{t=1}^{p} x_{it}\alpha_t \mid;$
4. $F_n(\alpha) = \frac{1}{n}\sum\limits_{i=1}^{n}\mid y_i - g\left(\mathbf{x}(i), \boldsymbol{\alpha}\right)\mid.$

Theorem 1 [13]. *Let the following conditions be satisfied:*

1. *for any* $c > 0$, $E\left(\max\limits_{\|\mathbf{u}\|\le c} f(\mathbf{u}, \xi_1)\right) < \infty$, *where* $\|\cdot\|$ *is a norm in* $\Re^l$
2. *for all* $z \in Y', P\{\xi_1 \in Y'\} = 1, f(\mathbf{u}, z) \to \infty$ *as* $\|u\| \to \infty$;
3. *there is a unique point* $\mathbf{u}_0$ *at which the function* $F(\mathbf{u})$ *attains its minimum.*

Then, for any n *and* $\omega \in \Omega', P(\Omega') = 1$, *there exists at least one vector* $\mathbf{u}_n = \mathbf{u}_n(\omega) \in I$ *for which the minimum value of* $F_n(\mathbf{u})$ *is attained, and for any* $n \ge 1$, *the vector* $\mathbf{u}_n$ *can be chosen to be* G_n'*-measurable, where* $G_n' = G_n \cap \Omega'$ *and* $G_n = \sigma\left\{\xi_i, \ i = \overline{1, n}\right\}$. *In this case, with probability 1,* $\mathbf{u}_n \to \mathbf{u}_0$ *and* $F_n(\mathbf{u}_n) \to F(\mathbf{u}_0)$, $n \to \infty$.

Theorem 2 [13]. *Let* $\{\xi(t), \ t \in \Re\}$ *be a random ergodic process stationary in the strict sense and defined on the probability space* $(\Omega, \Im, P)$ *with values in* $(Y, L(Y))$. *Suppose that the trajectories of the process are continuous and that the function f, described above, is continuous. Let the following conditions be satisfied:*

1. *for any* $c > 0$, $E\left\{\max\limits_{\|\mathbf{u}\|\le c} f(\mathbf{u}, \xi(0))\right\} < \infty$;
2. *if* I *is an unbounded set then for any* $z \in Y'$ *and* $P\{\xi(t) \in Y' \ \ \forall t \ge 0\} = 1$, *one has* $f(\mathbf{u}, z) \to \infty$ *as* $\|\mathbf{u}\| \to \infty$;
3. *there is a unique element* $\mathbf{u}_0 \in I$ *for which the minimal value of the function* $F(\mathbf{u}) = Ef(\mathbf{u}, \xi(0))$ *is attained.*

Then, for all $T > 0$ and $\omega \in \Omega', P(\Omega') = 1$, there is at least one vector $\mathbf{u}(T) \in I$ for which the minimal value of the function

$$F_T(\mathbf{u}) = \frac{1}{T} \int_0^T f(\mathbf{u}, \xi(t))\, dt$$

is attained and measurable, and we have

$$P\left\{\lim_{T\to\infty} \mathbf{u}(T) = u_0\right\} = 1, \quad P\left\{\lim_{T\to\infty} F_T(\mathbf{u}(T)) = F(\mathbf{u_0})\right\} = 1.$$

Remark The condition of the ergodic in the Theorems 1 and 2 will be fulfilled if the random stationary sequence or the random stationary process satisfies a condition of strong intermixing.

3 The Method of Empirical Means Is Applied to Nonstationary Models

Consider a more general case of a nonstationary model. We assume that the criterion function also depends on the temporal parameter, i.e., it is a function of three variables. For example, in discrete time, the criterion function has the form

$$F_n(u) = \frac{1}{n} \sum_{i=1}^{n} f(i, u, \xi_i).$$

For example, one can take

$$F_n(u) = \frac{1}{n} \sum_{i=1}^{n} [y_i - g(i, u)]^2$$

$$F_n(u) = \frac{1}{n} \sum_{i=1}^{n} | y_i - g(i, u) |$$

for the model of the observations

$$y_i = g(i, u_0) + \xi_i.$$

The unknown parameter is also assumed to be an element of some functional space. For example, one can consider the problem of the estimation of the unknown function $u(t) \in K$, where K is the compact set of functions defined on [0,1], by observations

$$y_i = u\left(\frac{i}{n}\right) + \xi_i, i = 0, \ldots, n$$

with some criterion function.

Theorem 3 [13]. *Let a stochastic function $f\left(i, \vec{u}, \xi_i\right)$ satisfy the following conditions:*

1. *for any $\vec{u} \in I$, there exists a function $F\left(\vec{u}\right)$ such that $F\left(\vec{u}\right) = \lim_{n\to\infty} F_n\left(\vec{u}\right)$ and a point $\vec{u}_0 \in I$ such that we have $F\left(\vec{u}_0\right) < F\left(\vec{u}\right)$ when $\vec{u} \neq \vec{u}_0$;*
2. *the function $f\left(i, \vec{u}, z\right)$ is continuous with respect to the second argument uniformly relative to i and z;*
3. *if the set I is unlimited, then $f\left(i, \vec{u}, z\right) \to \infty$ as $\|u\| \to \infty$ for fixed i and z;*
4. *there is a function $c(\gamma) \to 0$ as $\gamma \to 0$, and for any $\delta > 0$, there exists γ_0 such that, for any element $\vec{u}' \in I$ and $0 < \gamma < \gamma_0$, the following relationship is true:*
   $$\overline{\lim_{n\to\infty}} \frac{1}{n} \sum_{i=1}^{n} E \sup_{\substack{\left\|\vec{u} - \vec{u}'\right\| < \gamma \\ \left\|\vec{u} - \vec{u}_0\right\| < \delta}} \left|f\left(i, \vec{u}, \xi_i\right) - f\left(i, \vec{u}', \xi_i\right)\right| < c\left(\gamma\right);$$
5. *the function $f\left(i, \vec{u}, \xi_i\right)$ satisfies the strong mixing condition.*
   $$\alpha(j) = \sup_i \sup_{\substack{A \in \sigma_{-\infty}^{i} \\ B \in \sigma_{i+j}^{\infty}}} |P(AB) - P(A)P(B)| \leq \frac{c}{1+j^{1+\varepsilon}},$$
   $$\varepsilon > 0, \sigma_n^m = \sigma\left\{f\left(i, \vec{u}, \xi_i\right), \ n \leq i \leq m, \quad \vec{u} \in I\right\};$$
6. $E\left(f\left(i, \vec{u}, \xi_i\right)\right)^{2+\delta} < \infty, \varepsilon\delta > 2.$

Let $\vec{u}_n = \arg\min_{\vec{u}\in I} F_n\left(\vec{u}\right)$

Then, we have

$$P\left\{\lim_{n\to\infty} \left\|\vec{u}_n - \vec{u}_0\right\| = 0\right\} = 1,$$

$$P\left\{\lim_{n\to\infty} F_n\left(\vec{u}_n\right) = F\left(\vec{u}_0\right)\right\} = 1.$$

A similar statement is also true for the case in which a continuous stochastic function $f\left(t, \vec{u}, \xi(t)\right)$ is considered in an interval $[0, T]$, i.e., the following functional is considered:

$$F_T\left(\vec{u}\right) = \frac{1}{T}\int_0^T f\left(t, \vec{u}, \xi(t)\right) dt,$$

where $\xi(t)$ *is a random process that is stationary in a strong sense. It is necessary to find* $\min_{\vec{u}\in I} F_T(\vec{u})$ *and investigate the asymptotic behaviors of* $\vec{u}_T = \arg\min_{\vec{u}\in I} F_T(\vec{u})$ *and* $F_T(\vec{u}_T)$ *as* $T \to \infty$.

4 Method of Empirical Means for the Models with Random Functions Depending on Several Variables or Random Fields

Let us consider a stochastic programming problem where the empirical function is constructed on the basis of observations of a homogeneous random field.

Let $\left\{\xi\left(\vec{t}\right), \vec{t} \in \mathbb{R}^m\right\}$ be an ergodic homogeneous in a strict sense random field with continuous trajectories, defined on a probabilistic space (Ω, G, P), with values in some metric space $(Y, L(Y))$. We assume that I is a closed subset in $\Re^l$, $l \geq 1$, possibly $I = \Re^l$, and that $f: I \times Y \to \Re$ is a nonnegative continuous function.

Theorem 4 [13]. *Suppose that for the random field* $\xi(t) \in Y,\ t \in R^m$, *the conditions below are fulfilled:*

1. *for any* $c > 0$
   $$E\left\{\max_{\|u\|<c}\left(f\left(\mathrm{u}, \xi(0)\right)^2\right)\right\} < \infty;$$
2. *if I is unbounded then for each* $z \in Y$
   $$f(\mathrm{u}, z) \to \infty,\ \|u\| \to \infty;$$
3. *there is a unique element* $u_0 \in I$ *for which the minimal value of the function* $F(u) = Ef(u, \xi(0))$ *is attained.*

Then, for all $T_i > 0,\ i = 1, \ldots, m$ *and* $\omega \in \Omega',\ P(\Omega') = 1$, *there is at least one vector* $\mathbf{u}\left(\vec{T}\right) \in I,\ T = (T_1, \ldots, T_m)$ *for which the minimal value of the function*

$$F_T(\mathbf{u}) = \frac{1}{\prod_{i=1}^{m} T_i} \int_{t\in[0,T_1]\times\cdots\times[0,T_m]} f(\mathbf{u}, \xi(t))\, dt$$

is attained and

$$P\left\{u(T) \to u_0, F_T(u(T)) \to F(u_0), T_i \to \infty, i = 1, \ldots, m\right\} = 1.$$

Remark The ergodic condition will be fulfilled if the homogeneous field satisfies the condition of strong intermixing.

Now, we consider the case in which we observe the random field in the ball. Let $\left\{\xi\left(\vec{t}\right) = \xi\left(\vec{t}, \omega\right), \vec{t} \in \mathcal{R}^m\right\}, m \geq 1$ be a homogeneous in a strict sense random field on a complete probabilistic space (Ω, G, P) with values in some metric space $(Y, \mathfrak{B}(Y))$. Suppose that the realizations of $\xi\left(\vec{t}\right)$ are continuous on $\mathscr{R}^m$ with probability 1. We have a continuous nonnegative function $f : J \times Y \to R$, where J is a closed subset of $\mathscr{R}^l$, $l \geq 1$.

One has the observations $\left\{\xi\left(\vec{t}\right) : \left\|\vec{t}\right\| < T\right\}, T > 0$. The problem is to find the minimum points and the minimal value of the function

$$F\left(\vec{u}\right) = Ef\left(\vec{u}, \xi\left(\vec{0}\right)\right), \vec{u} \in J. \tag{1}$$

Let us investigate problem (1). We approximate it via minimization of the function

$$F_T\left(\vec{u}\right) = \frac{\int\limits_{\left\|\vec{t}\right\|<T} f\left(\vec{u}, \xi\left(\vec{t}\right)\right) d\vec{t}}{\int\limits_{\left\|\vec{t}\right\|<T} d\vec{t}}, \vec{u} \in J. \tag{2}$$

Denote

$$b_1\left(\vec{t}\right) = b_1\left(\vec{t}, c\right) = \frac{E\left(\inf\limits_{\|\vec{u}\|>c} f\left(\vec{u},\xi\left(\vec{t}\right)\right) - E\inf\limits_{\|\vec{u}\|>c} f\left(\vec{u},\xi\left(\vec{0}\right)\right)\right)\left(\inf\limits_{\|\vec{u}\|>c} f\left(\vec{u},\xi\left(\vec{0}\right)\right) - E\inf\limits_{\|\vec{u}\|>c} f\left(\vec{u},\xi\left(\vec{0}\right)\right)\right)}{E\left(\inf\limits_{\|\vec{u}\|>c} f\left(\vec{u},\xi\left(\vec{0}\right)\right) - E\inf\limits_{\|\vec{u}\|>c} f\left(\vec{u},\xi\left(\vec{0}\right)\right)\right)^2};$$

$$b_2\left(\vec{t}\right) = b_2\left(\vec{t}, \vec{u}\right) = \frac{E\left(f\left(\vec{u}, \xi\left(\vec{t}\right)\right) - F\left(\vec{u}\right)\right)\left(f\left(\vec{u}, \xi\left(\vec{0}\right)\right) - F\left(\vec{u}\right)\right)}{E\left(f\left(\vec{u}, \xi\left(\vec{0}\right)\right) - F\left(\vec{u}\right)\right)^2};$$

$$b_3\left(\vec{t}\right) = b_3\left(\vec{t}, K, \gamma\right) = \frac{E\left(\Psi\left(K,\gamma,\xi\left(\vec{t}\right)\right) - E\Psi\left(K,\gamma,\xi\left(\vec{0}\right)\right)\right)\left(\Psi\left(K,\gamma,\xi\left(\vec{0}\right)\right) - E\Psi\left(K,\gamma,\xi\left(\vec{0}\right)\right)\right)}{E\left(\Psi\left(K,\gamma,\xi\left(\vec{0}\right)\right) - E\Psi\left(K,\gamma,\xi\left(\vec{0}\right)\right)\right)^2},$$

where

$$\Psi\left(K, \gamma, z\right) = \sup_{\vec{u}, \vec{v} \in K: \|\vec{u} - \vec{v}\| < \gamma} \left|f\left(\vec{u}, z\right) - f\left(\vec{v}, z\right)\right|.$$

The following theorem takes place.

Theorem 5 [13]. *Let the next conditions be fulfilled:*

1. *for any* $c > 0$
$$E\left\{\max_{\|\vec{u}\|<c}\left(f\left(\vec{u},\xi\left(\vec{0}\right)\right)\right)^2\right\} < \infty;$$
2. *if* J *is unbounded then for each* $z \in Y$
$$f\left(\vec{u}, z\right) \to \infty, \text{ as } \|\vec{u}\| \to \infty;$$
3. *function* (1) *has a unique minimum point* u_0*;*
4. *for all* $c > 0$, $\vec{u} \in J$, *and any compact* $K \subset J$, $\gamma > 0$
$$\int_0^1 \frac{(\ln\rho)^2}{\rho^m}\left(\int_0^\rho \frac{1}{\tau^2}\left|B_i(\tau)\right| d\tau\right) d\rho < \infty, i = \overline{1,3},$$
where $B_i(\tau) = \int\limits_{\|\vec{t}\|<\tau} b_i\left(\vec{t}\right) d\vec{t}, i = \overline{1,3}.$

Then for all $T > 0$, $\omega \in \Omega'$, $P(\Omega') = 1$, there exists at least one minimum point $\vec{u}(T) = \vec{u}(T,\omega)$ of function (2). For any $T > 0$, the map $\vec{u}(T,\omega)$ can be chosen G-measurable, where $G = \{A \in G : A \subset \Omega'\}$.

For any minimum point $\vec{u}(T)$

$$P\left\{\vec{u}(T) \to \vec{u}_0, F_T\left(\vec{u}(T)\right) \to F\left(\vec{u}_0\right), T \to \infty\right\} = 1.$$

Let us consider a model with nonhomogeneous observations for a random field.

Let $\{\xi(t_1,t_2), (t_1,t_2) \in \mathbb{R}^2\}$ be a homogeneous in a strict sense real random field with continuous trajectories, defined on a complete probabilistic space (Ω, G, P), $X = [a; b] \subset \mathbb{R}$; $h : \mathbb{R}^2 \times X \times \mathbb{R} \to \mathbb{R}$ be a continuous function, convex on $x \in X$.

Investigating the problem

$$F_{T_1T_2}(x) = \frac{1}{T_1T_2}\int_0^{T_1}\int_0^{T_2} h\left(t_1, t_2, x, \xi\left(t_1, t_2\right)\right) dt_1 dt_2 \to \min, x \in X. \tag{3}$$

Let the next conditions be fulfilled:

$$\sup\left\{E\left[\max\left|h\left(t_1, t_2, x, \xi\left(t_1, t_2\right)\right)\right|, x \in X\right], t_1, t_2 \geq 0\right\} < \infty;$$

For any $x \in X$,

$$F(x) = \lim EF_{T_1T_2}(x), T_1, T_2 \to \infty;$$

there exist such $x_0 \in X$, $c > 0$, that

$$F(x) \geq F\left(x_0\right) + c\left|x - x_0\right|, x \in X. \tag{4}$$

Condition (3) implies that x_0 is a unique solution of the problem

$$F(x) \to \min, x \in X. \tag{5}$$

Convexity on $x \in X$ of a function h implies convexity of $F_{T_1T_2}(x)$ for any T_1, T_2, ω; convexity $EF_{T_1T_2}(x)$ for all T_1, T_2 and convexity $F(x)$.

For an arbitrary function $g : \mathbb{R} \to \mathbb{R}$, denote

$$g_+{}'(x) = \lim \frac{g(x+\Delta) - g(x)}{\Delta}, \Delta \to +0, \tag{6}$$

$$g_-{}'(x) = \lim \frac{g(x-\Delta) - g(x)}{\Delta}, \Delta \to +0, \tag{7}$$

if the limits exist.

The next lemma is evidently implied by properties of expectation.

Lemma 1 *Let the function $u : X \times \Omega \to \mathbb{R}$, be convex on the first argument and measurable on the second one, and*

$$E|u(x,\omega)| < \infty, x \in X.$$

Denote $v(x) = Eu(x, \omega)$. Then,

$$v_+{}'(x) = Eu_+{}'(x,\omega), v_-{}'(x) = Eu_-{}'(x,\omega).$$

Denote

$$g_{T_1T_2}(x) = EF_{T_1T_2}(x), x \in X.$$

The convexity of a function implies the existence of limits (6) and (7) for the function. Therefore, such limits exist:

- *for all t_1, t_2, y for $h(t_1, t_2, \cdot, y)$;*
- *for any t_1, t_2 for $Eh(t_1, t_2, \cdot, \xi(t_1, t_2))$;*
- *for all T_1, T_2, ω for $F_{T_1T_2}(\cdot)$;*
- *for all T_1, T_2 for $g_{T_1T_2}(\cdot)$;*
- *for $F(\cdot)$.*

By Lemma 1 for all $t_1, t_2 \in \mathbb{R}$, $T_1, T_2 > 0$, $x \in X$

$$(Eh)_+{}'(t_1, t_2, x, \xi(t_1, t_2)) = E\left\{h_+{}'(t_1, t_2, x, \xi(t_1, t_2))\right\},$$

$$(Eh)_-{}'(t_1, t_2, x, \xi(t_1, t_2)) = E\left\{h_-{}'(t_1, t_2, x, \xi(t_1, t_2))\right\},$$

$$\left(g_{T_1T_2}\right)_+'(x) = E\left\{\left(F_{T_1T_2}\right)_+'(x)\right\}, \left(g_{T_1T_2}\right)_-'(x) = E\left\{\left(F_{T_1T_2}\right)_-'(x)\right\}.$$

Lemma 2 *In addition to the conditions formulated above, the next conditions are fulfilled:*

1. *The field* $\xi(t_1, t_2)$ *satisfies a strong mixing condition, i.e., such a function* $a(d)$, $d \geq 0$; $a(d) \to 0$, $d \to \infty$ *exists such that for any* $H_1, H_2 \subset R^2$, *one has*
   $\sup\{|P(A \cap B) - P(A)P(B)|; A \in \sigma(H_1), B \in \sigma(H_2)\} \leq a(d(H_1H_2))$,
   where
   $\sigma(H) = \sigma\{\xi(t_1, t_2), (t_1, t_2) \in H\}$,
   $d(H_1, H_2) = \inf\{\|(t_1, t_2) - (s_1, s_2)\|; (t_1, t_2) \in H_1, (s_1, s_2) \in H_2\}$;
2. $a(d) \leq \frac{c_0}{1+d^{2+\varepsilon}}$, $\varepsilon > 0$,
3. *There exists* $L > 0$, *such that for all* t_1, t_2, ω
   $|h_+'(t_1, t_2, x_0, \xi(t_1, t_2))| \leq L$, $|h_-'(t_1, t_2, x_0, \xi(t_1, t_2))| \leq L$;
4. *For some* $\delta > \frac{8}{\varepsilon}$
   $E\{|h(t_1, t_2, x, \xi(t_1, t_2))|^{4+\delta}\} < \infty$, $x \in X$, t_1, $t_2 \in \mathbb{R}$;
5. $\left(g_{T_1T_2}\right)_+'(x_0) \to F_+'(x_0)$, $\left(g_{T_1T_2}\right)_-'(x_0) \to F_-'(x_0)$, $T \to \infty$;
6. *It exists* $c'' > 0$, *such that for any* $t_2 \in R^+$ $\int_0^{+\infty}\int_0^{+\infty} E\left|\beta_{t_1t_2}\beta_{s_1t_2}\right| dt_1 ds_1 \leq c''$; *where*
   $\beta_{t_1t_2} = h_+'(t_1, t_2, x_0, \xi(t_1, t_2)) - Eh_+'(t_1, t_2, x_0, \xi(t_1, t_2))$;
7. *It exists* $c''' > 0$, *such that for all* $t_1 \in \mathbb{R}^+$, *one has* $\int_0^{+\infty}\int_0^{+\infty} E\left|\beta_{t_1t_2}\beta_{t_1s_2}\right| dt_2 ds_2 \leq c'''$;.
8. *Analogous to (6) and (7), the conditions are fulfilled for the left derivative.*

Then, with probability 1,

$$\left(F_{T_1T_2}\right)_+'(x_0) \to F_+'(x_0), T_1, T_2 \to \infty; \tag{8}$$

$$\left(F_{T_1T_2}\right)_-'(x_0) \to F_-'(x_0), T_1, T_2 \to \infty. \tag{9}$$

The lemma is by standard means implied by the Borel–Cantelli lemma.

Theorem 6 *With probability 1, there exist such* $T_{01} = T_{01}(\omega)$, $T_{02} = T_{02}(\omega)$ *that for all* $T_1 > T_{01}$, $T_2 > T_{02}$, *problem (3) has a unique solution* $x(T_1, T_2) = x_0$.

Proof By (4)

$$F_+'(x_0) \geq c, F_-'(x_0) \geq c.$$

Owing to Lemma 2 with probability 1 beginning from some T_{01}, T_{02}, one has

$$\left(F_{T_1T_2}\right)_+'(x_0) > 0, \left(F_{T_1T_2}\right)_-'(x_0) > 0. \tag{10}$$

Now, (10) and the convexity of $F_{T_1T_2}$ imply the theorem.

Another important property of estimates is their limit distributions. It is important to know if the true value is an interior point of the domain of admissible values or if it belongs to the boundary of this domain. We will not formulate all the conditions under which one can prove the statement on the limit distribution of the estimate because these conditions are indeed very complicated. However, under some conditions on the smoothness of the criterion function and strong mixing condition on the respective random processes (or random fields), the asymptotic distribution is Gaussian.

For example, if we have the observations of the random field in area $\left\|\vec{t}\right\| < T$, then the normed variable has the form

$$\left(\int\limits_{\left\|\vec{t}\right\|<T} d\vec{t}\right)^{\frac{1}{2}} \left(\vec{u}(T) - \vec{u_0}\right)$$

and

$$\left(\int\limits_{\left\|\vec{t}\right\|<T} d\vec{t}\right)^{\frac{1}{2}} \left(F_T\left(\vec{u}(T)\right) - F\left(\vec{u_0}\right)\right).$$

5 Method of Empirical Means Under the Restrictions of Unknown Parameters, Described in the Form of Equalities and Inequalities

Furthermore, we consider a case where the restrictions are of the form

$$J = \{u : g(u) = (g_1(u), \ldots, g_n(u)) \leq 0\}.$$

Then, the normed sequence of vectors estimates converges weakly to the random vector η, which is the solution to the problem

$$\frac{1}{2}\vec{u}\,\Phi\left(\vec{u_0}\right)\vec{u} + \varsigma\vec{u} \to \min,$$

$$\nabla g^T\left(\vec{u_0}\right)\vec{u} \leq 0,$$

where ς is a Gaussian vector.

These models have been studied in detail in [14].

6 Large Deviations in the Method of Empirical Means

The following problem consists of obtaining some theorems of large deviations for a method of empirical means for the dependent observations. We formulate them for random fields. The results for random processes are analogous.

Let $\{\xi(t_1,t_2),(t_1,t_2)\in\mathbb{R}^2\}$ be a homogeneous in a strict sense random field with continuous trajectories defined on a full probabilistic space (Ω, G, P), with values in some metric space (Y,ρ).

Considering the problem

$$F(x) = E\,f\,(x,\xi\,(0,0)) \to \min,\quad x\in X, \tag{11}$$

where X is a nonempty compact subset of $\mathbb{R}$ and where $f : X\times Y\to\mathbb{R}$ is a continuous nonnegative function.

The problem can be approximated as follows:

$$F_{T_1T_2}(x) = \frac{1}{T_1T_2}\int_0^{T_1}\int_0^{T_2} f\,(x,\xi\,(t_1,t_2))\,dt_1dt_2 \to \min,\quad x\in X, \tag{12}$$

where $T_1>0$, $T_2>0$.

Owing to the properties of continuous functions at least one solution $x(T_1,T_2)$ of problem (12) exists, which is a measurable function of ω.

Suppose that

$$E\max\{|f\,(x,\xi\,(0,0))|\,,x\in X\}<\infty.$$

Then, $F(\cdot)$ is continuous, and at least one solution x_0 of problem (11) exists. Suppose that it is unique.

Let the field $\xi(t_1,t_2)$ satisfy the strong mixing condition, i.e., there exists such a function $a(d)$, $d\geq 0$; $a(d)\searrow 0$, $d\to\infty$ such that for all H_1, $H_2\subset\mathbb{R}^2$, one has

$$\sup\{|P\,(A\cap B)-P(A)P(B)|\,;A\in\sigma\,(H_1)\,,B\in\sigma\,(H_2)\}\leq a\,(d\,(H_1,H_2))\,,$$

where $\sigma(H)=\sigma\{\xi(t_1,t_2),(t_1,t_2)\in H\}$, $d(H_1,H_2)=\inf\,\{\|(t_1,t_2)-(s_1,s_2)\|;$ $(t_1,t_2)\in H_1,(s_1,s_2)\in H_2\}$.

Suppose that $a(d)=O(d^{-2-\varepsilon})$, $d\to\infty$, for some $\varepsilon>0$, and for some $\delta>\frac{8}{\varepsilon}$

$$E\left\{|f\,(x,\xi\,(0,0))|^{4+\delta}\right\}<\infty, x\in X.$$

If the conditions are fulfilled, then by [13] with probability 1

$$x\,(T_1,T_2)\to x_0,\,F_{T_1T_2}\,(x\,(T_1,T_2))\to F\,(x_0)\,;\,T_1,T_2\to\infty.$$

Consider the probability of large deviations $x(T_1, T_2)$ from x_0 and $F_{T_1 T_2}(x(T_1, T_2))$ from $F(x_0)$.

For any fixed y, one can consider $f(\cdot\ ,y)$ as an element of the space of continuous functions $C(X)$. Suppose that a convex compact set $K \subset C(X)$ exists such that for all $y \in Y$, one has $f(\cdot\ ,y) - F(\cdot) \in K$. Consider $F_{T_1 T_2} - F$ as random elements on (Ω, G, P) with values in K.

Let us use some results from functional analysis.

Definition 1 [15]. Let $(V, \|\circ\|)$ be a linear norm space; $B(x, r)$ is a closed ball with a radius r and a center x; $f : V \to [-\infty, +\infty]$ is some function; and x_f is its minimum point on V. Conditioning function ψ for f in x_f is a monotone nondecreasing function $\psi : [0, +\infty) \to [0, +\infty]$, $\psi(0) = 0$, such that $r > 0$, exists for any $x \in B(x_f, r)$ one has

$$f(x) \geq f\left(x_f\right) + \psi\left(\left\|x - x_f\right\|\right).$$

Let $V_0 \subset V$. Denote

$$\delta_{V_0}(x) = 0, x \in V_0,$$

$$\delta_{V_0}(x) = +\infty, x \notin V_0.$$

Theorem 6 [15]. *Let $(V, \|\circ\|)$ be a linear normed space, where $V_0 \subset V$ is closed and where $f_0, g_0 : V \to \mathbb{R}$ are continuous on V functions. Suppose*

$$\varepsilon = \sup\left\{\left|f_0(x) - g_0(x)\right|, x \in V_0\right\}.$$

Introduce functions $f, g : V \to (-\infty, +\infty]$:

$$f = f_0 + \delta_{V_0}, g = g_0 + \delta_{V_0}.$$

Then,

$$|\inf\{f(x), x \in V\} - \inf\{g(x), x \in V\}| \leq \varepsilon.$$

Let x_f be a minimum point of f on V, ψ is a conditioning function for f in x_f with a coefficient r. If ε is small enough for

$$\psi\left(\left\|x - x_f\right\|\right) \leq 2\varepsilon \Rightarrow \left\|x - x_f\right\| \leq r,$$

then, for any $x_g \in \arg\ \min\{g(x), x \in B(x_f, r)\}$, one has $\psi(\|x_f - x_g\|) \leq 2\varepsilon$. When ψ is convex and strictly increasing on $[0, r]$,

$$\psi^{-1}(2\varepsilon) \leq r \Rightarrow \left\|x_f - x_g\right\| \leq \psi^{-1}(2\varepsilon)$$

$$\forall x_g \in \arg\min \left\{ g(x), x \in B\left(x_f, r\right) \right\}.$$

Let us use some results from large deviation theory.

Definition 2 [8]. Let Σ be a separable Banach space, where $\{\zeta(t_1, t_2), (t_1, t_2) \in \mathbb{R}^2\}$ is a homogeneous in a strict sense random field on (Ω, G, P) with values in Σ. For $\tau > 0$, random values $\eta_1, \ldots, \eta_p; p \geq 2$ are called τ-measurably separated if η_j is $\sigma\{\zeta(t_1, t_2), (t_1, t_2) \in H_j\}$—measurable for all $j \in \{1, \ldots, p\}$, where $d(H_i, H_j) \geq \tau$, $i \neq j$; $H_j, j = 1, \ldots, p$ are Borel sets in $\mathbb{R}^2$; and $d(\cdot, \cdot)$ is the distance between the sets.

Definition 3 [8]. One says that the random field from Definition 2 satisfies the first hypothesis of hypermixing if such $\tau_0 \in \mathbb{N} \cup \{0\}$ and a nonincreasing function $\alpha : \{\tau > \tau_0\} \to [1, +\infty)$ exist,

$$\lim_{\tau \to \infty} \alpha(\tau) = 1; \left\| \eta_1 \ldots \eta_p \right\|_{L^1} \leq \prod_{j=1}^{p} \left\| \eta_j \right\|_{L^{\alpha(\tau)}}$$

for all $p \geq 2$, $\tau > \tau_0$; $\eta_1, \ldots, \eta_p$ τ-measurably separated, where

$$\|\eta\|_{L^r} = \left(E\left\{|\eta|^r\right\}\right)^{\frac{1}{r}}.$$

Suppose that $X = [a, b] \subset \mathbb{R}$. As is known [16], one has $(C(X))^* = M(X)$—a space of bounded signed measures on X, and

$$\langle g, Q \rangle = \int_X g(x) Q(dx); \; g \in C(X), \; Q \in M(X).$$

The next assertion takes place.

Theorem 7 *Let $\zeta(t_1, t_2)$, $(t_1, t_2) \in R^2$ be a homogeneous in a strict sense random field satisfying the hypermixing condition on (Ω, G, P), with values in a compact convex set $K \subset C(X)$. Then for all $Q \in M(X)$, there exists a finite limit*

$$\Lambda(Q) = \lim_{T_1, T_2 \to \infty} \frac{1}{T_1 T_2} \ln \left(E \exp \int_X \int_0^{T_1} \int_0^{T_2} \zeta\left(t_1, t_2\right)(x) dt_1 dt_2 Q(dx) \right),$$

and for any closed $A \subset K$,

$$\limsup_{T_1, T_2 \to \infty} \frac{1}{T_1 T_2} \ln P \left\{ \frac{1}{T_1 T_2} \int_0^{T_1} \int_0^{T_2} \zeta\left(t_1, t_2\right) dt_1 dt_2 \in A \right\} \leq - \inf\left\{ \Lambda^*(g), g \in A \right\},$$

where $\Lambda^*(g) = \sup\left\{\int_X g(x)Q(dx) - \Lambda(Q),\ Q \in M(X)\right\}$ *is a nonnegative convex lower semicontinuous function.*

Proof By the partition method analogous to that of [6], the following exists:

$$\Lambda(Q) = \lim_{T_1,T_2\to\infty} \frac{f_{T_1T_2}}{T_1T_2} \in [-\infty, +\infty].$$

The function under the limit is bounded, so the limit is finite.

Let us use the theorem from large deviations theory [6]. One has

$$H = K,\ J = C(X),\ J^* = M(X),\ \langle Q, g\rangle = \int_X g(x)Q(dx),\ \varepsilon_1 = \frac{1}{T_1},\ \varepsilon_2 = \frac{1}{T_2}.$$

Furthermore, $\mu\left(\frac{1}{T_1}, \frac{1}{T_2}\right)$ is a probability measure on $C(X)$, which is defined by the distribution

$$\frac{1}{T_1T_2}\int_0^{T_1}\int_0^{T_2} \zeta\,(t_1, t_2)\,dt_1dt_2.$$

Then,

$$\lim_{\varepsilon_1,\varepsilon_2\to 0} \varepsilon_1\varepsilon_2\Lambda_{\mu(\varepsilon_1,\varepsilon_2)}\left(\frac{Q}{\varepsilon_1\varepsilon_2}\right) =$$

$$= \lim_{T_1,T_2\to\infty} \frac{1}{T_1T_2} \ln\left(\int_{C(X)} \exp\left\{\int_X g(x)T_1T_2Q(dx)\right\}\mu\left(\frac{1}{T_1}, \frac{1}{T_2}\right)(dg)\right) =$$

$$= \lim_{T_1,T_2\to\infty} \frac{1}{T_1T_2} \ln\left(E\exp\left\{\int_X\left(\frac{1}{T_1T_2}\int_0^{T_1}\int_0^{T_2} \zeta\,(t_1, t_2)\,dt_1dt_2\right)(x)T_1T_2Q(dx)\right\}\right)$$

$$= \lim_{T_1,T_2\to\infty} \frac{f_{T_1T_2}}{T_1T_2} = \Lambda(Q).$$

The theorem is proved.

Return to problems (11) and (12).

Theorem 8 *Let $\xi(t_1, t_2)$ satisfy the hypermixing condition. Then, for any* $\varepsilon > 0$

$$\limsup_{T_1,T_2\to\infty} \frac{1}{T_1T_2} \ln P\left\{\left\|F_{T_1T_2} - F\right\| \geq \varepsilon\right\} \leq -\inf\{I(z), z \in A_\varepsilon\},$$

where $I(z) = \Lambda^*(z) = \sup\left\{\int_X z(x)Q(dx) - \Lambda(Q), Q \in M(X)\right\}$ *is a nonnegative lower semicontinuous convex function,*

$$\Lambda(Q) = \lim_{T_1,T_2\to\infty} \frac{1}{T_1T_2} \ln\left(E\exp\left\{\int_X\int_0^{T_1}\int_0^{T_2}[f(x,\xi(t_1,t_2)) - F(x)]\,dt_1dt_2Q(dx)\right\}\right),$$

$$A_\varepsilon = \{z \in K : \|z\| \geq \varepsilon\}.$$

Proof A_ε is a closed subset of K. The field

$$\zeta(t_1, t_2) = f(\cdot, \xi(t_1, t_2)) - F(\cdot),$$

with values in K, is a continuous function of $\xi(t_1, t_2)$, and then it satisfies the conditions of Theorem 7. Therefore, the theorem is a consequence of Theorem 7.

Theorem 9 *Let the conditions of Theorem 8 be satisfied. Then,*

$$\limsup_{T_1,T_2\to\infty} \frac{1}{T_1T_2} \ln P\left\{\left|\min\{F(x), x \in X\} - \min\left\{F_{T_1T_2}(x), x \in X\right\}\right| \geq \varepsilon\right\} \\ \leq -\inf\{I(z), z \in A_\varepsilon\}, \tag{13}$$

where $I(\cdot)$ *and* A_ε *are defined in Theorem 8.*

Suppose that a conditioning function ψ exists for F in x_0 with some constant r. Let $x(T_1, T_2)$ be a minimum point of (2) on $B(x_0, r)$. If ε is sufficiently small, then

$$\psi(|x - x_0|) \leq 2\varepsilon \Rightarrow |x - x_0| \leq r,$$

Then, one has

$$\limsup_{T_1,T_2\to\infty} \frac{1}{T_1T_2} \ln P\{\psi(|x(T_1, T_2) - x_0|) \geq 2\varepsilon\} \leq -\inf\{I(z), z \in A_\varepsilon\}. \tag{14}$$

Proof By Theorem 6 for all ω

$$\left|\min\{F(x), x \in X\} - \min\left\{F_{T_1T_2}(x), x \in X\right\}\right| \leq \left\|F_{T_1T_2} - F\right\|.$$

Then, Theorem 8 implies (13).

Furthermore, by Theorem 6 for allω

$$\psi\left(|x_0 - x(T_1, T_2)|\right) \le 2\left\|F_{T_1T_2} - F\right\|,$$

In addition, Theorem 8 implies (14).

Remark If, in addition to the conditions of Theorem 9, ψ is convex and strictly increasing on $[0, r]$,, then

$$\limsup_{T_1,T_2\to\infty} \frac{1}{T_1T_2} \ln P\left\{|x(T_1, T_2) - x_0| \ge \psi^{-1}(2\varepsilon)\right\} \le -\inf\{I(z), z \in A_\varepsilon\}. \tag{15}$$

Proof By Theorem 6 for allω

$$|x_0 - x(T_1, T_2)| \le \psi^{-1}\left(2\left\|F_{T_1T_2} - F\right\|\right).$$

Then,

$$P\left\{|x(T_1, T_2) - x_0| \ge \psi^{-1}(2\varepsilon)\right\} \le P\left\{\psi^{-1}\left(2\left\|F_{T_1T_2} - F\right\|\right) \ge \psi^{-1}(2\varepsilon)\right\} =$$

$$= P\left\{\left\|F_{T_1T_2} - F\right\| \ge \varepsilon\right\},$$

In addition, Theorem 8 implies (15).

Theorem 10 *Let the field $\{\xi(t_1, t_2), (t_1, t_2) \in R^2\}$ has hypermixing. Suppose that the function h does not depend on t_1, t_2. Let the function*

$$\min\left[h_+'(x_0, y), h_-'(x_0, y)\right], y \in R$$

be continuous.

Then,

$$\limsup_{T_1,T_2\to\infty} \frac{1}{T_1T_2} \ln P\left(A^c_{T_1T_2}\right) \le -\inf\left\{V^*(z), z \in [-L; 0]\right\}, \tag{16}$$

where $V^*(z) = \sup\{zQ(X) - V(Q), Q \in M(X)\}$,

$$V(Q) = \lim \frac{1}{T_1T_2} \ln E \exp\{Q(X) \times$$

$$\times \int_0^{T_1} \int_0^{T_2} \min\left[h_+{}'\left(x_0, \xi\left(t_1, t_2\right)\right), h_-{}'\left(x_0, \xi\left(t_1, t_2\right)\right)\right] dt_1 dt_2 \Big\}, T_1, T_2 \to \infty,$$

$$A_{T_1T_2} = \left\{\omega : \arg\min F_{T_1T_2}(x) = \{x_0\}, x \in X\right\}, A^c_{T_1T_2} = \Omega - A_{T_1T_2}.$$

Proof One has

$$P\left(A^c_{T_1T_2}\right) = P\left\{\min\left[\left(F_{T_1T_2}\right)_+{}'(x_0), \left(F_{T_1T_2}\right)_-{}'(x_0)\right] \in [-L; 0]\right\} \le$$

$$\le P\left\{\frac{1}{T_1T_2}\int_0^{T_1}\int_0^{T_2} \min\left[h_+{}'\left(x_0, \xi\left(t_1, t_2\right)\right), h_-{}'\left(x_0, \xi\left(t_1, t_2\right)\right)\right] dt_1 dt_2 \in [-L; 0]\right\}. \tag{17}$$

Denote

$$K = \{\alpha(x) = \alpha, x \in X; \alpha \in [-L; L]\}.$$

Evidently, K is a compact convex subset of $C(X)$.

Investigating the function

$$a_{t_1t_2} = a_{t_1t_2}(x) = \min\left[h_+{}'\left(x_0, \xi\left(t_1, t_2\right)\right), h_-{}'\left(x_0, \xi\left(t_1, t_2\right)\right)\right], x \in X.$$

For all t_1, t_2, ω, one has $a_{t_1t_2}(\cdot) \in K$. Let

$$K_1 = \{\alpha(x) = \alpha, x \in X; \alpha \in [-L; 0]\}.$$

Then, K_1 is a closed subset of K. Furthermore,

$$P\left\{\frac{1}{T_1T_2}\int_0^{T_1}\int_0^{T_2} \min\left[h_+{}'\left(x_0, \xi\left(t_1, t_2\right)\right), h_-{}'\left(x_0, \xi\left(t_1, t_2\right)\right)\right] dt_1 dt_2 \in [-L; 0]\right\} =$$

$$= P\left\{\left(\frac{1}{T_1T_2}\int_0^{T_1}\int_0^{T_2} a_{t_1t_2} dt_1 dt_2 = \frac{1}{T_1T_2}\int_0^{T_1}\int_0^{T_2} a_{t_1t_2} dt_1 dt_2(x), x \in X\right) \in K_1\right\}. \tag{18}$$

Let us use the theorem from large deviations theory. One has

$$\limsup_{T_1, T_2 \to \infty} \frac{1}{T_1T_2} \ln P\left\{\frac{1}{T_1T_2}\int_0^{T_1}\int_0^{T_2} a_{t_1t_2} dt_1 dt_2 \in K_1\right\} \le -\inf\left\{V^*(z), z \in K_1\right\}, \tag{19}$$

where $V^*(z) = \sup\{zQ(X) - V(Q), Q \in M(X)\}$,

$$V(Q) = \lim_{T_1,T_2\to\infty} \frac{1}{T_1 T_2} \ln E \exp\left\{\int_0^{T_1}\int_0^{T_2}\int_X a_{t_1t_2} Q(dx)dt_1dt_2\right\} =$$

$$= \lim_{T_1,T_2\to\infty} \frac{1}{T_1 T_2} \ln Eexp\left\{Q(X)\int_0^{T_1}\int_0^{T_2} a_{t_1t_2}dt_1dt_2\right\}.$$

Now, (17), (18), and (19) imply (16). The proof is complete.

7 Consequences

Large deviation theory is used in many spheres of science and practice, such as mathematics, physics, informatics, and economics. It continues to develop prospectively, and many new works on large deviations have appeared.

The results can be used for solving different stochastic optimization problems, which appear in recognition theory and regressive analysis, when one needs to find optimal solutions via observations of a stationary random process or a homogeneous random field [17–25].

Acknowledgments This work was supported by a joint project between the International Institute for Applied Systems Analysis (IIASA) and the National Academy of Sciences of Ukraine (NASU) on "Integrated robust modeling and management of food-energy-water–land use nexus for sustainable development" and by the project "Comprehensive analysis of robust preventive and adaptive measures of food, energy, water and social management in the context of systemic risks and consequences of COVID-19" (0122U000552) and the project № 2.3/25-П of the National Academy of Sciences of Ukraine.

References

1. Mikhalevich, V.S., Knopov, P.S., Golodnikov, A.N.: Mathematical models and methods of risks assessment in ecologically hazardous industries. Cybern. Syst. Anal. **30**(2), 259–273 (1994)
2. Knopov, P.S., Pardalos, P.: Simulation and Optimization Methods in Risk and Reliability Theory. Nova Science Publishers, 285 p (2009)
3. Ermoliev, Y.M., Knopov, P.S.: Method of empirical means in stochastic programming problems//cybernetics and systems. Analysis. **42**(6), 773–785 (2006)
4. Knopov, P.S.: Asymptotic properties of some classes of m-estimates. Cybern. Syst. Anal. **33**(4), 468–481 (1997)
5. Knopov, P.S., Kasitskaya, E.I.: Large deviations of empirical estimates in stochastic programming problems. Kibernetika i Sistemnyj Analiz. **40**(4), 52–60 (2004)
6. Knopov, P.S., Kasitskaya, E.J.: On large deviations of empirical estimates in a stochastic programming problem with time-dependent observations. Cybern. Syst. Anal. **46**(5), 724–728 (2010)

7. Knopov, P.S., Kasitskaya, E.J.: Properties of empirical estimates in stochastic optimization and identification problems. Ann. Oper. Res. **56**(1), 225–239 (1995)
8. Deuscel, J.-D., Stroock, D.W.: Large Deviations, 310 p. Academic Press (1989)
9. Dembo, A., Zeitouni, O.: Large Deviations Techniques and Applications, 397 p. Springer-Verlag, New York (1998)
10. Hollander, F.: Large Deviations, 142 p. American Mathematical Society (2000)
11. Sanov, I.N.: On the probability of large deviations of random variables. Math. Collect. **42**(84), 11–44. (in Russian)
12. Bryc, W.: On large deviations for uniformly strong mixing sequences. Stoc. Proc. Appl. **41**, 191–202 (1992)
13. Knopov, P.S., Kasitskaya, E.J.: Empirical Estimates in Stochastic Optimizastion and Identification, 250 p. Kluwer Academic Publishers (2002)
14. Knopov, P.S., Korkhin, A.S.: Regression Analysis Under A Priory Parameter Restrictions, 234 p. Springer (2012)
15. Kaniovski, Y.M., King, A.J., Wets, R.J.-B.: Probabilistic bounds (via large deviations) for the solutions of stochastic programming problems. Ann. Oper. Res. **56**, 189–208 (1995)
16. Dunford, N., Schwartz, J.: Linear Operators. P.I: General Theory, 896 p. Interscience, New York (1957)
17. Prekopa, A.: Stochastic Programming, 600 p. ©ordreehfc RIover ШA. РиЫ. (1995)
18. Atoyev, A., Knopov, P., Pepeliaev, V., Kisala, P., Romaniuk, R., Kalimoldayev, M.: Chapter 6. The mathematical problems of complex systems investigation under uncertainties. In: Wojcik, W., Sikora, J. (eds.) Recent advances in information technology, pp. 135–171. CRC Press/Balkema, Taylor & Francis Group, London, UK (2018)
19. Golodnikov, A.N., Knopov, P.S., Pepelyaev, V.A.: Estimation of reliability parameters under incomplete primary information. Theory Decis. **57**(4), 331–344 (2004)
20. Rychlik March, I., Rydén, J.: Probability and Risk Analysis, 286 p. Springer (2006)
21. Asmussen, S., Albrecher, H.: Ruin Probabilities, 621 p. World Scientific (2010)
22. Ermoliev, Y., Wets, R.J.-B. Numerical Techniques for Stochastic Optimization, 571 p. Springer
23. Marti, K., Ermoliev, Y., Macovski, M. (eds): Coping with Uncertainty: Robust Solutions. Lecture Notes in Economics and Mathematical Systems 633, 1st edn, 207 p (2010)
24. Pepelyaev, V.A., Golodnikov, A.N., Golodnikova, N.A.: Reliability optimization method alternative to bPOE. Cybern. Syst. Anal. **58**(4), 593–597 (2022). https://doi.org/10.1007/s10559-022-00492-9
25. Pepelyaev, V.A., Golodnikov, A.N., Golodnikova, N.A.: A new method of reliability optimization in the classical problem statement. Cybern. Syst. Anal. **58**(6), 917–922 (2022). https://doi.org/10.1007/s10559-023-00525-x

Automatic Tuning of the SkewedKruskal Algorithm

Mattia Lecchi and Giovanni Righini

1 Introduction

We consider the minimum cost spanning tree problem, that is, the problem of finding a minimum cost spanning tree (MST) of a connected weighted graph. A classical algorithm for this problem is due to Kruskal [4].

In its most basic implementation, Kruskal algorithm starts by fully sorting the edge list by nondecreasing edge weights. Then, starting from an empty set, a forest F is grown by scanning the edge list and inserting one edge at a time in F, discarding edges that produce cycles, until a spanning tree is obtained. The resulting time complexity is $O(m \log n)$, where m is the number of edges and n is the number of vertices.

Janson et al. [3] proved that the largest weight edge in an MST of a random weighted graph is expected to lie within the $\frac{1}{2}n \log n$ smallest weight edges. This suggests to save computing time, sorting only the relevant part of the edge list.

Following this idea, Paredes and Navarro [7] proposed the QuickKruskal algorithm, where edge sorting is interleaved with edge selection. For this purpose, edge sorting is done as in QuickSort, i.e., partitioning the unsorted edge list in two lists by comparing all weights with a suitably chosen pivot element; if the MST is found within the edges of the first list, then the edges in the second list are not sorted. This procedure is recursively executed for each list. From a theoretical viewpoint, the average time complexity of QuickKruskal is $O(m + n \log^2 n)$ for random graphs with randomly generated weights (Paredes and Navarro, [7]). From an experimental viewpoint, QuickKruskal is substantially faster than Kruskal algorithm.

M. Lecchi · G. Righini (✉)
Department of Computer Science, University of Milan, Milan, Italy
e-mail: mattia.lecchi@studenti.unimi.it; giovanni.righini@unimi.it

B. Goldengorin (ed.), *Theory, Algorithms, and Experiments in Applied Optimization*, Springer Optimization and Its Applications 226,
https://doi.org/10.1007/978-3-031-91357-0_9

A further improvement, named FilterKruskal, was proposed by Osipov et al. [6]. When an edge list is recursively partitioned into two parts as in QuickKruskal, after processing the first list and before processing the second one, all edges that would close a cycle in the current forest are filtered out from the second list, thus reducing its size and the computing time needed to sort it. As shown in [6], the average time complexity of the FilterKruskal algorithm for random graphs with randomly generated edge weights is $O(m + n \log n \log \frac{m}{n})$.

In this context, the recursive partition of the edge lists should better be done in an unbalanced way, contrary to QuickSort that achieves the best performance by splitting each list into two parts of equal size. The unbalanced partition is especially useful in the first recursive call, when the whole edge list is partitioned the first time. Following this observation, Righini and Righini [8] presented a variation of the algorithm, called SkewedKruskal, where a random sample is extracted from the complete edge list and the pivot element is suitably chosen among the samples. The size of the sample is $\lceil \sqrt{m} \rceil$, and the pivot is selected in position $\lceil \frac{n \log n}{2\sqrt{m}} \rceil$ in the sorted list of samples. This corresponds to the expected position of the *critical edge*, i.e., the largest weight edge in the MST according to the result of Janson et al., which refers to complete graphs with random weights generated according to a uniform distribution of probability.

However, for different types of graphs, different choices of the sample size and the pivot position can lead to better results. The purpose of this study is to optimize the time performance of the SkewedKruskal algorithm by estimating how to partition the sorted edge list on the basis of some graph characteristics, so that the MST is likely to be completely contained in one of the two initial edge lists and the size of such a list is minimized.

Paper Outline In Sect. 2, we describe two classes of randomly generated weighted graphs we studied, namely, random graphs and KNN graphs. Section 3 describes the main steps that can be followed to estimate the position or the weight of the critical edge starting from an automatic classification of the graph and the estimation of its parameters. Section 4 describes an alternative and more general technique, based on graph density, that does not require to classify the graph. In Sect. 5, we illustrate the results of computational tests and comparisons between different versions of the SkewedKruskal algorithm.

2 Randomly Generated Graphs

As a preliminary step, in this study, we initially consider two classes of randomly generated graphs, namely, random graphs and k-nearest neighbor (KNN in the remainder) graphs. Hereafter, we describe the details of the graph generation procedure.

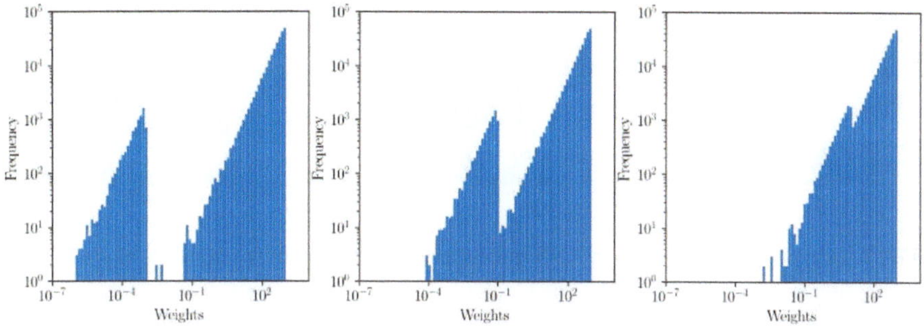

Fig. 1 Distribution of the edge weights in a random graph as a function of the clustering index γ (logarithmic scale)

Random Graphs In our tests, random graphs are generated according to the Erdős–Rényi model [2]: it allows to generate graphs $G_{n,\rho}$, where n is the number of vertices and $0 < \rho \leq 1$ is the graph density, i.e., the probability of including any edge (self-loops are excluded). The edge weights are generated according to a uniform probability distribution.

Besides size and density, we also consider a third parameter, that is, the clustering degree, indicated by γ. To generate random graphs with different clustering degree, we partition the vertex set into $\lceil\sqrt{n}\rceil$ disjoint subsets $C_1, C_2, ...C_{\lceil\sqrt{n}\rceil}$, such that $\lfloor\sqrt{n}\rfloor \leq |C_i| \leq \lceil\sqrt{n}\rceil\ \forall i = 1, \ldots, \lceil\sqrt{n}\rceil$. We define a parameter $0 < \gamma \leq 1$ representing the clustering degree. Then, the weight of all edges whose endpoints belong to the same cluster is multiplied by γ, while the weight of the other edges is divided by γ. For small values of γ, the resulting graph is strongly clustered.

According to this method to produce clustered random graphs, the number of intra-cluster edges grows as $\rho n\sqrt{n}$, while the number of inter-cluster edges grows as ρn^2. Figure 1 shows the distribution of the edge weights as a function of γ in random graphs $G_{1000,0.5}$ using a logarithmic scale, where the frequency of small weights can be better appreciated although they are relatively few.

KNN Graphs A KNN graph is obtained by generating its vertices at random as points in a circle of radius R. Each vertex is adjacent to its k-nearest neighbors, and the edge weights are the Euclidean distances between the endpoints. In our tests, these graphs were generated using $K - D$-trees and ball trees by the algorithms illustrated in [1, 5]) and implemented in the Python module `sklearn.neighbours`.

By construction, KNN graphs are clustered, with a clustering degree depending on k. Figure 2 shows the edge weight distribution of KNN graphs with $n = 1000$ for different values of k.

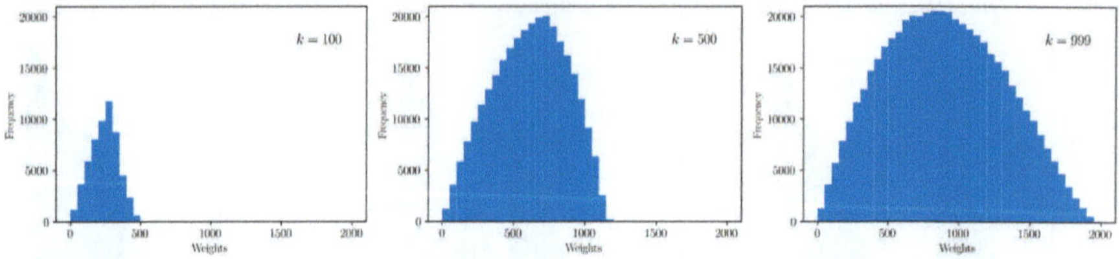

Fig. 2 Edge weight distribution in KNN graphs for $n = 1000$ and different values of k

3 Graph Classification

In this section, we describe a technique to estimate the position or the weight of the critical edge, starting from an automatic classification of the graph and an estimation of its relevant parameters. In Sect. 3.1, we describe an automatic classifier; in Sect. 3.2, we describe how the relevant parameters can be estimated once the graph has been classified; in Sect. 3.3, we analyze the relationship between the graph parameters and the position or weight of the critical edge; in Sect. 3.4, we consider the problem of suitably selecting a sample size from the edge list to obtain a reliable estimate of the position or the weight of the critical edge; in Sect. 3.5, we consider the selection of the pivot, i.e., the edge weight value that is used to partition the edge list into two parts.

3.1 Automatic Classification

We set up an automatic classifier to distinguish the two classes of randomly generated graphs shown above. The two classes are characterized by different distributions of edge weights. Therefore, the mean value and the variance of (a sample of) the edge weights are good candidates for an effective classifier.

Random Graphs By definition, the distribution of the edge weights is uniform when $\gamma = 1$. Hence, it is expected that the mean value and the variance of a sample subject to min-max scaling be close to the expected value $\mathbb{E}(X)$ and the variance $\mathrm{Var}(X)$ of a random variable X uniformly distributed between $a = 0$ and $b = 1$, i.e.,

$$\mathbb{E}(X) = \frac{a+b}{2} = \frac{1}{2}$$

and

$$\mathrm{Var}(X) = \frac{(b-a)^2}{12} = \frac{1}{12}.$$

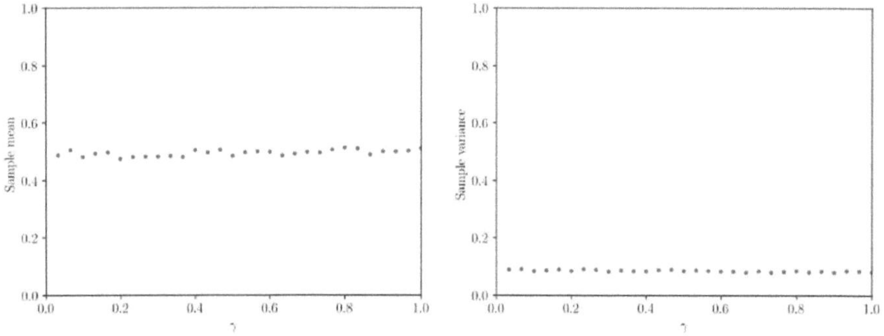

Fig. 3 Mean value and variance of edge weights in random graphs for different values of the clustering degree γ ($n = 2000$)

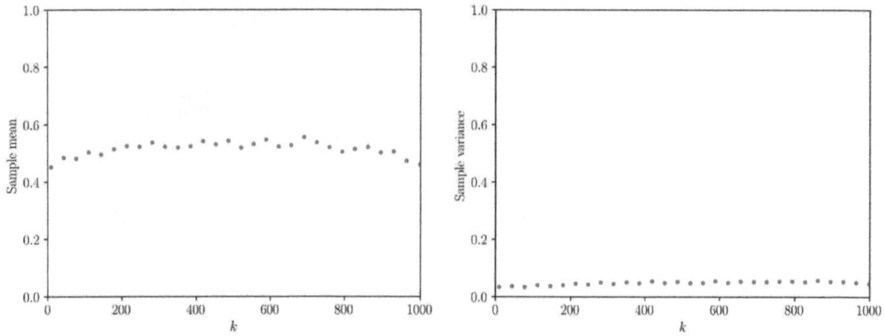

Fig. 4 Mean and average of the edge weights in KNN graphs for different values of k ($n = 1000$)

For clustered graphs, with low value of γ, deviations from these values are possible. However, the number of intra-cluster (small weight) edges is very small. Figure 3 shows the values of the mean and the variance for $n = 2000$ and different values of γ.

It is apparent that the value of the clustering degree does not significantly affect the mean and the variance.

KNN Graphs The mean of the edge weights in KNN graphs is very close to the mean of random graphs, when k is neither very small nor very large. However, the variance is slightly smaller, thus allowing to distinguish the two types of graphs. Figures 4 and 5 show the mean and the variance for $n = 1000$ and different values of k as well as for $k = 400$ and different values of n.

Similar to random graphs, the mean and the variance are almost independent on γ also for KNN graphs.

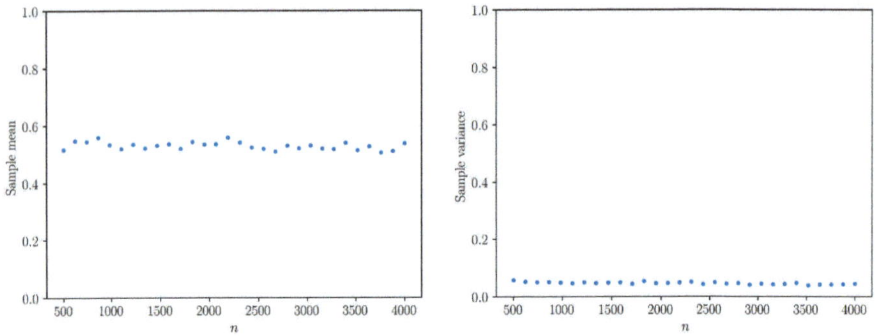

Fig. 5 Mean and average of the edge weights in KNN graphs for different values of n ($k = 400$)

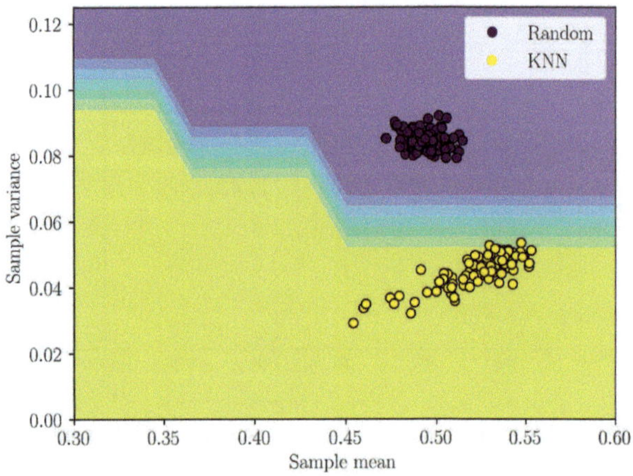

Fig. 6 Borders of the trained classifier

A Classifier From the observations above, it is possible to construct, train, and evaluate an automatic graph classifier, which takes the mean and the variance of (a sample of) the edge weights in input and automatically determines the type of graph.

We used a k-nearest neighbor classifier with $K = 5$. The model was trained with 120 random graphs and 120 KNN graphs. Its reliability has been tested on 200 randomly generated graphs, 100 for each type, taking a sample of 1000 edges. Figure 6 shows the results: each area corresponds to a type of graph after training the classifier. Each point lies in the area of its own type, i.e., all classifications are correct.

Selection of the Sample Size To minimize the impact of the classification on the computing time, one wants to use a sample size s as small as possible. Figure 7 shows the degradation in the accuracy of the classifier when s decreases. The results have been obtained from 70 graphs for each type, with random size and parameters. Keeping $s = 1000$, no classification errors were observed. This is the reason why the value $s = 1000$ was used to train the classifier.

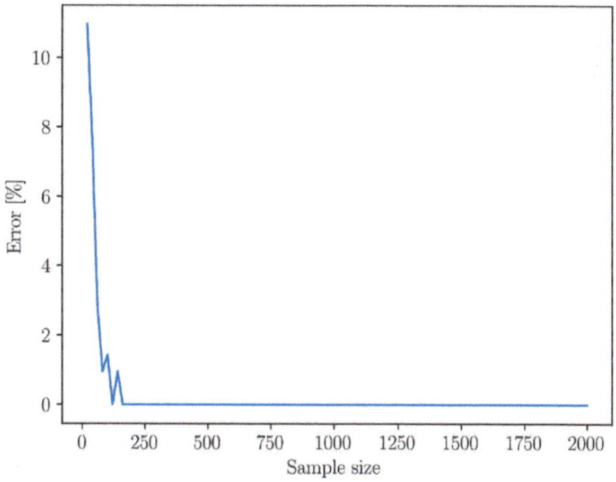

Fig. 7 Percentage error of the classifier for different values of the sample size

3.2 *Parameter Estimation*

Random Graphs In a random graph, the size of the graph and its density are immediately observable by counting the number n of vertices and the number m of edges. The clustering degree γ can be also estimated from the distribution of the edge weights. For this purpose, we analyzed the difference between theoretical quantiles and actual quantiles in the list of edge weights, and we studied its dependency on γ, n, and ρ. By a suitable regression on some randomly generated datasets, we could reliably estimate γ, especially for large values of n and ρ. However, further tests illustrated in Sect. 3.3 showed that γ is not a good predictor for the position of the critical edge in random graphs, and this is consistent with the results observed in Sect. 3.1. Therefore, we did not make further efforts to estimate γ.

KNN Graphs The parameter k in KNN graphs plays a similar role to the density ρ in random graphs, although the distribution of the edge weights in KNN graphs is not uniform as it is in random graphs. It is fair to assume a correlation between k and the number of edges m. This is confirmed by the analysis illustrated in Fig. 8 which shows an approximately linear correlation.

Hence, a simple way to estimate k is to take a fraction η of the ratio $\frac{m}{n}$. To calibrate η, we compute $\frac{nk}{m}$ for 50 graphs with n randomly selected between 500 and 5000 and k between 50 and $0.9n$. We obtain a mean value $\eta = 1.76258$ with a standard deviation of 0.04482. The average percentage error in the estimate of k is only 0.02455%.

Owing to the slight inflection of the curve when k tends to n, the estimate tends to be worse for large k. However, as observed in the remainder, a precise estimate is required only for small values of k.

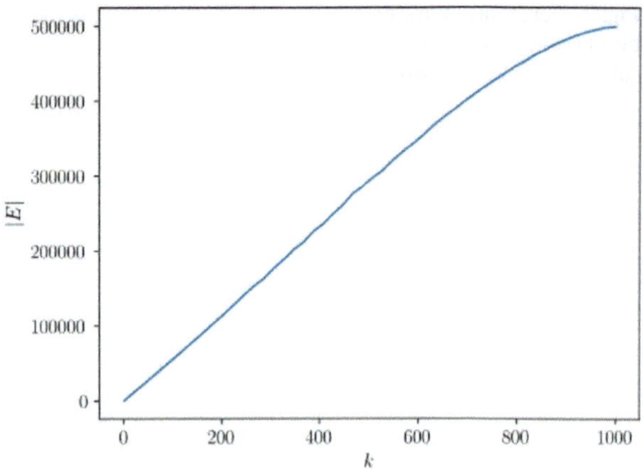

Fig. 8 Correlation between the parameter k and the number of edges in KNN graphs

3.3 Critical Edge and Graph Parameters

To optimize the time performance of the SkewedKruskal algorithm, the goal is to estimate the position or the weight of the critical edge. Let $\hat{w}$ be the critical edge weight, after a min-max scaling procedure which is executed to eliminate the dependency on the range of the weights. Alternatively, one can estimate the critical edge position in the sorted edge list: let $n - 1 \geq \hat{p} \leq m$ be such position in a sorted list of m edges. This estimate is more indirect, but it does not require any preprocessing to scale the weights.

In this section, we analyze the distribution of the values of $\hat{w}$ and $\hat{p}/m$ for different types of randomly generated graphs. This analysis allows to estimate a minimum size of the edge list fraction that is needed to include the MST.

Since the position and the weight of the critical edge depend on some parameters of the graphs, we search for the correlations that can be a base for a reliable prediction.

Random Graphs We generated random graphs with different values of size n, density ρ, and clustering degree γ, and we observed the values of $\hat{w}$ and $\hat{p}/m$.

Dependency on γ Figure 9 shows the results obtained from 200 random graphs with $n = 1000$, $\rho = 0.5$ and different values of γ. No significant correlation can be observed between $\hat{p}/m$ and the clustering degree. A weak correlation between $\hat{w}$ and the clustering degree is observed only for small values of γ.

Dependency on n Figure 10 shows the results obtained from 200 random graphs with $\gamma = 0.5$, $\rho = 0.5$ and different values of n from 50 to 3000. When n grows, both $\hat{p}/m$ and $\hat{w}$ decrease (both in average and in standard deviation).

Dependency on ρ Figure 11 shows the results obtained from 200 graphs with $n = 1000$, $\gamma = 0.5$ and different values of ρ. Both the average and the standard deviation of $\hat{p}/m$ decrease when ρ increases, and the same is observed for $\hat{w}$.

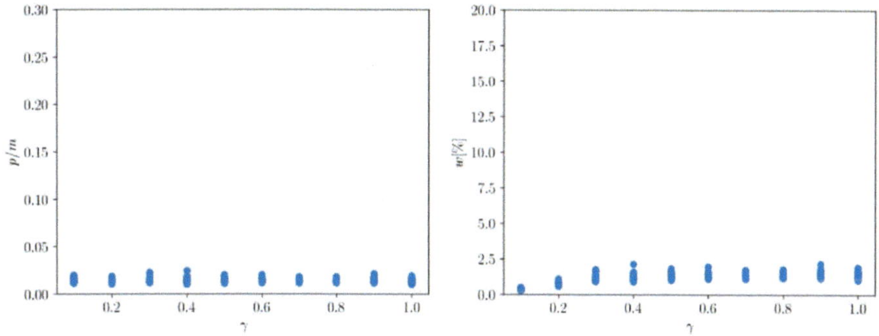

Fig. 9 Position $\hat{p}/m$ and weight $\hat{w}$ in random graphs for different values of the clustering degree γ

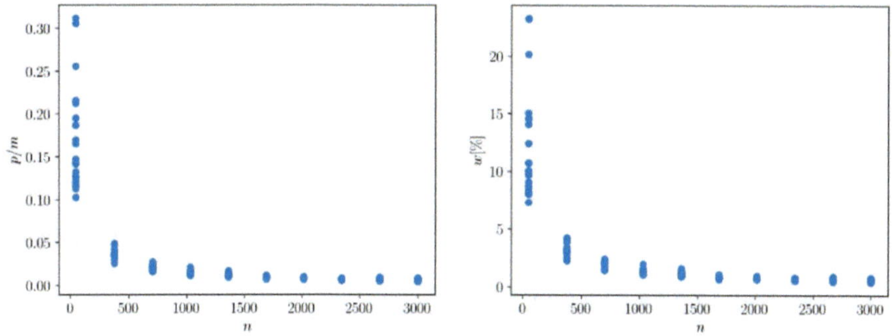

Fig. 10 Position $\hat{p}/m$ and weight $\hat{w}$ in random graphs for different values of the size n

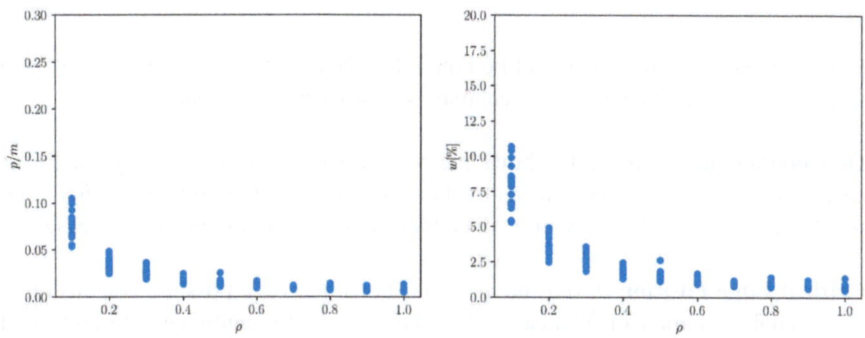

Fig. 11 Position $\hat{p}/m$ and weight $\hat{w}$ in random graphs for different values of the density ρ

KNN Graphs We generated KNN graphs with different values of n and k, and we observed the corresponding values of $\hat{w}$ and $\hat{p}/m$.

Dependency on k Figure 12 shows the results obtained from 400 graphs with $n = 3000$ and different values of k. A strong correlation is observed between $\hat{p}/m$ and

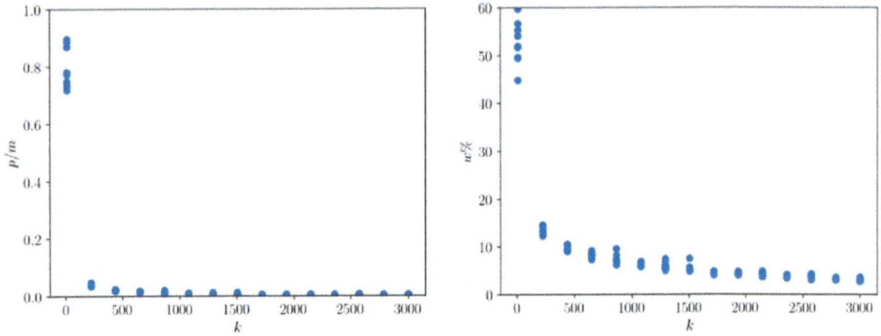

Fig. 12 Position $\hat{p}/m$ and weight $\hat{w}$ in random KNN graphs for different values of k

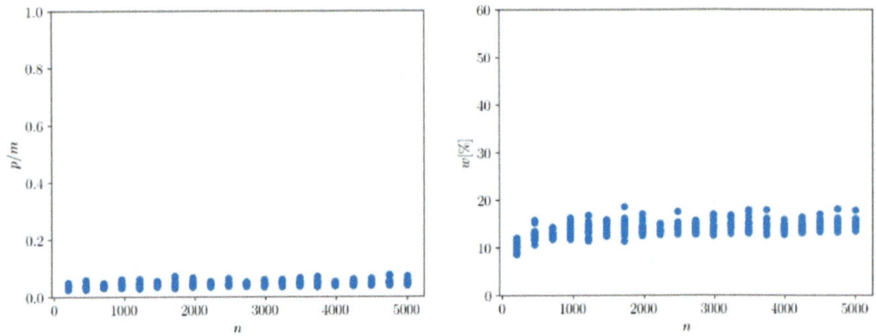

Fig. 13 Position $\hat{p}/m$ and weight $\hat{w}$ in random KNN graphs for different values of n.

k and between $\hat{w}$ and k. It should be noted that the connectivity of the graphs, and then the existence of a spanning tree, may be lost when k is too small.

Dependency on n Figure 13 shows the results obtained from 200 graphs, 10 for each size, with $k = 80$ and different values of n from 250 to 5000. Both $\hat{p}/m$ and $\hat{w}$ show a weak correlation with n, since they slightly increase when n increases.

Critical Edge Position From the analysis shown above, a possible conclusion is that a reliable estimate of $\hat{p}/m$ can be based on n and ρ for random graphs and n and k for KNN graphs. The type of graph at hand can be identified with the automatic classifier illustrated in Sect. 3.1.

We trained our models with 50 graphs for each type and different values of the relevant parameters. Then we observed the absolute errors in estimating $\hat{p}/m$ on 20 graphs for each type with random parameters. The results illustrated in Fig. 14 show that $\hat{p}/m$ can be reliably estimated, since the error is kept below 1.5% of the number of edges in the graph. In this representation, we do not distinguish between errors in

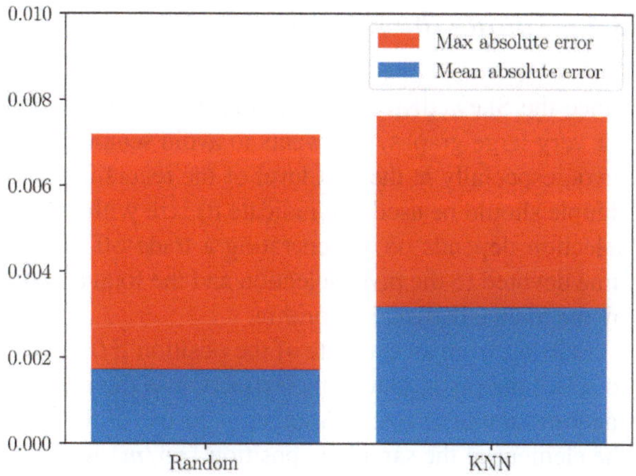

Fig. 14 Average and maximum absolute error in estimating $\hat{p}/m$ in different types of graphs

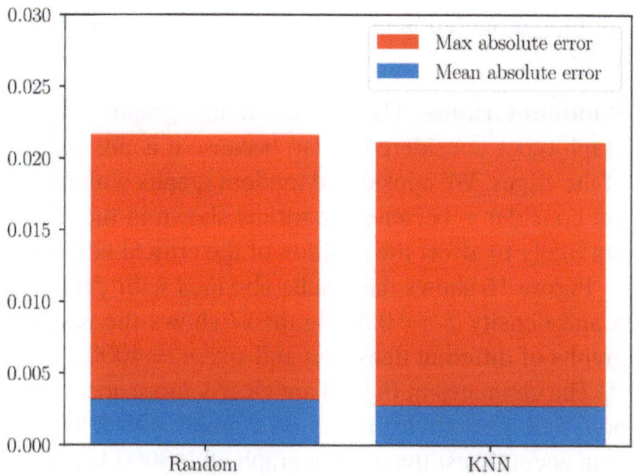

Fig. 15 Absolute error in estimating $\hat{p}/m$ from $\hat{w}$

excess and in defect. In the next section, we differentiate, because the two types of error are likely to produce different effects on the SkewedKruskal algorithm.

Critical Edge Weight Estimating $\hat{w}$ would be even more useful for the purpose of optimizing the SkewedKruskal algorithm, since the value $\hat{w}$ could be directly used as a pivot value, to partition the edge list. However, the effect of a wrong choice of the pivot value does not depend on the value itself but rather on its position in the edge list, since its position affects the number of missing MST edges in the left part of the partition at the first level.

Figure 15 shows the error in estimating $\hat{p}/m$ from $\hat{w}$. The largest errors occur with random graphs, especially when γ is close to 0, owing to the existence of many edges with small weight.

3.4 Tuning the Edge Sample Size

Since the SkewedKruskal algorithm is meant as a tool to solve the MST problem for very large graphs, one wants to avoid scanning the whole edge list to select the pivot, especially at the first level of the recursion. Instead, a suitably sized random sample should be used: we indicate its size with s. Clearly, the accuracy in the pivot selection depends on s, generating a trade-off between the amount of computing time devoted to the pivot selection and the total amount of computing time required by the SkewedKruskal algorithm.

Starting from an estimate of the position $\hat{p}/m$, we consider two main techniques to select the pivot. One is the extraction of the edge in position $\lceil s\hat{p}/m \rceil$ in a sorted random sample of size s. Another is the use of the IntroSelect algorithm to compute the element of the sample in position $\lceil s\hat{p}/m \rceil$ without sorting it.

We remark that this analysis applies to the technique based on the estimate of the critical edge position; when the critical edge weight is estimated, instead, no additional operation is required, because the estimated value $\hat{w}$ can be directly used as a pivot.

Random Graphs The case of random graphs is the most favorable among the two graph types considered so far, because it is possible to discard a very large fraction of the edges. We considered random graphs with different values of n and ρ. We did not consider γ because the results shown in the previous sections suggest that it is not likely to affect the position of the critical edge significantly.

Figure 16 shows the results obtained with $four$ random graphs of different size n and density $\rho = 0.5$. Figure 17 shows the results obtained with $four$ random graphs of different density ρ and size $n = 4000$.

The estimates in Fig. 16 are clearly more accurate for larger n. It is interesting to note that accurate estimates on a graph with 2000 vertices require a larger sample than accurate estimates on a graph with 8000 vertices.

Density turns out to be relevant: estimates in sparse graphs are unstable, i.e., the variability in the estimate of $\hat{p}/m$ is larger than in dense graphs, as shown in Fig. 17.

KNN Graphs The critical edge position in KNN graphs heavily depends on parameter k. Here, we want to analyze its dependency on the sample size s.

Figure 18 shows the results obtained with random KNN graphs of different size and $k = 500$, while Fig. 19 shows the results obtained with random KNN graphs with different values of k and $n = 5000$.

The quality of estimates is robust with respect to the size n. When k is low (for instance, $k = 50$), the estimate is less accurate. This was expected, because for small k the variance of the critical edge position tends to be large.

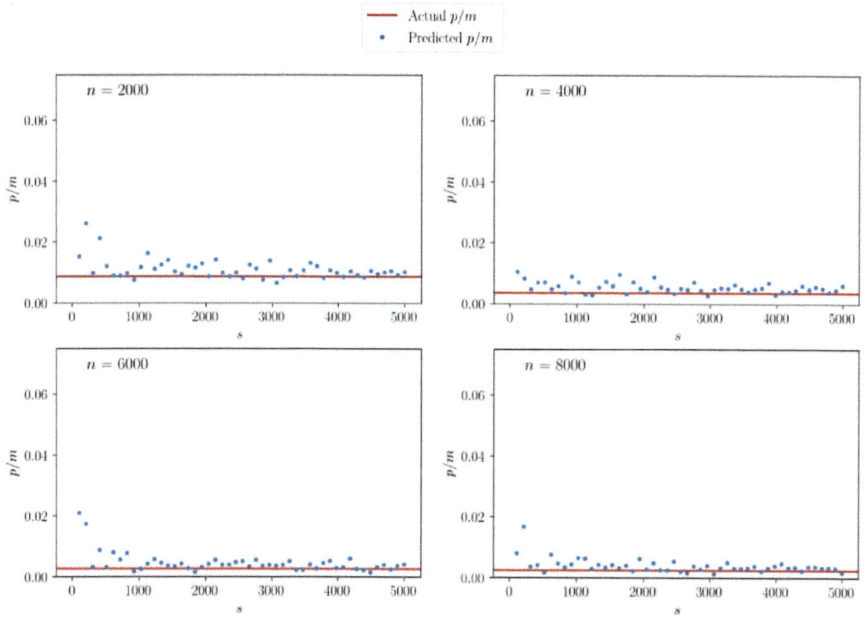

Fig. 16 Pivot selection on *four* random graphs of different size. The horizontal line represents the position of the critical edge. The dots represent the estimated position of the critical edge as a function of the sample size *s*

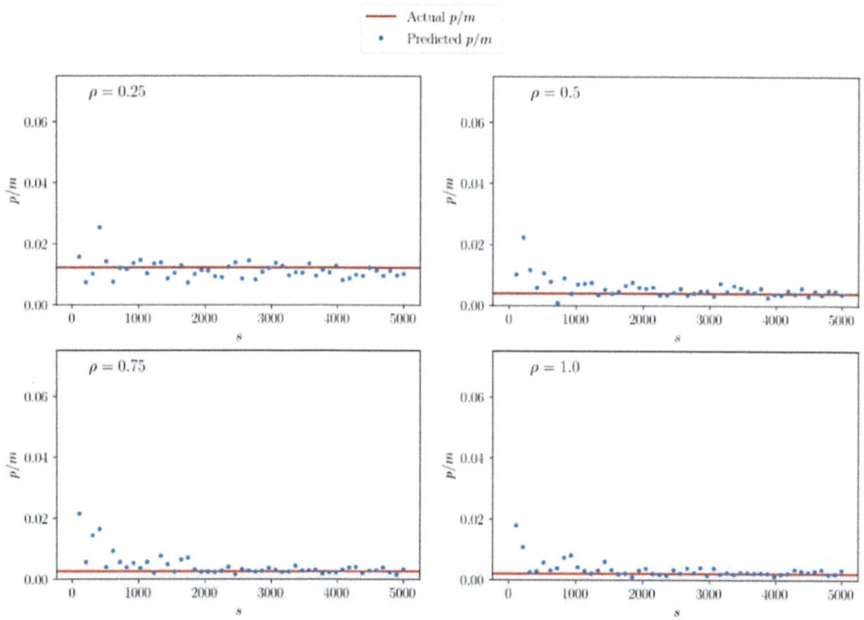

Fig. 17 Pivot selection on four random graphs of different density. The horizontal line represents the position of the critical edge. The dots represent the estimated position of the critical edge as a function of the sample size *s*

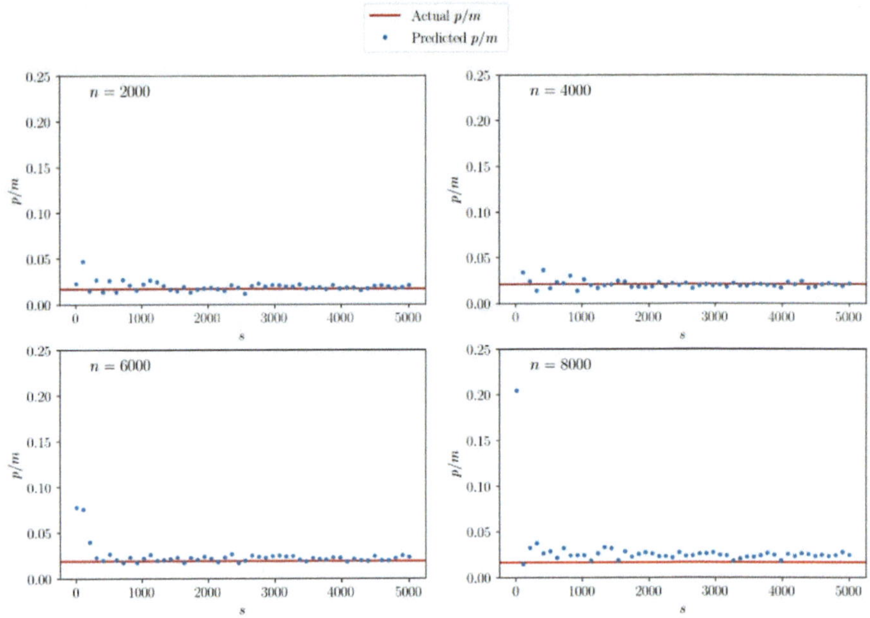

Fig. 18 Pivot selection in random KNN graphs of different size

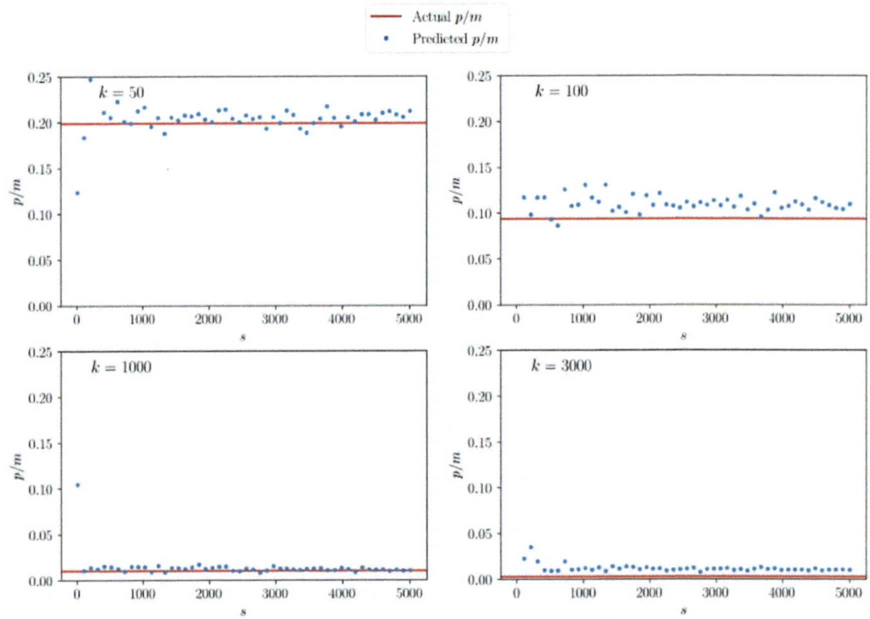

Fig. 19 Pivot selection in random KNN graphs with different values of k

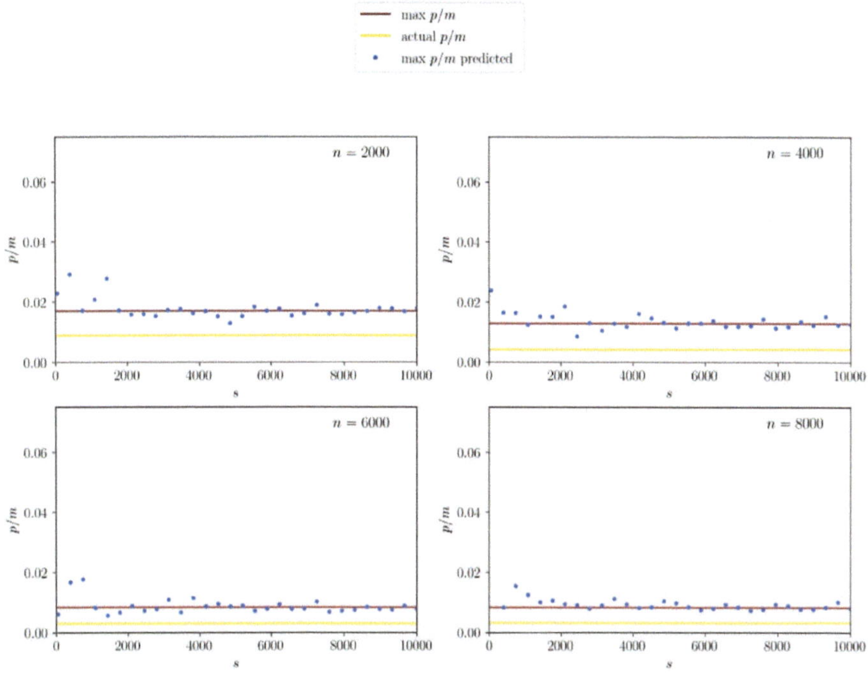

Fig. 20 Critical edge range for different values of the random sample size s (random graphs of four different size n) Each box represents a single instance of a graph

3.5 Selection of the Pivot

The outcome of the search for the critical edge position reported in Sect. 3.3 shows a significant standard deviation of $\hat{p}/m$ for all graph types. This suggests that the pivot selection should be done in a rather conservative way to increase the probability that the pivot is chosen in a position beyond that of the critical edge.

To search for such a pivot, we generated many graphs with the same parameters, and we observed the maximum values of $\hat{p}/m$. Then, with a nonlinear regression model, we tuned a threshold $\overline{p}/m$. To make this estimate even more reliable, we added 20% of the thresholds, artificially shifting the pivot to larger weight edges.

Once computed the threshold value, it is also necessary to revise the cardinality s of the sample. In the remainder, we show some graphical representations, where two horizontal straight lines indicate the threshold $\overline{p}/m$ obtained from the regression model, including the artificial shift described above, and the actual value of $\hat{p}/m$ for the graph, while the dots show the outcome of the estimate of $\hat{p}/m$ for different values of s.

Random Graphs Figure 20 shows that a reliable estimate for $\overline{p}/m$ requires a larger size of the random sample compared to the sample size needed to estimate $\hat{p}/m$.

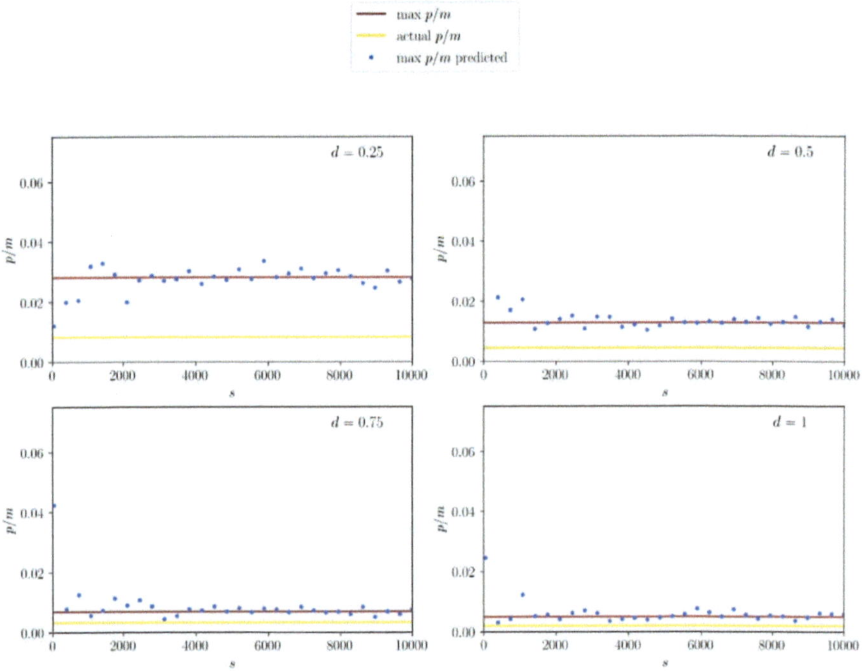

Fig. 21 Critical edge range for different values of the random sample size s (random graphs of four different density values ρ)

Similarly, for random graphs of different density shown in Fig. 21, the precision of the threshold estimates turns out to be affected by the density. For low values of ρ, it is necessary to further increase the random sample size s.

In all cases, the estimated conservative threshold $\overline{p}/m$ always yields an effective selection of the pivot, since the selected pivot position is systematically beyond the position of the critical edge.

KNN Graphs Figure 22 shows the results obtained with KNN graphs with $k = 400$. As with random graphs, it is necessary to consider a larger sample to estimate the thresholds compared to the sample needed to estimate $\hat{p}/m$. Furthermore, for small n, the precision is more sensitive to s than it is for large n, as expected. Figure 23 shows similar results with KNN graphs with different values of k and $n = 4000$. The width of the estimated range is heavily dependent on k; the larger the value of k, the better the precision of the estimated thresholds.

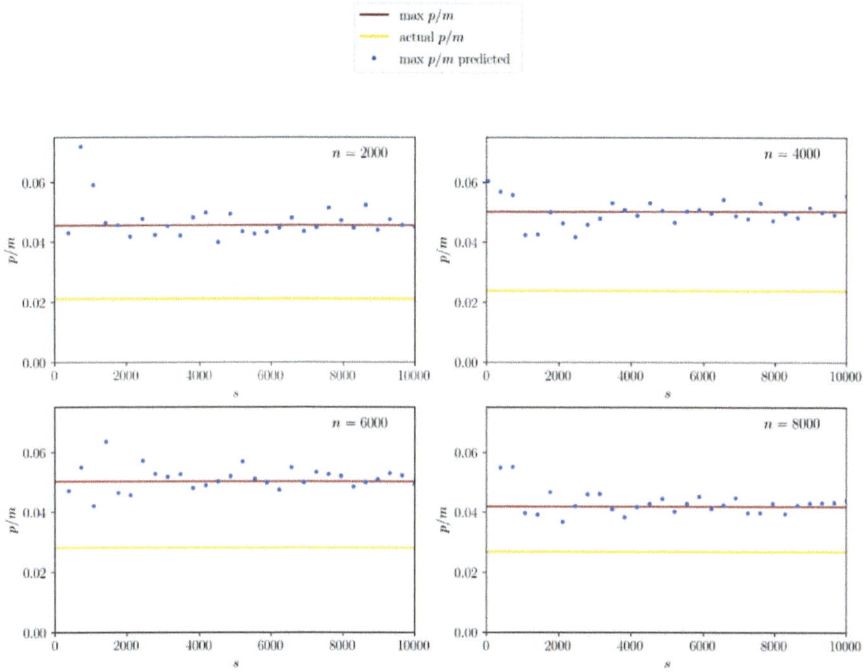

Fig. 22 Critical edge range for different values of the random sample size s (KNN graphs of four different size n)

4 A More General Technique

The classifier presented in Sect. 2 allows to identify the two types of randomly generated graphs considered so far; in turn, this allows to use suitable indicators to estimate $\hat{p}/m$ or $\hat{w}$. However, for the sake of general applicability, one wants to be able to estimate $\hat{p}/m$ or $\hat{w}$ for any type of graph and without any prior knowledge about it. For this purpose, we search for some significant correlation between $\hat{p}/m$ or $\hat{w}$ and some general characteristic of graphs, independent of their type.

The most important advantage of estimating the critical edge weight is that it is not necessary to extract a random sample and to suitably tune its size s. When the graph is classified and hence some assumptions can be done on its edge weight distribution, one can estimate a range that is likely to contain the critical edge weight $\hat{w}$ instead of the critical edge position $\hat{p}/m$, as illustrated in Sect. 3. On the contrary, generic models like those illustrated in this section cannot predict $\hat{w}$, because the edge weight distribution is unknown. Therefore, in this section, we only consider the problem of estimating $\hat{p}/m$ and to select a pivot value from it.

Standard Deviation of Weight Distribution From the observation of Fig. 6, one can note that the standard deviation of the distribution of the edge weights in random

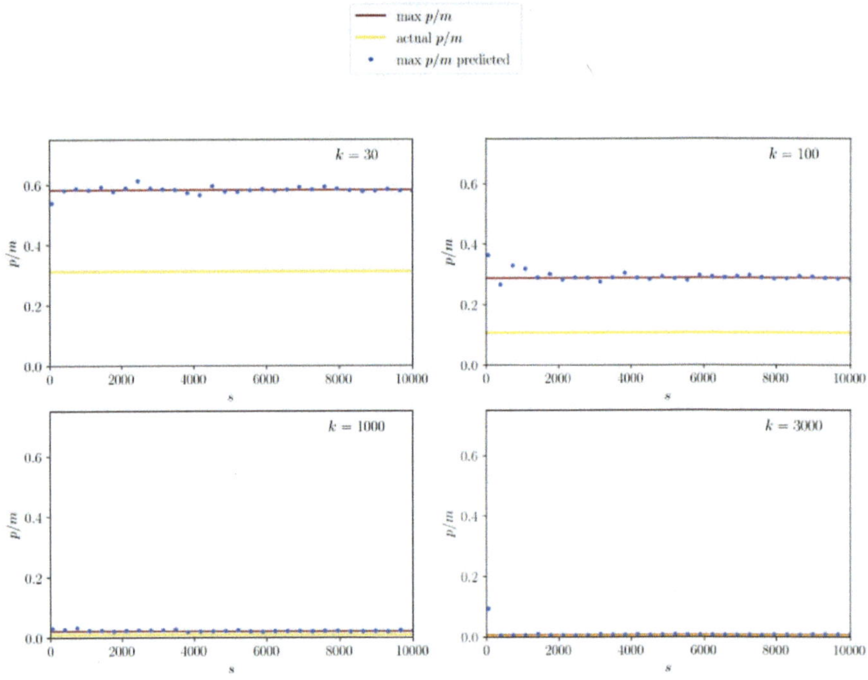

Fig. 23 Critical edge range for different values of the random sample size s (KNN graphs of four different values of k)

Fig. 24 Position of the critical edge as a function of the standard deviation of the weights distribution

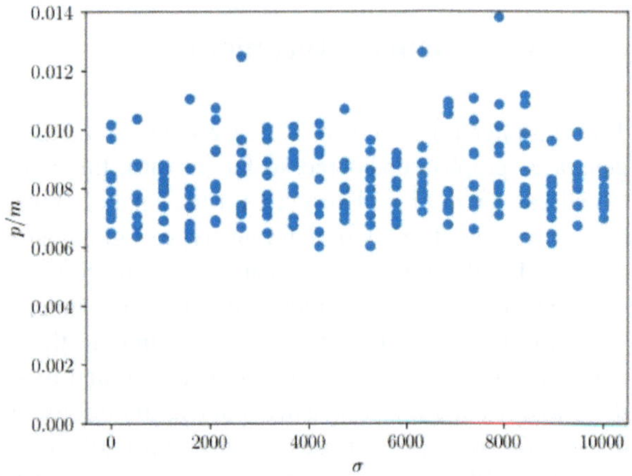

graphs is larger than in KNN graphs. Hence, a search direction is for a possible correlation between such a standard deviation and $\hat{p}/m$. To test this hypothesis, we generated random graphs whose weights follow the normal distribution, with different values of the standard deviation. Unfortunately, from the results shown in Fig. 24, no correlation is visible.

Fig. 25 Position of the critical edge as a function of graph density

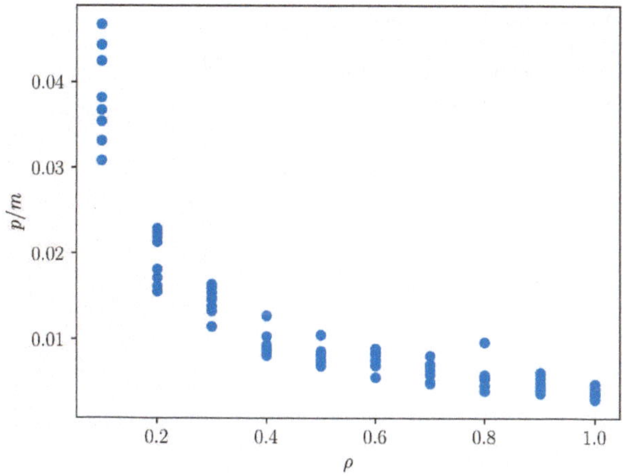

Density Another potentially useful characteristic is density: in random graphs and KNN graphs, when ρ and k (respectively) increase, $\hat{p}/m$ decreases. To search for a possible correlation between $\hat{p}/m$ and the graph density independent of the weight distribution, we generated random graphs with 8 different distribution models, including Gaussian, exponential, and gamma.

The results shown in Fig. 25 show a rather strong correlation between $\hat{p}/m$ and the graph density, while the weight distribution turns out to be unlikely to affect $\hat{p}/m$. This confirms the observations outlined in Sect. 3.3: the parameters that affect the weights, such as γ in random graphs, do not significantly affect $\hat{p}/m$, while the parameters that affect the graph structure, such as k in KNN graphs, also affect the position of the critical edge.

Therefore, we defined a nonlinear regression model, trained with 500 randomly generated graphs with different values of n and ρ and different weight distributions. We used 12 values of ρ to obtain a good approximation for low density: 6 values are equally spaced in a logarithmic scale between 0 (excluded) and 0.3 (included), while the other 6 values are equally spaced in a linear scale between 0.3 (excluded) and 1 (included). Figure 26 shows the quality of the estimate obtained with 20 graphs for each type, generated with random parameters.

Sample Size The technique based on the estimate of the critical edge position requires to extract a pivot element in a given position from the edge list. Since the IntroSelect procedure is time-consuming, in spite of its theoretically linear worst-case time complexity, it is advisable to extract the pivot from a random sample of suitable size s. Figure 27 shows the error in the estimate of $\hat{p}/m$ with different graph types of different size n as a function of the sample size s.

For all graph types, the method is reliable, and s can be tuned to a relatively small value, compared to the size of the whole edge list.

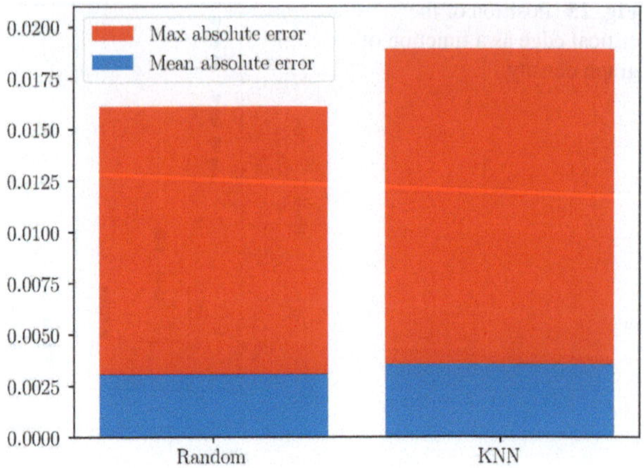

Fig. 26 Errors in estimating the position of the critical edge based on the density for different graph types

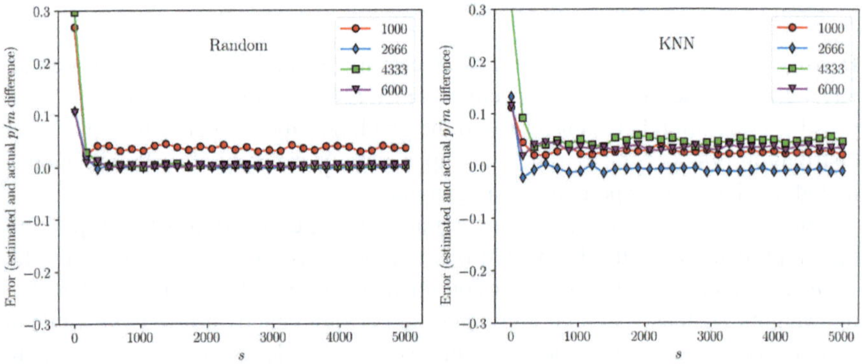

Fig. 27 Density-based estimate of the critical edge position

5 Computational Results

Computational Environment All computational tests reported in this section have been carried out on an AMD Ryzen 5 4500U processor. Algorithms have been coded in Python, and the learning algorithms were taken from the open-source library `Scikit-learn`. Input and output files are available from the authors upon request, as well as the code.

Versions of the SkewedKruskal Algorithm From the previous analysis, it is possible to design different techniques to drive the SkewedKruskal algorithm. We considered six of them, identified here as follows:

- Algorithm 1a: estimate the position of the critical edge, based on graph classification
- Algorithm 1b: estimate the maximum position of the critical edge, based on graph classification

- Algorithm 2a: estimate the weight of the critical edge, based on graph classification
- Algorithm 2b: estimate the maximum weight of the critical edge, based on graph classification
- Algorithm 3a: estimate the position of the critical edge, based on graph density
- Algorithm 3b: estimate the maximum position of the critical edge, based on graph density

Optimal Size of the Random Sample The methods based on an estimate of $\hat{p}/m$ require a suitable sizing of a random sample of the edge set.

In [8], it was suggested to select a random sample of size $s = \lceil\sqrt{m}\rceil$ from the edge list. However, in some of the techniques shown so far, the random sample is used not only for extracting the pivot but also for the purpose of classification and parameter estimation. Therefore, a larger value of s could work better. For this reason, we introduce a multiplicative factor α (to be suitably tuned), so that $s = \lceil\alpha\sqrt{m}\rceil$.

5.1 Calibrating the Algorithm by Graph Classification

Hereafter, we present the results of some computational tests in which graphs of different types (random graphs, KNN graphs) were generated, they were (correctly) classified, their critical edge position or critical edge maximum position or critical edge weight were estimated, and such an estimate was finally used to drive the SkewedKruskal algorithm, by suitably partitioning the edge list. Hence, this technique, corresponding to Algorithms 1a, 1b, 2a, and 2b, exploits the assumption that the input graphs belong to a restricted and known subset of randomly generated graph types.

Random Graphs Figure 28 shows how the computing time of SkewedKruskal depends on the size of the random sample, when the pivot element is selected after estimating its position $\hat{p}/m$ in the edge list (Algorithm 1a). The tests have been done on random graphs with density $\rho = 0.5$ and different size n. Each point is the average value from five graphs.

The computing time is highly variable: the illustration shows several peaks for some values of α. We also observed a rather high variance of the computing time for the same value of α.

When the pivot is selected in a more conservative way (Algorithm 1b), better results are achieved. Figure 29 shows the computing time of SkewedKruskal for the same graphs as in Fig. 28 when the pivot is selected in position $\overline{p}/m$, instead of $\hat{p}/m$.

The observed computing time is smaller in average and also more stable. Concerning the optimal calibration of the sample size, values of α close to 2.5 yielded the best outcome.

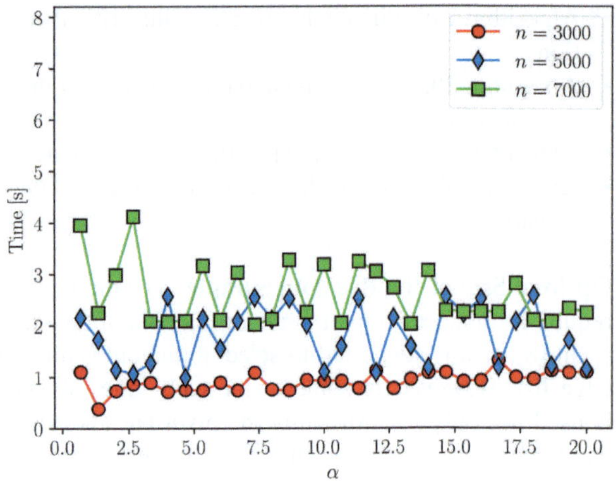

Fig. 28 Computing time of SkewedKruskal driven by the classification of the graph type and an estimate of $\hat{p}/m$ (random graphs with $\rho = 0.5$)

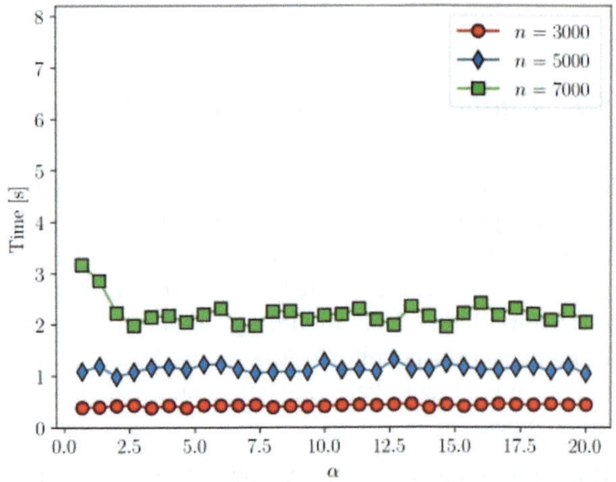

Fig. 29 Computing time of SkewedKruskal driven by the classification of the graph type and an estimate of $\overline{p}/m$ (random graphs with $\rho = 0.5$)

Better results were obtained by estimating the maximum value of the critical edge weight instead of the critical edge position (Algorithms 2a and 2b), as illustrated in Fig. 30.

KNN Graphs Similar results were obtained with KNN random graphs. Figure 31 shows the computing time of SkewedKruskal when the pivot element is selected in position $\hat{p}/m$ (Algorithm 1a) in random KNN graphs with $k = n/10$.

The outcome is rather unstable, with some peaks due to graphs where the critical edge position was underestimated. This happened in 15% of the graphs.

As shown in Fig. 32, more stable average computing time is observed when the pivot is selected in position $\overline{p}/m$ (Algorithm 1b). However, the average computing time is slightly worse in this case.

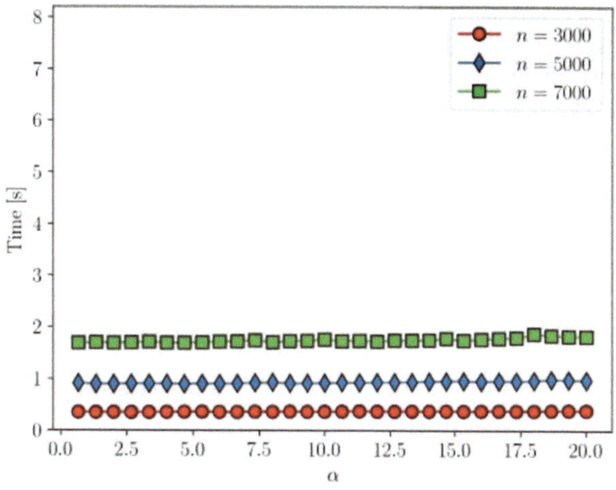

Fig. 30 Computing time of SkewedKruskal driven by the classification of the graph type and an estimate of $\hat{w}$ (random graphs with $\rho = 0.5$)

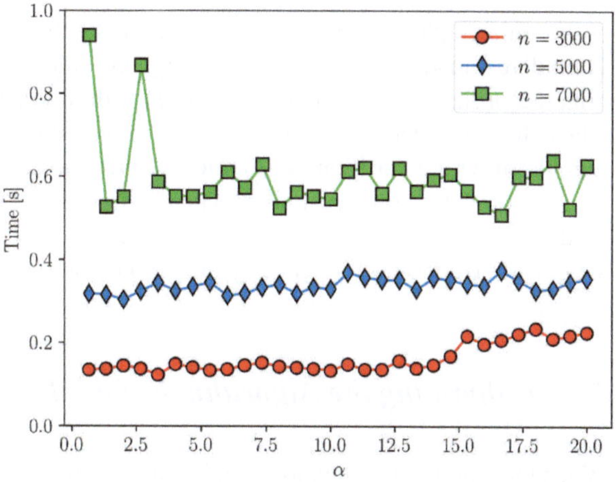

Fig. 31 Computing time of SkewedKruskal driven by the classification of the graph type and an estimate of $\hat{p}/m$ (KNN random graphs)

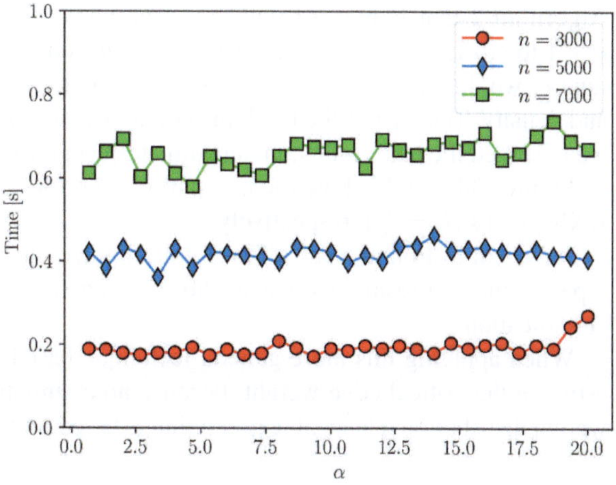

Fig. 32 Computing time of SkewedKruskal driven by the classification of the graph type and an estimate of $\overline{p}/m$ (KNN random graphs)

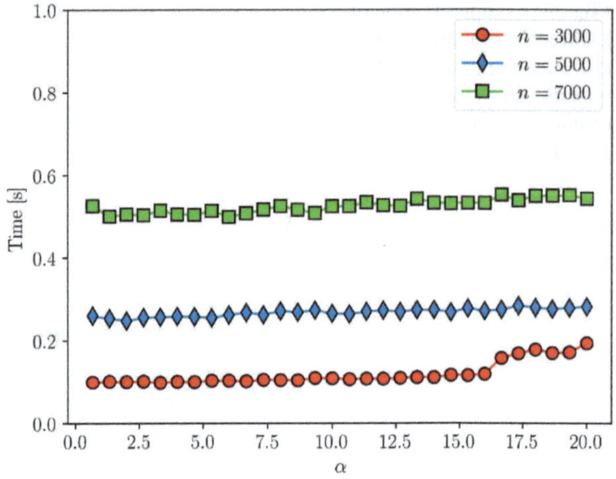

Fig. 33 Computing time of SkewedKruskal driven by the classification of the graph type and an estimate of $\overline{w}$ (KNN random graphs)

For this graph type, it is effective to estimate the critical edge weight. However, 20 underestimates were observed among 100 graphs when the pivot selection was done according to an estimate of $\hat{w}$ (Algorithm 2a). Better results were achieved when the pivot was selected according to an estimate of $\overline{w}$ (Algorithm 2b): the number of wrong partitions was almost reduced to zero, as shown in Fig. 33.

The computing time is similar to that observed for Algorithm 1b with $\alpha \approx 3$, but its dependency on s looks more predictable, as expected: in this case, the random sample of size s is used to classify the graph, not to compute $\overline{w}$.

5.2 *Calibrating the Algorithm by Graph Density*

When the graph type is unknown and no assumption can be done about it, it is harder to estimate $\hat{p}/m$, since the distribution of the edges is unknown. For this reason, in Algorithm 3 that is meant to address this more general settings, we use a factor $c > 1$ to artificially increase the value of $\hat{p}/m$ estimated from the graph density. To tune c, we did some preliminary tests with random graphs of different type, size, and density. The size of the random sample was set to $s = 10\sqrt{m}$. Setting $c = 1.4$, we observed the best results; underestimates occurred in only 5% of the graphs.

Figures 34 and 35 show the computing time for random graphs ($\rho = 0.5$) and KNN graphs ($k = \frac{n}{10}$), respectively.

Values of α around $five$ allowed to minimize the computing time for all graph types, achieving results similar to those obtained with algorithms based on graph classification.

When applying this more general technique based on graph density, we cannot estimate the critical edge weight, because no assumption can be done on the edge weight distribution. Hence, the prediction is based only on the critical edge position.

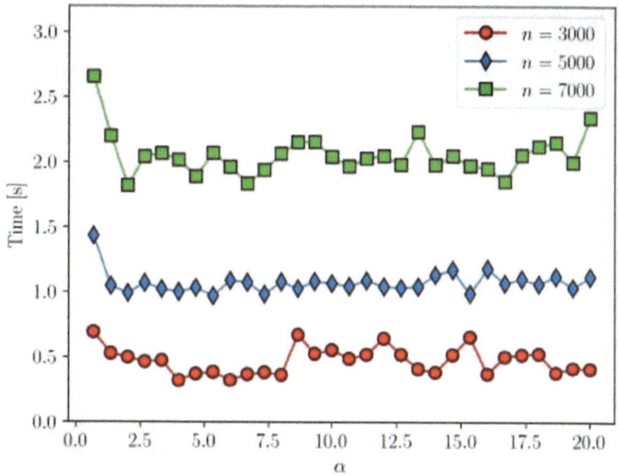

Fig. 34 Computing time of SkewedKruskal driven by graph density and an estimate of $c\overline{p}/m$ (random graphs)

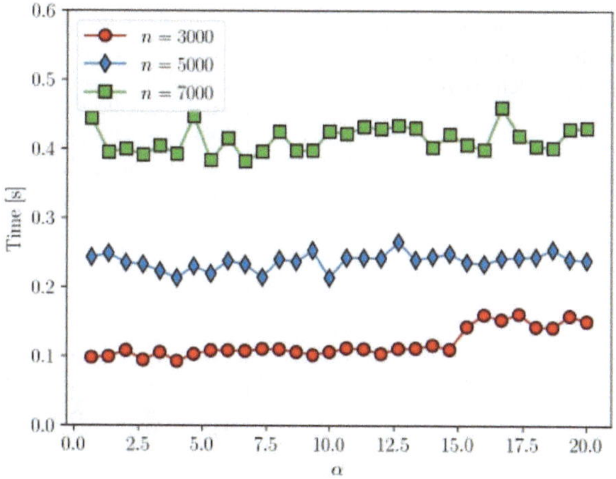

Fig. 35 Computing time of SkewedKruskal driven by graph density and an estimate of $c\overline{p}/m$ (KNN random graphs)

5.3 *Comparison Between Algorithms*

To compare the versions of the SkewedKruskal algorithm, we observe the computing time for different values of the graph types and parameters. Following the results outlined in the previous subsections, the SkewedKruskal algorithm driven by graph classification uses the estimate of the maximum position of the critical edge $\overline{p}/m$ or the maximum weight of the critical edge $\overline{w}$ tuning the sample size with $\alpha = 3$, while the algorithm driven by graph density uses $\alpha = 5$.

Random Graphs Figure 36 shows the results with random graphs with $\rho = 0.5$, $\gamma = 1$, and some values of n.

All SkewedKruskal variants have definitely smaller computing time compared with Kruskal algorithm, and the gap grows with the size n. The variants based on classification and density are faster and more robust than the original Skewed-

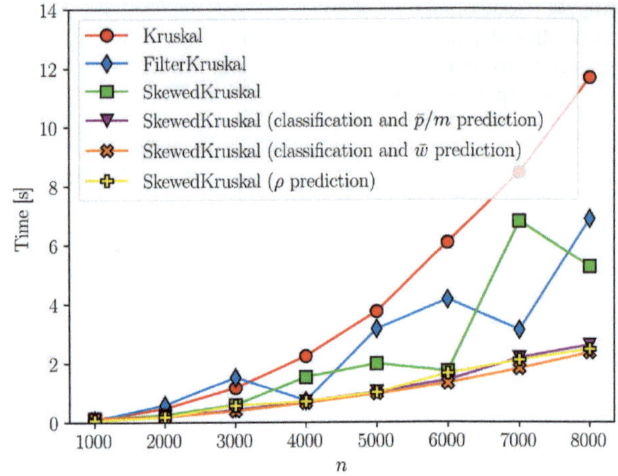

Fig. 36 Computing time with random graphs of different size n

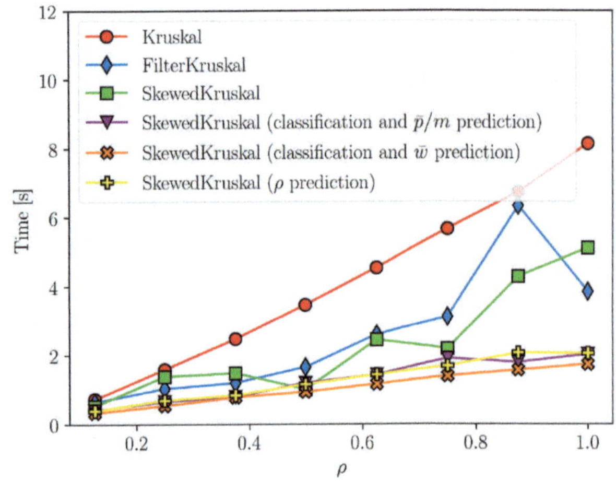

Fig. 37 Computing time with random graphs of different density ρ

Kruskal algorithm, which is sometimes forced to make a recursive call on the second list at the first recursion level due to an underestimation of the critical arc weight.

The same observations hold when the size is fixed ($n = 5000$ in our tests), and computing time is measured on graphs with different density, as shown in Fig. 37. Varying the degree of clustering γ, for fixed size $n = 5000$ and density $\rho = 0.5$, the computing time reported in Fig. 38 is observed. In general, the computing time turns out to be smaller for all variants of SkewedKruskal when γ is very close to 0, i.e., the degree of clustering is large. For the other values of γ, the algorithms based on classification and density have almost identical computing time.

KNN Graphs The comparison between the algorithms on KNN graphs shows even larger gaps between the computing time taken by SkewedKruskal and that of the new variants we have analyzed. Figure 39 shows the computing time for KNN graphs of different size n and $k = \frac{n}{10}$.

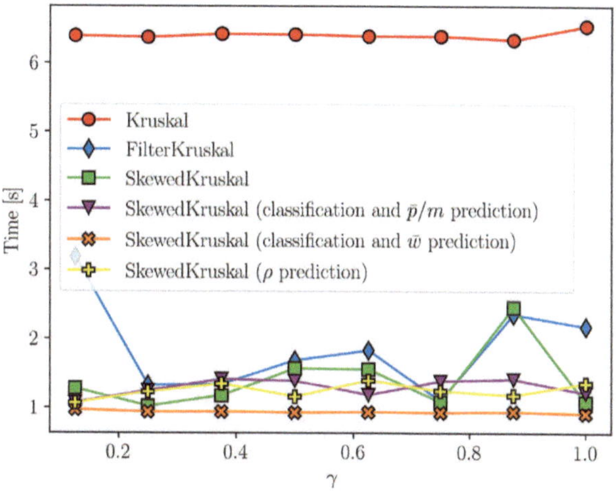

Fig. 38 Computing time with random graphs of different clustering degree γ

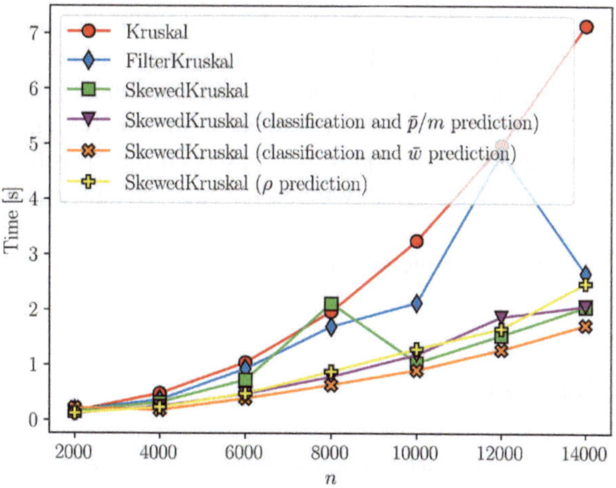

Fig. 39 Computing time with KNN graphs of different size

As with random graphs, the computing time growth with n looks also more predictable for the new versions compared with the original one. The same observation holds for the results shown in Fig. 40, obtained with KNN graphs with fixed size $n = 5000$ and different values of k.

6 Conclusions

In this study, we have applied some techniques based on the analysis of the data of each specific instance of the minimum cost spannning tree problem to optimize the computing time taken by the SkewedKruskal algorithm, a variation of the classical Kruskal algorithm in which the underlying idea is to minimize the edge subset

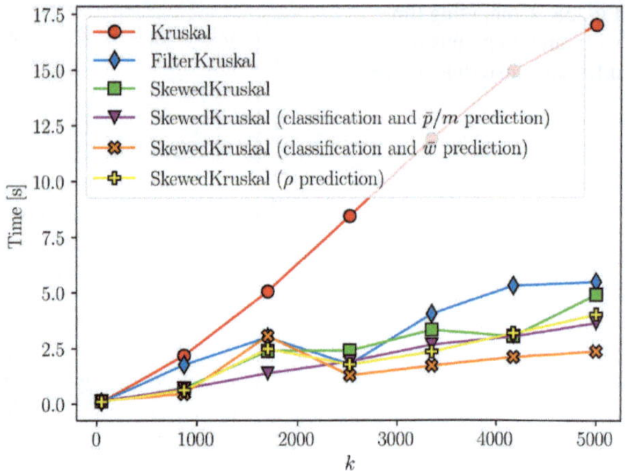

Fig. 40 Computing time with KNN graphs of different values of k

that needs to be examined and sorted to find a minimum spanning tree. This can be viewed as an application of (heuristic) learning algorithms to the performance improvement of (exact) optimization algorithms.

The results we have obtained show that the approach is promising; we have analyzed two techniques, one based on the classification of graphs in which one can use models suitably tuned for each specific graph type and the other based on graph density, which can be evaluated in any general case with no assumption on the structure of the graph.

References

1. Bentley, J.L.: Multidimensional binary search trees used for associative searching. Commun. ACM **18**, 509–517 (1975)
2. Erdös, P., Rényi, A.: On random Graphs I. Publicationes Mathematicae Debrecen, Debrecen (1959)
3. Janson, S., Knuth, D.E., Łuczak, T., Pittel, B.: The birth of the giant component. Random Struct. Algorithms **4**(3), 233–358 (1993)
4. Kruskal, J.B.: On the shortest spanning subtree of a graph and the traveling salesman problem. Proc. Am. Math. Soc. **7**(1), 48–50 (1956)
5. Omohundro, S.: Five balltree construction algorithms. ICSI Technical Report TR-89-063 (1989)
6. Osipov, V., Sanders, P., Singler, J.: The filter-kruskal minimum spanning tree algorithm. In: Proceedings of the Meeting on Algorithm Engineering & Experiments (2009)
7. Paredes, R., Navarro, G.: Optimal incremental sorting. In: Proceedings of the Meeting on Algorithm Engineering & Experiments (2006)
8. Righini, E., Righini, G.: The skewed kruskal algorithm. Learning and Intelligent Optimization, pp. 130–135 (2022)

Approximating the Feasible Criterion Set in Multi-criteria Optimization Problems Described by Equations in Partial Derivatives

Alexander V. Lotov

1 Introduction

The multi-criteria optimization (also known as multi-objective optimization) is an area of decision making with multiple criteria which deals with mathematical optimization problems involving more than one criterion to be optimized (see [29] and [36]). In practical problems, situation with several criteria is met fairly often. For a nontrivial multi-criteria optimization problem, there does not exist a single decision that simultaneously optimizes each criterion. In that case, the criteria are said to be conflicting. A decision is called Pareto optimal (or Pareto efficient or nondominated) if none of the criterion values can be improved without worsening some other criterion values. In the case of conflicting criteria, an infinite number of Pareto optimal decisions do usually exist. Without additional information about subjective preferences, which must be obtained from decision maker, all Pareto optimal decisions are considered equally good. Although the set of Pareto optimal decisions is considered as the mathematical solution to a multi-criteria optimization problem, in real-life decision problems, a single decision is requested that satisfies the subjective preferences of decision maker.

The objective functions define mapping from the decision space into criterion space. The set of feasible decisions (feasible set) is transformed into the feasible criterion set (FCS). The Pareto optimal decisions are mapped to the FCS vectors (points), which are called Pareto optimal, Pareto efficient, or nondominated. Normally, decision maker is interested in maximizing or minimizing the value of a

A. V. Lotov (✉)
Federal Research Center “Computer Science and Control” of Russian Academy of Sciences, Moscow, Russia
e-mail: avlotov@yandex.ru

B. Goldengorin (ed.), *Theory, Algorithms, and Experiments in Applied Optimization*, Springer Optimization and Its Applications 226,
https://doi.org/10.1007/978-3-031-91357-0_10

criterion. For this reason, the Pareto optimal criterion vectors provide a part of the frontier of the FCS, which is known as the Pareto frontier (see, for example, [28]).

Researchers in the field of decision making do use studying multi-criteria optimization problems from different viewpoints. There exist different solution philosophies and goals. Many different approaches have been proposed for supporting decision maker in the process of searching for the preferred decision in the multi-criteria case. One of the approaches is based on providing decision maker with information on the Pareto frontier. Since the Pareto frontier is the image of the set of Pareto optimal decisions, in a rational decision making, the preferred criterion point should be sought among the Pareto frontier points. After exploring the Pareto frontier, decision maker can express his/her subjective preferences by selecting the preferred criterion point that belongs to the Pareto frontier.

There are two main ways of informing decision maker about the Pareto frontier. First of all, it is possible to deal with a large number of Pareto frontier points, say, to provide a large list of them. However, this way of informing has proven to be difficult for decision makers [17]. To improve the situation, a widely known graphic method called the parallel coordinate graphs can be used (see, for example, [16]). In this method, Pareto frontier points are displayed on parallel vertical axes associated with the values of particular criteria. A specific criterion point is displayed then by a polyline that connects the criterion values located on different axes. The polylines associated with all selected criterion points are displayed simultaneously. In the case of a large number of criterion points, the picture may look very complicated. Moreover, the parallel coordinate plotting method does not display trade-offs between criteria.

Secondly, one can visualize the Pareto frontier as a whole. General discussion of Pareto frontier visualization is given in [21]. This idea was proposed in 1954 by S. Gass and T. Saati for two criteria [9]. Visualization of the Pareto frontier in the bi-criterion case was approved by B. Roy [31]. The development of this idea for the case of more than two criteria based on interactive visualization of bi-criterion slices (cross sections) of the FCS was proposed in 1975 [19]. Later it turned to be more convenient to visualize bi-criterion slices of the Edgeworth-Pareto hull (EPH) of the FCS, that is, the maximum set in the criterion space that has the same Pareto frontier as the FCS. This approach has got the name the Interactive Decision Maps (IDM) technique [23].

To visualize the Pareto frontier, you first need to approximate it. However, the problem of approximating the Pareto frontier is ill-posed, since the Pareto frontier is unstable with respect to perturbations of the system parameters [32]. In contrast, the problem of approximating the FCS (or its EPH) is usually stated correctly. Using this evidence, an approximation of the FCS (or its EPH) is computed in the framework of the IDM technique. For the approximation, simple shapes are used such as polyhedrons (in the convex case) or systems of balls, rectangles, and cones (in the non-convex case). This ensures a fast interactive display of bi-criterion slices of the FCS (or its EPH) at the request of decision maker.

As already mentioned, information about the Pareto frontier is provided by the IDM technique in the graphical form of bi-criterion slices of the FCS (or its EPH). A bi-criterion slice is specified by selecting two criteria that define the plane on which the slice will be displayed, while the values of the other criteria should be fixed. A collection of slices is usually called *the decision map* if one of the other criteria (the so-called third or color criterion) changes from one slice to another, while the values of the rest of the criteria are fixed.

The nondominated frontier of a bi-criterion slice of the FCS (or its EPH) displays the bi-objective trade-off curve between the two criteria, while the values of the other criteria are fixed. The curves for different values of the third criterion do not intersect, although sometimes they may coincide. Because of this, sets of trade-off curves look like sets of contour lines on topographic maps, while the value of the third criterion plays the role of the height level related to a contour line. For this reason, such collections of trade-off curves have got the name decision maps. If the distance between the trade-off curves is small, this may mean that the value of the third criterion improves significantly due to a small worsening of the values of the first two criteria. By changing the values of the remaining criteria, decision maker can get a series of decision maps or even animate them. Decision maps provide decision maker with information about the feasible values of the criteria and the criterion (objective) trade-offs. This knowledge helps decision maker to specify the preferred nondominant criterion point (feasible goal) which is used to find the preferred effective decision.

From computational point of view, the most difficult task when using the IDM technique is approximation of the FCS. In the 1990s, the approximation problem was successfully solved for linear models for which the FCS and its EPH are convex. Effective algorithms have been developed for it. It allowed the use of the IDM technique in a number of applied decision-making tasks. Many applications are presented, for example, in [23]. Approximation and visualization of the EPH in the nonlinear non-convex case began to develop intensively in the twenty-first century. In this paper, we consider the IDM technique in the case of multi-criteria optimization problems described by linear partial differential equations.

We start with the mathematical definition of the IDM technique. Then the approach is formulated for an abstract system of linear partial differential equations. In this approach, the right-hand sides of the equations and the boundary values are the control functions that need to be selected. An example is provided by the evolutionary equation. The FCS approximation for a linear system of partial differential equations consists in replacing the original system with a finite-dimensional system and approximating its FCS by polyhedrons. Since the approximation methods of the FCS and its EPH are very close in the convex case, we have limited ourselves to numerical methods for approximating the FCS.

2 Mathematical Introduction of IDM

We start with the general mathematical formalization of the IDM technique. Let decision variables $\mathbf{x}$ belong to a linear functional space $\mathbb{W}$ called the decision space.

Let the variety of feasible decisions (the feasible set in decision space) be denoted by $X \subset \mathbb{W}$. Let us assume that the criteria $\mathbf{z}$ are given by points (vectors) of m-dimensional linear space R^m. Criterion vector $\mathbf{z}$ is related to the decisions $\mathbf{x}$ by the objective vector functions given by the mapping

$$f : \mathbb{W} \to R^m.$$

Then, the variety of feasible criterion vectors Z named the feasible criterion set (FCS) is defined as $f(X)$, i.e.,

$$Z = \{\mathbf{z} \in R^m : \mathbf{z} = f(\mathbf{x}),\ \mathbf{x} \in X\}.$$

Let us assume that preferences of decision maker are described by an irreflexive and transitive binary relation $\mathbf{z} \succ \mathbf{z}'$ that is defined on pairs of elements of R^m and means that $\mathbf{z}$ is better than $\mathbf{z}'$. Let us consider a special case of such binary relation related to the assumption that decision maker is interested in decreasing the values of the criteria. The binary relation is called the Pareto domination if $\mathbf{z} \succ \mathbf{z}'$ means $\mathbf{z} \leq \mathbf{z}'$ and $\mathbf{z} \neq \mathbf{z}'$. Under the nondominated (Pareto optimal) criterion vector, one understands such a vector $\mathbf{z} \in Z$, for which does not exist another element $\mathbf{z}' \in Z$ that dominates it, that is,

$$\{\mathbf{z}' \in Z : \mathbf{z}' \leq \mathbf{z},\ \mathbf{z}' \neq \mathbf{z}\} = \varnothing.$$

The variety $P(Z)$ of nondominated elements of Z, that is,

$$P(Z) = \{\mathbf{z} \in Z : \{\mathbf{z}' \in Z : \mathbf{z}' \leq \mathbf{z},\ \mathbf{z}' \neq \mathbf{z}\} = \varnothing\},$$

is called the nondominated criterion set or Pareto-efficient set as well as Pareto optimal criterion set. The set $P(Z)$ is the part of the frontier of the FCS and is called the nondominated frontier (or Pareto-efficient frontier or Pareto optimal frontier). The feasible decisions that give rise to the elements of the set $P(Z)$ are called the nondominated, or Pareto-efficient, or Pareto optimal decisions.

If decision maker is interested in the Pareto frontier only, it may be reasonable to approximate and visualize another variety that is simpler than the variety Z but has the same Pareto frontier. The broadest of such varieties denoted by Z^* is named the Edgeworth–Pareto Hull (EPH) of the set Z. Along with the vectors from Z, the EPH contains the non-feasible criterion vectors dominated by vectors of Z. Since the vectors dominated by any vector $\mathbf{z}$ from R^m are given by $\mathbf{z} + R^m_+$, where R^m_+ is the nonnegative cone of R^m, the EPH is given by

$$Z^* = \{\mathbf{z}^* \in R^m : \mathbf{z}^* = \mathbf{z}_1 + \mathbf{z}_2, \mathbf{z}_1 \in Z, \mathbf{z}_2 \in R^m_+\}$$

or, in short, $Z^* = Z + R^m_+$. It is important that the nondominated frontiers of the sets Z and Z^* coincide and the frontiers of the EPH have a simpler shape than the frontiers of the FCS. Therefore, the nondominated frontier can be understood easier.

Once again, the IDM technique consists in approximating the set Z (or the set Z^*) using simple figures. Then, the approximation is displayed by using various decision maps, i.e., collections of its bi-criterion slices. The decision maps are used for informing the decision maker and specification of the goal. Computing a decision that results in the goal finalizes the decision support procedure (feasible goal method). Details of the algorithms for interactive visualization of the nondominated frontier are given in the book [23] that contains multiple examples of real-life application.

3 Control Problems for Equations in Partial Derivatives

Let us formulate the multi-criteria decision problem for a decision situation described by a mathematical model that includes equations in partial derivatives. In this case, decisions are given by functions (usually of space and/or time). The researchers use to call them control functions. Let us consider a typical example of such decision problem.

Let the set D be a confined set of the finite-dimensional linear space R^r. Let ∂D be the boundary of the set D. Let the linear metric functional spaces Φ, U, and V be given. The spaces Φ, U, and V contain the functions

$$\varphi \in \Phi, \quad u \in U, \quad v \in V$$

defined at $D + \partial D$, D, and ∂D, correspondingly. Let us consider the linear equation in partial derivatives

$$A\varphi = u, \tag{1}$$

defined on D where A is a linear operator from Φ into U, and let us assume that the boundary condition

$$a\varphi = v \tag{2}$$

is given on ∂D where a is a linear operator that maps Φ into V. The relations (1)–(2) form the boundary problem which is assumed to be stated correctly. It means, firstly, that there exists the only function $\varphi \in \Phi$ that satisfies (1)–(2) for any pair $u \in U$ and $v \in V$ and, secondly, the function $\varphi \in \Phi$ depends continuously on $u \in U$ and $v \in V$.

For the boundary problem under study, the control problem is considered, i.e., the problem of selecting the functions $u \in U$ and $v \in V$. The decision space in this case may be not only composed of U and V but include Φ, too. It means that the decision space $\mathbb{W}$ is given by $\Phi \times U \times V$. This fact is related to constraints imposed on the right-hand functions u and v—it is assumed that the constraints are imposed on functions u and v as well as on the solution of the boundary problem (1)–(2). In the general case, the constraints may have the form

$$\{\varphi, u, v\} \in \Psi \tag{3}$$

where Ψ is a subset of the decision space $\Phi \times U \times V$. The feasible set in decision space is given in this case by the relations (1)–(3).

Let us introduce the multi-criteria decision problem for the system (1)–(3). Let linear continuous operators F_1, $F_2, and F_3$ be given that map the spaces Φ, U, and V into the criterion space R^m. Let us denote

$$f(\varphi, u, v) = F_1\varphi + F_2u + F_3v. \tag{4}$$

Then, the FCS is defined in the usual way as vectors of R^m that can be obtained by the mapping (4) from the vectors of the feasible decision set given by (1)–(3). Let us assume that the set Ψ is convex. In this case, the set Z is also convex.

Since the boundary problem (1)–(2) is stated correctly, we can require that the multi-criteria decision problem given by the relations (1)–(4) is also stated correctly. It means, first of all, that there exists the only feasible set in decision space, and it depends continuously on the set Ψ. Since the operators F_1, F_2, F_3 are linear and continuous, the FCS has the same properties.

Let us consider an example of the described problem—the multi-criteria decision problem that includes the evolutionary equation

$$\frac{\partial\varphi}{\partial t} + A\varphi = u \quad \text{on} \quad D \equiv H \times (0, T) \tag{5}$$

with the boundary condition

$$a\varphi = v \quad \text{on} \quad \partial H \times (0, T) \tag{6}$$

and the initial condition

$$\varphi(0) = w \quad \text{at} \quad H \tag{7}$$

where H is a confined set of a finite-dimensional linear space. Here, a is a linear operator specifying the boundary condition. In this system, controls are given by u, v, and w.

Let the constraint imposed at φ, u, v have the form

$$\left\{\varphi|_{t=\tau}, u|_{t=\tau}, v|_{t=\tau}\right\} \in \Psi_\tau, \quad \tau \in (0, T), \tag{8}$$

where Ψ_τ is a convex subset of $\Phi \times U \times V$ for $\tau \in (0, T)$ and

$$w \in W_0 \subset \Psi_0 \tag{9}$$

where Ψ_0 is the space of initial conditions. The set W_0 is assumed to be convex. Then, the FCS is convex, too.

4 Approximating the Decision Problem

Since the problems of approximating the FCS and its EPH are closely related, here, we limit ourselves to methods for approximating the FCS for the system (1)–(4). The methods are based on substitution of the system by its finite-dimensional analogue that can be based on using, say, finite differences (the finite element methods can be used as well). Polyhedral approximation of the FCS for the finite-dimensional analogue is discussed in the next section.

Let us consider uniform meshes with the step h in the vicinity of the set D. Let us denote by ∂D_h the set of mesh points, which are used for approximating the boundary condition (2), and by D_h the internal mesh points. Analogous to the definition of the spaces Φ, U, and V, we can define the spaces of mesh functions Φ_h, U_h and V_h on $D_h + \partial D_h$, D_h, and ∂D_h correspondingly. The following equations are considered in the space of such functions,

$$A^h \varphi_h = u_h \quad \text{on} \quad D_h, \tag{10}$$

$$a^h \varphi_h = v_h \quad \text{on} \quad \partial D_h, \tag{11}$$

where A^h and a^h are the linear operators, $\varphi_h \in \Phi_h, u_h \in U_h, v_h \in V_h$.

Let the finite-dimensional system (10)–(11) approximate boundary problem (1)–(2). Then, since (1)–(2) is stated correctly, according to the Lax equivalence theorem [18], the solution of (10)–(11) approximates the solution of (1)–(2).

Let us assume that φ_h, u_h, and v_h satisfy

$$\{\varphi_h, u_h, v_h\} \in \Psi_h \tag{12}$$

where Ψ_h is a polyhedral subset of $\Phi_h \times U_h \times V_h$. Let F_1^h, F_2^h, and F_3^h be linear operators from Φ_h, U_h, and V_h into R^m. Let

$$\mathbf{z} = F_1^h \varphi_h + F_2^h u_h + F_3^h v_h. \tag{13}$$

The FCS for the system (10)–(13) is given by

$$Z^h = \Big\{ \mathbf{z} \in R^m : \mathbf{z} = F_1^h \varphi_h + F_2^h u_h + F_3^h v_h, \tag{14}$$

$$A^h \varphi_h = u_h, \ \ a^h \varphi_h = v_h, \ \ \{\varphi_h, u_h, v_h\} \in \Psi_h \Big\}.$$

The scheme for proving proximity of the FCS for the finite-dimensional system (10)–(13) to the FCS for the initial decision problem (1)–(4) is based on the assumption that the initial decision problem is stated correctly. Then, it is possible to prove that if the system (10)–(13) approximates the system (1)–(4), then the FCS for the system (10)–(13) approximates the FCS for the system (1)–(4) (see [20] for details).

Now, let us consider the problem of approximating the set Z^h by a polyhedral set. The relations (10)–(12) provide a finite system of linear equations and inequalities imposed on the finite-dimensional vectors φ_h, u_h, v_h. Let us give a simpler description of the system under consideration. Let us denote $\mathbf{x} = (\varphi_h, u_h, v_h)$. Let the total number of elements of vector $\mathbf{x}$ be N, that is, it belongs to the linear metric space R^N. Then, the relations (10)–(12) can be provided in the form $\mathbf{x} \in X$ where

$$X = \left\{\mathbf{x} \in R^N : H_1\mathbf{x} = \mathbf{h_1}, H_2\mathbf{x} \leq \mathbf{h_2}\right\} \tag{15}$$

and H_1, H_2, h_1, and h_2 are given matrices and vectors. Let us denote

$$F^h\mathbf{x} = F_1^h\varphi_h + F_2^h u_h + F_3^h v_h. \tag{16}$$

Then, the set Z^h is given by

$$Z^h = \left\{\mathbf{z} \in R^m : \mathbf{z} = F^h\mathbf{x},\ H_1\mathbf{x} = \mathbf{h_1}, H_2\mathbf{x} \leq \mathbf{h_2}\right\}. \tag{17}$$

Note that the set Z^h is convex and can be approximated by a polyhedral set.

Several approaches can be proposed for polyhedral approximating the m-dimensional set given by (17). In the next section, we describe one of them that is based on the theory and methods for approximating finite-dimensional convex compact bodies, that is, closed confined convex sets that belong to a finite-dimensional linear metric space and contain internal points. The convex compact bodies are approximated by convex polyhedrons (polytopes), that is, convex hulls of a finite number of points.

5 Polyhedral Approximating the FCS

A convex polyhedron can be given in two forms, namely, by its vertices and by a finite system of linear inequalities, which solution set coincides with it [37]. The modern development of the theory of polyhedral approximation of the convex sets was started in 1975 by D.E. McClure and R. Vitale [27]. In their paper, approximating of two-dimensional sets was considered, and the one concept of the polygon of best approximation was introduced. P.M. Gruber proved in 1983 that, for any finite-dimensional convex compact body, the polyhedron of best approximation (PBA) does exist that results in the minimal Hausdorff distance from the body among the inscribed polyhedrons, that is, polyhedrons with vertices that belong to the boundary of the approximated body, with a given number of vertices. Moreover, this distance converges to zero with the number of vertices tending to infinity [10].

One is able to construct PBA in the case of simple two-dimensional bodies, but PBA have not been constructed for $m > 2$. However, properties of PBA can be used for evaluating the quality of sequences of approximating polyhedrons constructed by various methods. In particular, the methods that construct inscribed polyhedrons that have the same order of convergence as PBA can be denoted as optimal methods for approximating the convex compact bodies by polyhedrons.

The following traditional method of approximating a compact convex body with inscribed polyhedrons has been well known for several decades. It is based on computing the support function of the body $C \subset R^m$ that is defined for a given direction $\mathbf{u}$ of the unit sphere of directions as

$$g_C(\mathbf{u}) = \max\{(\mathbf{u}, \mathbf{z}) : \mathbf{z} \in C\}$$

where $(\mathbf{u}, \mathbf{z})$ is the scalar product of vectors $\mathbf{u}$ and $\mathbf{z}$. The point $\mathbf{z}^*$ that belongs to the frontier of the set C and provides maximum of the scalar product is found as well. The convex hull of the points $\mathbf{z}^*$ provides an internal approximation of the set C, while intersection of the half-spaces that contact the set C at the points $\mathbf{z}^*$ provides an external approximation. In the framework of the traditional method, values of the support function are computed for a finite number of directions $\mathbf{u}$ given in advance.

It is clear that such a system of directions $\mathbf{u}$ does not take into account the actual shape of the approximated body. For this reason, the traditional method requires too many calculations of the support function. Polyhedrons constructed by this method have too many vertices and faces. Therefore, they are not optimal [35].

Let's consider a class of iterative methods for polyhedral approximation of convex compact bodies that construct sequences of inscribed polyhedrons:

$$P^0, P^1, \ldots, P^l, \ldots. \tag{18}$$

As the number of vertices increases, the old vertices are not excluded, and only one new vertex is added during the iteration. Specific methods of this class can be characterized by various algorithms for selecting a new vertex and constructing a new approximation. Such methods are known as incremental methods.

In the paper [27], several ideas were proposed aimed at improving the performance of incremental procedures in the two-dimensional case. These ideas were developed for multidimensional bodies in various theoretical studies (see, for example, papers by R. Schneider [33] and P.M. Gruber [11, 12]). However, these studies did not contain numerical methods that can be implemented as computer software.

Interesting ideas aimed at improving the approximation procedures were proposed by J.L. Cohon (see [7]). The problem of approximating the bi-criterion Pareto frontier in a linear water management problem was studied. In the framework of his method, J.L. Cohon approximated the Pareto frontier, which is not convex even in the linear case, by a piecewise linear curve that does belong the FCS. The peculiarity of the method is computing the FCS support function for the directions that are perpendicular to the linear segments of the approximation. The point of the Pareto frontier that is most distanced from the corresponding linear segment is selected as the new break point of the piecewise linear approximation. Thus, the approximation process is adapted to the shape of the Pareto frontier. Though such improvement of the approximation can be easily performed in the two-dimensional case, it turned out to be too complicated in the case of more than two criteria [34]. We can assume that this result is related to the fact that the Pareto frontier, which is non-convex even in the linear problems, is unstable with respect to rounding errors. Nevertheless, the ideas proposed by J.L. Cohon of using directions orthogonal to the linear faces of the approximation and selecting a single new point of the Pareto frontier to expand the approximation turned out to be very valuable.

These ideas were used in the adaptive iterative incremental method for polyhedral approximation of multidimensional compact convex bodies, developed at the Computing Center of the USSR Academy of Sciences in the 1980s. During its gradual development, the method has received several different names. At first, it was simply called *adaptive method.* Later, an abbreviated name began to be used, that is, *A-method.* Since there exist a large number of adaptive methods (see, for example, the method for Pareto frontier approximation described in [29]), it was decided to change its name. Unfortunately, the new name has been translated into English in several different ways. In the book [23], it was called *estimation refinement (ER) method.* We will use this name in this paper.

Unlike the methods of R. Schneider and P.M. Gruber, the computational scheme of the ER method is based on standard computing the support function of the approximated body $C \subset R^m$ and on a numerically realizable algorithm for constructing a convex hull of a polyhedron and a point. Although, at first glance, the polyhedrons constructed by the incremental methods have nothing in common with PBA, since the vertices of PBA with l vertices differ from the vertices of PBA with $l-1$ vertices, the optimality of the ER method has been proved [13].

Let's take a general look at the iteration of the ER method. In the ER method, sequences of polyhedrons (18) are created, as usually in the incremental methods, with an increase in the number of vertices by one. The vertices must belong to the boundary of the approximated body. Before the l-th iteration, the polyhedron P^{l-1} must be represented in two forms: by a list of its vertices and as the intersection of several half-spaces, i.e., by a system of linear inequalities

$$(\mathbf{u^i}, \mathbf{z}) \leq h_i, i = 1, \ldots, M^{l-1} \quad (19)$$

where $\mathbf{u^i}$, $i = 1, \ldots, M^{l-1}$, are vectors that belong to the unit sphere. Let us denote this set of vectors by $U(P^{l-1})$.

Before the first iteration, the initial bodily polyhedron P^0 must be prepared. Its vertices must belong to the boundary of the approximated set C. The number of vertices must be equal to $m+1$. The polyhedron P^0 must be provided by the list of these vertices as the convex hull of them. At the same time, P^0 should be described by a finite system of linear inequalities as its solution set. The number of inequalities must be equal to $m+1$ as well. Such a polyhedron can be constructed by optimizing several linear functions over the convex body C.

The l-th iteration of the ER method is as follows:

Step 1. Among $\mathbf{u} \in U(P^{l-1})$, find a vector $\mathbf{u}^*$ with maximal increment, that is, the maximal value

$$g_C(\mathbf{u^i}) - h_i, i = 1, \ldots, M^{l-1}. \tag{20}$$

The values $g_C(\mathbf{u})$ are found by solving the related optimization problems. If the maximal increment is sufficiently small, stop; else, select such a point $\mathbf{z}^*$ on the boundary of C that $(\mathbf{u}^*, \mathbf{z}^*) = g_C(\mathbf{u}^*)$. Let the point $\mathbf{z}^*$ be the new vertex.

Step 2. Let $\mathrm{conv}\{z^*, P^{l-1}\}$ be the new approximating polyhedron P^l. Insert $\mathbf{z}^*$ into the list of vertices. Construct the new linear inequality system, which solution set coincides with the polyhedron P^l. Start next iteration.

After a finite number of iterations, a polyhedron P^L with the desired small increment value will be constructed. Note that the method has an important feature: in addition to the internal polyhedron P^l, an external polyhedron

$$\hat{P}^l = \{\mathbf{z} : (\mathbf{u}, \mathbf{z}) \le g_C(\mathbf{u}), \mathbf{u} \in U(P^l)\}$$

that contains the body C is also provided. Due to this, internal and external estimates of the body C are obtained at any iteration. If needed, one can use the values of the increments (20) for estimating the Hausdorff distance between the polyhedron P^L and the body C.

Let us discuss several features of the ER method. To improve the current approximation, one selects the point of the set C that is most distant from the half-spaces describing the approximation. Of course, including the point of the set C that is furthest from the approximation would presumably accelerate the convergence of the method. However, it requires, even in the case of a linear system, solving a non-convex optimization problem. At the same time, finding a point of C furthest from a half-space requires computing the support function for the direction given by this half-space. Since the set C is convex, it can be done by solving a convex optimization problem. In the linear case, it is the linear programming problem.

At the second stage of the method, it is necessary to construct the convex hull of a polyhedron and a point in the m-dimensional space in the form of the solution set of a linear inequality system. This was the main computational obstacle of the method in the case of $m > 2$. The problem was solved by V.A. Bushenkov in 1981 in [2, 3] where one of the first algorithms of this type was proposed. The algorithm developed by V.A. Bushenkov was based on a general approach, currently known as the *under-beyond scheme* (see, for example, [30]). It was used in 1982 by V.A. Bushenkov and A.V. Lotov in [4] as a part of the new adaptive iterative method for approximating compact convex bodies, which later became known as the ER method. A new algorithm for constructing the convex hull of a multidimensional polyhedron and a point that proved to be robust to rounding errors was proposed in 1986 by O.L. Chernykh [6]. It is also based on the under-beyond scheme. The algorithm has been incorporated in the ER method software.

Thanks to the algorithm developed by O.L. Chernykh, the ER method has been implemented as reliable software for approximating compact convex bodies in spaces from three to ten dimensions. In turn, the software of the ER method was used in convex decision problems in the framework of the IDM technique. A detailed description of the ER method is given, for example, in Chapter 6 of the book [23].

6 Conclusion

Several real-world applications of the IDM technique in linear multi-criteria decision problems are described in the book [23]. An example of application of the IDM method in multi-criteria decision problems described by finite-dimensional approximations of linear boundary value control problems in partial derivatives is given in the paper [26] devoted to the forest plantation growth control.

For the case of the non-convex FCS, several different methods for approximating and visualizing the FCS and its EPH have been developed (see [1, 22], etc.). They have been applied in several real-world tasks (see, for example, [5, 8, 24], and [25]).

Based on the ER method, G.K. Kamenev proposed several new optimal methods for approximating the multidimensional convex compact bodies. He collected the results of his theoretical and experimental studies of the methods in the monographs [14] and [15]. Unlike the ER method, the new methods have not yet been implemented in software and have not been studied experimentally.

Competing Interests The author has no conflicts of interest to declare that are relevant to the content of this paper.

References

1. Berezkin, V.E., Kamenev, G.K., Lotov, A.V.: Hybrid adaptive methods for approximating a nonconvex multi-dimensional Pareto frontier. Comput. Math. Math. Phys. **46**(11), 1918–1931 (2006)
2. Bushenkov, V.A.: A numerical algorithm for constructing the projections of polyhedral sets. In: Proceedings of the Conference M.Ph.T.I.–1980: Aerophysics and Applied Mathematics. M.Ph.T.I.: Moscow (1981). In Russian
3. Bushenkov, V.A.: An iteration method of constructing orthogonal projections of covex polyhedral sets. U.S.S.R. Comput. Maths. Math. Phys. **25**(5), 1–5 (1985)
4. Bushenkov, V.A., Lotov A.V.: Methods for the construction and application of generalized reachable sets. Computing Center of USSR Academy of Sciences, Moscow (1982). In Russian
5. Castelletti A., Lotov A., Soncini-Sessa R.: Visualization-based multi-criteria improvement of environmental decision-making using linearization of response surfaces. Environ. Model. Softw. **25**, 1552–1564 (2010)

6. Chernykh, O.L.: Construction of the convex hull of a finite set of points when the computations are approximate. U.S.S.R. Comput. Maths. Math. Phys. **28**(5), 71–77 (1988)
7. Cohon, J.L.: Multi-objective Programming and Planning. Academic Press, Cambridge (1978)
8. Dietrich J., Schumann A.H., Lotov A.V.: Workflow oriented participatory decision support for integrated river basin planning. In: Topics on System Analysis and Integrated Water Resource Management, pp. 207–221. Elsevier, Amsterdam (2006)
9. Gass, S., Saaty, T.: The computational algorithm for the parametric objective function. Nav. Res. Logist. Q. **2**, 39 (1955)
10. Gruber, P.M.: Approximation of convex bodies. In: Convexity and Its Applications, pp. 131–162. Birkhauser, Basel (1983)
11. Gruber, P.M.: Volume approximation of convex bodies by inscribed polytopes. Math. Ann. **281**(2), 229–245 (1988)
12. Gruber, P.M.: Volume approximation of convex bodies by circumscribed polytopes. In: Applied Geometry and Discrete Mathematics. DIMACS Ser., vol. 4, pp. 309–317 American Mathematical Society, Providence (1991)
13. Kamenev, G.K.: Investigation of an algorithm for the approximation of convex bodies. Comput. Maths. Math. Phys. **34**(4), 608–616 (1994)
14. Kamenev, G.K.: Optimal Adaptive Methods for Polyhedral Approximation of Convex Bodies. Computing Center of Russian Academy of Sciences, Moscow (2007) (in Russian)
15. Kamenev, G.K.: Numerical Study of Convex Bodies Polyhedral Approximation Methods Efficiency. Computing Center of Russian Academy of Sciences, Moscow (2010) (in Russian)
16. Korhonen, P.J., Wallenius, J.: Visualization in the multiple objective decision-making framework. In: Multiobjective Optimization: Interactive and Evolutionary Approaches, pp. 195–212. Springer, Berlin (2008)
17. Larichev, O.I.: Cognitive validity in design of decision aiding techniques. J. Multi-Criteria Decision Anal. **1**(3), 127–138 (1992)
18. Lax, P.D., Richtmyer, R.D.: Survey of the stability of linear finite difference equations. Comm. Pure Appl. Math. **9**(2), 267–293 (1956)
19. Lotov, A.V.: Exploration of economic systems with the help of feasible sets. In: Proceedings of the International Conference on the Modeling of Economic Processes (Erevan, 1974), pp. 132–137. Computing Center of USSR Academy of Sciences, Moscow (1975) (in Russian)
20. Lotov, A.V.: On stability and approximation of generalized reachable sets. Sov. Phys. Dokl. **30**, 66–69 (1985)
21. Lotov, A.V., Miettinen, K.: Visualizing the Pareto Frontier. In: Multiobjective Optimization: Interactive and Evolutionary Approaches, pp. 213–244. Springer, Berlin (2008)
22. Lotov, A.V., Ryabikov, A.I.: Launch pad method in multiextremal multiobjective optimization problems. Comput. Math. Math. Phys. **59**(12), 2041–2056 (2019)
23. Lotov, A.V., Bushenkov, V.A., Kamenev, G.K.: Interactive Decision Maps: Approximation and Visualization of Pareto Frontier. Kluwer, Alphen aan den Rijn (2004)
24. Lotov, A.V., Kamenev, G.K., Berezkin, V.E., Miettinen, K.: Optimal control of cooling process in continuous casting of steel using a visualization-based multi-criteria approach. Appl. Math. Model. **29**(7), 653–672 (2005)
25. Lotov, A.V., Ryabikov, A.I., Bolgov, M.V., Buber, A.L.: Application of the Pareto Front to Finding Compromise Rules for Lake Baikal Water Level Management. Sci. Tech. Inf. Process. **50**(6), 582–594 (2023)
26. Marques, S., Bushenkov, V.A., Lotov, A.V., Marto, M., Borges, J.G.: Bi-level participatory forest management planning supported by Pareto Frontier visualization. Forest Sci. **66**(4), 490–500 (2020)
27. McClure, D.E., Vitale, R.: Polygonal approximation of plane convex bodies. J. Math. Analys Appl. **51**(2), 326–358 (1975)
28. Miettinen, K.M.: Nonlinear Multiobjective Optimization. Kluwer, Alphen aan den Rijn (1999)
29. Pardalos, P.M., Zilinskas, A., Zilinskas, J.: Non-convex Multi-objective Optimization. Springer, Berlin (2018)

30. Preparata, F.P., Shamos, M.I.: Computational Geometry: An Introduction. Springer, Berlin (1985)
31. Roy, B.: Decisions avec criteres multiples. Metra Int. **11**(1), 121–151 (1972)
32. Sawaragi, Y., Nakayama, H., Tanino, T.: Theory of Multiobjective Optimization. Academic Press, Cambridge (1985)
33. Schneider, R.: Polyhedral approximation of smooth convex bodies. J. Math. Anal. Appl. **128**(2), 470–474 (1987)
34. Solanki, R.S., Appino, P.A., Cohon, J.L.: Approximating the noninferior set in multiobjective linear programming problems. Eur. J. Oper. Res. **68**, 356–373 (1993)
35. Sonnevend, G.: An optimal sequential algorithm for uniform approximation of convex functions on $[0, 1]^2$. Appl. Math. **23**(10), 127–142 (1983)
36. Steuer, R.E.: Multiple Criteria Optimization: Theory, Computation, and Application. Wiley, Hoboken (1986)
37. Weyl, H.: Elementare Theorie der konvexen Polyeder. Comm. Math. Helv. **6**, 290–306 (1935)

The (In)Famous Big-M Technique for Solving Bilevel Programs

Saeed Mohammadi, Mohammad Reza Hesamzadeh, and Dina Khastieva

1 Introduction

The single-level reformulation approach for solving linear bilevel programs (linear BLPs) requires finding disjunctive parameters (big-M parameters) that do not cut off any bilevel-optimal solution. Heuristic techniques are often used to find the big-M parameters [3] although it is well-known that these techniques might fail. Recent papers [12, 21, 28, 29], and [20, 26, 30], and [27] show the computational challenges of finding the correct big-M parameters (and solving BLPs) which is shown to be strongly NP-hard [15]. This in turn makes finding the global optimal solution of linear BLPs quite challenging. Accordingly, we propose a disjunctive-based decomposition (DBD) solution algorithm for solving the linear BLPs which (1) does not require finding big-M parameters, and thus (2) it is computationally advantageous. Accordingly, using our proposed DBD algorithm, no heuristic technique is needed to find big-M parameters, and the obtained solution is guaranteed to be optimal to the given linear BLP under reasonable assumptions. Table 1 compares our proposed DBD algorithm with the relevant solution algorithms in the literature.

The rest of this chapter is organized as follows: Sect. 2 explains the proposed DBD algorithm to solve a general linear BLP. Section 3 gives an illustrative example. Section 4 presents several general linear BLPs, and Sect. 5 contains large-scale case studies. Section 9 concludes the chapter.

S. Mohammadi (✉) · M. R. Hesamzadeh · D. Khastieva
School of Electrical Engineering and Computer Science, KTH Royal Institute of Technology, Stockholm, Sweden
e-mail: saeedmoh@kth.se; mrhesa@kth.se; dinak@kth.se

B. Goldengorin (ed.), *Theory, Algorithms, and Experiments in Applied Optimization*, Springer Optimization and Its Applications 226,
https://doi.org/10.1007/978-3-031-91357-0_11

Table 1 A comparative study based on the existing literature

Solution algorithm	[21]	[29]	[11]	[12]	[30]	DBD
Require tuning disjunctive parameters	No	No	Yes	Yes	Yes	No
Find global optimal solution	No	No	No	No	Yes	Yes

2 The Proposed Disjunctive-Based Decomposition (DBD) Algorithm

The general linear BLP is given in (1) with matrices $A \in \mathbb{R}^{(n_x \times 1)}$, $B, G \in \mathbb{R}^{(n_y \times 1)}$, $C \in \mathbb{R}^{(n_u \times n_x)}$, $D \in \mathbb{R}^{(n_u \times n_y)}$, $E \in \mathbb{R}^{(n_u \times 1)}$, $H \in \mathbb{R}^{(n_l \times n_x)}$, $J \in \mathbb{R}^{(n_l \times n_y)}$, and $N \in \mathbb{R}^{(n_l \times 1)}$:

$$\underset{\mathbf{x}}{\text{Minimize}} \quad f(\mathbf{x}, \mathbf{y}) = A^{\mathsf{T}}\mathbf{x} + B^{\mathsf{T}}\mathbf{y} \tag{1a}$$

$$\text{Subject to: } C\mathbf{x} + D\mathbf{y} \leq E \tag{1b}$$

$$\mathbf{x} \geq 0 \tag{1c}$$

$$\mathbf{x} \in \mathbb{X} \tag{1d}$$

$$\mathbf{y} \in S_1(\mathbf{x}) \tag{1e}$$

$$S_1(\mathbf{x}) := \arg\min_{\mathbf{y}} g(\mathbf{x}, \mathbf{y}) = G^{\mathsf{T}}\mathbf{y} \tag{1f}$$

$$\text{Subject to: } H\mathbf{x} + J\mathbf{y} \leq N : \boldsymbol{\lambda} \tag{1g}$$

$$\mathbf{y} \geq 0 \tag{1h}$$

$$\mathbf{y} \in \mathbb{Y} \tag{1i}$$

The sets $\mathbb{X} \subseteq \mathbb{R}^{(n_x \times 1)}$ and $\mathbb{Y} \subseteq \mathbb{R}^{(n_y \times 1)}$ are polyhedral sets. Since the lower-level problem ($\overset{\min}{\mathbf{y}}\{G^{\mathsf{T}}\mathbf{y} | H\mathbf{x} + J\mathbf{y} \leq N, \mathbf{y} \geq 0, \mathbf{y} \in \mathbb{Y}\}$) is a linear programming (LP) problem, the Karush-Kuhn-Tucker (KKT) conditions are necessary and sufficient, and the linear BLP (1) can be reformulated as the following single-level equivalent model SL:

$$\mathbf{SL} := \underset{\mathbf{x},\mathbf{y},\boldsymbol{\lambda}}{\text{Minimize}} \quad A^{\mathsf{T}}\mathbf{x} + B^{\mathsf{T}}\mathbf{y} \tag{2a}$$

$$\text{Subject to: } C\mathbf{x} + D\mathbf{y} \leq E \tag{2b}$$

$$0 \leq (N - H\mathbf{x} - J\mathbf{y}) \perp \boldsymbol{\lambda} \geq 0 \tag{2c}$$

$$0 \leq \left(G + J^{\mathsf{T}}\boldsymbol{\lambda}\right) \perp \mathbf{y} \geq 0 \tag{2d}$$

$$\mathbf{x} \geq 0 \tag{2e}$$

$$\boldsymbol{\lambda} \in \mathbb{R}^{(n_l \times 1)} \tag{2f}$$

$\boldsymbol{\lambda} \in \mathbb{R}^{(n_l \times 1)}$ is the vector of Lagrange multipliers for $H\mathbf{x} + J\mathbf{y} \leq N$. Accordingly, if $(\mathbf{x}, \mathbf{y}, \boldsymbol{\lambda})$ is a global optimal solution of SL problem, then $(\mathbf{x}, \mathbf{y})$ is a global optimal solution of the original linear BLP (1). We assume that the upper- and lower-level variables $\mathbf{x}$ and $\mathbf{y}$ are bounded, consistent with practical real-life problems, and that disjunctive parameters $\{M_1, M_2, M_3, M_4\}$ for the lower-level complementarity conditions exist and they are sufficiently large.

Definition 1 A set of positive scalars $\{M_1, M_2, M_3, M_4\}$ is **LP-correct** if the vector of solutions $\mathbf{y}$ to the lower-level problem and vector of solutions $\boldsymbol{\lambda}$ to its dual problem are the same as vectors of solutions $\{\mathbf{y}, \boldsymbol{\lambda}\}$ to problem (3e)-(3h).

The complementary slackness condition in SL (designated by $\perp$) have disjunctive property. Therefore, it can be linearized with the big-M technique [17] with LP-correct big-M parameters $\{M_1, M_2, M_3, M_4\}$: This results an equivalent problem for SL formulated in (3).

$$\underset{\mathbf{x},\mathbf{y},\boldsymbol{\lambda},\mathbf{u}_1,\mathbf{u}_2}{\text{Minimize}} \quad f(\mathbf{x}, \mathbf{y}) = A^\mathsf{T}\mathbf{x} + B^\mathsf{T}\mathbf{y} \tag{3a}$$

$$\text{Subject to: } C\mathbf{x} + D\mathbf{y} \leq E : \boldsymbol{\mu}_1 \tag{3b}$$

$$H\mathbf{x} + J\mathbf{y} \leq N : \boldsymbol{\mu}_2 \tag{3c}$$

$$-J^\mathsf{T}\boldsymbol{\lambda} \leq G : \boldsymbol{\mu}_3 \tag{3d}$$

$$N - H\mathbf{x} - J\mathbf{y} \leq M_1(\mathbf{e}_1 - \mathbf{u}_1) : \boldsymbol{\nu}_1 \tag{3e}$$

$$\boldsymbol{\lambda} \leq M_2\mathbf{u}_1 : \boldsymbol{\gamma}_1 \tag{3f}$$

$$G + J^\mathsf{T}\boldsymbol{\lambda} \leq M_3(\mathbf{e}_2 - \mathbf{u}_2) : \boldsymbol{\nu}_2 \tag{3g}$$

$$\mathbf{y} \leq M_4\mathbf{u}_2 : \boldsymbol{\gamma}_2 \tag{3h}$$

$$\mathbf{x} \in \mathbb{X}, \mathbf{y} \in \mathbb{Y} \tag{3i}$$

$$\mathbf{x} \geq 0, \mathbf{y} \geq 0, \boldsymbol{\lambda} \geq 0 \tag{3j}$$

where $\mathbf{u}_1 \in \{0, 1\}^{(n_l \times 1)}$ and $\mathbf{u}_2 \in \{0, 1\}^{(n_y \times 1)}$ are vectors of binary variables, the parameters $\{M_1, \ldots, M_4\}$ are LP-correct, and $\mathbf{e}_1 \in \mathbb{R}^{(n_l \times 1)}$ and $\mathbf{e}_2 \in \mathbb{R}^{(n_y \times 1)}$ are vectors of ones. In the proposed DBD approach, (3) is replaced by a subproblem (SP) in (5) and a master problem (MP) in (6) where MP and SP do not depend on any disjunctive parameter (big-M). In effect, the role of SP is to find an optimal solution for a given $\mathbf{u}$. The goal of MP is to find an optimal value for a relaxation of (3) with feasibility/optimality cuts added to MP iteratively. The procedure then alternates between SP which provides an upper bound (UB) for the optimal objective value to (3) and MP which provides a lower bound (LB) for that, in order to solve (3). From this perspective, our approach has some similarities to the Benders decomposition [6].

Next, we fix the binary variables $\mathbf{u}_1 = \hat{\mathbf{u}}_1$ and $\mathbf{u}_2 = \hat{\mathbf{u}}_2$ so that (3) is an LP with its dual program as (4):

$$\mathbf{DP} := \underset{\mu_1,\mu_2,\nu_1,\nu_2,\gamma_1,\gamma_2}{\text{Maximize}} \; (-\mu_1^\top E - \mu_2^\top N - \mu_3^\top G + \nu_1^\top (N - M_1(\mathbf{e}_1 - \hat{\mathbf{u}}_1)) - \gamma_1^\top M_2 \hat{\mathbf{u}}_1 + \nu_2^\top (G - M_3(\mathbf{e}_2 - \hat{\mathbf{u}}_2)) - \gamma_2^\top M_4 \hat{\mathbf{u}}_2) \tag{4a}$$

$$\text{Subject to:} \quad -\mu_1^\top C + \left(\nu_1^\top - \mu_2^\top\right) H \le A^\top : \mathbf{x} \tag{4b}$$

$$-\mu_1^\top D + (\nu_1^\top - \mu_2^\top) J - \gamma_2^\top \le B^\top : \mathbf{y} \tag{4c}$$

$$-\mu_3^\top J^\top - \gamma_1^\top - \nu_2^\top J^\top \le 0 : \lambda \tag{4d}$$

$$\mu_1, \mu_2, \nu_1, \nu_2, \gamma_1, \gamma_2 \ge 0 \tag{4e}$$

The DP is equivalent to the SP below as it is shown in Proposition 1. The SP is solved in Step 1 of the proposed DBD algorithm in Fig. 1. It shows that the dependence on the disjunctive parameters can be dropped under certain circumstances (Proposition 1):

$$\mathbf{SP}: \quad \underset{\mu_1,\mu_2,\nu_1,\nu_2,\gamma_1,\gamma_2}{\text{Maximize}} \; h(\mu, \nu) \tag{5a}$$

$$\text{Subject to:} \quad -\mu_1^\top C + \left(\nu_1^\top - \mu_2^\top\right) H \le A^\top : \mathbf{x} \tag{5b}$$

$$-\mu_1^\top D + (\nu_1^\top - \mu_2^\top) J - \gamma_2^\top \le B^\top : \mathbf{y} \tag{5c}$$

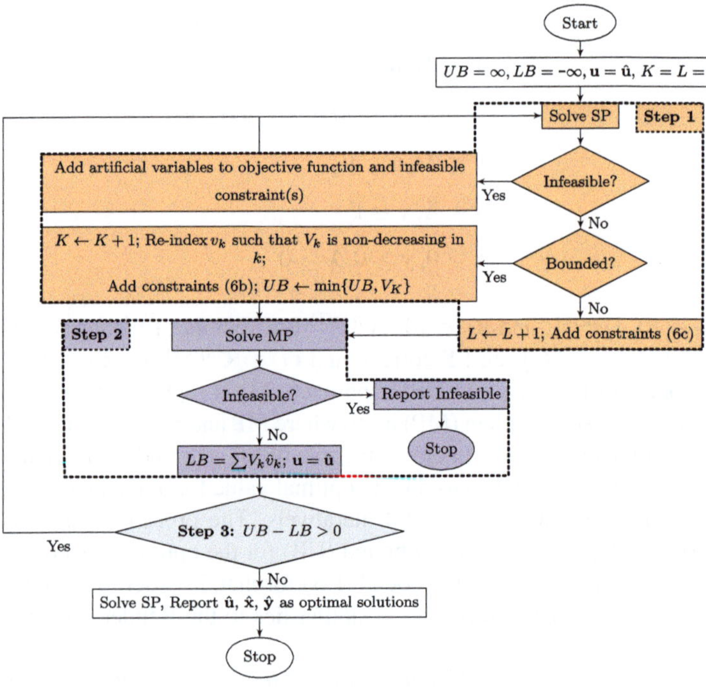

Fig. 1 The proposed DBD algorithm

$$-\boldsymbol{\mu}_3^{\mathsf{T}} J^{\mathsf{T}} - \boldsymbol{\gamma}_1^{\mathsf{T}} - \boldsymbol{\nu}_2^{\mathsf{T}} J^{\mathsf{T}} \leq 0 : \boldsymbol{\lambda} \tag{5d}$$

$$\boldsymbol{\nu}_1^{\mathsf{T}}(\mathbf{e}_1 - \hat{\mathbf{u}}_1) = \boldsymbol{\nu}_2^{\mathsf{T}}(\mathbf{e}_2 - \hat{\mathbf{u}}_2) = 0 \tag{5e}$$

$$\boldsymbol{\gamma}_1^{\mathsf{T}}\hat{\mathbf{u}}_1 = \boldsymbol{\gamma}_2^{\mathsf{T}}\hat{\mathbf{u}}_2 = 0 \tag{5f}$$

$$\boldsymbol{\mu}_1, \boldsymbol{\mu}_2, \boldsymbol{\nu}_1, \boldsymbol{\nu}_2, \boldsymbol{\gamma}_1, \boldsymbol{\gamma}_2 \geq 0 \tag{5g}$$

Proposition 1 *Suppose LP-correct parameters* $\{M_1, M_2, M_3, M_4\}$ *exist. Furthermore, assume that these parameters are large enough so that (i)* $(N - H\mathbf{x} - J\mathbf{y})_j < M_1, \forall \mathbf{x}, \mathbf{y}, j$, *and (ii)* $(G + J^{\mathsf{T}}\boldsymbol{\lambda})_j < M_3, \forall \boldsymbol{\lambda}, j$ *(element* j *of each vector). Then, the terms* $\boldsymbol{\nu}_1^{\mathsf{T}}(\mathbf{e}_1 - \hat{\mathbf{u}}_1)$, $\boldsymbol{\gamma}_1^{\mathsf{T}}\hat{\mathbf{u}}_1$, $\boldsymbol{\nu}_2^{\mathsf{T}}(\mathbf{e}_2 - \hat{\mathbf{u}}_2)$, *and* $\boldsymbol{\gamma}_2^{\mathsf{T}}\hat{\mathbf{u}}_2$ *in the objective function of the DP are zero at an optimal point of* (3). *This means that the DP and the SP have the same solution set.*

Proof If the binary variable $(\hat{\mathbf{u}}_1)_j = 0$, then row j of constraints (3e) $\big((N - H\mathbf{x} - J\mathbf{y})_j \leq M_1\big)$ is not binding, and the corresponding Lagrange multiplier $(\boldsymbol{\nu}_1)_j = 0$. Accordingly, the term $(\boldsymbol{\nu}_1)_j(\mathbf{e}_1 - \hat{\mathbf{u}}_1)_j$ is zero. Otherwise, the binary variable is $(\hat{\mathbf{u}}_1)_j = 1$, and it is clear that $(\boldsymbol{\nu}_1)_j(\mathbf{e}_1 - \hat{\mathbf{u}}_1)_j$ is zero. Therefore, we have $\boldsymbol{\nu}_1^{\mathsf{T}}(\mathbf{e}_1 - \hat{\mathbf{u}}_1) = 0$. The same arguments are valid for the other three terms in Proposition 1. Hence, these terms can be removed from the objective function of DP and represented as constraints (5e)–(5f) in the SP. □

In the case of infeasibility of the SP, an alternative always-feasible version is formulated and solved employing artificial variables in the objective function and infeasible constraint(s) [11]. Observe that SP provides an upper bound to the optimal solution of the mixed-integer LP (MILP) problem in (3) and it does not have any big-M parameters.

In Step 2, the MP (6) is solved using the optimal values (V_k) of the SP and extreme points $k \in \{0, \ldots, K\}$ and extreme rays $l \in \{1, \ldots, L\}$ found by solving the SP in previous iterations. If the SP has an extreme point, then index of the binary variables corresponding to $\boldsymbol{\nu}_1 > 0$ and $\boldsymbol{\nu}_2 > 0$ are stored in the set Ω_k, and the set Ω'_k stores index of the binary variables corresponding to $\boldsymbol{\gamma}_1 > 0$ and $\boldsymbol{\gamma}_2 > 0$. If the SP has an extreme ray, then the set Ψ_l stores indices of the binary variables corresponding to $\boldsymbol{\nu}_1 > 0$ and $\boldsymbol{\nu}_2 > 0$, and the set Ψ'_l stores indices of the binary variable corresponding to $\boldsymbol{\gamma}_1 > 0$ and $\boldsymbol{\gamma}_2 > 0$. $|\Omega_k|$, $|\Omega'_k|$, $|\Psi'_l|$, and $|\Psi_l|$ are the cardinality of these sets, and (6e) is used to enforce the tree search algorithm. Lastly, in the objective function (6a), V_ks should be reordered such that V_k is nondecreasing in k:

$$\textbf{MP} : \underset{u_i, v_k}{\text{Minimize}} \quad \sum_{k=0}^{K} V_k v_k \tag{6a}$$

$$\text{Subject to: } \sum_{i \in \Omega_k} u_i + \sum_{i \in \Omega'_k} u_i \leq |\Omega_k| + |\Omega'_k| - 1 + \sum_{k' \geq k} v_{k'}; \quad \forall k \geq 1 \tag{6b}$$

$$\sum_{i\in\Psi_l} u_i + \sum_{i\in\Psi'_l}(1-u_i) \le |\Psi_l| + |\Psi'_l| - 1;\ \ \forall l \ge 1 \tag{6c}$$

$$\sum_{k=0}^{K} v_k = 1 \tag{6d}$$

$$u_i, v_k \in \{0, 1\} \tag{6e}$$

As it is proved in [5], the MP provides a lower bound for the optimal solution of the MILP problem (3). In Step 3, if $UB - LB = 0$, then the DBD algorithm terminates. Otherwise, the next iteration starts from the new values obtained in Step 2. Proposed DBD algorithm is detailed in Fig. 1 and applied to a simple example in Sect. 3 to clearly explain the process.

3 Illustrative Example

In this section, the proposed DBD algorithm for illustrative example is implemented in Julia 1.4.0 [7], and LP problems are solved with GNU Linear Programming Kit (GLPK) 0.12.1 [23]. Detailed operation of our proposed DBD algorithm is shown using the illustrative linear BLP (7) below:

$$\underset{x\in\mathbb{R}^+}{\text{Minimize}}\quad f(\mathbf{x},\mathbf{y}) = 0.01x - y \tag{7a}$$

$$\text{Subject to: } 0 \le x \le 1 \tag{7b}$$

$$y \in S_2(x) \tag{7c}$$

$$S_2(x) := \underset{y\in\mathbb{R}^+}{\arg\min}\quad g(\mathbf{x},\mathbf{y}) = y \tag{7d}$$

$$\text{Subject to: } 0.01y - x \ge -0.5 \tag{7e}$$

$$y + x \ge 1 \tag{7f}$$

λ_1 and λ_2 are Lagrange variables of constraints $0.01y - x \ge -0.5$ and $y + x \ge 1$, respectively. The single-level MILP reformulation of (7) is derived in (8):

$$\underset{x,y,\lambda_1,\lambda_2,u_1,u_2,u_3}{\text{Minimize}}\quad f(\mathbf{x},\mathbf{y}) = 0.01x - y \tag{8a}$$

$$\text{Subject to: } -x \ge -1 : \mu_1 \tag{8b}$$

$$1 - 0.01\lambda_1 - \lambda_2 \ge 0 : \mu_2 \tag{8c}$$

$$0.01y - x \ge -0.5 : \mu_3 \tag{8d}$$

$$y + x \ge 1 : \mu_4 \tag{8e}$$

$$x - 0.01y \geq 0.5 - M_1(1 - u_1) : \nu_1 \tag{8f}$$

$$-x - y \geq -1 - M_2(1 - u_2) : \nu_2 \tag{8g}$$

$$0.01\lambda_1 + \lambda_2 \geq 1 - M_5(1 - u_3) : \nu_3 \tag{8h}$$

$$-\lambda_1 \geq -M_3 u_1 : \gamma_1 \tag{8i}$$

$$-\lambda_2 \geq -M_4 u_2 : \gamma_2 \tag{8j}$$

$$-y \geq -M_6 u_3 : \gamma_3 \tag{8k}$$

$$x, y, \lambda_1, \lambda_2 \in \mathbb{R}^+ \tag{8l}$$

$$u_1, u_2, u_3 \in \{0, 1\} \tag{8m}$$

The SP for a given $\hat{\mathbf{u}} = [\hat{u}_1, \hat{u}_2, \hat{u}_3]^\intercal$ is derived in SP-1 below where $\boldsymbol{\mu} = [\mu_1, \mu_2, \mu_3, \mu_4]^\intercal$, $\boldsymbol{\nu} = [\nu_1, \nu_2, \nu_3]^\intercal$, and $\boldsymbol{\gamma} = [\gamma_1, \gamma_2, \gamma_3]^\intercal$ are vectors of Lagrange multipliers:

$$\textbf{SP-1} : \underset{\boldsymbol{\mu},\boldsymbol{\nu},\boldsymbol{\gamma}}{\text{Maximize}} \quad -\mu_1 - \mu_2 - 0.5\mu_3 + \mu_4 + 0.5\nu_1 - \nu_2 + \nu_3 \tag{9a}$$

$$\text{Subject to: } -\mu_1 - \mu_3 + \mu_4 + \nu_1 - \nu_2 \leq 0.01 : x \tag{9b}$$

$$0.01\mu_3 + \mu_4 - 0.01\nu_1 - \nu_2 - \gamma_3 \leq -1 : y \tag{9c}$$

$$-0.01\mu_2 + 0.01\nu_3 - \gamma_1 \leq 0 : \lambda_1 \tag{9d}$$

$$-\mu_2 + \nu_3 - \gamma_2 \leq 0 : \lambda_2 \tag{9e}$$

$$\nu_1(1 - \hat{u}_1) = \nu_2(1 - \hat{u}_2) = \nu_3(1 - \hat{u}_3) = 0 \tag{9f}$$

$$\gamma_1\hat{u}_1 = \gamma_2\hat{u}_2 = \gamma_3\hat{u}_3 = 0 \tag{9g}$$

$$\mu_1, \mu_2, \mu_3, \mu_4, \nu_1, \nu_2, \nu_3, \gamma_1, \gamma_2, \gamma_3 \in \mathbb{R}^+ \tag{9h}$$

SP-1 is solved for all combinations of $\hat{u}_1$ to $\hat{u}_3$ as shown in Table 2 only for reference and the final result is shown in bold. The linear BLP problem (7) is solved in four iterations as follows:

Initialization Initialize the proposed DBD algorithm with $UB = 10^4$, $LB = -10^4$, $\hat{\mathbf{u}} = [0, 0, 0]^\intercal$, $K = L = 0$, and $V_0 = -10^4$.

Iteration 1 The SP-1 is solved in Step 1 which is unbounded. This results in $L \leftarrow L + 1 = 1$ and $\gamma_3 = 1.01 > 0$ from row 1 of Table 2. Index sets are $\Psi_L = \{\}$ and $\Psi'_L = \{3\}$ with cardinalities $|\Psi_L| = 0$ and $|\Psi'_L| = 1$. In Step 2 employing (6c), we generate the constraint $1 \leq u_3$ and formulate the MP-1 in (10):

$$\textbf{MP-1} : \underset{u_3, v_0}{\text{Minimize}} \quad V_0 v_0 \tag{10a}$$

$$\text{Subject to: } 1 \leq u_3 \tag{10b}$$

Table 2 SP-1 solutions for all combinations of $\{\hat{u}_1, \hat{u}_2, \hat{u}_3\}$

Row	$\hat{u}_1$	$\hat{u}_2$	$\hat{u}_3$	$\hat{\mu}_1$	$\hat{\mu}_2$	$\hat{\mu}_3$	$\hat{\mu}_4$	$\hat{\nu}_1$	$\hat{\nu}_2$	$\hat{\nu}_3$	$\hat{\gamma}_1$	$\hat{\gamma}_2$	$\hat{\gamma}_3$	$\hat{V}_K$	Status
1	0	0	0	0	0	0	0.01	0	0	0	0	0	1.01	$-\infty$	UNB
2	0	0	1	×	×	×	×	×	×	×	×	×	×	×	INF
3	0	1	0	0	0	0	0.01	0	0	0	0	0	1.01	$-\infty$	UNB
4	0	1	1	0	0	0	0	0	1	0	0	0	0	−1	OPT
5	1	0	0	0	0	0	0.01	0	0	0	0	0	1.01	$-\infty$	UNB
6	1	0	1	99.99	0	0	0	100	0	0	0	0	0	**−49.99**	OPT
7	1	1	0	0	0	0	0.01	0	0	0	0	0	1.01	$-\infty$	UNB
8	1	1	1	0	0	0	0	1	0.99	0	0	0	0	−0.49	OPT

UNB, unbounded; INF, infeasible; OPT, optimal

$$v_0 = 1 \tag{10c}$$

$$u_3, v_0 \in \{0, 1\} \tag{10d}$$

u_1 and u_2 are free and initialized by zero. Optimal solutions are $v_0 = 1$ and $\hat{\mathbf{u}} = [0, 0, 1]^\intercal$. Step 3 is to go to the next iteration since $LB = V_0 v_0 = -10^4$ and $UB - LB > 0$.

Iteration 2 The problem SP-1 is infeasible with $\hat{\mathbf{u}} = [0, 0, 1]^\intercal$. Therefore, we add the term $(-s - t)$ to the objective function (9a) and change the infeasible constraint (9c) to $0.01\mu_3 + \mu_4 - 0.01\nu_1 - \nu_2 - \gamma_3 + s - t \leq -1$. In this case, SP-1 becomes feasible but unbounded. This results in $L \leftarrow L+1 = 2$, $\nu_3 = 100 > 0$, $\gamma_1 = 1 > 0$, and $\gamma_2 = 100 > 0$. Index sets are $\Psi_L = \{3\}$ and $\Psi'_L = \{1, 2\}$ with cardinalities $|\Psi'_L| = 2$ and $|\Psi_L| = 1$. In Step 2 employing (6c), the constraint $u_3 \leq u_1 + u_2$ is generated in MP-2:

$$\textbf{MP-2}: \underset{u_1,u_2,u_3,v_0}{\text{Minimize}} \quad V_0 v_0 \tag{11a}$$

$$\text{Subject to: } 1 \leq u_3 \tag{11b}$$

$$v_0 = 1 \tag{11c}$$

$$u_3 \leq u_1 + u_2 \tag{11d}$$

$$u_1, u_2, u_3, v_0 \in \{0, 1\} \tag{11e}$$

Optimal solutions are $v_0 = 1$ and $\hat{\mathbf{u}} = [1, 0, 1]^\intercal$. Then, Step 3 is to go to the next iteration since $LB = V_0 v_0 = -10^4$ and $UB - LB > 0$.

Iteration 3 First, we solve the SP-1 with $\hat{\mathbf{u}} = [1, 0, 1]^\intercal$. It is bounded where $\nu_1 = 100 > 0$, $K \leftarrow K + 1 = 1$, and $V_K = -49.99$. Then, the UB is updated to $UB \leftarrow \min\{UB = 10^4, V_K = -49.499\} = -49.99$. Index sets are $\Omega_K = \{1\}$ and $\Omega'_K = \{\}$ with cardinalities $|\Omega_K| = 1$ and $|\Omega'_K| = 0$. In Step 2 employing (6b), the constraint $u_1 \leq v_1$ is formulated in MP-3. Indices are not changed since V_ks are already nondecreasing in k ($V_0 \leq V_1$):

$$\textbf{MP-3}: \underset{u_1,u_2,u_3,v_0,v_1}{\text{Minimize}} \quad V_0v_0 - 49.99v_1 \tag{12a}$$

$$\text{Subject to: } v_0 + v_1 = 1 \tag{12b}$$

$$1 \leq u_3 \leq u_1 + u_2 \tag{12c}$$

$$u_1 \leq v_1 \tag{12d}$$

$$u_1, u_2, u_3, v_0, v_1 \in \{0, 1\} \tag{12e}$$

Optimal solutions are $v_0 = 1$, $v_1 = 0$, and $\hat{\mathbf{u}} = [0, 1, 1]^\mathsf{T}$ which result in $LB = V_0v_0 + V_1v_1 = -10^4$. We go to the next iteration since $UB - LB > 0$.

Iteration 4 The SP-1 is solved with $\hat{\mathbf{u}} = [0, 1, 1]^\mathsf{T}$ which is bounded so $K \leftarrow K + 1 = 2$, $V_K = -1$, $\nu_2 = 1 > 0$, and $UB \leftarrow \min\{UB = -49.99, V_K = -1\} = -49.99$. The index sets are $\Omega_K = \{2\}$ and $\Omega'_K = \{\}$ with cardinalities $|\Omega_K| = 1$ and $|\Omega'_K| = 0$. The constraint $u_2 \leq v_2$ is generated using (6b). V_ks are nondecreasing in k which results in MP-4:

$$\textbf{MP-4}: \underset{u_1,u_2,u_3,v_0,v_1,v_2}{\text{Minimize}} \quad V_0v_0 - 49.99v_1 - v_2 \tag{13a}$$

$$\text{Subject to: } v_0 + v_1 = 1 \tag{13b}$$

$$1 \leq u_3 \leq u_1 + u_2 \tag{13c}$$

$$u_1 \leq v_1 + v_2 \tag{13d}$$

$$u_2 \leq v_2 \tag{13e}$$

$$u_1, u_2, u_3, v_0, v_1, v_2 \in \{0, 1\} \tag{13f}$$

Optimal solutions are $v_0 = v_2 = 0$, $v_1 = 1$, and $\hat{\mathbf{u}} = [1, 0, 1]^\mathsf{T}$. This results in $LB = -49.99$ and $UB - LB = 0$. Therefore, there is no need for more iterations and $\mathbf{u} = [1, 0, 1]^\mathsf{T}$ is optimal. The optimal solutions of MILP problem (8) (or the bilevel-optimal solutions of linear BLP (7)) are $\hat{\mathbf{u}} = [1, 0, 1]^\mathsf{T}$, $x = 1$, and $y = 50$. Note that in iterations 1 to 4 we did not have to find the LP-correct big-M parameters at all.

4 General Case Studies

Numerical results for eight general case studies are presented in Tables 3, 4, 5, and 6. Number of variables and constraints in the upper- and lower-level problems is shown in Table 3. Values of the parameter matrices A to N are available in the corresponding references.

Table 3 General case studies

Row	Problem	n_x	n_y	n_u	n_l	Reference
1	*bard*1	2	3	2	6	[4]
2	*bard*2	2	2	2	5	[4]
3	*candler*	2	6	2	6	[8]
4	*anan*	1	1	1	6	[1]
5	*bard*3	1	2	1	5	[2]
6	*bard*4	1	2	1	5	[3]
7	*clark*	1	2	1	5	[9]
8	*colson*	2	3	1	3	[10]

n_x, number of variables in the upper-level problem; n_y, number of variables in the lower-level problem; n_u, number of constraints in the upper-level problem; n_l, number of constraints in the lower-level problem

Table 4 Solutions

Row	Problem	$\hat{f}(\mathbf{x}, \mathbf{y})$	$\hat{g}(\mathbf{x}, \mathbf{y})$	$\hat{\mathbf{x}}$	$\hat{\mathbf{y}}$
1	*bard*1	−26	3.2	[0,0.9]	[0,0.6,0.4]
2	*bard*2	−3.25	−4	[2,0]	[1.5,0]
3	*candler*	−29.2	3.2	[0,0.9]	[0,0.6,0.4,0,0,0]
4	*anan*	−49	17	[16]	[11]
5	*bard*3	−1	0	[1]	[0,0]
6	*bard*4	−2	−1	[0]	[0,1]
7	*clark*	−13	−4	[5]	[4,2]
8	*colson*	−14.6	0.3	[0,0.65]	[0,0.3,0]

Table 5 Solution time (wall time) and number of iterations for different initial points $\hat{\mathbf{u}}$

		Iterations				Time (seconds)			
Row	Problem	Min	Median	Mean	Max	Min	Median	Mean	Max
1	*bard*1	7	8	8.7	12	0.67	0.79	0.84	1.13
2	*bard*2	1	6	7.2	11	0.18	0.62	0.73	1.1
3	*candler*	7	11	10.99	20	0.73	1.08	1.1	1.91
4	*anan*	1	5	5.46	8	0.18	0.57	0.61	0.91
5	*bard*3	1	3	2.65	4	0.19	0.37	0.35	0.51
6	*bard*4	1	6	5.72	8	0.17	0.6	0.58	0.84
7	*clark*	1	12	10.78	14	0.17	1.15	1.06	1.45
8	*colson*	1	4	3.79	7	0.17	0.43	0.42	0.77

These problems are solved with both the big-M algorithm and our proposed DBD algorithm. Optimal values of objective functions ($\hat{f}(\mathbf{x}, \mathbf{y})$ and $\hat{g}(\mathbf{x}, \mathbf{y})$) and optimal solutions of the upper- and lower-level decision variables ($\hat{\mathbf{x}}$ and $\hat{\mathbf{y}}$) are shown in Table 4. As we can see for these small case studies, both the big-M algorithm and our proposed DBD algorithm achieve the same results.

Table 6 Number of iterations in the big-M algorithm and the proposed DBD algorithm

		Big-M algorithm			Tuned big-M		DBD		
Row	Problem	Big-M	Iterations	Solution	Big-M	Iterations	Iterations	%	$\hat{f}(\mathbf{x}, \mathbf{y})$
1	*bard*1	5	Inf.	Inf.	10	**25**	**8**	**68**	− 26
2	*bard*2	6	13	1.75	10	**15**	**6**	**60**	− 3.25
3	*candler*	5	28	− 23	20	**29**	**11**	**62**	− 29.2
4	*anan*	10	Inf.	Inf.	50	**21**	**5**	**76**	− 49
5	*bard*3	5	Inf.	Inf.	10	**14**	**3**	**79**	− 1
6	*bard*4	5	Inf.	Inf.	10	**15**	**6**	**60**	− 2
7	*clark*	10	31	− 11	100	**10**	**12**	**−20**	−13
8	*colson*	10	59	− 14.6	20	**57**	**4**	**93**	−14.6

Inf., Infeasible

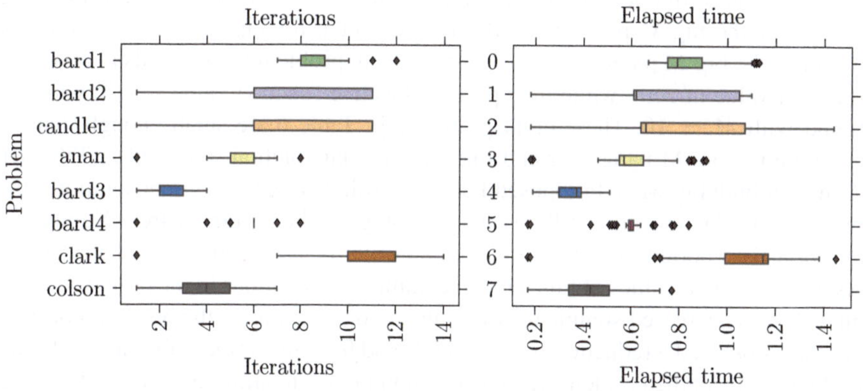

Fig. 2 Distribution of solution time and number of iterations for different initial points

The vector of binary variables $\mathbf{u}$ is initialized using an arbitrary binary vector in the proposed DBD algorithm in Fig. 1. To study sensitivity of the proposed DBD algorithm to the initial binary vector $\hat{\mathbf{u}}$, eight problems listed in Table 3 are solved with different initial binary vectors. This requires solving each problem up to $2^{(n_y+n_l)}$ times. For instance, the problem *candler* should be solved up to $2^{6+6} = 4{,}096$ times. To compare results of different problems, 100 random initial binary vectors are selected for each problem. Number of iterations in the proposed DBD algorithm for different initial points $\hat{\mathbf{u}}$ is demonstrated in Table 5 and Fig. 2. These calculations are performed on a computer with 2.4 GHz 8-core processor and memory of 32 GB. Solution time is elapsed real time or wall time in seconds.

Number of iterations in the big-M algorithm and the proposed DBD algorithm are shown in Table 6 for problems *bard*1 to *colson*. The big-M algorithm requires tuning the big-M parameter for each problem. We observe two issues in tuning

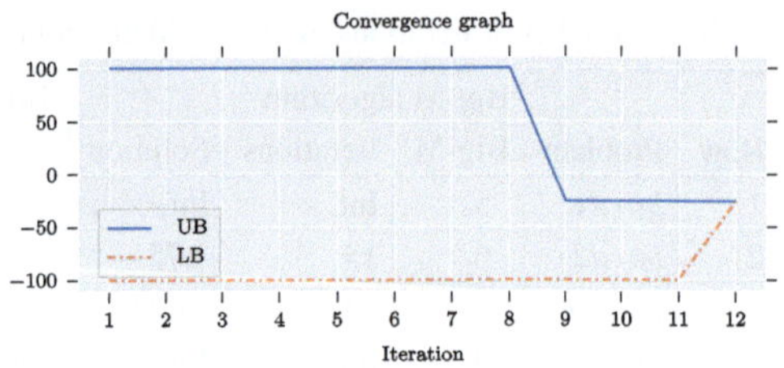

Fig. 3 The convergence graph of the proposed DBD algorithm

the big-M parameters: First, some problems may become infeasible for some big-M parameters. For example, problems $bard1$, $anan$, $bard3$, and $bard4$ become infeasible with big-M parameters 5, 10, 5, and 5, respectively. Second, sub-optimal solutions are found with sub-optimal big-M parameters. For instance, employing the big-M algorithm, objective function value for problem $bard2$ with big-M parameter 6 is 1.75 instead of optimal objective value which is −3.25. In addition, using large values for the big-M parameters, do not guarantee finding optimal solution. For instance, the big-M parameter equal to 5×10^5 in problem $bard1$ leads to objective function value of zero, while the optimal objective function value is −26 which is found with $M = 10$. These eight problems in Table 6 are small enough that we can tune the big-M parameter and find the optimal solution (reported in Table 6). Although both big-M and proposed DBD algorithms find the optimal solutions, the proposed DBD algorithm still has the advantage of being big-M free. Number of iterations is reported in Table 6 for both algorithms in problems $bard1$ to $colson$. As we see in bold values of this table, the number of iterations in the proposed DBD algorithm is decreased significantly as compared to the one in the big-M algorithm. We have observed a reduction between 60% and 93% in number of iterations for all problems except the problem $clark$ where number of iterations are close (10 and 12 iterations).

The convergence graph of the proposed DBD algorithm for problem $bard1$ is shown in Fig. 3. The UB and LB values are shown for iterations 1 to 12. The UB and LB are initialized with 100 and −100, respectively. UB is decreased to −26 at iteration 9, and LB is increased to −26 at iteration 12 where the proposed DBD algorithm stops.

5 Large-Scale Case Studies

Performance of the proposed approach in large-scale case studies is investigated in this section. First, linear BLPs are defined with random numbers. Then, these problems are solved both with the big-M algorithm [17] and our proposed DBD algorithm.

5.1 Case Study Parameters

Four large-scale case studies (Case 1 to Case 4) are defined with $n_x = n_y = 25, 50, 75,$ and 100, respectively, to study the effect of problem size on performance of the big-M and the DBD algorithms. Numbers of constraints are $n_u = n_l = 15, 25, 50,$ and 75, respectively. The proposed DBD algorithm does not have any disjunctive parameter (such as the big-M). The big-M parameter is set to 100 in all case studies for the big-M algorithm. The parameter matrices are defined with random numbers as shown in (14). Normal(*Mean*, *Std.*) is a function which gives a random number from normal distribution with the given mean (*Mean*) and standard deviation (*Std.*). Uniform(*Min*, *Max*) is a function which gives a random number from the uniform distribution with the given minimum and maximum values (*Min* and *Max*):

$$A = [\text{Uniform}(0, 1)]^{(n_x \times 1)}; \; B = [\text{Uniform}(0, 1)]^{(n_y \times 1)}; \tag{14a}$$

$$C = [\text{Normal}(0, 1)]^{(n_x \times n_u)}; \; D = [0]^{(n_y \times n_u)}; \tag{14b}$$

$$E = [\text{Normal}(0, 1)]^{(n_u \times 1)}; \; G = [\text{Uniform}(0, 1)]^{(n_y \times 1)}; \tag{14c}$$

$$H = [\text{Normal}(0, 1)]^{(n_x \times n_l)}; \; J = [\text{Normal}(0, 1)]^{(n_y \times n_l)}; \tag{14d}$$

$$N = [\text{Normal}(0, 1)]^{(n_l \times 1)}; \tag{14e}$$

The matrix D is set to zero to avoid infeasible problems. Uniform distribution with values between zero and one is employed in A, B, and G in the objective functions to avoid unbounded problems. Other parameters are determined from a normal distribution with mean and standard deviation equal to zero and one, respectively. Since all parameters (A to N) are obtained using random numbers, 100 problems are defined for each case to mitigate the effect of the random values. This allows us to focus mainly on the performance of our proposed DBD algorithm for different problem sizes.

All problems in Sects. 4 and 5 are modeled in GAMS 25.1.3 [14] and solved with CPLEX 12.8.0.0 solver [16]. The optimality gap for CPLEX solver (optCR option) is set to zero. The branch-and-cut algorithm of the CPLEX solver is used for MIP problems. Pre-solving option of the CPLEX solver are disabled to have a fair comparison between results of the CPLEX solver and the proposed DBD algorithm [16]. All simulations for the large-scale case studies are performed using Tegner processing systems available at KTH PDC Center for High Performance Computing (HPC).

Average number of iterations and average of optimal objective values for *Case 1* to *Case 4* are shown in Table 7. This table demonstrates significant reduction in number of iterations in our proposed DBD algorithm. For *Case 1*, the number of iterations in the DBD algorithm is decreased by 65% compared with the big-

Table 7 The large-scale case studies

Problem	n_x	n_y	n_u	n_l	Average number of iterations			Average optimal value		
					Big-M	DBD	%	Big-M	DBD	%
Case 1	25	25	15	15	840	287	65.9	2.4912	2.6007	4.39
Case 2	50	50	25	25	8,947	371	95.85	2.4619	2.5022	1.64
Case 3	75	75	50	50	457,002	457	99.9	4.6533	4.7974	3.09
Case 4	100	100	75	75	714,711	688	99.9	11.6428	11.6735	0.26

M algorithm. This is why the big-M algorithm finds a sub-optimal solution with optimality gap of 4.39% (the DBD algorithm finds the optimal solution with zero optimality gaps in all cases). For *Case 2*, the DBD algorithm finds the optimal solution in 371 iterations as compared to the 8947 iterations from the big-M algorithm (a reduction of 95.85%). Also, the big-M algorithm solution has an optimality gap of 1.64%. For *Case 3* and *Case 4*, the reduction in the number of iterations is significant (99.9%). The big-M algorithm finds sub-optimal solutions with optimality gaps of 3.09% and 0.26%, respectively. As mentioned before, the proposed DBD algorithm finds the optimal solutions for both *Case 3* and *Case 4* with zero duality gap.

5.2 Comparison with Existing Bilevel Approaches

Existing approaches to solve bilevel problems are listed below:

- **Nonlinear**
  Replacing the lower-level problem with the necessary and sufficient KKT conditions which provides a single-level problem. This problem is nonlinear due to the complementarity condition which could be solved with global solvers that can handle nonlinear non-convex problems.
- **Big-M Technique + Mixed-Integer Solver**
  Reformulating the single-level nonlinear problem with the big-M technique. This results to a mixed-integer problem which requires finding the big-M parameters. Finding the right value for the big-M parameters is challenging as mentioned in literature (see [16]). This problem could be solved with enough iterations on the integer variables.
- **Decomposition**
  Decomposition approaches such as Benders decomposition [6] can divide this problem into two parts with two sets of variables. The first-stage master problem (MP) deals with finding the optimal value for the first set of variables. Optimal values for the second set of variables are found in the second-stage subproblem (SP) employing solutions of the MP. Mostly, the integer variables are used in the MP, and continuous variables are used in the SP. These approaches also require the right values for the big-M parameters.

- **Branching on Complementarity**
  Solve the bilevel problem with branching on the complementarity. This could be done by introducing logic constraints and writing an algorithm to deal with the complementarity between the upper-level and lower-level problems. Here, the bilevel problem is solved as it is. Some solvers are able to solve two-level problems without rewriting them as a single-level problem which are not efficient and not applicable for general problems.
- **Built-in Solvers**
  Solving the bilevel problem with high-level KKT operators. Some solvers such as YALMIP (see [22]) has a built-in bilevel solver. This can be done either by branching on the complementarity or by writing the KKT conditions and finding the right values for the big-M parameters internally. These approaches are not essentially optimal, not applicable for general problems, and also have the same problems of the big-M technique.

  In this research, we will avoid the big-M parameters by introducing a parameter-free decomposition approach. A disjunctive-based decomposition (DBD) approach is presented which cuts out the big-M parameters completely. The rest of this chapter is organized as follows: Sect. 6 formulates the bilevel programming problems. This includes LP-LP (6.2), LP-QP (8.1), LP-QCQP (8.2), and LP-LCP (8.3). The proposed DBD algorithm is explained in Sect. 2. Section 7 provides numerical results and Sect. 9 concludes the chapter.

6 General Problem Formulation

General bilevel programming problem is formulated in (15) for $\mathbf{x} \in \mathbb{X} \subseteq \mathbb{R}^{(n_x \times 1)}$ and $\mathbf{y} \in \mathbb{Y} \subseteq \mathbb{R}^{(n_y \times 1)}$. (15a) and (15b) propose upper-level problem, and (15c) and (15d) is called lower-level problem,

$$\underset{\mathbf{x}\in\mathbb{X}}{\text{Minimize}} \quad f(\mathbf{x},\mathbf{y}) \tag{15a}$$

$$\text{Subject to: } F(\mathbf{x},\mathbf{y}) \le 0 \tag{15b}$$

$$\underset{\mathbf{y}\in\mathbb{Y}}{\text{Minimize}} \quad g(\mathbf{x},\mathbf{y}) \tag{15c}$$

$$\text{Subject to: } G(\mathbf{x},\mathbf{y}) \le 0 : \boldsymbol{\gamma} \tag{15d}$$

where $f,\ g{:}\mathbb{R}^{(n_x\times 1)} \times \mathbb{R}^{(n_y\times 1)} \rightarrow \mathbb{R}$, $F{:}\mathbb{R}^{(n_x\times 1)} \times \mathbb{R}^{(n_y\times 1)} \rightarrow \mathbb{R}^{(n_u\times 1)}$, and $G{:}\mathbb{R}^{(n_x\times 1)} \times \mathbb{R}^{(n_y\times 1)} \rightarrow \mathbb{R}^{(n_l\times 1)}$.

Assumption 1 The lower-level problem in (15c) to (15d) is convex, and Slater constraint qualification [13] holds for this problem.

The Lagrangian function for the general problem (15) is defined as $\mathcal{L}(\mathbf{x}, \mathbf{y}, \boldsymbol{\gamma}) = g(\mathbf{x}, \mathbf{y}) - \boldsymbol{\gamma} G(\mathbf{x}, \mathbf{y})$. The lower-level problem (15c) to (15d) could be replaced with KKT conditions (first-order optimality conditions) as in (16) (see [25]):

$$\underset{\mathbf{x}\in\mathbb{X},\mathbf{y}\in\mathbb{Y},\boldsymbol{\gamma}}{\text{Minimize}} \;\; f(\mathbf{x}, \mathbf{y}) \tag{16a}$$

$$\text{Subject to: } F(\mathbf{x}, \mathbf{y}) \leq 0 : \boldsymbol{\lambda}_1 \tag{16b}$$

$$G(\mathbf{x}, \mathbf{y}) \leq 0 : \boldsymbol{\lambda}_2 \tag{16c}$$

$$\nabla_{\mathbf{y}} \mathcal{L}(\mathbf{x}, \mathbf{y}, \boldsymbol{\gamma}) = 0 : \boldsymbol{\mu} \tag{16d}$$

$$\boldsymbol{\gamma}^{\mathsf{T}} G(\mathbf{x}, \mathbf{y}) = 0 : \boldsymbol{\nu} \tag{16e}$$

$$\boldsymbol{\gamma} \geq 0 : \boldsymbol{\xi} \tag{16f}$$

The constraint (16e) is called complementarity condition which has a disjunctive property. This implies in row i, either $\boldsymbol{\gamma}_i' = 0$ or the constraint is active $G(\mathbf{x}, \mathbf{y})_i = 0$ or both. This constraint could be linearized employing the big-M technique. Problem (16) is reformulated in (18) using the duality theorem. First, the Lagrangian function is written as $\mathcal{L}(\mathbf{x}, \mathbf{y}, \boldsymbol{\gamma}, \boldsymbol{\lambda}_1, \boldsymbol{\lambda}_2, \boldsymbol{\mu}, \boldsymbol{\nu}) = f(\mathbf{x}, \mathbf{y}) + \boldsymbol{\lambda}_1^{\mathsf{T}} F(\mathbf{x}, \mathbf{y}) + \boldsymbol{\lambda}_2^{\mathsf{T}} G(\mathbf{x}, \mathbf{y}) + \boldsymbol{\mu}^{\mathsf{T}} \nabla_{\mathbf{y}} \mathcal{L}(\mathbf{x}, \mathbf{y}, \boldsymbol{\gamma}) + \boldsymbol{\nu}^{\mathsf{T}} \boldsymbol{\gamma}^{\mathsf{T}} G(\mathbf{x}, \mathbf{y}) - \boldsymbol{\xi}^{\mathsf{T}} \boldsymbol{\gamma}$. Then the dual function will be

$$\inf_{\mathbf{x}\in\mathbb{X},\mathbf{y}\in\mathbb{Y}} \{\mathcal{L}(\mathbf{x}, \mathbf{y}, \boldsymbol{\gamma}, \boldsymbol{\lambda}_1, \boldsymbol{\lambda}_2, \boldsymbol{\mu}, \boldsymbol{\nu})\} =$$

$$\begin{cases} \mathcal{L}(\mathbf{x}, \mathbf{y}, \boldsymbol{\gamma}, \boldsymbol{\lambda}_1, \boldsymbol{\lambda}_2, \boldsymbol{\mu}, \boldsymbol{\nu}) \,|_{\mathbf{x}=0,\mathbf{y}=0,\boldsymbol{\gamma}=0} & \text{if } \nabla_{[\mathbf{x}^{\mathsf{T}},\mathbf{y}^{\mathsf{T}},\boldsymbol{\gamma}^{\mathsf{T}}]^{\mathsf{T}}} \mathcal{L}(\mathbf{x}, \mathbf{y}, \boldsymbol{\gamma}, \boldsymbol{\lambda}_1, \boldsymbol{\lambda}_2, \boldsymbol{\mu}, \boldsymbol{\nu}) = 0 \\ -\infty & \text{else.} \end{cases} \tag{17a}$$

We can rewrite the dual function (17) as an optimization problem in (18) which is called dual problem:

$$\underset{\boldsymbol{\lambda}_1\geq 0,\boldsymbol{\lambda}_2\geq 0,\boldsymbol{\mu}\geq 0,\boldsymbol{\nu}\geq 0}{\text{Maximize}} \;\; \mathcal{L}(\mathbf{x}, \mathbf{y}, \boldsymbol{\gamma}, \boldsymbol{\lambda}_1, \boldsymbol{\lambda}_2, \boldsymbol{\mu}, \boldsymbol{\nu}) \,|_{\mathbf{x}=0,\mathbf{y}=0,\boldsymbol{\gamma}=0} \tag{18a}$$

$$\text{Subject to: } \nabla_{[\mathbf{x}^{\mathsf{T}},\mathbf{y}^{\mathsf{T}},\boldsymbol{\gamma}^{\mathsf{T}}]^{\mathsf{T}}} \mathcal{L}(\mathbf{x}, \mathbf{y}, \boldsymbol{\gamma}, \boldsymbol{\lambda}_1, \boldsymbol{\lambda}_2, \boldsymbol{\mu}, \boldsymbol{\nu}) = 0 \tag{18b}$$

This method is employed in Sect. 6.1 for general LP, quadratic program (QP), quadratically constrained quadratic program (QCQP), and linear complementarity problem (LCP) problems to formulate bilevel problems as single-level problems.

6.1 General Problem Formulation and Strong Duality Result

Types of the upper-level and lower-level problems in the general bilevel programming problem in (15) vary from problem to problem. General formulation of bilevel problems for four cases are studied in this research including LP-LP, LP-QP, LP-QCQP, and LP-LCP. After formulating the bilevel problem as a single-level problem, a proposition is proposed to remove the big-M parameters.

6.2 LP-LP Bilevel Problem

The general LP-LP bilevel problem is formulated in (19) for the same $\mathbf{x}$ and $\mathbf{y}$,

$$\underset{\mathbf{x}\geq 0}{\text{Minimize}} \quad A^{\mathsf{T}}\mathbf{x}+B^{\mathsf{T}}\mathbf{y} \tag{19a}$$

$$\text{Subject to: } C\mathbf{x}+D\mathbf{y} \leq H \tag{19b}$$

$$\underset{\mathbf{y}\geq 0}{\text{Minimize}} \quad J^{\mathsf{T}}\mathbf{x}+K^{\mathsf{T}}\mathbf{y} \tag{19c}$$

$$\text{Subject to: } L\mathbf{x}+N\mathbf{y} \leq P : \boldsymbol{\gamma} \tag{19d}$$

where $A, J \in \mathbb{R}^{(n_x \times 1)}$, $B, K \in \mathbb{R}^{(n_y \times 1)}$, $C \in \mathbb{R}^{(n_u \times n_x)}$, $D \in \mathbb{R}^{(n_u \times n_y)}$, $H \in \mathbb{R}^{(n_u \times 1)}$, $L \in \mathbb{R}^{(n_l \times n_x)}$, $N \in \mathbb{R}^{(n_l \times n_y)}$, $P \in \mathbb{R}^{(n_l \times 1)}$, and $\boldsymbol{\lambda} \in \mathbb{R}^{(n_l \times 1)}$ is vector of Lagrange multipliers. The lower-level problem is reformulated into a single-level problem in (20) using the necessary and sufficient Karush-Kuhn-Tucker conditions of the LP problem in (19c) to (19d):

$$\underset{\mathbf{x}\geq 0,\mathbf{y},\boldsymbol{\gamma}}{\text{Minimize}} \quad A^{\mathsf{T}}\mathbf{x}+B^{\mathsf{T}}\mathbf{y} \tag{20a}$$

$$\text{Subject to: } C\mathbf{x}+D\mathbf{y} \leq H \tag{20b}$$

$$0 \leq \left(K+N^{\mathsf{T}}\boldsymbol{\gamma}\right) \perp \mathbf{y} \geq 0 \tag{20c}$$

$$0 \leq (P-L\mathbf{x}-N\mathbf{y}) \perp \boldsymbol{\gamma} \geq 0 \tag{20d}$$

Disjunctive property of the complementary slackness conditions (allocated by $\perp$ sign) in (20c) and (20d) is linearized in the MILP problem (21) by the big-M technique where $\{M_1, M_2, M_3, M_4\}$ is an LP-correct set (refer to Definition 3):

$$\underset{\mathbf{x}\geq 0,\mathbf{y}\geq 0,\boldsymbol{\gamma}\geq 0,\mathbf{u}_1,\mathbf{u}_2}{\text{Minimize}} \quad A^{\mathsf{T}}\mathbf{x}+B^{\mathsf{T}}\mathbf{y} \tag{21a}$$

$$\text{Subject to: } C\mathbf{x}+D\mathbf{y} \leq H : \boldsymbol{\lambda}_1 \tag{21b}$$

$$L\mathbf{x}+N\mathbf{y} \leq P : \boldsymbol{\lambda}_2 \tag{21c}$$

$$-N^{\top}\boldsymbol{\gamma} \leq K : \boldsymbol{\lambda}_3 \quad (21d)$$

$$K+N^{\top}\boldsymbol{\gamma} \leq M_1 \left(\mathbf{e}_1-\mathbf{u}_1\right) : \boldsymbol{\mu}_1 \quad (21e)$$

$$P-L\mathbf{x}-N\mathbf{y} \leq M_2 \left(\mathbf{e}_2-\mathbf{u}_2\right) : \boldsymbol{\mu}_2 \quad (21f)$$

$$\mathbf{y} \leq M_3\mathbf{u}_1 : \boldsymbol{\nu}_1 \quad (21g)$$

$$\boldsymbol{\gamma} \leq M_4\mathbf{u}_2 : \boldsymbol{\nu}_2 \quad (21h)$$

Definition 1 *LP-correct*: Under Assumption 1, a set of positive numbers $\{M \,|M \in \mathbb{R}^{+}\}$ is LP-correct if the set of optimal solutions $\{\hat{\mathbf{y}}, \hat{\boldsymbol{\gamma}}\}$ to the primal/dual of the LP problem in (19c) to (19d) is the same as the set of optimal solutions to (21e) to (21h).

$\mathbf{e}_1 \in \mathbb{R}^{(n_y\times 1)}$ and $\mathbf{e}_2 \in \mathbb{R}^{(n_l\times 1)}$ are vectors of ones. $\mathbf{u}_1 \in \{0, 1\}^{(n_y\times 1)}$ and $\mathbf{u}_2 \in \{0, 1\}^{(n_l\times 1)}$ are vectors of binary variables. For a given vector of binary variables $\hat{\mathbf{u}}_1$ and $\hat{\mathbf{u}}_2$, (21) is an LP problem. Therefore, we may replace this problem with (22) employing the duality theorem:

$$\underset{\boldsymbol{\lambda}_1,\boldsymbol{\lambda}_2,\boldsymbol{\lambda}_3,\boldsymbol{\mu}_1,\boldsymbol{\mu}_2,\boldsymbol{\nu}_1,\boldsymbol{\nu}_2}{\text{Maximize}} \quad -\boldsymbol{\lambda}_1^{\top} H-\boldsymbol{\lambda}_2^{\top} P-\boldsymbol{\lambda}_3^{\top} K+\boldsymbol{\mu}_1^{\top} \left(K-M_1 \left(\mathbf{e}_1-\mathbf{u}_1\right)\right)$$

$$+\boldsymbol{\mu}_2^{\top} \left(P-M_2 \left(\mathbf{e}_2-\mathbf{u}_2\right)\right)$$

$$-\boldsymbol{\nu}_1^{\top} M_3\mathbf{u}_1-\boldsymbol{\nu}_2^{\top} M_4\mathbf{u}_2 \quad (22a)$$

$$\text{Subject to: } -\boldsymbol{\lambda}_1^{\top} C-\boldsymbol{\lambda}_2^{\top} L+\boldsymbol{\mu}_2^{\top} L \leq A^{\top} : \mathbf{x} \quad (22b)$$

$$-\boldsymbol{\lambda}_1^{\top} D-\boldsymbol{\lambda}_2^{\top} N+\boldsymbol{\mu}_2^{\top} N-\boldsymbol{\nu}_1^{\top} \leq B^{\top} : \mathbf{y} \quad (22c)$$

$$\boldsymbol{\lambda}_3^{\top} N^{\top}-\boldsymbol{\mu}_1^{\top} N^{\top}-\boldsymbol{\nu}_2^{\top} \leq \mathbf{0} : \boldsymbol{\gamma} \quad (22d)$$

Proposition 1 *If Assumption 1 holds for* (19) *and LP-correct set* $\{M_1, M_2, M_3, M_4\}$ *exists and if* $M_1, M_2, M_3,$ *and* M_4 *are large enough so that for row k:* $(K+N^{\top}\boldsymbol{\gamma})_k < M_1$, $(P-L\mathbf{x}-N\mathbf{y})_k < M_2$, $(\mathbf{y})_k < M_3$, *and* $(\boldsymbol{\gamma})_k < M_4$, $\forall\boldsymbol{\gamma}, \mathbf{x}, \mathbf{y}$, *then the terms* $\boldsymbol{\mu}_1^{\top}\left(\mathbf{e}_1-\hat{\mathbf{u}}_1\right)$, $\boldsymbol{\mu}_2\left(\mathbf{e}_2-\hat{\mathbf{u}}_2\right)$, $\boldsymbol{\nu}_1^{\top}\hat{\mathbf{u}}_1$, *and* $\boldsymbol{\nu}_2^{\top}\hat{\mathbf{u}}_2$ *are zero in the optimal solution of* (22). *Therefore, optimal solutions of* (22) *and* (23) *are the same.*

Proof *(Proof of Proposition 1)* First, consider the row k of the constraints $(K+N^{\top}\boldsymbol{\gamma})_k < M_1$. (i) If the binary variable $(\mathbf{u})_k = 1$ is one, from (21e), we have $(K+N^{\top}\boldsymbol{\gamma})_k = 0$, and the term $\left(\boldsymbol{\mu}_1^{\top}\left(\mathbf{e}_1-\hat{\mathbf{u}}_1\right)\right)_k$ is zero. (ii) If the binary variable $(\mathbf{u})_k = 0$ is zero, from (21e), we have a large enough M_1 so that $(K+N^{\top}\boldsymbol{\gamma})_k \leq M_1$ is not binding. Therefore, the corresponding Lagrange multiplier $(\boldsymbol{\mu}_1)_k = 0$ is zero which cause the term $\left(\boldsymbol{\mu}_1^{\top}\left(\mathbf{e}_1-\hat{\mathbf{u}}_1\right)\right)_k$ to be zero. Therefore, for all $\mathbf{u}, \boldsymbol{\gamma}, \mathbf{x}$, and $\mathbf{y}$, the term $\left(\boldsymbol{\mu}_1^{\top}\left(\mathbf{e}_1-\hat{\mathbf{u}}_1\right)\right)_k = 0$ is zero. We can write similar arguments for the other three terms. Therefore, we can remove these terms from objective of (22) and add constraints enforcing these terms to be zero in (23). □

The dual problem (22) is rewritten in (23) employing Proposition 1. This is the SP in the proposed DBD algorithm:

$$\textbf{SP-LP}: \underset{\lambda_1,\lambda_2,\mu_1,\mu_2,\nu_1,\nu_2}{\text{Maximize}} \quad -\lambda_1^{\mathsf{T}} H - \lambda_2^{\mathsf{T}} P - \lambda_3^{\mathsf{T}} K + \mu_1^{\mathsf{T}} K + \mu_2^{\mathsf{T}} P \tag{23a}$$

$$\text{Subject to: } -\lambda_1^{\mathsf{T}} C - \lambda_2^{\mathsf{T}} L + \mu_2^{\mathsf{T}} L \leq A^{\mathsf{T}} : \mathbf{x} \tag{23b}$$

$$-\lambda_1^{\mathsf{T}} D - \lambda_2^{\mathsf{T}} N + \mu_2^{\mathsf{T}} N - \nu_1^{\mathsf{T}} \leq B^{\mathsf{T}} : \mathbf{y} \tag{23c}$$

$$\lambda_3^{\mathsf{T}} N^{\mathsf{T}} - \mu_1^{\mathsf{T}} N^{\mathsf{T}} - \nu_2^{\mathsf{T}} \leq \mathbf{0} : \gamma \tag{23d}$$

$$\mu_1^{\mathsf{T}} \left(\mathbf{e}_1 - \hat{\mathbf{u}}_1\right) = \mu_2 \left(\mathbf{e}_2 - \hat{\mathbf{u}}_2\right) = \nu_1^{\mathsf{T}} \hat{\mathbf{u}}_1 = \nu_2^{\mathsf{T}} \hat{\mathbf{u}}_2 = 0 \tag{23e}$$

7 Numerical Results

Numerical results for general case studies are presented below. The size of the problems is shown in Table 8. The optimal solutions for each problem are shown in Table 9. Solution time and number of iterations to solve each problem are shown in Table 10.

Table 8 Case studies

Problem	n_x	n_y	n_u	n_l	Reference
bard1	2	3	2	6	[4]
bard2	2	2	2	5	[4]
candler	2	6	2	6	[8]
anan	1	1	1	6	[1]
bard3	1	2	1	5	[2]
bard4	1	2	1	5	[3]
clark	1	2	1	5	[9]
colson	2	3	1	3	[10]

Table 9 Solutions

Problem	$\hat{f}$	$\hat{g}$	$\hat{\mathbf{x}}$	$\hat{\mathbf{y}}$
bard1	− 26	3.2	[0,0.9]	[0,0.6,0.4]
bard2	− 3.25	− 4	[2,0]	[1.5,0]
candler	− 29.2	3.2	[0,0.9]	[0,0.6,0.4,0,0,0]
anan	− 49	17	[16]	[11]
bard3	− 1	0	[1]	[0,0]
bard4	− 2	− 1	[0]	[0,1]
clark	− 13	− 4	[5]	[4,2]
colson	− 14.6	0.3	[0,0.65]	[0,0.3,0]

Table 10 Solution time (wall time) and number of iterations

	Iterations				Time (seconds)			
Problem	Min	Median	Mean	Max	Min	Median	Mean	Max
bard1	7	8	8.7	12	0.67	0.79	0.84	1.13
bard2	1	6	7.2	11	0.18	0.62	0.73	1.1
candler	7	11	10.99	20	0.73	1.08	1.1	1.91
anan	1	5	5.46	8	0.18	0.57	0.61	0.91
bard3	1	3	2.65	4	0.19	0.37	0.35	0.51
bard4	1	6	5.72	8	0.17	0.6	0.58	0.84
clark	1	12	10.78	14	0.17	1.15	1.06	1.45
colson	1	4	3.79	7	0.17	0.43	0.42	0.77

8 Extended Problems

8.1 LP-QP Bilevel Model

The general LP-QP bilevel problem is formulated in (24) for the same **x** and **y**,

$$\underset{\mathbf{x}\geq 0}{\text{Minimize}} \ \ A^{\mathsf{T}}\mathbf{x}+B^{\mathsf{T}}\mathbf{y} \tag{24a}$$

$$\text{Subject to: } C\mathbf{x}+D\mathbf{y} \leq H \tag{24b}$$

$$\underset{\mathbf{y}\geq 0}{\text{Minimize}} \ \ J^{\mathsf{T}}\mathbf{x}+K^{\mathsf{T}}\mathbf{y}+\frac{1}{2}\mathbf{y}^{\mathsf{T}}Q\mathbf{y} \tag{24c}$$

$$\text{Subject to: } L\mathbf{x}+N\mathbf{y} \leq P : \boldsymbol{\gamma}, \tag{24d}$$

where $Q \in \mathbb{R}^{(n_y \times n_y)}$ and $K \in \mathbb{R}^{(n_y \times 1)}$. We assume that the lower-level problem in (29c) to (29d) is convex. Therefore, Q is positive semidefinite ($Q \succeq 0$). This problem is reformulated into a single-level problem in (25) employing the necessary and sufficient KKT conditions of the lower-level QP problem in (24c) to (24d):

$$\underset{\mathbf{x}\geq 0,\mathbf{y},\boldsymbol{\gamma}}{\text{Minimize}} \ \ A^{\mathsf{T}}\mathbf{x}+B^{\mathsf{T}}\mathbf{y} \tag{25a}$$

$$\text{Subject to: } C\mathbf{x}+D\mathbf{y} \leq H \tag{25b}$$

$$0 \leq \left(Q\mathbf{y}+K+N^{\mathsf{T}}\boldsymbol{\gamma}\right) \perp \mathbf{y} \geq 0 \tag{25c}$$

$$0 \leq (P-L\mathbf{x}-N\mathbf{y}) \perp \boldsymbol{\gamma} \geq 0 \tag{25d}$$

Similarly, (25c) and (25d) are linearized in the MILP problem (26) with an LP-correct set $\{M_1, M_2, M_3, M_4\}$ and the big-M technique:

$$\underset{\mathbf{x}\geq 0,\mathbf{y}\geq 0,\boldsymbol{\gamma}\geq 0,\mathbf{u}_1,\mathbf{u}_2}{\text{Minimize}} \quad A^\top\mathbf{x}+B^\top\mathbf{y} \tag{26a}$$

$$\text{Subject to: } C\mathbf{x}+D\mathbf{y} \leq H : \boldsymbol{\lambda}_1 \tag{26b}$$

$$L\mathbf{x}+N\mathbf{y} \leq P : \boldsymbol{\lambda}_2 \tag{26c}$$

$$-Q\mathbf{y}-N^\top\boldsymbol{\gamma} \leq K : \boldsymbol{\lambda}_3 \tag{26d}$$

$$Q\mathbf{y}+K+N^\top\boldsymbol{\gamma} \leq M_1\left(\mathbf{e}_1-\mathbf{u}_1\right) : \boldsymbol{\mu}_1 \tag{26e}$$

$$P-L\mathbf{x}-N\mathbf{y} \leq M_2\left(\mathbf{e}_2-\mathbf{u}_2\right) : \boldsymbol{\mu}_2 \tag{26f}$$

$$\mathbf{y} \leq M_3\mathbf{u}_1 : \boldsymbol{\nu}_1 \tag{26g}$$

$$\boldsymbol{\gamma} \leq M_4\mathbf{u}_2 : \boldsymbol{\nu}_2 \tag{26h}$$

Definition 2 *QP-correct*: Under Assumption 1, a set of positive numbers $\{M \mid M \in \mathbb{R}^+\}$ is QP-correct if the set of optimal solutions $\{\hat{\mathbf{y}}, \hat{\boldsymbol{\gamma}}\}$ to the primal/dual of the QP problem in (24c) to (24d) is the same as the set of optimal solutions to (26e) to (26h).

$\mathbf{e}_1 \in \mathbb{R}^{(n_y\times 1)}$ and $\mathbf{e}_2 \in \mathbb{R}^{(n_l\times 1)}$ are vectors of ones. $\mathbf{u}_1 \in \{0, 1\}^{(n_y\times 1)}$ and $\mathbf{u}_2 \in \{0, 1\}^{(n_l\times 1)}$ are vectors of binary variables. For given binary vectors $\hat{\mathbf{u}}_1$ and $\hat{\mathbf{u}}_2$, (26) becomes an LP problem which is rewritten in (27) employing the duality theorem:

$$\underset{\boldsymbol{\lambda}_1,\boldsymbol{\lambda}_2,\boldsymbol{\mu}_1,\boldsymbol{\mu}_2,\boldsymbol{\nu}_1,\boldsymbol{\nu}_2}{\text{Maximize}} \quad -\boldsymbol{\lambda}_1^\top H-\boldsymbol{\lambda}_2^\top P-\boldsymbol{\lambda}_3^\top K+\boldsymbol{\mu}_1^\top\left(K-M_1\left(\mathbf{e}_1-\hat{\mathbf{u}}_1\right)\right)+\boldsymbol{\mu}_2^\top\left(P-M_2\left(\mathbf{e}_2-\hat{\mathbf{u}}_2\right)\right)$$

$$-\boldsymbol{\nu}_1^\top M_3\hat{\mathbf{u}}_1-\boldsymbol{\nu}_2^\top M_4\hat{\mathbf{u}}_2 \tag{27a}$$

$$\text{Subject to: } -\boldsymbol{\lambda}_1^\top C-\boldsymbol{\lambda}_2^\top L+\boldsymbol{\mu}_2^\top L \leq A^\top : \mathbf{x} \tag{27b}$$

$$-\boldsymbol{\lambda}_1^\top D-\boldsymbol{\lambda}_2^\top N+\boldsymbol{\lambda}_3^\top Q-\boldsymbol{\mu}_1^\top Q+\boldsymbol{\mu}_2^\top N-\boldsymbol{\nu}_1^\top \leq B^\top : \mathbf{y} \tag{27c}$$

$$\boldsymbol{\lambda}_3^\top N^\top-\boldsymbol{\mu}_1^\top N^\top-\boldsymbol{\nu}_2^\top \leq 0 : \boldsymbol{\gamma} \tag{27d}$$

Corollary 1 *If Assumption 1 holds for* (24) *and a QP-correct set* $\{M_1, M_2, M_3, M_4\}$ *exists and if* $M_1, M_2, M_3,$ *and* M_4 *are large enough so that for row* k $(Q\mathbf{y}+K+N^\top\boldsymbol{\gamma})_k < M_1$, $(P-L\mathbf{x}-N\mathbf{y})_k < M_2$, $(\mathbf{y})_k < M_3$, *and* $(\boldsymbol{\gamma})_k < M_4$, $\forall\boldsymbol{\gamma}, \mathbf{x}, \mathbf{y}$, *then the terms* $\boldsymbol{\mu}_1^\top\left(\mathbf{e}_1-\hat{\mathbf{u}}_1\right)$, $\boldsymbol{\mu}_2\left(\mathbf{e}_2-\hat{\mathbf{u}}_2\right)$, $\boldsymbol{\nu}_1^\top\hat{\mathbf{u}}_1$, *and* $\boldsymbol{\nu}_2^\top\hat{\mathbf{u}}_2$ *are zero in the optimal solution of* (27). *Therefore, optimal solutions of* (27) *and* (28) *are the same.*

Proof *(Proof of Corollary 1)* First consider the row k of the constraints $(Q\mathbf{y}+K+N^\top\boldsymbol{\gamma})_k < M_1$. (i) If the binary variable $(\mathbf{u})_k = 1$ is one, from (26e), we have $(Q\mathbf{y}+K+N^\top\boldsymbol{\gamma})_k = 0$, and the term $\left(\boldsymbol{\mu}_1^\top\left(\mathbf{e}_1-\hat{\mathbf{u}}_1\right)\right)_k$ is zero. (ii) If the binary variable $(\mathbf{u})_k = 0$ is zero, from (26e), we have a large enough M_1 so that $(Q\mathbf{y}+K+N^\top\boldsymbol{\gamma})_k \leq M_1$ is not binding. Therefore, the corresponding Lagrange multiplier $(\boldsymbol{\mu}_1)_k = 0$ is zero which cause the term $\left(\boldsymbol{\mu}_1^\top\left(\mathbf{e}_1-\hat{\mathbf{u}}_1\right)\right)_k$ to be zero. Therefore, for all $\mathbf{u}, \boldsymbol{\gamma}, \mathbf{x}$, and $\mathbf{y}$, the term $\left(\boldsymbol{\mu}_1^\top\left(\mathbf{e}_1-\hat{\mathbf{u}}_1\right)\right)_k = 0$ is zero. We can write similar arguments for the other three terms. Therefore, we can remove these terms from objective of (27) and add constraints enforcing these terms to be zero in (28). □

The dual problem (27) is rewritten in (28) employing Corollary 1. This is the SP in the proposed DBD algorithm:

$$\textbf{SP-QP}: \underset{\boldsymbol{\lambda}_1,\boldsymbol{\lambda}_3,\boldsymbol{\lambda}_2,\boldsymbol{\mu}_1,\boldsymbol{\mu}_2,\boldsymbol{\nu}_1,\boldsymbol{\nu}_2}{\text{Maximize}} \quad -\boldsymbol{\lambda}_1^{\mathsf{T}}H-\boldsymbol{\lambda}_2^{\mathsf{T}}P-\boldsymbol{\lambda}_3^{\mathsf{T}}K+\boldsymbol{\mu}_1^{\mathsf{T}}K+\boldsymbol{\mu}_2^{\mathsf{T}}P \tag{28a}$$

$$\text{Subject to: } -\boldsymbol{\lambda}_1^{\mathsf{T}}C-\boldsymbol{\lambda}_2^{\mathsf{T}}L+\boldsymbol{\mu}_2^{\mathsf{T}}L \leq A^{\mathsf{T}} : \mathbf{x} \tag{28b}$$

$$-\boldsymbol{\lambda}_1^{\mathsf{T}}D-\boldsymbol{\lambda}_2^{\mathsf{T}}N+\boldsymbol{\lambda}_3^{\mathsf{T}}Q-\boldsymbol{\mu}_1^{\mathsf{T}}Q+\boldsymbol{\mu}_2^{\mathsf{T}}N-\boldsymbol{\nu}_1^{\mathsf{T}} \leq B^{\mathsf{T}} : \mathbf{y} \tag{28c}$$

$$\boldsymbol{\lambda}_3^{\mathsf{T}}N^{\mathsf{T}}-\boldsymbol{\mu}_1^{\mathsf{T}}N^{\mathsf{T}}-\boldsymbol{\nu}_2^{\mathsf{T}} \leq 0 : \boldsymbol{\gamma} \tag{28d}$$

$$\boldsymbol{\mu}_1^{\mathsf{T}}\left(\mathbf{e}_1-\hat{\mathbf{u}}_1\right) = \boldsymbol{\mu}_2^{\mathsf{T}}\left(\mathbf{e}_2-\hat{\mathbf{u}}_2\right) = \boldsymbol{\nu}_1^{\mathsf{T}}\hat{\mathbf{u}}_1 = \boldsymbol{\nu}_2^{\mathsf{T}}\hat{\mathbf{u}}_2 = 0 \tag{28e}$$

8.2 LP-QCQP Bilevel Model

The general LP-QCQP bilevel problem is formulated in (29) for the same $\mathbf{x}$ and $\mathbf{y}$.

$$\underset{\mathbf{x}\geq 0}{\text{Minimize}} \quad A^{\mathsf{T}}\mathbf{x}+B^{\mathsf{T}}\mathbf{y} \tag{29a}$$

$$\text{Subject to: } C\mathbf{x}+D\mathbf{y} \leq H \tag{29b}$$

$$\underset{\mathbf{y}\geq 0}{\text{Minimize}} \quad \frac{1}{2}\mathbf{y}^{\mathsf{T}}Q\mathbf{y}+q^{\mathsf{T}}\mathbf{y} \tag{29c}$$

$$\text{Subject to: } \frac{1}{2}\mathbf{y}^{\mathsf{T}}Q_i\mathbf{y}+q_i^{\mathsf{T}}\mathbf{y} \leq R_i : \boldsymbol{\gamma}_i' \tag{29d}$$

where $Q_i \in \mathbb{R}^{(n_y \times n_y)}$, $q_i \in \mathbb{R}^{(n_y \times 1)}$, and $R_i \in \mathbb{R}$. $\boldsymbol{\gamma}_i' \in \mathbb{R}$ are vectors of Lagrange multipliers. We assume that the lower-level problem in (29c) to (29d) is convex. Therefore, Q and $Q_i, \forall i \in \{1, \ldots, n_l\}$ are positive semidefinite ($Q \succeq 0$ and $Q_i \succeq 0, \forall i \in \{1, \ldots, n_l\}$). The lower-level QCQP problem is replaced by its KKT conditions in (30) for all $i \in \{1, \ldots, n_l\}$:

$$\underset{\mathbf{x}\geq 0}{\text{Minimize}} \quad A^{\mathsf{T}}\mathbf{x}+B^{\mathsf{T}}\mathbf{y} \tag{30a}$$

$$\text{Subject to: } C\mathbf{x}+D\mathbf{y} \leq H \tag{30b}$$

$$0 \leq \left(R_i-\frac{1}{2}\mathbf{y}^{\mathsf{T}}Q_i\mathbf{y}-q_i^{\mathsf{T}}\mathbf{y}\right) \perp \boldsymbol{\gamma}_i' \geq 0 \tag{30c}$$

$$0 \leq \left(Q\mathbf{y}+q+\sum_i \left(\boldsymbol{\gamma}_i' Q_i\mathbf{y}+q_i\boldsymbol{\gamma}_i'\right)\right) \perp \mathbf{y} \geq 0 \tag{30d}$$

The complementarity conditions in (30c) and (30d) are linearized in (31) with the big-M technique and a bilevel-correct set $\{M_1, M_2, M_3, M_4\}$:

$$\underset{\mathbf{x}\geq 0,\mathbf{y}'\geq 0,\gamma_i'\geq 0}{\text{Minimize}} \quad A^{\mathsf{T}}\mathbf{x}+B'^{\mathsf{T}}\mathbf{y}' \tag{31a}$$

$$\text{Subject to: } C\mathbf{x}+D'\mathbf{y}' \leq H \tag{31b}$$

$$\frac{1}{2}\mathbf{y}^{\mathsf{T}\prime}Q'_i\mathbf{y}'+Q_i'^{\mathsf{T}}\mathbf{y}' \leq R_i \tag{31c}$$

$$-Q'\mathbf{y}'-\sum_i\left(\gamma_i'Q'_i\mathbf{y}'+q'_i\gamma_i'\right) \leq \tag{31d}$$

$$R_i-\frac{1}{2}\mathbf{y}^{\mathsf{T}\prime}Q'_i\mathbf{y}'-Q_i'^{\mathsf{T}}\mathbf{y}' \leq M_1\left(\mathbf{e}_1'-\mathbf{u}_1'\right) \tag{31e}$$

$$Q'\mathbf{y}'+q+\sum_i\left(\gamma_i'Q'_i\mathbf{y}'+q'_i\gamma_i'\right) \leq M_2\left(\mathbf{e}_2'-\mathbf{u}_2'\right) \tag{31f}$$

$$\gamma_i' \leq M_3\mathbf{u}_1' \tag{31g}$$

$$\mathbf{y}' \leq M_4\mathbf{u}_2' \tag{31h}$$

Definition 3 *QCQP-correct*: Under Assumption 1, a set of positive numbers $\{M \mid M \in \mathbb{R}^+\}$ is QCQP-correct if the set of optimal solutions $\{\hat{\mathbf{y}}, \hat{\gamma}\}$ to the primal/dual of the QCQP problem in (29c) to (29d) is the same as the set of optimal solutions to (31e) to (31h).

$\mathbf{e}_1 \in \mathbb{R}$ and $\mathbf{e}_2 \in \mathbb{R}^{(n_y\times 1)}$ are vectors of ones. $\mathbf{u}_1 \in \{0, 1\}^{(1\times 1)}$ and $\mathbf{u}_2 \in \{0, 1\}^{(n_y\times 1)}$ are vectors of binary variables. For given vectors of binary variables $\hat{\mathbf{u}}_1$ and $\hat{\mathbf{u}}_2$, (31) could be rewritten as a QCLP problem in (32). For instance, $\gamma_i'Q'_i\mathbf{y} = [\,\mathbf{x}\ \mathbf{y}'\ \gamma_i'\,]\begin{bmatrix} 0 & 0 & 0 \\ 0 & 0 & \frac{1}{2}Q'_i \\ 0 & \frac{1}{2}Q'_i & 0 \end{bmatrix}\begin{bmatrix} \mathbf{x} \\ \mathbf{y}' \\ \gamma_i' \end{bmatrix} = \mathbf{y}^{\mathsf{T}}Q_i\mathbf{y}$, where $\mathbf{y}^{\mathsf{T}} = [\,\mathbf{x}\ \mathbf{y}'\ \gamma_i'\,]$:

$$\underset{\mathbf{x}\geq 0,\mathbf{y}\geq 0}{\text{Minimize}} \quad B^{\mathsf{T}}\mathbf{y} \tag{32a}$$

$$\text{Subject to: } D\mathbf{y} \leq H : \boldsymbol{\lambda}_1 \tag{32b}$$

$$\frac{1}{2}\mathbf{y}^{\mathsf{T}}Q_i\mathbf{y}+q_i^{\mathsf{T}}\mathbf{y} \leq R_i : \boldsymbol{\lambda}_i' \tag{32c}$$

$$-Q\mathbf{y}-\sum_i\left(\mathbf{y}^{\mathsf{T}}Q_i\mathbf{y}+q_i\mathbf{y}\right) \leq : \boldsymbol{\lambda}_2 \tag{32d}$$

$$R_i-\frac{1}{2}\mathbf{y}^{\mathsf{T}}Q_i\mathbf{y}-q_i^{\mathsf{T}}\mathbf{y} \leq M_1\left(\mathbf{e}_1-\mathbf{u}_1\right) : \boldsymbol{\mu}_i' \tag{32e}$$

$$Q\mathbf{y}+q+\sum_i\left(\mathbf{y}^{\mathsf{T}}Q_i\mathbf{y}+q_i\mathbf{y}\right) \leq M_2\left(\mathbf{e}_2-\mathbf{u}_2\right) : \boldsymbol{\mu} \tag{32f}$$

$$\mathbf{y} \leq M_3\mathbf{u}_1 : \boldsymbol{\nu}_i' \tag{32g}$$

$$\mathbf{y} \leq M_4 \mathbf{u}_2 : \boldsymbol{\nu} \tag{32h}$$

We may replace (32) with its dual problem (33) employing the duality theorem:

$$\begin{aligned}
&\text{Maximize } -\frac{1}{2}\Big(+\boldsymbol{\lambda}_1^{\mathsf{T}} D^{\mathsf{T}} + \boldsymbol{\lambda}_i'^{\mathsf{T}} q_i - \boldsymbol{\lambda}_2^{\mathsf{T}} \Big(Q + \sum_i q_i^{\mathsf{T}}\Big) - \boldsymbol{\mu}_i'^{\mathsf{T}} q_i \\
&+\boldsymbol{\mu}^{\mathsf{T}} \Big(Q + \sum_i q_i^{\mathsf{T}}\Big) + \boldsymbol{\nu}_i'^{\mathsf{T}} + \boldsymbol{\nu}^{\mathsf{T}}\Big)^{\mathsf{T}} \\
&\left(\boldsymbol{\lambda}_i'^{\mathsf{T}} Q_i - 2\boldsymbol{\lambda}_2^{\mathsf{T}} Q_i - \boldsymbol{\mu}_i'^{\mathsf{T}} Q_i + 2\boldsymbol{\mu}^{\mathsf{T}} Q_i\right)^{-1} \\
&\Big(B + \boldsymbol{\lambda}_1^{\mathsf{T}} D^{\mathsf{T}} + \boldsymbol{\lambda}_i'^{\mathsf{T}} q_i - \boldsymbol{\lambda}_2^{\mathsf{T}} \Big(Q + \sum_i q_i^{\mathsf{T}}\Big) - \boldsymbol{\mu}_i'^{\mathsf{T}} q_i + \boldsymbol{\mu}^{\mathsf{T}} \Big(Q + \sum_i q_i^{\mathsf{T}}\Big) + \boldsymbol{\nu}_i'^{\mathsf{T}} + \boldsymbol{\nu}^{\mathsf{T}}\Big) \\
&+\left(-\boldsymbol{\lambda}_1^{\mathsf{T}} H - \boldsymbol{\lambda}_i'^{\mathsf{T}} R_i - \boldsymbol{\lambda}_2^{\mathsf{T}} q + \boldsymbol{\mu}_i'^{\mathsf{T}} \left(R_i - M_1 \left(\mathbf{e}_1 - \hat{\mathbf{u}}_1\right)\right) + \boldsymbol{\mu}^{\mathsf{T}} \left(q - M_2 \left(\mathbf{e}_2 - \hat{\mathbf{u}}_2\right)\right)\right. \\
&\left.-\boldsymbol{\nu}_i'^{\mathsf{T}} M_3 \hat{\mathbf{u}}_1 - \boldsymbol{\nu}^{\mathsf{T}} M_4 \hat{\mathbf{u}}_2\right)
\end{aligned} \tag{33a}$$

$$\text{Subject to: } \boldsymbol{\lambda}_1 \geq 0, \boldsymbol{\lambda}_2 \geq 0, \boldsymbol{\lambda}_i' \geq 0, \boldsymbol{\mu} \geq 0, \boldsymbol{\mu}_i' \geq 0, \boldsymbol{\nu} \geq 0, \boldsymbol{\nu}_i' \geq 0 \tag{33b}$$

Corollary 2 *If Assumption 1 holds for* (29) *and a QCQP-correct set* $\{M_1, M_2, M_3, M_4\}$ *exists and if* $M_1, M_2, M_3,$ *and* M_4 *are large enough so that for row k:* $\left(R_i - \frac{1}{2}\mathbf{y}^{\mathsf{T}} Q_i \mathbf{y} - q_i^{\mathsf{T}} \mathbf{y}\right)_k < M_1$, $\left(Q\mathbf{y} + q + \sum_i (\mathbf{y}^{\mathsf{T}} Q_i \mathbf{y} + q_i \mathbf{y})\right)_k < M_3$, *and* $(\boldsymbol{\gamma})_k < M_4$, $\forall \boldsymbol{\gamma}, \mathbf{x}, \mathbf{y}$, *then the terms* $\boldsymbol{\mu}_1^{\mathsf{T}} (\mathbf{e}_1 - \hat{\mathbf{u}}_1)$, $\boldsymbol{\mu}_2 (\mathbf{e}_2 - \hat{\mathbf{u}}_2)$, $\boldsymbol{\nu}_1^{\mathsf{T}} \hat{\mathbf{u}}_1$, *and* $\boldsymbol{\nu}_2^{\mathsf{T}} \hat{\mathbf{u}}_2$ *are zero in the optimal solution of* (33). *Therefore, optimal solutions of* (33) *and* (34) *are the same.*

Proof *(Proof of Corollary 2.)* First consider the row k of the constraints $\left(R_i - \frac{1}{2}\mathbf{y}^{\mathsf{T}} Q_i \mathbf{y} - q_i^{\mathsf{T}} \mathbf{y}\right)_k < M_1$. (i) If the binary variable $(\mathbf{u})_k = 1$ is one, from (31e), we have $\left(R_i - \frac{1}{2}\mathbf{y}^{\mathsf{T}} Q_i \mathbf{y} - q_i^{\mathsf{T}} \mathbf{y}\right)_k = 0$, and the term $\left(\boldsymbol{\mu}_1^{\mathsf{T}} (\mathbf{e}_1 - \hat{\mathbf{u}}_1)\right)_k$ is zero. (ii) If the binary variable $(\mathbf{u})_k = 0$ is zero, from (31e), we have a large enough M_1 so that $\left(R_i - \frac{1}{2}\mathbf{y}^{\mathsf{T}} Q_i \mathbf{y} - q_i^{\mathsf{T}} \mathbf{y}\right)_k \leq M_1$ is not binding. Therefore, the corresponding Lagrange multiplier $(\boldsymbol{\mu}_1)_k = 0$ is zero which cause the term $\left(\boldsymbol{\mu}_1^{\mathsf{T}} (\mathbf{e}_1 - \hat{\mathbf{u}}_1)\right)_k$ to be zero. Therefore, for all $\mathbf{u}$, $\boldsymbol{\gamma}$, $\mathbf{x}$, and $\mathbf{y}$, the term $\left(\boldsymbol{\mu}_1^{\mathsf{T}} (\mathbf{e}_1 - \hat{\mathbf{u}}_1)\right)_k = 0$ is zero. We can write similar arguments for the other three terms. Therefore, we can remove these terms from objective of (33) and add constraints enforcing these terms to be zero in (34). □

The dual problem (33) is rewritten in (34) employing Corollary 2. This is the SP in the proposed DBD algorithm:

$$\begin{aligned}
\mathbf{SP}: \text{Maximize } &-\frac{1}{2}\Big(+\boldsymbol{\lambda}_1^{\mathsf{T}} D^{\mathsf{T}} + \boldsymbol{\lambda}_i'^{\mathsf{T}} q_i - \boldsymbol{\lambda}_2^{\mathsf{T}} \Big(Q + \sum_i q_i^{\mathsf{T}}\Big) - \boldsymbol{\mu}_i'^{\mathsf{T}} q_i \\
&+\boldsymbol{\mu}^{\mathsf{T}} \Big(Q + \sum_i q_i^{\mathsf{T}}\Big) + \boldsymbol{\nu}_i'^{\mathsf{T}} + \boldsymbol{\nu}^{\mathsf{T}}\Big)^{\mathsf{T}}
\end{aligned}$$

$$\begin{aligned}
&\left(\boldsymbol{\lambda}_i'^{\mathsf{T}} Q_i - 2\boldsymbol{\lambda}_2^{\mathsf{T}} Q_i - \boldsymbol{\mu}_i'^{\mathsf{T}} Q_i + 2\boldsymbol{\mu}^{\mathsf{T}} Q_i\right)^{-1} \\
&\Big(B + \boldsymbol{\lambda}_1^{\mathsf{T}} D^{\mathsf{T}} + \boldsymbol{\lambda}_i'^{\mathsf{T}} q_i - \boldsymbol{\lambda}_2^{\mathsf{T}} \Big(Q + \sum_i q_i^{\mathsf{T}}\Big) - \boldsymbol{\mu}_i'^{\mathsf{T}} q_i \\
&+ \boldsymbol{\mu}^{\mathsf{T}} \Big(Q + \sum_i q_i^{\mathsf{T}}\Big) + \boldsymbol{\nu}_i'^{\mathsf{T}} + \boldsymbol{\nu}^{\mathsf{T}}\Big) \\
&+ \left(-\boldsymbol{\lambda}_1^{\mathsf{T}} H - \boldsymbol{\lambda}_i'^{\mathsf{T}} R_i - \boldsymbol{\lambda}_2^{\mathsf{T}} q + \boldsymbol{\mu}_i'^{\mathsf{T}} R_i + \boldsymbol{\mu}^{\mathsf{T}} q\right) \qquad (34a)
\end{aligned}$$

$$\text{Subject to: } \boldsymbol{\lambda}_1 \geq 0,\ \boldsymbol{\lambda}_2 \geq 0,\ \boldsymbol{\lambda}_i' \geq 0,\ \boldsymbol{\mu} \geq 0,\ \boldsymbol{\mu}_i' \geq 0,\ \boldsymbol{\nu} \geq 0,\ \boldsymbol{\nu}_i' \geq 0 \qquad (34b)$$

$$\boldsymbol{\mu}_i'^{\mathsf{T}} \left(\mathbf{e}_1 - \hat{\mathbf{u}}_1\right) = \boldsymbol{\mu}^{\mathsf{T}} \left(\mathbf{e}_2 - \hat{\mathbf{u}}_2\right) = \boldsymbol{\nu}_i'^{\mathsf{T}} \hat{\mathbf{u}}_1 = \boldsymbol{\nu}^{\mathsf{T}} \hat{\mathbf{u}}_2 = 0 \qquad (34c)$$

8.3 LP-LCP Bilevel Model

The general LP-LCP bilevel problem is formulated in (35) for the $\mathbf{x}$ and $\mathbf{y}$,

$$\underset{\mathbf{x}\geq 0,\mathbf{y}\geq 0}{\text{Minimize}} \quad A^{\mathsf{T}}\mathbf{x} + B^{\mathsf{T}}\mathbf{y} \qquad (35a)$$

$$\text{Subject to: } C\mathbf{x} + D\mathbf{y} \leq H \qquad (35b)$$

$$0 \leq (Q\mathbf{y} + q) \perp \mathbf{y} \geq 0, \qquad (35c)$$

where $Q \in \mathbb{R}^{(n_y \times n_y)}$ and $q \in \mathbb{R}^{(n_y \times 1)}$.

8.3.1 Big-M Technique

The complementarity condition in (35c) could be reformulated with the big-M technique as written in (36),

$$\underset{\mathbf{x}\geq 0,\mathbf{y}\geq 0,\mathbf{u}}{\text{Minimize}} \quad A^{\mathsf{T}}\mathbf{x} + B^{\mathsf{T}}\mathbf{y} \qquad (36a)$$

$$\text{Subject to: } C\mathbf{x} + D\mathbf{y} \leq H : \boldsymbol{\lambda}_1 \qquad (36b)$$

$$-Q\mathbf{y} \leq : \boldsymbol{\lambda}_2 \qquad (36c)$$

$$Q\mathbf{y} + q \leq M_1(\mathbf{e} - \mathbf{u}) : \boldsymbol{\nu} \qquad (36d)$$

$$\mathbf{y} \leq M_2\mathbf{u} : \boldsymbol{\gamma}, \qquad (36e)$$

where $\mathbf{u} \in \{0, 1\}^{(n_y \times 1)}$ is vector of binary variables, $\mathbf{e} \in \mathbb{R}^{(n_y \times 1)}$ is vector of ones, and $\{M_1, M_2\}$ is LP-correct set of big-M parameters. (36) is an MILP problem, and we can solve this problem with the off-the-shelf optimization software with given disjunctive parameters M_1 and M_2 (big-M technique). For a given vector of binary variables, $\hat{\mathbf{u}}$, (36) is an LP problem. Therefore, we may replace this problem with

(37) employing the duality theorem:

$$\underset{\boldsymbol{\lambda}_1\geq 0,\boldsymbol{\lambda}_2\geq 0,\boldsymbol{\nu}\geq 0,\boldsymbol{\gamma}\geq 0}{\text{Maximize}} \quad -\boldsymbol{\lambda}_1^{\top} - \boldsymbol{\lambda}_2^{\top} q + \boldsymbol{\nu}^{\top}\left(q - M_1\left(\mathbf{e}-\hat{\mathbf{u}}\right)\right) - \boldsymbol{\gamma}^{\top} M_2 \hat{\mathbf{u}} \tag{37a}$$

$$\text{Subject to: } -\boldsymbol{\lambda}_1^{\top} C \leq A^{\top} : \mathbf{x} \tag{37b}$$

$$-\boldsymbol{\lambda}_1^{\top} D + \boldsymbol{\lambda}_2^{\top} Q - \boldsymbol{\nu}^{\top} Q - \boldsymbol{\gamma}^{\top} \leq B^{\top} : \mathbf{y} \tag{37c}$$

The dual problem in (37) is rewritten in (38) employing Proposition 1. This is the SP in the proposed DBD algorithm:

$$\textbf{SP}: \underset{\boldsymbol{\lambda}_1\geq 0,\boldsymbol{\lambda}_2\geq 0,\boldsymbol{\nu}\geq 0,\boldsymbol{\gamma}\geq 0}{\text{Maximize}} \quad -\boldsymbol{\lambda}_1^{\top} - \boldsymbol{\lambda}_2^{\top} q + \boldsymbol{\nu}^{\top} q \tag{38a}$$

$$\text{Subject to: } -\boldsymbol{\lambda}_1^{\top} C \leq A^{\top} : \mathbf{x} \tag{38b}$$

$$-\boldsymbol{\lambda}_1^{\top} D + \boldsymbol{\lambda}_2^{\top} Q - \boldsymbol{\nu}^{\top} Q - \boldsymbol{\gamma}^{\top} \leq B^{\top} : \mathbf{y} \tag{38c}$$

$$\boldsymbol{\nu}^{\top}\left(\mathbf{e}-\hat{\mathbf{u}}\right) = \boldsymbol{\gamma}^{\top}\hat{\mathbf{u}} = 0 \tag{38d}$$

QP: The LCP problem in (35) could be converted to a QP problem as in (39):

$$\underset{\mathbf{x}\geq 0}{\text{Minimize}} \quad A^{\top}\mathbf{x} + B^{\top}\mathbf{y} \tag{39a}$$

$$\text{Subject to: } C\mathbf{x} + D\mathbf{y} \leq H \tag{39b}$$

$$\underset{\mathbf{y}\geq 0}{\text{Minimize}} \quad \mathbf{y}^{\top} Q\mathbf{y} + \mathbf{y}^{\top} q \tag{39c}$$

$$\text{Subject to: } -Q\mathbf{y} \leq : \boldsymbol{\lambda} \tag{39d}$$

Problem (39) has the same structure as the LP-QP problem presented in Sect. 8.1, and we may solve it with the same approach.

9 Conclusion

This chapter proposes a disjunctive-based decomposition (DBD) approach to solve linear bilevel programs (linear BLPs). The proposed DBD algorithm does not need selecting LP-correct big-M parameters which is a hard (if not impossible) process. Accordingly, the obtained solutions are guaranteed to be bilevel optimal without checking the LP-correctness of the big-M parameters as long as the conditions of the Proposition 1 are met. Effectiveness of the proposed DBD algorithm is shown through 12 different case studies. All these studies confirm promising performance of the proposed DBD algorithm. Also, different applications of the proposed DBD algorithm are reported in [19] and [18] addressing electricity network investment problem and in [24] addressing optimal bidding problem of a hydropower generator.

10 Declarations

The authors would like to thank Prof. Steve A. Gabriel for his insightful and valuable comments on earlier versions of this book chapter. This work was financially supported by the Swedish Energy Agency (Energimyndigheten) under Grant number 3233. The required computation is performed by computing resources from the Swedish National Infrastructure for Computing (SNIC) at PDC Center for High Performance Computing at KTH Royal Institute of Technology which was supported by the Swedish Research Council under Grant 2018-05973.

References

1. Anandalingam, G., White, D.J.: A solution method for the linear static Stackelberg problem using penalty functions. IEEE Trans. Autom. Control **35**(10), 1170–1173 (1990)
2. Bard, J.F.: Some properties of the bilevel programming problem. J. Optim. Theory Appl. **68**(2), 371–378 (1991)
3. Bard, J.F.: Practical Bilevel Optimization: Algorithms and applications. Springer, Berlin (1998)
4. Bard, J.F., Falk, J.E.: An explicit solution to the multi-level programming problem. Comput. Oper. Res. **9**(1), 77–100 (1982)
5. Bean, J.C.: A bender's approach to disjunctive programming (1992). Technical report available at http://hdl.handle.net/2027.42/3479
6. Benders, J.F.: Partitioning procedures for solving mixed-variables programming problems. Numer. Math. **4**, 238–252 (1962)
7. Bezanson, J., Edelman, A., Karpinski, S., Shah, V.B.: Julia: a fresh approach to numerical computing. SIAM Rev. **59**(1), 65–98 (2017)
8. Candler, W., Townsley, R.: A linear two-level programming problem. Comput. Oper. Res. **9**(1), 59–76 (1982)
9. Clark, P.A., Westerberg, A.W.: Bilevel programming for steady-state chemical process design—I. Fundamentals and algorithms. Comput. Chem. Eng. **14**(1), 87–97 (1990)
10. Colson, B.: BIPA (bilevel programming with approximation methods): Software guide and test problems. Centre de Recherche sur les Transports (CRT), Publication CRT-2002-38, Université de Montréal (2002). https://researchportal.unamur.be/en/publications/bipa-bilevel-programming-with-approximation-methods-software-guid
11. Conejo, A.J., Castillo, E., Minguez, R., Garcia-Bertrand, R.: Decomposition Techniques in Mathematical Programming: Engineering and Science Applications. Springer Science & Business Media, Berlin (2006)
12. Davarikia, H., Barati, M.: A tri-level programming model for attack-resilient control of power grids. J. Mod. Power Syst. Clean Energy **6**(5), 918–929 (2018)
13. Dhara, A., Dutta, J.: Optimality Conditions in Convex Optimization: A Finite-Dimensional View. CRC Press, Boca Raton (2011)
14. GAMS Development Corporation: General Algebraic Modeling System (GAMS) Release 25.1.3. Washington, DC (2021)
15. Hansen, P., Jaumard, B., Savard, G.: New branch-and-bound rules for linear bilevel programming. SIAM J. Scient. Stat. Comput. **13**(5), 1194–1217 (1992)
16. IBM ILOG: Cplex optimizer 12.8 (2018)
17. José, F.-A., McCarl, B.: A representation and economic interpretation of a two-level programming problem. J. Oper. Res. Soc. **32**(9), 783–792 (1981)

18. Khastieva, D., Hesamzadeh, M.R., Vogelsang, I., Rosellon, J.: Transmission network investment using incentive regulation: a disjunctive programming approach. Netw. Spatial Econ. **20**(4), 1029–1068 (2020)
19. Khastieva, D., Mohammadi, S., Hesamzadeh, M.R., Bunn, D.: Optimal transmission investment with regulated incentives based upon forward considerations of firm and intermittent resources with batteries. IEEE Trans. Power Syst. **36**(5), 4420–4434 (2021)
20. Kleinert, T., Labbé, M., Plein, F., Schmidt, M.: There's No Free Lunch: On the Hardness of Choosing a Correct Big-M in Bilevel Optimization (2019). Preprint
21. Kovács, A.: Bilevel programming approach to demand response management with day-ahead tariff. J. Mod. Power Syst. Clean Energy **7**(6), 1632–1643 (2019)
22. Lofberg, J.: Yalmip: a toolbox for modeling and optimization in MATLAB. In: 2004 IEEE International Conference on Robotics and Automation (IEEE Cat. No.04CH37508), pp. 284–289 (2004)
23. Makhorin, A.: GLPK (GNU Linear Programming Kit) (2021)
24. Moiseeva, E., Hesamzadeh, M.R.: Strategic bidding of a hydropower producer under uncertainty: modified benders approach. IEEE Trans. Power Syst. **33**(1), 861–873 (2018)
25. Nocedal, J., Wright, S.: Numerical Optimization. Springer Science & Business Media, Berlin (2006)
26. Pineda, S., Morales, J.M.: Solving linear bilevel problems using big-ms: not all that glitters is gold. IEEE Trans. Power Syst. **34**(3), 2469–2471 (2019)
27. Pineda, S., Bylling, H., Morales, J.M.: Efficiently solving linear bilevel programming problems using off-the-shelf optimization software. Optim. Eng. **19**(1), 187–211 (2018)
28. Sinha, A., Malo, P., Deb, K.: A review on bilevel optimization: from classical to evolutionary approaches and applications. IEEE Trans. Evol. Comput. **22**(2), 276–295 (2018)
29. Xu, F., Guo, Q., Sun, H., Zhang, B., Jia, L.: A two-level hierarchical discrete-device control method for power networks with integrated wind farms. J. Mod. Power Syst. Clean Energy **7**(1), 88–98 (2019)
30. Zeng, B., An, Y.: Solving bilevel mixed integer program by reformulations and decomposition. Optim. Online 1–34 (2014). https://optimization-online.org/wp-content/uploads/2014/07/4455.pdf

Optimization Techniques for Twin Support Vector Machines in Primal Space

Hossein Moosaei, Fatemeh Bazikar, and Milan Hladík

1 Twin Support Vector Machine for Binary Data Classification

Support vector machines are powerful supervised learning tools with proven effectiveness across diverse applications. Traditional SVMs create two parallel support hyperplanes to separate two classes (positive and negative data points) by maximizing the distance between them [22]. This approach follows the principle of structural risk minimization, where the final separating hyperplane is positioned equidistantly between the two support hyperplanes. The proximal support vector machine (PSVM) differs considerably from the SVM in its classification procedure. While the SVM classifies points by separating them into two disjoint half-spaces, the PSVM assigns points to the closest of two parallel planes, pushing these planes as far apart as possible. This formulation simplifies the computational process by reducing the classification problem to the solution of a single system of linear equations. In contrast, SVMs require the solution of quadratic programming

H. Moosaei (✉)
Department of Informatics, Faculty of Science, Jan Evangelista Purkyně University, Ústí nad Labem, Czech Republic
e-mail: hossein.moosaei@ujep.cz

F. Bazikar
Department of Computer Science, Faculty of Mathematical Sciences, Alzahra University, Tehran, Iran
e-mail: F.Bazikar@alzahra.ac.ir

M. Hladík
Department of Applied Mathematics, Faculty of Mathematics and Physics, Charles University, Prague, Czech Republic
e-mail: hladik@kam.mff.cuni.cz

B. Goldengorin (ed.), *Theory, Algorithms, and Experiments in Applied Optimization*, Springer Optimization and Its Applications 226,
https://doi.org/10.1007/978-3-031-91357-0_12

Fig. 1 Binary classification by using TWSVM

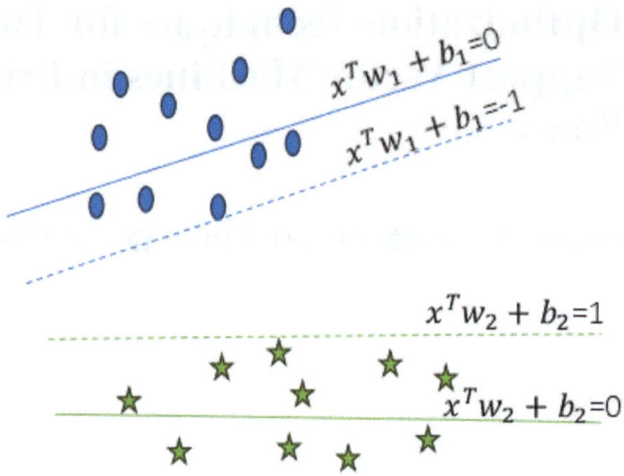

problems, which are computationally more demanding. The PSVM framework can also be interpreted as a regularized least squares problem, which further emphasizes its computational efficiency. Building on this foundation, the generalized eigenvalue proximal support vector machine (GEPSVM) extends the PSVM model by introducing nonparallel hyperplanes for classification. In GEPSVM, each hyperplane is designed to be proximal to one of the two data classes while maintaining maximum distance from the other, providing a more flexible approach to handling complex datasets. Mangasarian and Wild [11] proposed this method, where nonparallel hyperplanes are derived from the eigenvectors associated with the smallest eigenvalues of two related generalized eigenvalue problems. Similarly, twin support vector machine, introduced by Jayadeva et al. [7], aims to generate two nonparallel hyperplanes where each is closer to one class and farther from the other. Unlike SVMs, which solve a single quadratic programming problem (QPP), TWSVM solves two smaller QPPs, making it computationally faster. TWSVM has shown particular effectiveness with datasets like the "Cross Planes" dataset. Figure 1 illustrates the classification of two classes using TWSVM.

Next, we will explain the TWSVM method for binary data classification.

The training set T for the given binary classification problem is defined as $T = \{(x_i, y_i)|x_i \in \mathbb{R}^n, y_i \in \{\pm 1\}, i = 1, \cdots, m\}$. In this section, m_1 represents the number of data points in class $+1$, and m_2 represents those in class -1, where $m_1 + m_2 = m$. The matrix A of dimensions $m_1 \times n$ is created with each row corresponding to $(x_i)^\top$, where $y_i = +1$. Similarly, matrix B of size $m_2 \times n$ is constructed using data points where $y_i = -1$. In this formulation of the TWSVM, the goal is to determine two hyperplanes, $x^\top w_1 + b_1 = 0$ and $x^\top w_2 + b_2 = 0$, by solving two quadratic programming problems:

$$\begin{aligned}
\min_{w_1, b_1, \xi_1} \quad & \frac{1}{2}\|Aw_1 + e_1 b_1\|^2 + \frac{C_1}{2} e_2^\top \xi_1 \\
s.t. \quad & -(Bw_1 + e_2 b_1) + \xi_1 \geq e_2, \\
& \xi_1 \geq 0,
\end{aligned} \tag{1}$$

and

$$\begin{aligned}
\min_{w_2, b_2, \xi_2} \quad & \frac{1}{2}\|Bw_2 + e_2 b_2\|^2 + \frac{C_2}{2} e_1^\top \xi_2 \\
s.t. \quad & (Aw_2 + e_1 b_2) + \xi_2 \geq e_1, \\
& \xi_2 \geq 0,
\end{aligned} \tag{2}$$

where $w_1, w_2 \in \mathbb{R}^n$, ξ_1, $\xi_2 \geq 0$, $b_1, b_2 \in \mathbb{R}$, $\xi_1 \in \mathbb{R}^{m_2}$, and $\xi_2 \in \mathbb{R}^{m_1}$. The constants $C_1 > 0$ and $C_2 > 0$ are regularization parameters, while $e_1 \in \mathbb{R}^{m_1}$ and $e_2 \in \mathbb{R}^{m_2}$ are vectors of ones.

To interpret the first problem (1), we observe that the objective function consists of two main components:

- The first term, $\frac{1}{2}\|Aw_1 + e_1 b_1\|^2$, minimizes the squared distance between the hyperplane defined by $x^\top w_1 + b_1 = 0$ and the data points of class $+1$. This encourages the hyperplane to be positioned close to the points of class $+1$, providing a tighter fit around them.
- The second term, $\frac{C_1}{2} e_2^\top \xi_1$, penalizes any margin violations by introducing slack variables ξ_1. These slack variables measure the degree to which points from class -1 violate the margin constraints. The constant C_1 is a regularization parameter that controls the trade-off between minimizing these violations and maintaining a tight fit to the points of class $+1$.

The constraints of the problem ensure the following:

- The first constraint, $-(Bw_1 + e_2 b_1) + \xi_1 \geq e_2$, guarantees that the points from class -1 lie on or beyond the margin boundary defined by the hyperplane. If a point violates this condition, the corresponding slack variable ξ_1 compensates for this violation, but the objective function will penalize such violations.
- The second constraint, $\xi_1 \geq 0$, ensures that the slack variables are nonnegative, meaning no point from class -1 can violate the margin by more than what is allowed by ξ_1.

In summary, the goal of the optimization problem (1) is to find the optimal hyperplane parameters w_1 and b_1 that best separate the two classes of data. The hyperplane is encouraged to be as close as possible to the points of class $+1$ while at the same time ensuring that the points from class -1 remain outside the margin. The slack variables ξ_1 allow for some misclassification or margin violations, but these are penalized by the second term in the objective function, with the trade-off controlled by the parameter C_1.

The second optimization problem (2), similar to problem (1), focuses on separating the data points of class -1. The objective function aims to minimize the squared distance to keep the hyperplane close to the points in this class while also penalizing misclassifications of class $+1$. The constraints ensure that the hyperplane maintains a minimum distance from the points in class $+1$, allowing for some errors captured by slack variables. Overall, this problem seeks to determine the parameters of the hyperplane that effectively separate class -1 while managing the likelihood of misclassifying class $+1$ points.

Thus, the TWSVM framework consists of solving two quadratic programming problems (QPPs) that are smaller in size compared to the single large QPP in a standard SVM [22]. Each QPP in TWSVM is designed to focus on a specific class, with constraints influenced by the opposing class. This results in two smaller optimization problems, making TWSVM considerably faster than traditional SVM approaches. Specifically, TWSVM is approximately four times faster, as it solves two problems of size $m/2$, compared to the standard SVM's single problem of size m. While the time complexity for standard SVM is proportional to m^3, TWSVM solves two problems, each with a complexity proportional to $(m/2)^3$, leading to a speedup factor of about 4.

Building on the foundational concepts of the standard SVM, we now proceed to derive the dual formulation for TWSVM by determining the dual forms of (1) and (2). We begin with (1), deriving its Wolfe dual through the Lagrangian function associated with the problem, which is expressed as follows:

$$\begin{aligned}\mathcal{L}(w_1, b_1, \xi_1, \alpha, \beta) = &\frac{1}{2}\|Aw_1 + e_1 b_1\|^2 + \frac{C_1}{2} e_2^\top \xi_1 \\ &- \alpha^\top(-Bw_1 - e_2 b_1 + \xi_1 - e_2) - \beta^\top \xi_1. \end{aligned} \tag{3}$$

Here, $\alpha = (\alpha_1, \alpha_2, \cdots, \alpha_{m_2})^\top$ and $\beta = (\beta_1, \beta_2, \cdots, \beta_{m_2})^\top$ represent the Lagrange multipliers. Since (1) is a convex optimization problem, the Karush-Kuhn-Tucker (KKT) optimal conditions are both necessary and sufficient for the solution [1, 9]. We can thus drive the KKT conditions for (1), which are as follows:

- Stationary conditions:

$$\frac{\partial \mathcal{L}}{\partial w_1} = A^\top(Aw_1 + e_1 b_1) + B^\top \alpha = 0, \tag{4}$$

$$\frac{\partial \mathcal{L}}{\partial b_1} = e_1^\top(Aw_1 + e_1 b_1) + e_2^\top \alpha = 0, \tag{5}$$

$$\frac{\partial \mathcal{L}}{\partial \xi_1} = C_1 e_2 - \alpha - \beta = 0. \tag{6}$$

- Primal feasibility:

$$-(Bw_1 + e_2 b_1) + \xi_1 \geq e_2, \tag{7}$$

$$\xi_1 \geq 0. \tag{8}$$

- Dual feasibility:

$$\alpha \geq 0, \quad \beta \geq 0.$$

- Complementary slackness:

$$\alpha^\top [-(Bw_1 + e_2 b_1) + \xi_1 - e_2] = 0, \tag{9}$$

$$\beta^\top \xi_1 = 0. \tag{10}$$

Given $\beta \geq 0$, from Eq. (6), it follows that $\alpha \leq C_1 e_2$. Additionally, since $\alpha \geq 0$, the constraint $0 \leq \alpha \leq C_1 e_2$ holds. By combining Eqs. (4) and (5), we obtain

$$[A \;\; e_1]^\top [A \;\; e_1] [w_1 \;\; b_1]^\top + [B \;\; e_2]^\top \alpha = 0. \tag{11}$$

We define $H = [A \;\; e_1]$, $G = [B \;\; e_2]$, and $u = [w_1^\top \;\; b_1]^\top$. With these definitions, Eq. (11) can be rewritten as

$$(H^\top H)u + G^\top \alpha = 0. \tag{12}$$

If $H^\top H$ is invertible, then Eq. (12) leads to

$$u = -(H^\top H)^{-1}(G^\top \alpha),$$

and using the KKT conditions (4)–(12) and incorporating equation (12), we derive the Wolfe dual of (1) as follows:

$$\begin{aligned} \max \;\; & \mathcal{L}(w_1, b_1, \xi_1, \alpha, \beta) \\ s.t. \;\; & \nabla_{w_1} \mathcal{L}(w_1, b_1, \xi_1, \alpha, \beta) = 0, \\ & \frac{\partial \mathcal{L}}{\partial b_1} = 0, \\ & \frac{\partial \mathcal{L}}{\partial \xi_1} = 0, \\ & \alpha \geq 0, \;\; \beta \geq 0. \end{aligned} \tag{13}$$

The Wolfe dual of problem (1) is defined as

$$\begin{aligned} \max_{\alpha} \;\; & e_2^\top \alpha - \frac{1}{2} \alpha^\top G (H^\top H)^{-1} G^\top \alpha \\ s.t. \;\; & 0 \leq \alpha \leq C_1 e_2. \end{aligned} \tag{14}$$

Similarly, for (2), we derive its dual as

$$\begin{aligned} \max_{\nu} \;\; & e_1^\top \nu - \frac{1}{2} \nu^\top J (Q^\top Q)^{-1} J^\top \nu \\ s.t. \;\; & 0 \leq \nu \leq C_2 e_1, \end{aligned} \tag{15}$$

where $J = [A \;\; e_1]$, $Q = [B \;\; e_2]$, and the vector $v = [w_2^\top \;\; b_2]^\top$ is given by

$$v = (Q^\top Q)^{-1} J^\top \nu. \tag{16}$$

To obtain the separating hyperplanes, we solve (14) and (15) for α and ν, respectively. These solutions are then substituted into Eqs. (12) and (16) to compute the vectors $u = [w_1^\top, b_1]^\top$ and $v = [w_2^\top, b_2]^\top$. The resulting vectors define the separating hyperplanes

$$x^\top w_1 + b_1 = 0, \quad \text{and} \quad x^\top w_2 + b_2 = 0. \tag{17}$$

Remark 1 In the derivation of the dual problems, a modification is introduced when computing inverse matrices, such as $(A^\top A)$, to address cases of singularity or ill-conditioning. Instead of directly inverting $(A^\top A)$, a small positive scalar, δ, is added to the matrix, resulting in $(A^\top A + \delta I)^{-1}$. This adjustment helps to prevent issues arising from near-singular matrices and ensures that the inversion process remains stable. This technique is commonly employed to maintain numerical robustness when working with potentially ill-conditioned matrices in optimization problems.

For a new data point $x \in \mathbb{R}^n$, class membership is determined by evaluating its proximity to these hyperplanes. Indeed, the classification decision is made by assigning the point x to the class corresponding to the hyperplane for which the normalized distance is minimized:

$$class\ (x) = \arg\min \frac{|x^\top w_r + b_r|}{\|w_r\|^2}, \quad r = 1, 2. \tag{18}$$

2 Twin Bounded Support Vector Machine for Binary Data Classification

One widely recognized advantage of SVM's lies in their foundation on the structural risk minimization principle. However, in the primal forms of TWSVM (1) and (2), only the empirical risk is accounted for moreover, the dual formulations of TWSVM explicitly involve the inverse matrices $(H^\top H)^{-1}$ and $(Q^\top Q)^{-1}$. The presence of these matrices suggests an implicit assumption that both $(H^\top H)$ and $(Q^\top Q)$ are nonsingular, a condition that cannot always be guaranteed. This issue was notably addressed by Jayadeva et al. [7], who proposed a regularization method to mitigate the problem of matrix singularity. Nonetheless, applying such regularization transforms the original TWSVM problem into an approximate version rather than preserving its exact formulation. In response to these limitations, Shao et al. [20] introduced an enhanced variant of TWSVM in 2011. Their model incorporates improvements that ensure alignment with the structural risk minimization principle and automatically generate matrices in the dual formulation with guaranteed invertibility. Shao et al. [20] named this model twin bounded support vector machine (TBSVM), which will be explored further in the following:

$$
\begin{aligned}
\min_{w_1,b_1,\xi_1} \quad & \frac{1}{2}\|Aw_1 + e_1 b_1\|^2 + \frac{C_1}{2} e_2^\top \xi_1 + \frac{C_3}{2}\left(\|w_1\|^2 + b_1^2\right) \\
s.t. \quad & -(Bw_1 + e_2 b_1) + \xi_1 \geq e_2, \\
& \xi_1 \geq 0,
\end{aligned} \tag{19}
$$

and

$$
\begin{aligned}
\min_{w_2,b_2,\xi_2} \quad & \frac{1}{2}\|Bw_2 + e_2 b_2\|^2 + \frac{C_2}{2} e_1^\top \xi_2 + \frac{C_4}{2}\left(\|w_2\|^2 + b_2^2\right) \\
s.t. \quad & (Aw_2 + e_1 b_2) + \xi_2 \geq e_1, \\
& \xi_2 \geq 0,
\end{aligned} \tag{20}
$$

where ξ_1 and ξ_2 denote the error variables and $C_i, i = 1, \cdots, 4$, are positive parameters. Comparing the primal optimization problems (1)–(2) with (19)–(20), it becomes apparent that (19) incorporates an additional term $\frac{C_3}{2}(\|w_1\|^2 + b_1^2)$, while (20) includes $\frac{C_4}{2}(\|w_2\|^2 + b_2^2)$. To understand the impact of these additional terms, we focus on (19), noting that the analysis for (20) follows a similar rationale.

It is argued in [20] that the extra term $\frac{C_3}{2}(\|w_1\|^2 + b_1^2)$ contributes to the minimization of structural risk as represented in Eq. (19). Recall that structural risk minimization in the SVM context is achieved by maximizing the margin, quantified by the Euclidean distance between the two supporting hyperplanes. For TWSVM, the margin between the two classes is measured by the distance between the two bounding hyperplanes for each class, i.e., $x^\top w_1 + b_1 = -1$ and $x^\top w_1 + b_1 = 1$. As a result, the distance between the two hyperplane boundaries in the first class is $\frac{2}{\|w_1\|^2}$, which must be maximized. This observation justifies including the term $\|w_1\|^2$ in the objective function in (19). Additionally, the term b_1^2 is introduced, drawing from the PSVM framework as discussed by Fung and Mangasarian [3]. If we define the augmented vectors,

$$
X = \left[x^\top, 1\right]^\top, \quad \text{and} \quad W_1 = \left[w_1^\top, b_1\right]^\top,
$$

where the decision boundary is represented by the equation $X^\top W_1 = 0$ in the $\mathbb{R}^{n+1}$ space. The distance between the two bounding hyperplanes in this space is expressed as $\frac{2}{\|W_1\|}$, which simplifies to $\frac{2}{\sqrt{\|w_1\|^2+b_1^2}}$. This further justifies the inclusion of the term $\|w_1\|^2 + b_1^2$ in the objective function of problem (19). Similarly, analogous reasoning supports the inclusion of $\|w_2\|^2 + b_2^2$ in the objective function of problem (20) for the second class. In the TWSVM framework, the margin for each class is optimized independently, so a similar approach is applied to both classes to ensure optimal separation. Extensive computational studies, such as those in [3, 4], show that these formulations perform similarly to the classical formulations (1) and (2), while offering additional benefits, as mentioned above, including the strong convexity of the objective functions.

To solve the problems (19) and (20), it is necessary to derive their dual formulations. For this purpose, we construct the Lagrangian dual of (19). The Lagrangian function for the primal problem (19) is given by

$$\begin{aligned}\mathcal{L}(w_1, b_1, \xi_1, \alpha, \beta) = &\frac{1}{2}\|Aw_1 + e_1 b_1\|^2 + \frac{C_1}{2} e_2^\top \xi_1 + \frac{C_3}{2}(\|w_1\|^2 + b_1^2)\\ &- \alpha_2^\top \xi_1(-(B^\top w_1 + e_2 b_1) + \xi_1 - e_2) - \beta \xi_1,\end{aligned}$$

where $\alpha = (\alpha_1, \cdots, \alpha_{m_2})^\top$ and $\beta = (\beta_1, \cdots, \beta_{m_2})^\top$ are vectors of Lagrange multipliers.

Similar to TWSVM, the Wolfe dual of problem (19) can be formulated as

$$\begin{aligned}\max_{\alpha}\ & e_2^\top \alpha - \frac{1}{2}\alpha^\top G(H^\top H + C_3 I)^{-1} G^\top \alpha\\ s.t.\ \ & 0 \le \alpha \le C_1 e_2.\end{aligned} \tag{21}$$

Analogously, the Wolfe dual of (20) has the form of

$$\begin{aligned}\max_{\nu}\ & e_1^\top \nu - \frac{1}{2}\nu^\top J(Q^\top Q + C_4 I)^{-1} J^\top \nu\\ s.t.\ \ & 0 \le \nu \le C_2 e_1.\end{aligned} \tag{22}$$

In these dual formulations, $G,\ H,\ J$, and Q are matrices similar to those used in problems (14) and (15). Note that the matrices $(H^\top H + C_3 I)$ and $(Q^\top Q + C_4 I)$ are inherently nonsingular. Consequently, their inverses $(H^\top H + C_3 I)^{-1}$ and $(Q^\top Q + C_4 I)^{-1}$ exist without the need for additional assumptions or modifications. Solving the dual problems (21) and (22) yields vectors α and ν, respectively, in a process similar to that described in Sect. 1. We may obtain vectors $[w_1, b_1]$ and $[w_2, b_2]$, respectively, once we know α and ν. Thus determined, the separation hyperplanes are given by the vectors $[w_1, b_1]$ and $[w_2, b_2]$:

$$x^\top w_1 + b_1 = 0, \quad \text{and} \quad x^\top w_2 + b_2 = 0.$$

For a new data point $x \in \mathbb{R}^n$, it can be classified using the same decision rule as in (18). The remaining details follow the established framework for TWSVM (see [7, 20]).

3 Generalized Newton Method for Twin Bounded Support Vector Machines

In this section, we introduce a Newton-based method for solving the optimization problems associated with the twin bounded support vector machine (TBSVM) in the primal space, which we refer to as NTBSVM. The traditional TBSVM approaches typically operate in the dual space, which can lead to increased computational complexity. By addressing the optimization problems in the primal space, NTBSVM aims to streamline the solution process while preserving high classification accuracy.

The original TBSVM model involves solving constrained quadratic programming problems (QPPs), as shown in the problems (19) and (20). These QPPs can be computationally expensive, especially for large datasets. To address this challenge, we convert these constrained problems into unconstrained optimization problems, thereby simplifying the optimization procedure. To achieve this, we leverage generalized Newton's method for solving unconstrained optimization problems. This approach is particularly challenging here, as the objective functions are only once-differentiable, not twice, which prevents the use of the standard Newton method, as it relies on the availability of a Hessian matrix.

By introducing 2-norm slack vectors, denoted as ξ_1 and ξ_2, into the objective functions of problems (19) and (20), we obtain the unconstrained minimization problems.

For the primal optimization problems in (19) and (20), our goal is to minimize

$$\begin{aligned}
\min_{w_1,b_1,\xi_1} \quad & \frac{1}{2}\|Aw_1 + e_1b_1\|^2 + \frac{C_1}{2}\|\xi_1\|^2 + \frac{C_3}{2}(\|w_1\|^2 + b_1^2) \\
s.t. \quad & -(Bw_1 + e_2b_1) + \xi_1 \geq e_2, \\
& \xi_1 \geq 0,
\end{aligned} \tag{23}$$

and

$$\begin{aligned}
\min_{w_2,b_2,\xi_2} \quad & \frac{1}{2}\|Bw_2 + e_2b_2\|^2 + \frac{C_2}{2}\|\xi_2\|^2 + \frac{C_4}{2}(\|w_2\|^2 + b_2^2) \\
s.t. \quad & (Aw_2 + e_1b_2) + \xi_2 \geq e_1, \\
& \xi_2 \geq 0.
\end{aligned} \tag{24}$$

For the optimal solution of problems (23) and (24), we have

$$\xi_1 = (e_2 + (Bw_1 + e_2b_1))_+, \tag{25}$$

$$\xi_2 = (e_1 - (Aw_2 + e_1b_2))_+, \tag{26}$$

where the plus function $(\cdot)_+$ replaces the negative components of a vector with zeros. Thus, we can replace ξ_1 and ξ_2 in (23) and (24) with (25) and (26), respectively, and convert the TBSVM problems (23) and (24) into an equivalent TBSVM that are an unconstrained optimization problems as follows:

$$\min_{w_1,b_1} F_1(w_1,b_1) = \frac{1}{2}\|Aw_1 + e_1b_1\|^2 + \frac{C_1}{2}\|(e_2 + (Bw_1 + e_2b_1))_+\|^2 + \frac{C_3}{2}(\|w_1\|^2 + b_1^2), \tag{27}$$

and

$$\min_{w_2,b_2} F_2(w_2,b_2) = \frac{1}{2}\|Bw_2 + e_2b_2\|^2 + \frac{C_2}{2}\|(e_1 - (Aw_2 + e_1b_2))_+\|^2 + \frac{C_4}{2}(\|w_2\|^2 + b_2^2). \tag{28}$$

The objective functions $F_1(w_1, b_1)$ and $F_2(w_2, b_2)$ are differentiable, and their gradients can be computed. Then, we have

$$\nabla F_1(w_1,b_1) = \begin{bmatrix} \frac{\partial F_1}{\partial w_1} \\ \frac{\partial F_1}{\partial b_1} \end{bmatrix} = \begin{bmatrix} A^\top(Aw_1 + e_1b_1) \\ e_1^\top(Aw_1 + e_1b_1) \end{bmatrix} + C_1 \begin{bmatrix} B^\top(e_2 + (Bw_1 + e_2b_1))_+ \\ e_2^\top(e_2 + (Bw_1 + e_2b_1))_+ \end{bmatrix} + C_3 \begin{bmatrix} w_1 \\ b_1 \end{bmatrix}, \tag{29}$$

and

$$\nabla F_2(w_2,b_2) = \begin{bmatrix} \frac{\partial F_2}{\partial w_2} \\ \frac{\partial F_2}{\partial b_2} \end{bmatrix} = \begin{bmatrix} B^\top(Bw_2 + e_2b_2) \\ e_2^\top(Bw_2 + e_2b_2) \end{bmatrix} + C_2 \begin{bmatrix} -A^\top(e_1 - (Aw_2 + e_1b_2))_+ \\ -e_1^\top(e_1 - (Aw_2 + e_1b_2))_+ \end{bmatrix} + C_4 \begin{bmatrix} w_2 \\ b_2 \end{bmatrix}. \tag{30}$$

The functions $\nabla F_1(w_1, b_1)$ and $\nabla F_2(w_2, b_2)$ are not differentiable, as they contain terms involving the plus function $(\cdot)_+$. Then, the objective functions of problems (27) and (28) are piecewise quadratic, convex, and differentiable. However, they are not twice differentiable, so we cannot apply standard Newton's method to solve these problems. To solve problems (27) and (28), we first introduce some definitions related to the objective functions of these problems.

Generalized Hessians

The class LC^1 of functions is defined as follows [5]:

Definition 1 A function $f : \mathbb{R}^n \to \mathbb{R}$ is said to be an LC^1 function on an open set A if:

- f is continuously differentiable on A.
- The gradient ∇f is locally Lipschitz continuous on A.

Although the ordinary second derivative of f may not exist everywhere, the condition that ∇f is locally Lipschitz on A ensures that ∇f is differentiable almost everywhere on A. In such cases, we can define a generalized derivative for ∇f in Clarke's sense [2], which allows us to extend the notion of the Hessian to a broader class of functions.

Definition 2 Let $f : \mathbb{R}^n \to \mathbb{R}$ be an LC^1 function on an open set A, and let $x \in A$. The generalized Hessian of f at x is the set of $n \times n$ matrices defined by

$$\partial^2 f(x) = \text{co}\left\{\widetilde{H} \in \mathbb{R}^{n\times n} : \exists x_k \to x \text{ where } \nabla f \text{ is differentiable at } x_k, \text{ and } \nabla^2 f(x_k) \to \widetilde{H}\right\},$$

where co$\{\cdot\}$ stands for the convex hull.

By construction, $\partial^2 f(x_0)$ is a nonempty, compact, convex set of symmetric matrices, which reduces to $\{\nabla^2 f(x_0)\}$ when f is differentiable at x_0.

Example 1 Let $h : A \subseteq \mathbb{R}^n \to \mathbb{R}$ be a function that is twice continuously differentiable on an open set A, and define a new function f on A as

$$f(x) = [(h(x))_+]^2.$$

Clearly, f is LC^1 function. For any point $x_0 \in A$, the second-order subdifferential of f, $\partial^2 f(x_0)$, is given by the following cases:

$$\partial^2 f(x_0) = \begin{cases} \left\{2h(x_0)\nabla^2 h(x_0) + 2\nabla h(x_0)\nabla h(x_0)^\top\right\} & \text{if } h(x_0) > 0, \\ \{0\} & \text{if } h(x_0) < 0, \\ \left\{2\alpha\nabla h(x_0)\nabla h(x_0)^\top \mid \alpha \in [0, 1]\right\} & \text{if } h(x_0) = 0. \end{cases}$$

In this expression, $\nabla h(x_0)$ and $\nabla^2 h(x_0)$ represent the gradient and Hessian of h evaluated at x_0, respectively, and α is a scalar in the interval $[0, 1]$.

To better understand the concept of the generalized Hessian, we now focus on a specific case derived from the general form above. This will help to illustrate the concept more clearly and provide concrete insight into how the generalized Hessian operates in practice.

Example 2 The function $f(x) = \|(Ax - b)_+\|^2$, where $A \in \mathbb{R}^{m\times n}$ and $b \in \mathbb{R}^m$, has the following properties:

- f is an LC^1 function since it is continuously differentiable on $\mathbb{R}^n$, and its gradient, ∇f, is Lipschitz continuous on $\mathbb{R}^n$ with constant $2\|A^\top\|\|A\|$.
- In particular, the gradient of f is given by

$$\nabla f = 2A^\top(Ax-b)_+ = A^\top(Ax-b) + A^\top|Ax-b|.$$

- The generalized Hessian of f at $x \in \mathbb{R}^n$, denoted by $\partial^2 f(x)$, exists. According to the above example, each element $\widetilde{H} \in \partial^2 f(x)$ takes the form:

$$\widetilde{H} = A^\top(I + \widetilde{D}(x))A,$$

where $\widetilde{D}(x)$ is a diagonal matrix with entries $\widetilde{D}_{ii}(x)$, determined as follows:

$$\widetilde{D}_{ii}(x) = \begin{cases} 1, & \text{if } (Ax-b)_i > 0, \\ -1, & \text{if } (Ax-b)_i < 0, \\ \in [-1, 1], & \text{if } (Ax-b)_i = 0. \end{cases} \tag{31}$$

For computational purposes, in the case of $(Ax-b)_i = 0$, $\widetilde{D}_{ii}(x)$ is usually chosen to be 0.

This form of $\widetilde{D}(x)$ will be used throughout the following sections.

The concept of the generalized Hessian enables us to handle functions whose second derivatives may not exist in the classical sense. This is particularly important for analyzing the behavior of nonsmooth or piecewise smooth functions, which frequently arise in optimization problems.

Now, we will revisit our problems and explore the Lipschitz continuity properties of the partial derivatives of F_1 and provide the generalized Hessian of F_1.

Theorem 1 *The functions $\nabla F_1(s)$ and $\nabla F_2(s)$ are globally Lipschitz.*

Proof From Eq. (29), we have the following expression for the gradient of F_1 with respect to $s = [w_1^\top, b_1]^\top$:

$$\nabla F_1(s) = T_1 s + C_1 T_2 (e_2 + T_3)_+ + C_3 T_4 s,$$

where

$$T_1 = \begin{bmatrix} A^\top A & A^\top e_1 \\ e_1^\top A & e_1^\top e_1 \end{bmatrix}, \quad T_2 = \begin{bmatrix} B^\top \\ e_2^\top \end{bmatrix}, \quad T_3 = \begin{bmatrix} B \; e_2 \end{bmatrix}, \quad \text{and } T_4 = I_{n+1}.$$

Now, consider the difference between the gradients evaluated at two points, s_1 and s_2:

$$\begin{aligned} &\|\nabla F_1(s_1) - \nabla F_1(s_2)\| \\ &= \|T_1 s_1 + C_1 T_2 (e_2 + T_3 s_1)_+ + C_3 T_4 s_1 - T_1 s_2 - C_1 T_2 (e_2 + T_3 s_2)_+ - C_3 T_4 s_2\| . \end{aligned}$$

We can now bound the difference above by applying the triangle inequality:

$$\|T_1(s_1 - s_2)\| + \|C_3 T_4(s_1 - s_2)\| + \left\| C_1 \, |T_2|^\top \left((e_2 + T_3 s_1) - (e_2 + T_3 s_2)\right) \right\|.$$

Since $(e_2 + T_3 s_1) - (e_2 + T_3 s_2) = T_3(s_1 - s_2)$, we can further simplify the bound:

$$\|\nabla F_1(s_1) - \nabla F_1(s_2)\| \leq \|T_1\| \cdot \|s_1 - s_2\| + C_3 \|T_4\| \cdot \|s_1 - s_2\| + C_1 \| |T_2|^\top T_3 \| \cdot \|s_1 - s_2\|.$$

Thus, we infer that the gradient ∇F_1 is globally Lipschitz continuous with constant

$$K = \|T_1\| + C_3 \|T_4\| + C_1 \| |T_2|^\top T_3 \|.$$

We can obtain a similar result for ∇F_2. □

Based on the aforementioned points and considering the above theorem, we know that $\nabla F_1(s)$ and $\nabla F_2(s)$ are almost always differentiable, and the generalized Hessian of F_1 and F_2 exists everywhere. For the generalized Hessian, we use the fact that the Clarke subdifferential for a piecewise smooth function can be computed by differentiating at points where the function is smooth. Applying this to F_1, we obtain an element of the generalized Hessian of the objective function F_1 as follows (i.e., $\widetilde{H}(s) \in \partial^2 F_1(s)$):

$$\widetilde{H}(s) = T_1 + C_1 T_2 D_1 T_3 + C_3 T_4,$$

where D_1 is a diagonal matrix with entries determined by the sign function as follows:

$$D_1 = \operatorname{diag}(\operatorname{sgn}(e_2 + T_3 s)_+).$$

The entries of D_1 depend on the values of $e_2 + T_3 s$, capturing the nonsmooth behavior of the function when the argument switches between positive and negative values.

These results establish the global Lipschitz continuity of the gradient and the structure of the generalized Hessian for F_1 and F_2, which are essential for analyzing the stability and convergence properties of optimization algorithms applied to these functions.

Generalized Newton Method

Now, we discuss the application of the Generalized Newton Method to solve the following optimization problem:

$$\min_{w_1, b_1} F_1(w_1, b_1) = \frac{1}{2}\|A w_1 + e_1 b_1\|^2 + \frac{C_1}{2}\|(e_2 + (B w_1 + e_2 b_1))_+\|^2 + \frac{C_3}{2}(\|w_1\|^2 + b_1^2), \tag{32}$$

where $w_1 \in \mathbb{R}^n$ and $b_1 \in \mathbb{R}$, and A, B, e_1, e_2, C_1, C_3 are given parameters.

To apply the Generalized Newton Method, we first compute the gradient and the generalized Hessian of the objective function $F_1(w_1, b_1)$. The gradient of the function can be computed as follows:

$$\nabla F_1(w_1, b_1) = \left[\frac{\partial F_1}{\partial w_1}, \frac{\partial F_1}{\partial b_1}\right]^\top.$$

From the previous results, we know that

$$\frac{\partial F_1}{\partial w_1} = A^\top(Aw_1 + e_1b_1) + C_1B^\top(e_2 + Bw_1 + e_2b_1)_+ + C_3w_1,$$

and

$$\frac{\partial F_1}{\partial b_1} = e_1^\top(Aw_1 + e_1b_1) + C_1e_2^\top(e_2 + Bw_1 + e_2b_1)_+ + C_3b_1.$$

The generalized Hessian of the function is given by

$$\widetilde{H}(w_1, b_1) = T_1 + C_1T_2D_1T_3 + C_3T_4,$$

where $D_1 = \text{diag}(\text{sgn}(e_2 + (Bw_1 + e_2b_1))_+)$.

The Generalized Newton Method updates the parameters w_1 and b_1 iteratively using the update rule

$$\begin{bmatrix} w_1^{k+1} \\ b_1^{k+1} \end{bmatrix} = \begin{bmatrix} w_1^k \\ b_1^k \end{bmatrix} - \alpha_k \left(\widetilde{H}(w_1^k, b_1^k)\right)^{-1} \nabla F_1(w_1^k, b_1^k),$$

where α_k is a suitable step size determined through line search to ensure convergence.

To address potential singularity in the generalized Hessian matrix, the direction d_k is modified by adding a small regularization term δ. Consequently, the generalized Newton method updates the parameter $s = [w_1^\top, b_1]^\top$ iteratively using the update rule

$$d_k = -(\widetilde{H}(s^{(k)}) + I\delta)^{-1}\nabla F_1(s^{(k)}),$$
$$s^{(k+1)} = s^{(k)} + \alpha_k d_k,$$

with the step length α_k and the initial vector $s^{(0)}$. Furthermore, the stopping criterion for this method is as follows:

$$\|s^{(k+1)} - s^{(k)}\| \leq tol.$$

Under appropriate conditions, such as the Lipschitz continuity of the gradient and the positive definiteness of the generalized Hessian, the Generalized Newton Method can guarantee convergence to a local minimum of the function $F_1(s)$. In optimization algorithms, choosing an appropriate step size is crucial for ensuring convergence and improving the performance of the algorithm. The Armijo method, also known as the backtracking line search, is a widely used technique to determine an appropriate step size that satisfies certain criteria.

The Armijo method aims to find a step size α_k such that the following condition is satisfied:

$$F_1(s^{(k)} + \alpha_k d_k) \leq F_1(s^{(k)}) + c\alpha_k \nabla F_1(s^{(k)})^\top d_k, \tag{33}$$

where $c \in (0, 1)$ is a predefined constant, often set to a small value such as $c = 10^{-4}$. This condition guarantees that the decrease in the objective function is proportional to the step size and the norm of the gradient, ensuring an appropriately balanced step size that is neither excessively large nor too small.

To solve the above problems, we now apply the Generalized Newton Method with an Armijo step size [6, 10, 12–15, 19].

Algorithm 1 Armijo Generalized Newton Algorithm for (32)

Start with any $s^{(0)} \in \mathbb{R}^{n+1}$. For $k = 0, 1, \ldots$:

1. Stop if $||\nabla F_1(s^{(k)})|| \leq tol$.
2. Set

$$s^{(k+1)} = s^{(k)} + \alpha_k d_k,$$

where $\alpha_k = \max\{1, \frac{1}{2}, \frac{1}{4}, \ldots\}$ such that:

$$F_1(s^{(k)} + \alpha_k d_k) \leq F_1(s^{(k)}) + c\alpha_k \nabla F_1(s^{(k)})^\top d_k,$$

for some $c \in (0, \frac{1}{2})$, and d_k is the Newton direction:

$$d_k = -(\widetilde{H}(s^{(k)}) + I\delta)^{-1} \nabla F_1(s^{(k)}).$$

3. Set $k = k + 1$, and go to step 1.

Remark on Complexity

The complexity of applying the Generalized Newton Method to solve the optimization problem in (32) is dominated by the steps involved in computing the gradient, Hessian, and solving the linear system at each iteration. Specifically, computing the gradient and Hessian involves matrix-vector multiplications and element-wise operations, which have a complexity of $O(n^2)$. However, the most computationally expensive part is solving a system of equations, which typically requires $O(n^3)$ operations. Therefore, the overall complexity of each iteration is $O(n^3)$, and the total complexity for k iterations is $O(kn^3)$, where k is the number of iterations required for convergence. This implies that the method's efficiency depends critically on both the number of iterations and the problem size.

4 Least Squares Twin Bounded Support Vector Machine for Binary Data Classification

Building on the least squares support vector machine (LS-SVM) framework introduced by Suykens and Vandewalle [21], Kumar and Gopal [8] developed the least squares twin support vector machine (LS-TWSVM). This variant enhances the traditional TWSVM by reformulating its primal problems using a least squares approach. A key modification is the replacement of inequality constraints in the standard TWSVM with equality constraints, which simplifies the optimization process considerably. Consequently, LS-TWSVM avoids complex QP problem-solving and reduces computational complexity by solving two systems of linear equations.

The main advantage of LS-TWSVM lies in its efficiency, especially for large-scale classification tasks where computational resources and time are critical. By eliminating the quadratic programming bottleneck, LS-TWSVM enables faster computation without compromising classification performance. Empirical results across various datasets demonstrate that LS-TWSVM not only matches but often outperforms conventional TWSVM models as well, while significantly reducing training and prediction times. This makes LS-TWSVM an attractive choice for binary classification tasks in resource-constrained environments. However, LS-TWSVM's reliance on empirical risk minimization, which focuses solely on minimizing training error, can make it prone to overfitting, particularly in noisy datasets. To address this limitation, an enhanced version called the Least Squares Twin Bound Support Vector Machine (LS-TBSVM) was proposed in [23]. This variant introduces improvements that enhance classification accuracy and reduce the risk of overfitting, making it more robust in real-world applications. In general, variants of the Least Squares TWSVM have garnered significant interest from researchers due to their simple structure and strong performance [16–18].

In the following, we introduce the least squares reformulation of the primal problem for TBSVM. The critical adjustment involves replacing the inequality constraints of the original TBSVM with equality constraints, thereby streamlining the optimization process. The modified primal problem (19) is expressed as follows:

$$\begin{aligned} \min_{w_1,b_1,\xi_1} \quad & \frac{1}{2}\|Aw_1 + e_1 b_1\|^2 + \frac{C_1}{2}\|\xi_1\|^2 + \frac{C_3}{2}(\|w_1\|^2 + b_1^2) \\ s.t. \quad & -(Bw_1 + e_2 b_1) + \xi_1 = e_2, \end{aligned} \tag{34}$$

where the objective function now includes the square of the 2-norm of the slack ξ_1 with weight $\frac{C_1}{2}$, replacing the 1-norm used in the original TBSVM formulation. This modification leverages the strong convexity advantage, enhancing the robustness and stability of the model. This modification renders the constraint $\xi_1 \geq 0$ redundant, as the inclusion of the squared 2-norm ensures that the slack variable is naturally nonnegative. The simplicity of this change enables us to express the solution of the quadratic programming problem (QPP) as a system of linear equations. By substituting the equality constraint into the objective function, we reformulate the problem as

$$\min_{w_1,b_1} \frac{1}{2}\|Aw_1 + e_1b_1\|^2 + \frac{C_1}{2}\|(Bw_1 + e_2b_1) + e_2\|^2 + \frac{C_3}{2}(\|w_1\|^2 + b_1^2). \tag{35}$$

Since (35) is an unconstrained convex optimization problem, its KKT conditions result in a system of linear equations. Furthermore, the strict convexity of the objective function ensures that (35) has a unique solution $[w_1^\top, b_1]^\top$. By taking the gradient of the objective function with respect to w_1 and b_1, and setting it to zero, we obtain

$$A^\top(Aw_1 + e_1b_1) + C_1B^\top(Bw_1 + e_2b_1 + e_2) + C_3w_1 = 0, \tag{36}$$

$$e_1^\top(Aw_1 + e_1b_1) + C_1e_2^\top(Bw_1 + e_2b_1 + e_2) + C_3b_1 = 0. \tag{37}$$

Hence, arranging (36) and (37) in matrix form and solving for w_1 and b_1 give

$$\begin{bmatrix} A^\top A & A^\top e_1 \\ e_1^\top A & e_1^\top e_1 \end{bmatrix}\begin{bmatrix} w_1 \\ b_1 \end{bmatrix} + C_3\begin{bmatrix} w_1 \\ b_1 \end{bmatrix} + C_1\begin{bmatrix} B^\top B & B^\top e_2 \\ e_2^\top B & e_2^\top e_2 \end{bmatrix}\begin{bmatrix} w_1 \\ b_1 \end{bmatrix} + C_1\begin{bmatrix} B^\top e_2 \\ e_2^\top e_2 \end{bmatrix} = 0. \tag{38}$$

Now, we have

$$\begin{bmatrix} w_1 \\ b_1 \end{bmatrix} = -C_1\left[[A^\top \ e_1^\top]^\top[A \ e_1] + C_1[B^\top \ e_2^\top]^\top[B \ e_2] + C_3I\right]^{-1}\begin{bmatrix} B^\top e_2 \\ e_2^\top e_2 \end{bmatrix}. \tag{39}$$

Denoting $H = [A \ e_1]$ and $G = [B \ e_2]$, the solution becomes

$$\begin{bmatrix} w_1 \\ b_1 \end{bmatrix} = -(H^\top H + C_1G^\top G + C_3I)^{-1}(C_1G^\top e_2). \tag{40}$$

Similarly, for the second hyperplane, we can define the following QPP:

$$\begin{aligned} \min_{w_2,b_2,\xi_2} \ & \frac{1}{2}\|Bw_2 + e_2b_2\|^2 + \frac{C_2}{2}\|\xi_2\|^2 + \frac{C_4}{2}(\|w_2\|^2 + b_2^2) \\ s.t. \ & (Aw_2 + e_1b_2) + \xi_2 = e_1, \end{aligned} \tag{41}$$

and the solution to this problem is

$$\begin{bmatrix} w_2 \\ b_2 \end{bmatrix} = (G^\top G + C_2 H^\top H + C_4 I)^{-1}(C_2 H^\top e_1). \tag{42}$$

We note that the above matrices are always invertible because $C_1 - C_4$ are positive, and the terms $C_3 I$ and $C_4 I$ ensure the matrices are positive definite. Thus, we obtain the hyperplanes $x^\top w_1 + b_1 = 0$ and $x^\top w_2 + b_2 = 0$. A new data point $x \in \mathbb{R}^n$ is classified based on its minimum perpendicular distance to either hyperplane. The classification rule can be applied using the same decision criterion as in (18).

Remark on Complexity The simplicity of this approach lies in the fact that by substituting the equality constraint into the objective function, we transform the quadratic programming problem (QPP) into an unconstrained convex optimization problem. This transformation reduces the problem to a system of linear equations, which can be efficiently solved. The complexity of solving the system primarily depends on the size of the matrix involved. Since the problem is convex and the objective function is strictly convex, the solution is guaranteed to be unique. Therefore, the computational complexity is dominated by solving a system of equations step, which typically scales as $O(n^3)$ for the $n \times n$ matrix, where n is the number of variables.

Acknowledgments The work of H. Moosaei and M. Hladík was supported by the Czech Science Foundation Grant 25-15714S.

References

1. Chandra, S., Jayadeva, Mehra, A.: Numerical optimization with applications. Alpha Sci. Int. (2009). https://books.google.cz/books?id=MogtPwAACAAJ&redir_esc=y
2. Clarke, F.H.: Optimization and Nonsmooth Analysis. SIAM, Philadelphia (1990)
3. Fung, G., Mangasarian, O.L.: Proximal support vector machine classifiers. In Proceedings of the Seventh ACM SIGKDD International Conference on Knowledge Discovery and Data Mining, pp. 77–86 (2001)
4. Lee, Y.J., Mangasarian, O.L.: RSVM: reduced support vector machines. In Proceedings of the 2001 SIAM International Conference on Data Mining, pp. 1–17. SIAM, Philadelphia (2001)
5. Hiriart-Urruty, J.B., Strodiot, J.J., Nguyen, V.H.: Generalized Hessian matrix and second-order optimality conditions for problems with $C_{1,1}$ data. Appl. Math. Optim. **11**(1), 43–56 (1984)
6. Ketabchi, S., Moosaei, H., Razzaghi, M., Pardalos, P.M.: An improvement on parametric $\nu-$support vector algorithm for classification. Ann. Oper. Res. **276**(1), 155–168 (2019)
7. Khemchandani, R., Chandra, S.: Twin support vector machines for pattern classification. IEEE Trans. Pattern Anal. Mach. Intell. **29**(5), 905–910 (2007)
8. Kumar, M.A., Gopal, M.: Least squares twin support vector machines for pattern classification. Expert Syst. Appl. **36**(4), 7535–7543 (2009)
9. Mangasarian, O.L.: Nonlinear Programming. Society for Industrial and Applied Mathematics, Philadelphia (1994)

10. Mangasarian, O.L.: A Newton method for linear programming. J. Optim. Theory Appl. **121**, 1–18 (2004)
11. Mangasarian, O.L., Wild, E.W.: Multisurface proximal support vector machine classification via generalized eigenvalues. IEEE Trans. Pattern Anal. Mach. Intell. **28**(1), 69–74 (2006)
12. Mangasarian, O.L.: A finite Newton method for classification. Optim. Methods Softw. **17**(5), 913–929 (2002)
13. Moosaei, H., Bazikar, F., Ketabchi, S., Hladík, M.: Universum parametric-margin $\nu-$support vector machine for classification using the difference of convex functions algorithm. Appl. Intell. **52**(3), 2634–2654 (2022)
14. Moosaei, H., Khosravi, S., Bazikar, F., Hladík, M., Rosario Guarracino, M.: A novel method for solving universum twin bounded support vector machine in the primal space. Ann. Math. Artif. Intell. (2023). https://doi.org/10.1007/s10472-023-09871-0
15. Moosaei, H., Hladík, M., Razzaghi, M., Ketabchi, S.: Newton-based approach to solving K-SVCR and Twin-KSVC multi-class classification in the primal space. Comput. Oper. Res. **160**, 106370 (2023)
16. Moosaei, H., Hladík, M.: Least squares approach to K-SVCR multi-class classification with its applications. Ann. Math. Artif. Intell. **90**(7), 873–892 (2022)
17. Moosaei, H., Bazikar, F., Hladík, M.: Multi-task twin support vector machine with Universum data. Eng. Appl. Artif. Intell. **132**, 107951 (2024)
18. Moosaei, H., Bazikar, F., Hladík, M., Pardalos, P.M.: Sparse least-squares Universum twin bounded support vector machine with adaptive L_p-norms and feature selection. Expert Syst. Appl. **248**, 123378 (2024)
19. Pardalos, P.M., Ketabchi, S., Moosaei, H.: Minimum norm solution to the positive semi-definite linear complementarity problem. Optimization **63**(3), 359–369 (2014)
20. Shao, Y.H., Zhang, C.H., Wang, X.B., Deng, N.Y.: Improvements on twin support vector machines. IEEE Trans. Neural Netw. **22**(6), 962–968 (1998)
21. Suykens, J.A., Vandewalle, J.: Least squares support vector machine classifiers. Neural Proces. Lett. **9**, 293–300 (1999)
22. Vapnik, V.: Statistical Learning Theory. Wiley, Hoboken (1998)
23. Xu, Y., Xi, W., Lv, X., Guo, R.: An improved least squares twin support vector machine. J. Inf. Comput. Sci. **9**(4), 1063–1071 (2012)

On the Calculation of Similarity Measures for Figures by Nonsmooth Global Optimization

Vladimir Norkin and Georg Pflug

1 Introduction

This paper considers the problem of measuring similarity between two objects/figures in Euclidean and metric spaces. The figures are represented as clouds of points and/or line segments in the space. Generally, similarity is understood as a binary relation satisfying reflexivity, symmetricity, and transitivity properties. In machine learning, the similarity of objects is understood as belonging to the same class and is based on feature analysis of the objects. In the present paper, the similarity is understood as a measure of the best alignment of one space figure upon another under a certain family of transformations. Alignment can include displacement, rotation, stretching, and other transformations dependent on some parameters. As a measure of closedness, the minimal Pompeiu-Hausdorff distance (or its robust analogues) is used between transformed figures, and thus the problem is reduced to minimization of this distance over a set of parameters. The obtained measure of similarity can be used further within machine leaning classification method like the nearest neighbor classifier. The use of the Pompeiu-Hausdorff distance to metricize spaces of sets is well known [6, 48, 51].

In particular, we use the minimal Hausdorff distance between all possible isometric transformations of the two compared figures, also known under the

V. Norkin (✉)
V.M. Glushkov Institute of Cybernetics of the National Academy of Sciences of Ukraine & National Technical University of Ukraine "Igor Sikorsky Kyiv Polytechnic Institute", Kyiv, Ukraine
e-mail: v.norkin@kpi.ua

G. Pflug
University of Vienna, Vienna, Austria
e-mail: georg.pflug@univie.ac.at

B. Goldengorin (ed.), *Theory, Algorithms, and Experiments in Applied Optimization*, Springer Optimization and Its Applications 226,
https://doi.org/10.1007/978-3-031-91357-0_13

name of Gromov-Hausdorff distance. Finding the corresponding similarity measure is reduced to a multidimensional global optimization problem with equality constraints that express isometricity conditions. In this paper, we show that the calculation of Hausdorf-based similarity measures between clouds of points and line segments is a Lipschitz nonsmooth global optimization problem, which is typically multi-extremal. We estimate the related Lipschitz constants and develop a branch-and-bound scheme combined with nonsmooth local minimization algorithms to solve this problem.

The problem of recognizing figures on a plane and in a space is an important component of the theory and practice of computer vision [43]. It includes the actual task of recognizing figures and the task of highlighting the figures in the image. Particular cases of this problem are the recognition of scanned printed and handwritten letters, texts, drawings, diagrams, maps, and other flat images. A more general problem is the recognition of three-dimensional shapes from their images on a plane.

As alternative, neural networks are successfully used for image recognition, but data preparation and training of networks are a laborious and nontrivial problem. Remark that classes of equivalence of objects with respect to a group of transformations are considered for designing neural networks invariant to the group transformations of inputs [16, 21]. However, we concentrate here on image recognition based on measures of similarity.

The problem of figure recognition is considered in the following setting. Let there be a 2D scanned black and white or gray image in some pixel format. First, it is transformed into a set of points in Euclidean space, the first coordinates represent the location, and the subsequent coordinates can describe the degree of pixel brightness in grayscale or other colors. It is assumed that there is a collection of samples of geometric figures, which are also represented by sets of points in two-dimensional space. Recognition consists of a sequential optimal alignment of the quested image with samples and finding the best match. Alignment can include displacement, rotation, and stretching of the specimen in a plane or space [3, 7, 33]. The alignment transformation depends on several parameters, the translation vector, the angle of rotation, the scaling coefficient, etc. For example, a general affine transformation of a two-dimensional Euclidean space is determined by six parameters, a two-dimensional displacement vector, and 2×2 linear transformation matrix. As an answer to the recognition task, the class label is returned for which the best match with the image is obtained.

Finding a pattern within an image is quite similar. The image is converted from a pixel format to a nonuniform point cloud in Euclidean space. Patterns are also represented as sets of points in space. The problem of searching for a pattern is formulated as finding the optimal alignment of the moved and transformed pattern within a fragment of the image. The one-sided Hausdorff distance (or its modifications) is used as a measure of closedness, i.e., deviation of the transformed sample from the image. To avoid the sample falling on the filled area of the image, where the one-sided distance is close to zero, the image may be pre-converted into a contour view.

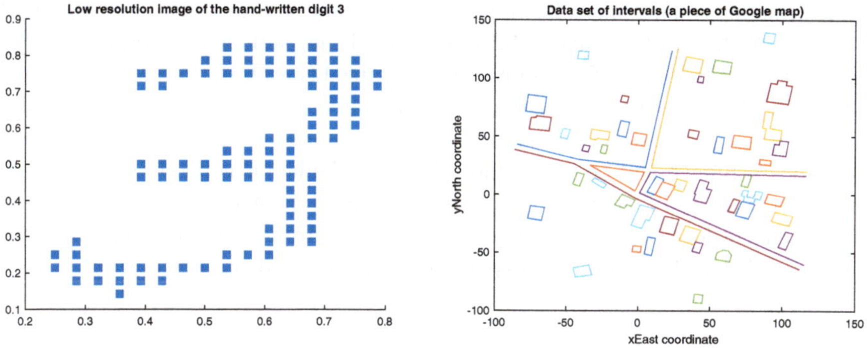

Fig. 1 Examples of figures represented by points and intervals in $\mathbb{R}^2$

Instead of points, the image can consist of a finite set of continuous line segments, joint or disjoint [14, 18, 28, 63, 64] (Fig. 1). The point images are just particular case of the segment images. The same technology based on Hausdorff type similarity measures can be applied to the recognition of segment images, but calculation of these similarity measures is more involved.

In this paper, the main attention is paid to the problem of finding the optimal transformation and solving the arising continuous optimization problems by non-smooth global optimization methods.

The problem of calculation of the image-to-class distance is a laborious task because of large instances in the class and the difficulty of calculation of distances to elements of the class [4]. In the paper, we also propose how to reduce the complexity of this task by exploiting a triangle inequality and the Lipschitz property of optimization problems involved.

Thus, the contribution of this paper includes:

- the description of geometric objects both as clouds of points and also as clouds of line segments,
- the introduction of new generalized Hausdorff-like similarity measures between movable objects,
- highlighting the Lipschitz nature of the figure similarity assessment problems, in particular in case of line segment representation of objects,
- the calculation of the corresponding Lipschitz constants,
- the development of the problem specific branch-and-bound algorithm for the optimal alignment of plane figures,
- the development a figure-to-class robust similarity measure,
- and complexity reduction techniques in calculations of similarity measures.

The paper is organized as follows. Section 2 reviews the literature. Section 3 formally defines similarity relations through pseudo-distances and transformations. Section 4 reviews Hausdorff distance and its modifications. Section 5 studies Lipschitz transformations of sets. Section 6 considers groups and semigroups of transformations on the plane. Section 7 introduces and studies transformation-based generalized Hausdorff distances. Section 8 establishes sufficient conditions for the existence of optimal transformations for generalized Hausdorff distances. Section 9 establishes Lipschitz properties for distance functions. Section 10 explores calculation of Hausdorff distance between figures containing line segments. Section 11 describes a branch-and-bound method for the calculation generalized Hausdorff distances. Section 12 suggests an optimization-based method for calculation of the Gromov-Hausdorff distance for bendable figures. Section 13 explores application of generalized Hausdorff distances to pattern recognition and classification.

2 Literature Review

In the described formulation, the problem of recognizing figures on a plane by transforming and comparing them to minimize the Hausdorff-distance (H-distance) was considered in [29, 34, 53, 54, 60]. Figures can be represented by clouds of points or lines segments [13, 14, 18, 28, 63, 64]. In these articles, the problem of minimizing the H-distance is solved by general global optimization techniques, e.g., by branch-and-bound [35], exhaustive and heuristic methods. In [57], various analogs of the H-distance are considered, and it is shown that the arising optimization problem concerning the parameters of the displacement is indeed multiextremal and nonsmooth. Robust analogs of the H-distance in the context of image recognition problems are proposed in [29, 30, 32, 36, 37, 50, 52–54, 62]. By itself, the problem of calculating the H-distance between large sets of points, not to mention the problem of its optimization, is very laborious. Papers [15, 26, 58, 65] are devoted to the problem of accelerated calculation of this distance. The applications of the H-distance and its analogs for solving problems of image analysis, recognition, and computer vision are considered in [31, 37, 50, 61, 62]. The estimation of the similarity of contour images has its specifics; for this, the strengthened Hausdorff metric, namely, the Frechet metric, is used [2, 55]. A comprehensive review of metric similarity measures for searching information in large data bases is given in [49].

For the evaluation of similarity between partially nonrigid objects based on Gromov-Hausdorff metrics [11, 24, 59], see [1, 9, 10, 22, 38, 39]. In this approach, objects (surfaces) are considered as compact metric spaces with some intrinsic metrics, e.g., geodesic ones. The similarity between objects is measured as Gromov-Hausdorff distance between the corresponding metric spaces. This is the minimal Hausdorff distance between all possible isometric embeddings of the two objects into common metric spaces. The rigidity of objects is limited to bending transformations of objects with preserving the internal distances between points in the object and not necessarily preserving angles in the transformed object.

Relation of the Paper to Results in the Literature We consider objects as equivalence classes of movable objects, represented by clouds of points and line segments as in [14, 18, 28, 63, 64]. Objects are considered as belonging to the same class, if they can be continuously transformed to each other by a transformation from some group as in [3, 7, 29, 34, 53, 54, 60]. In these works, the degree of similarity is measured by the minimal achieved Hausdorff distance or by stable to outliers its robust modifications. We extend this approach to include parametric semigroups of nonexpanding transformations and by using for similarity remeasuring so-called generalized Hausdorff metrics. For this, we develop and investigate new Hausdorff-like distances, which resemble Gromov-Hausdorff metrics [11, 24]; in particular, we establish triangle inequalities for these distances. We also suggest robust modifications of these distances to include grayscale information. We show that optimal matching problem is a nonsmooth global Lipschitz optimization problem with respect to transformation parameters. We calculate the corresponding (variable) Lipschitz constants and adapt a branch-and-bound method for solving the optimal matching problem. To improve bounds in the method, we use the so-called tangent minorants, i.e., functions, which are less than the minimized function and touch its graph at some points. It is worth to remark that the calculation of Hausdorff distance for line segment figures is a Lipschitz global one-dimensional optimization problem, which is solved within a branch-and-bound scheme. We discuss the complexity reduction of the nearest neighbor classification rule by exploiting triangle inequality and Lipschitz nature of the proposed distances.

3 Similarity Relations

Definition 1 (Similarity Relation) Let $\mathcal{C}$ be a set (of objects). A binary relation on $\mathcal{C}$ is called a similarity relation, if it is reflexive, symmetric, and transitive.

In what follows, objects can be (a) finite sets of points or/and bounded linear segments in $\mathbb{R}^n$, $n = 2, 3, \ldots$, with fixed distances between them; (b) a connected graph in $\mathbb{R}^n$ with fixed lengths of edges. This graph can be considered as a finite metric space of its vertexes with distances between them as shortest paths in the graph. In the case (a) the set of objects is metrizable, in particular, with Hausdorff distance and in the case (b) the set of objects is metrizable with Gromov-Hausdorff distance (see Sect. 12).

Definition 2 (Pseudo-Distance) A function $dist(\cdot, \cdot)$ is called a pseudo-distance on a space $\mathcal{C}$, if $dist(A, B) \geq 0$, $dist(A, A) = 0$, $dist(A, B) = dist(B, A)$, and $dist(A, B) \leq dist(A, C) + dist(C, B)$ for any $A, B, C \in \mathcal{C}$, although $dist(A, B) = 0$ does not necessarily mean that $A = B$.

Lemma 1 (Similarity Through a Pseudo-Distance) *Pseudo-distance defines a similarity relation:* $A \simeq B$ *if* $dist(A, B) = 0$. *This relation is reflexive, symmetric and transitive.*

Definition 3 (Similarity Through Transformation) Let $\mathfrak{I}$ be a group of transformations in $\mathcal{C}$. Let us define the $\mathfrak{I}$-relation $\sim_{\mathfrak{I}}$ (or simply $\sim$) in the space $\mathcal{C}$: sets $A, B \in \mathcal{C}$ are $\mathfrak{I}$-related, in notation: $A \sim B$, if there is some $\mathbf{T} \in \mathfrak{I}$ such that $\mathbf{T}A = B$.

Lemma 2 (Transformation-Based Similarity) *Relation* $\sim_{\mathfrak{I}}$ *is a similarity relation; i.e., it is reflexive, symmetric, and transitive.*

4 Hausdorff Distance and Its Robust Modifications

Definition 4 (Pompeiu-Hausdorff Distance) [48], see, e.g., [6, 51]. The Pompeiu-Hausdorff deviation of a set $A \subset \mathbb{R}^n$ from $B \subset \mathbb{R}^n$ is the quantity

$$\Delta(A, B) = \sup_{a\in A} \inf_{b\in B} \|a - b\|,$$

and the Pompeiu-Hausdorff distance between A and B is the quantity

$$H(A, B) = \max\left(\Delta(A, B), \Delta(B, A)\right).$$

The distance $H(A, B)$ is a metric in the space of compact subsets $\mathbb{C}$ of $\mathbb{R}^n$. For a survey of various metrics in the set space, see [17]; for metrics in the space of convex sets, see [8].

In particular, the Pompeiu-Hausdorff deviation and distance satisfy the triangle inequality:

$$\Delta(A, B) \le \Delta(A, C) + \Delta(C, B),$$
$$H(A, B) \le H(A, C) + H(C, B).$$

Due to the triangle inequality, the set-defined functions $\Delta(\cdot, B)$ and $H(\cdot, B)$ satisfy the following Lipschtz property.

$$\begin{aligned} |\Delta(A_1, B) - \Delta(A_2, B)| &\le \max\{\Delta(A_1, A_2), \Delta(A_2, A_1)\} = H(A_1, A_2), \\ |H(A_1, B) - H(A_2, B)| &\le \max\{\Delta(A_1, A_2), \Delta(A_2, A_1)\} = H(A_1, A_2). \end{aligned} \tag{1}$$

A special case is the point-to-set distance function $d(x, C) = \Delta(\{x\}, C) = \inf_{c\in C} \|x - c\|$. It satisfies the following Lipschitz property; see [51, Example 9.6]: For any nonempty, closed set $C \subset \mathbb{R}^k$, the distance function $x \mapsto d(x, C)$ is Lipschitz continuous on $\mathbb{R}^k$ with Lipschitz constant 1, i.e.,

$$|d(x_1, C) - d(x_2, C)| \le \|x_1 - x_2\|$$

for any $x_1, x_2 \in \mathbb{R}^n$.

The Case of Finite Sets In the case of finite sets A and B, one may adopt a statistical language [44, 50, 52, 66]. Let the figures A and B in space $\mathbb{R}^k$ be given by sets of points (vectors), $A = \{x^1, , \ldots, x^n\} \subset \mathbb{R}^k$ and $B = \{y^1, \ldots y^m\} \subset \mathbb{R}^k$. For the number of elements in these sets, we will use the notation $|A| = n,\ |B| = m$.

The "distribution function" of the numerical set $\{d(x, y^j), y^j \in B\}$ is the function

$$F_{x,B}(t) = \frac{1}{|B|}\left\{\# y^j \in B : d(x, y^j) \leq t\right\},$$

where # means the number of elements such that..., and the quantile $q_\alpha(x, B)$ of level α of the numerical set $\{d(x, y^j), y^j \in B\}$ is the quantity:

$$q_\alpha(x, B) = \min\left\{t : F_{x,B}(t) \geq \alpha\right\}, \alpha \geq 1/|B|. \tag{2}$$

In other words, $q_\alpha(x, B)$ is the smallest t such that at least $\lceil \alpha m \rceil$ points of B are in a distance from x not larger than t.

Let $F^\alpha_{A,B}(t) = \frac{1}{|A|}\{\# x \in A : q_\alpha(x, B) \leq t\}$ be the distribution function of the numerical set $\{q_\alpha(x^i, B), x^i \in A\}$. The (α, β)-quantile $Q_{\alpha,\beta}(A, B)$ of the deviation of a finite set A from a finite set B is the $(1 - \beta)$-quantile of the numerical set $\{q_\alpha(x^i, B), x^i \in A\}$:

$$Q_{\alpha,\beta}(A, B) = \min\left\{t : F^\alpha_{A,B}(t) \geq 1 - \beta\right\}, \beta \geq 1/|A|. \tag{3}$$

In other words, $Q_{\alpha,\beta}(A, B)$ is the smallest t such that at least $\rceil n(1 - \beta)\lceil$ points from A are such that each of them is in a distance smaller than t from at least $\lceil m\alpha \rceil$ points from B.

Defining the quantity

$$D_{\alpha,\beta}(A, B) = \max\left\{Q_{\alpha,\beta}(A, B), Q_{\beta,\alpha}(B, A)\right\}$$

one can see that it satisfies the augmented triangle inequality [44]:

$$D_{\alpha,\beta}(A, B) \leq D_{\alpha,0}(A, Z) + D_{\alpha,0}(Z, B) + \min\left\{D_{\alpha,0}(A, A), D_{\alpha,0}(B, B)\right\}.$$

For $\alpha = 1/|B|$ and $\beta = 1/|A|$, it holds: $q_\alpha(x, B) = d(x, B)$, $Q_{\alpha,\beta}(A, B) = \Delta(A, B)$, $D_{\alpha,\beta}(A, B) = H(A, B)$.

The average quantile deviation of a set A from B is the quantity

$$\bar{\Delta}_\alpha(A, B) = \frac{1}{n}\sum_{i=1}^{n} q_\alpha(x^i, B),$$

and the averaged Hausdorff quantile distance between A and B is the quantity

$$\bar{D}_{\alpha,\beta}(A, B) = \max\left\{\bar{\Delta}_\alpha(A, B), \bar{\Delta}_\beta(B, A)\right\}.$$

The quantile distances $\bar{\Delta}_\alpha(A, B)$ and $\bar{D}_{\alpha,\beta}(A, B)$ are more robust with respect to outliers in sets A and B than the initial values $\Delta(A, B)$ and $H(A, B)$.

Instead of $d(x, C) = \min_{y \in C} d(x, y)$, one may use an approximate value $M_\gamma(x, C) = -\gamma \ln\left(\sum_{y \in C} \exp(-d(x, y)/\gamma)\right)$ with $\gamma > 0$. When γ tends to zero, $M_\gamma(x, C)$ tends to $d(x, C)$.

It was proved in [23] that the function

$$d(A, B) = \frac{1}{|A \cup B|}\left(\frac{1}{|A|}\sum_{a \in A}\sum_{b \in B \setminus A} d(a, b) + \frac{1}{|B|}\sum_{b \in B}\sum_{a \in A \setminus B} d(a, b)\right)$$

is a metric in the space of finite sets.

Remark 2 (Incorporation of the Grayscale Information into a Distance Formation) Suppose that the points b_j from a figure B carry some extra information $s_j > 0$ like brightness/darkness in some gray scale. Then, we can rescale this information by attributing to each point of B a number $p_j = s_j / \sum_j s_j$, which is interpreted as probability of point b_j to belong to figure B. Then the cumulative distribution function of the distance between a point x and B is defined as

$$F_{x,B}(t) = \left\{\sum_{b_j \in B:\, d(x,b_j) < t} p_j\right\}.$$

The corresponding quantiles are defined as in (2).

The deviation of the set A having grayscale information r_i on each point a_i from the set B is defined in an analogous way. First, we normalize this information; i.e., calculate $q_i = r_i / \sum_i r_i$. The cumulative distribution function for the set of numbers $\{q_\alpha(a_i, B), a_i \in A\}$ is

$$F^\alpha_{A,B}(t) = \left\{\sum_{a_i \in A: q_\alpha(a_i, B) \le t} q_i\right\}.$$

The (α, β)-quantile distance between A and B is defined as in (3).

5 Lipschitz Transformations of Sets

Let $\mathbb{C}$ be the family of compact subsets in $\mathbb{R}^n$ and $H(\cdot, \cdot)$ be the Hausdorff distance on $\mathbb{C}$, and let the "norm" of a set be $\|A\| = \sup_{a \in A} \|a\|$. For a set $A \subset \mathbb{R}^n$ and operator $\mathbf{T} : \mathbb{R}^n \to \mathbb{R}^n$ we define the image set $\mathbf{T}A = \{\mathbf{T}a : a \in A\}$.

Assume now that a parameterized family of operators T_z is given, where the parameter z belongs to some metric space $(\mathbb{Z}, \rho)$,

Definition 5 An operator family $\{\mathbf{T}_z : \mathbb{C} \to \mathbb{C}, z \in \mathbb{Z}\}$ is called Lipschitz one, if for any set $A \in \mathbb{C}$ and any points $z_1, z_2 \in \mathbb{Z}$ there is a constant L_A such that it holds that

$$H\left(\mathbf{T}_{z_1} A, \mathbf{T}_{z_2} A\right) \le L_A \cdot \rho(z_1, z_2).$$

Lemma 3 *For a Lipschitz operator family* $\mathbf{T}_z : \mathbb{C} \to \mathbb{C}$, $z \in \mathbb{Z}$, *and any* $A, B \in \mathbb{C}$, *the functions* $\delta_{AB}(z) = \Delta(\mathbf{T}_z A, B)$ *and* $h_{AB}(z) = H(\mathbf{T}_z A, B)$ *are Lipschitz ones on* $\mathbb{Z}$*; i.e., for any* $z_1, z_2 \in \mathbb{Z}$, *it holds that* $|\delta_{AB}(z_1) - \delta_{AB}(z_2)| \le L_A \cdot \rho(z_1, z_2)$ *and* $|h_{AB}(z_1) - h_{AB}(z_2)| \le L_A \cdot \rho(z_1, z_2)$.

Proof By (1) and the Lipschitz property of $\mathbf{T}_z$, we have

$$\begin{aligned}
\left|\delta_{AB}(z) - \delta_{AB}(z')\right| &= \left|\Delta(\mathbf{T}_z A, B) - \Delta\left(\mathbf{T}_{z'} A, B\right)\right| \\
&\le H\left(\mathbf{T}_z A, \mathbf{T}_{z'} A\right) \le L_A \cdot \rho(z, z'); \\
\left|h_{AB}(z) - h_{AB}(z')\right| &= \left|H(\mathbf{T}_z A, B) - H\left(\mathbf{T}_{z'} A, B\right)\right| \\
&\le H\left(\mathbf{T}_z A, \mathbf{T}_{z'} A\right) \le L_A \cdot \rho(z, z').
\end{aligned}$$

□

Let us consider several examples of transformations in $\mathbb{R}^2$ [20]. Denote $x = (x_1, x_2)^T$ an arbitrary point in $\mathbb{R}^2$ and $\hat{x} = (x_1, x_2, 1)^T$ its homogeneous coordinates, $(\cdot)^T$ denotes transposition of a matrix or a vector. The homogeneous coordinates allow to represent transformations in a convenient matrix multiplicative form.

Example 3 A rigid motion (translation-rotation) transformation has the form,

$$\mathbf{T}_z x = \begin{pmatrix} \cos z_1 & \sin z_1 \\ -\sin z_1 & \cos z_1 \end{pmatrix} \begin{pmatrix} x_1 \\ x_2 \end{pmatrix} + \begin{pmatrix} z_2 \\ z_3 \end{pmatrix}, \quad z = (z_1, z_2, z_3) \in \mathbb{Z}.$$

In the computationally convenient homogenous coordinates, this transformation has the following representation:

$$\hat{\mathbf{T}}_z \hat{x} = \begin{pmatrix} \cos z_1 & \sin z_1 & z_2 \\ -\sin z_1 & \cos z_1 & z_3 \\ 0 & 0 & 1 \end{pmatrix} \begin{pmatrix} x_1 \\ x_2 \\ 1 \end{pmatrix} = \begin{pmatrix} 1 & 0 & z_2 \\ 0 & 1 & z_3 \\ 0 & 0 & 1 \end{pmatrix} \begin{pmatrix} \cos z_1 & \sin z_1 & 0 \\ -\sin z_1 & \cos z_1 & 0 \\ 0 & 0 & 1 \end{pmatrix} \begin{pmatrix} x_1 \\ x_2 \\ 1 \end{pmatrix}.$$

Example 4 Scaled translation-rotation transformation has the form

$$\mathbf{T}_z x = z_2 \begin{pmatrix} \cos z_1 & \sin z_1 \\ -\sin z_1 & \cos z_1 \end{pmatrix} \begin{pmatrix} x_1 \\ x_2 \end{pmatrix} + \begin{pmatrix} z_3 \\ z_4 \end{pmatrix}, \quad z = (z_1, z_2, z_3, z_4) \in \mathbb{Z},$$

and in the extended form

$$\hat{\mathbf{T}}_z\hat{x} = \begin{pmatrix} z_2\cos z_1 & z_2\sin z_1 & z_3 \\ -z_2\sin z_1 & z_2\cos z_1 & z_4 \\ 0 & 0 & 1 \end{pmatrix} \begin{pmatrix} x_1 \\ x_2 \\ 1 \end{pmatrix}$$

$$= \begin{pmatrix} 1 & 0 & z_3 \\ 0 & 1 & z_4 \\ 0 & 0 & 1 \end{pmatrix} \begin{pmatrix} z_2 & 0 & 0 \\ 0 & z_2 & 0 \\ 0 & 0 & 1 \end{pmatrix} \begin{pmatrix} \cos z_1 & \sin z_1 & 0 \\ -\sin z_1 & \cos z_1 & 0 \\ 0 & 0 & 1 \end{pmatrix} \begin{pmatrix} x_1 \\ x_2 \\ 1 \end{pmatrix}.$$

Example 5 An affine transformation has the form

$$\mathbf{T}_z x = \begin{pmatrix} z_1 & z_2 \\ z_3 & z_4 \end{pmatrix} \begin{pmatrix} x_1 \\ x_2 \end{pmatrix} + \begin{pmatrix} z_5 \\ z_6 \end{pmatrix}, \quad z = (z_1, \ldots, z_6) \in \mathbb{Z}, \quad \hat{\mathbf{T}}_z\hat{x} = \begin{pmatrix} z_1 & z_2 & z_5 \\ z_3 & z_4 & z_6 \\ 0 & 0 & 1 \end{pmatrix} \begin{pmatrix} x_1 \\ x_2 \\ 1 \end{pmatrix}.$$

Example 6 And finally, a projective transformation has the form

$$\mathbf{T}_z x = \begin{pmatrix} (z_{11}x_1 + z_{12}x_2 + z_{13})/(z_{31}x_1 + z_{32}x_2 + 1) \\ (z_{21}x_1 + z_{22}x_2 + z_{23})/(z_{31}x_1 + z_{32}x_2 + 1) \end{pmatrix}, \quad z = \begin{pmatrix} z_{11} & z_{12} & z_{13} \\ z_{21} & z_{22} & z_{23} \\ z_{31} & z_{32} & 1 \end{pmatrix} \in Z,$$

$$\hat{\mathbf{T}}_z\hat{x} = (z_{31}x_1 + z_{32}x_2 + 1)^{-1} \begin{pmatrix} z_{11} & z_{12} & z_{13} \\ z_{21} & z_{22} & z_{23} \\ z_{31} & z_{32} & 1 \end{pmatrix} \begin{pmatrix} x_1 \\ x_2 \\ 1 \end{pmatrix}.$$

6 Groups of Transformations and Equivalence Classes of Sets

Recall that a set $\mathfrak{I}$ of transformations with composition $\mathbf{T}_1\mathbf{T}_2$ (juxtaposition) is called a *semigroup*, if the following properties are met:

- If $\mathbf{T}_1, \mathbf{T}_2 \in \mathfrak{I}$, then $\mathbf{T}_1\mathbf{T}_2 \in \mathfrak{I}$;
- For $\mathbf{T}_1, \mathbf{T}_2, \mathbf{T}_3 \in \mathfrak{I}$, it holds $\mathbf{T}_1(\mathbf{T}_2\mathbf{T}_3) = (\mathbf{T}_1\mathbf{T}_2)\mathbf{T}_3$ (associativity).

$\mathfrak{I}$ is called a *group*, if it is a semigroup and the following additional properties hold:

- There is an identity element $\mathbf{I} \in \mathfrak{I}$ such that $\mathbf{IT} = \mathbf{TI} = \mathbf{T}$ for any $\mathbf{T} \in \mathfrak{I}$;
- For each $\mathbf{T} \in \mathfrak{I}$, there is an inverse element $\mathbf{T}^{-1} \in \mathfrak{I}$ such that $\mathbf{TT}^{-1} = \mathbf{T}^{-1}\mathbf{T} = \mathbf{I}$.

Example 7 (Rotation and Translation Group in $\mathbb{R}^2$) Consider the isometric transformation $\mathbf{T}_{z_0,z_1}$ of points on the plane $\mathbb{R}^2$, given by

$$\begin{pmatrix} x_1 \\ x_2 \end{pmatrix} \mapsto \begin{pmatrix} \cos\varphi & \sin\varphi \\ -\sin\varphi & \cos\varphi \end{pmatrix} \begin{pmatrix} x_1 \\ x_2 \end{pmatrix} + \begin{pmatrix} z_{11} \\ z_{12} \end{pmatrix}.$$

It is convenient to identify the $\mathbb{R}^2$ with the complex plane. To do so, one may set $z_0 = e^{i\phi}$, and $z_1 = z_{11} + iz_{12}$ with $i = \sqrt{-1}$. The pair (z_0, z_1) of complex numbers is the parameter of the transformation $\mathbf{T}_{z_0,z_1}$. This transformation maps the vector $(x_1, x_2)^T$, now written as $x = x_1 + ix_2$, into

$$\mathbf{T}_{z_0,z_1}(x) = z_0 x + z_1.$$

The set of transformations $\mathfrak{I} = \left\{\mathbf{T}_{z_0,z_1} : \mathbf{T}_{z_0,z_1} x = z_0 x + z_1,\ |z_0| = 1\right\}$ constitutes a group of isometric transformations of a plane. Notice that $\mathbf{T}_{1,0}(x) = x$; hence, it is the identity transformation.

Moreover, if $\mathbf{T}_{z_0,z_1}, \mathbf{T'}_{z'_0,z'_1} \in \mathfrak{I}$, then

$$\begin{aligned} \mathbf{T}_{z_0,z_1}\mathbf{T}_{z'_0,z'_1} &= \mathbf{T}_{z_0 z'_0, z_0 z'_1 + z_1}, \quad \left|z_0 z'_0\right| = 1; \\ \mathbf{T}^{-1}_{z_0,z_1} &= \mathbf{T}_{1/z_0, -z_1/z_0} \\ \mathbf{T}_{z_0,z_1}\mathbf{T}^{-1}_{z_0,z_1} &= \mathbf{T}_{1,0} \\ \mathbf{T}^{-1}_{z_0,z_1}\mathbf{T}_{z_0,z_1} &= \mathbf{T}_{1,0} \end{aligned}$$

The group is not commutative in general, since $z_0 z'_1 + z_1$ is not identical to $z'_0 z_1 + z'_1$.

Transformations $\mathbf{T}_{z_0,z_1} \in \mathfrak{I}$ preserve distances between complex numbers. Let u, w be two complex numbers, then

$$\left|\mathbf{T}_{z_0,z_1}u - \mathbf{T}_{z_0,z_1}w\right| = \left|\mathbf{T}_{z_0,z_1}(u - w)\right| = |z_0(u - w)| = |z_0| \cdot |u - w| = |u - w|.$$

Example 8 (scaling, rotation, and translation group). The family of transformations $\mathfrak{I} = \left\{\mathbf{T}_{z_0,z_1} : \mathbf{T}_{z_0,z_1} x = z_0 x + z_1,\ z_0 \neq 0\right\}$, where z_0, z_1 are complex parameters, x is a complex variable, constitute a group of transformations of a plane, which do not preserve distances.

Example 9 (Nonexpanding Scaling, Rotation, and Translation Group) The family $\mathfrak{I} = \left\{\mathbf{T}_{z_0,z_1} : \mathbf{T}_{z_0,z_1} x = z_0 x + z_1,\ |z_0| \leq 1\right\}$ constitutes a semigroup of nonexpanding transformations of the plane $\mathbb{R}^2$.

Definition 6 (Classes of Equivalent Sets) Let $\mathfrak{I}$ be a group of transformations in $\mathbb{R}^n$ and $\mathbb{C}$ be the set of compacts in $\mathbb{R}^n$. We define a $\mathfrak{I}$-equivalence relation $\sim_{\mathfrak{I}}$ (or simply $\sim$) in the space $\mathbb{C}$. Sets $A, B \in \mathbb{C}$ are $\mathfrak{I}$-equivalent, $A \sim B$, if there is some $\mathbf{T} \in \mathfrak{I}$ such that $\mathbf{T}A = B$.

The relation $\sim_{\mathfrak{I}}$ is reflexive, symmetric, and transitive:

- $A \sim A$,
- $A \sim B$ implies $B \sim A$,
- $A \sim B$ and $B \sim C$ imply $A \sim C$.

Thus , the set of compacts $\mathbb{C}$ decomposes into equivalence classes. For a set A, we denote $\tilde{A}$ the family of subsets, which are equivalent to A.

Concerning transformations $\mathfrak{I}$, we will use the following assumptions.

Assumption 1 (Attainability of Infinum) For $\mathfrak{I}$ and any compact sets $A, B \in \mathbb{C}$, there exist $\mathbf{T}_1, \mathbf{T}_2 \in \mathfrak{I}$ such that

(a)

$$\inf_{\mathbf{T} \in \mathfrak{I}} \Delta(\mathbf{T}A, B) = \Delta(\mathbf{T}_1 A, B);$$

(b)

$$\inf_{\mathbf{T} \in \mathfrak{I}} H(\mathbf{T}A, B) = H(\mathbf{T}_2 A, B).$$

Later in Sect. 8, we give sufficient conditions for Assumption 1 to be fulfilled.

Assumption 2 (Isometric Transformation Group $\mathfrak{I}$) For any compact sets $A, B \in \mathbb{C}$ and $\mathbf{T} \in \mathfrak{I}$, it holds that $\Delta(\mathbf{T}A, \mathbf{T}B) = \Delta(A, B)$, and hence $H(\mathbf{T}A, \mathbf{T}B) = H(A, B)$.

Assumption 3 (Nonexpanding Transformation Semigroup $\mathfrak{I}$) For any compact sets $A, B \in \mathbb{C}$ and $\mathbf{T} \in \mathfrak{I}$, it holds $\Delta(\mathbf{T}A, \mathbf{T}B) \leq \Delta(A, B)$, and hence $H(\mathbf{T}A, \mathbf{T}B) \leq H(A, B)$.

For a particular nonexpanding transformation semigroup, see Example 9.

Assumption 4 Transformations $\mathbf{T} \in \mathfrak{I}$ preserve lengths and Lebesgue volumes in $\mathbb{R}^n$.

In what follows, we cosider a subfamily $\mathfrak{D}$ of compacts $\mathbb{C}$. Define $\mathfrak{D}$ as the family of all compact sets A in $\mathbb{R}^n$ such that they are the closure of their open interior A°, $(A^\circ)^c = A$.

It is easy to see that for each $x \in A \in \mathfrak{D}$ and each $\varepsilon > 0$, there exists an open ball contained in $\{y \in A : d(y, x) < \varepsilon\}$. Notice that with λ being the Lebesgue measure, $\lambda(A) > 0$ for all $A \in \mathfrak{D}$.

7 Group and Semigroup Distance Between Equivalence Classes of Sets

As it was mentioned in Sect. 2, the Hausdorff distance and its modifications are widely used in image analysis and recognition. For comparison of movable objects, generalizations of the Hausdorff distance are used. In [19, Section 1.5], the following generalized Hausdorff distance (GH-distance) was defined:

$$GH(A, B) = \inf_{\mathbf{T}_1, \mathbf{T}_2 \in \mathfrak{I}} H(\mathbf{T}_1 A, \mathbf{T}_2 B), \tag{4}$$

where A, B are compact sets in $\mathbb{R}^n$ and $\mathfrak{I}$ is a group of transformations in $\mathbb{R}^n$. In the case of the isometric transformation group $\mathfrak{I}$, it holds

$$\begin{aligned} GH(A, B) &= \inf\nolimits_{\mathbf{T}_1, \mathbf{T}_2 \in \mathfrak{I}} H(\mathbf{T}_1 A, \mathbf{T}_2 B) \\ &= \inf\nolimits_{\mathbf{T}_1, \mathbf{T}_2 \in \mathfrak{I}} H(\mathbf{T}_2 \mathbf{T}_2^{-1} \mathbf{T}_1 A, \mathbf{T}_2 B) \\ &= \inf\nolimits_{\mathbf{T}_1, \mathbf{T}_2 \in \mathfrak{I}} H(\mathbf{T}_2^{-1} \mathbf{T}_1 A, B) \\ &= \inf\nolimits_{\mathbf{T} \in \mathfrak{I}} H(\mathbf{T} A, B) = \inf\nolimits_{\mathbf{T} \in \mathfrak{I}} H(A, \mathbf{T} B). \end{aligned}$$

Below, we introduce a new variant of the generalized Hausdorff distance, the SH-distance, which measures similarity between objects up to a semigroup of nonexpanding transformations.

Definition 7 (SH-Deviation and SH-Distance Between Sets) Let $\mathfrak{I} = \{\mathbf{T}_z, z \in \mathbb{Z}\}$ be a *semigroup* of nonexpanding operators $\mathbf{T} : \mathbb{C} \to \mathbb{C}$. Define a SH-deviation and SH-distance between sets A and B as

$$\begin{aligned} \Delta_{SH}(A, B) &= \inf_{\mathbf{T} \in \mathfrak{I}} \Delta(\mathbf{T} A, B) = \inf_{z \in \mathbb{Z}} \Delta(\mathbf{T}_z A, B), \\ SH(A, B) &= \max \left\{ \inf_{\mathbf{T} \in \mathfrak{I}} \Delta(\mathbf{T} A, B), \inf_{\mathbf{T} \in \mathfrak{I}} \Delta(\mathbf{T} B, A) \right\} \\ &= \max \left\{ \inf_{z \in \mathbb{Z}} \Delta(\mathbf{T}_z A, B), \inf_{z \in \mathbb{Z}} \Delta(\mathbf{T}_z B, A) \right\}. \end{aligned}$$

If $\mathfrak{I}$ is an isometric group, we can compare SH and GH distances,

$$\begin{aligned} SH(A, B) &= \max \left\{ \inf_{\mathbf{T} \in \mathfrak{I}} \Delta(\mathbf{T} A, B), \inf_{\mathbf{T} \in \mathfrak{I}} \Delta(\mathbf{T} B, A) \right\} \\ &= \max \left\{ \inf_{\mathbf{T} \in \mathfrak{I}} \Delta(\mathbf{T} A, B), \inf_{\mathbf{T} \in \mathfrak{I}} \Delta(\mathbf{T}^{-1} B, A) \right\} \\ &\leq \inf_{\mathbf{T} \in \mathfrak{I}} \max \left\{ \Delta(\mathbf{T} A, B), \Delta(\mathbf{T}^{-1} B, A) \right\} \\ &= \inf_{\mathbf{T} \in \mathfrak{I}} \max \{ \Delta(\mathbf{T} A, B), \Delta(B, \mathbf{T} A) \} \\ &= \inf_{\mathbf{T} \in \mathfrak{I}} H(\mathbf{T} A, B) = GH(A, B). \end{aligned}$$

SH and GH distances do not coincide in general.

In the next two lemmas, we show that GH and SH quantities are pseudo-distances and thus define certain similarity relations between classes of equivalent sets; A and A' are equivalent if $GH(A, A') = 0$ or $SH(A, A') = 0$, respectively.

Lemma 4 (Properties of the GH-Distance) *Under Assumption 2 (isometricity), the GH-distance has the following properties.*

(i) $GH(A, B) = GH(B, A)$.
(ii) The triangle inequalities hold:

$$\Delta_{GH}(A, B) \leq \Delta_{GH}(A, C) + \Delta_{GH}(C, B),$$

$$GH(A, B) \leq GH(A, C) + GH(C, B).$$

(iii) If $GH(A, B) = 0$ *and* $GH(B, C) = 0$, *then* $GH(A, C) = 0$.
(iv) If $GH(A, A') = 0$ *and* $GH(B, B') = 0$, *then* $GH(A, B) = GH(A', B')$.

Proof (i)

$$\begin{aligned} GH(A, B) &= \inf_{\mathbf{T}\in\mathfrak{I}} H(\mathbf{T}A, B) \\ &= \inf_{\mathbf{T}\in\mathfrak{I}} \max\{\Delta(\mathbf{T}A, B)\Delta(B, \mathbf{T}A)\} \\ &= \inf_{\mathbf{T}\in\mathfrak{I}} \max\left\{\Delta(A, \mathbf{T}^{-1}B)\Delta(\mathbf{T}^{-1}B, A)\right\} \\ &= \inf_{\mathbf{T}\in\mathfrak{I}} \max\{\Delta(A, \mathbf{T}B)\Delta(\mathbf{T}B, A)\} \\ &= \inf_{\mathbf{T}\in\mathfrak{I}} H(\mathbf{T}B, A) = GH(B, A). \end{aligned}$$

(ii) Take any $C \in \mathbb{C}$. We need to prove that

$$\inf_{\mathbf{T}\in\mathfrak{I}} H(\mathbf{T}A, B) \;\leq\; \inf_{\mathbf{T}\in\mathfrak{I}} H(\mathbf{T}A, C) + \inf_{\mathbf{T}\in\mathfrak{I}} H(\mathbf{T}B, C).$$

For any $\varepsilon > 0$, there exist $\mathbf{T}_1 \in \mathfrak{I}$ and $\mathbf{T}_2 \in \mathfrak{I}$ such that

$$H(\mathbf{T}_1 A, C) \leq \inf_{\mathbf{T}} H(\mathbf{T}A, C) + \varepsilon, \quad H(\mathbf{T}_2 B, C) \leq \inf_{\mathbf{T}} H(\mathbf{T}B, C) + \varepsilon.$$

Then, accounting for group property $\mathbf{T}_2\mathbf{T}_1 \in \mathfrak{I}$, the triangle inequality for the Hausdorff distance, and Assumption 2, we have

$$\begin{aligned} \inf_{\mathbf{T}} H(\mathbf{T}A, B) &\leq H(\mathbf{T}_2\mathbf{T}_1 A, B) = H(\mathbf{T}_2\mathbf{T}_1 A, \mathbf{T}_2 C) + H(\mathbf{T}_2 C, B) \\ &\leq H(\mathbf{T}_1 A, C) + H(\mathbf{T}_2 B, C) \\ &\leq \inf_{\mathbf{T}} H(\mathbf{T}A, C) + \inf_{\mathbf{T}} H(\mathbf{T}B, C) + 2\varepsilon. \end{aligned}$$

Since ε can be arbitrarily small, the statement follows.

(iii) From $GH(A, B) = GH(B, C) = 0$ by the triangle inequality (ii), it follows

$$0 \le GH(A, C) \le GH(A, B) + GH(B, C) = 0.$$

(iv) Let A, A', B, B' be compact sets in $\mathbb{R}^n$ such that $GH(A, A') = 0$ and $GH(B, B') = 0$. Let us show that $GH(A, B) = GH(A', B')$. By the triangle inequality,

$$\begin{aligned} GH(A, B) &\le GH(A, A') + GH(A', B) = GH(A', B) \\ &\le GH(A', B') + GH(B', B) = GH(A', B'). \end{aligned}$$

Similarly,

$$\begin{aligned} GH(A', B') &\le GH(A', A) + GH(A, B') = GH(A, B') \\ &\le GH(A, B) + GH(B, B') = GH(A, B). \end{aligned}$$

Hence, $GH(A, B) = GH(A', B')$.

□

The SH-distance has similar properties as the GH one.

Lemma 5 (Properties of the SH-Distance) *Under Assumption 3 (nonexpansion), the SH-distance has the following properties.*

(i) $SH(A, B) = SH(B, A)$.
(ii) The triangle inequalities hold:

$$\Delta_{SH}(A, B) \le \Delta_{SH}(A, C) + \Delta_{SH}(C, B),$$

$$SH(A, B) \le SH(A, C) + SH(C, B).$$

(iii) If $SH(A, B) = 0$ *and* $SH(B, C) = 0$*, then* $SH(A, C) = 0$.
(iv) If $SH(A, A') = 0$ *and* $SH(B, B') = 0$*, then* $SH(A, B) = SH(A', B')$.

Proof (i) For any $A, B \in \mathbb{C}$ by construction,

$$\begin{aligned} SH(A, B) &= \max\left\{ \inf_{\mathbf{T}\in\mathfrak{I}} \Delta\left(\mathbf{T}A, B\right), \inf_{\mathbf{T}\in\mathfrak{I}} \Delta\left(\mathbf{T}B, A\right) \right\} \\ &= \max\left\{ \inf_{\mathbf{T}\in\mathfrak{I}} \Delta\left(\mathbf{T}B, A\right), \inf_{\mathbf{T}\in\mathfrak{I}} \Delta\left(\mathbf{T}A, B\right) \right\} = SH(B, A). \end{aligned}$$

(ii) Let A, B, C be compact sets in $\mathbb{R}^n$. Take an arbitrary $\varepsilon > 0$, and find $\mathbf{T}_1$, $\mathbf{T}_2 \in \mathfrak{I}$ such that

$$\Delta(\mathbf{T}_1 A, C) \le \inf_{\mathbf{T}\in\mathfrak{I}} \Delta(\mathbf{T}A, C) + \varepsilon, \quad \Delta(\mathbf{T}_2 C, B) \le \inf_{\mathbf{T}\in\mathfrak{I}} \Delta(\mathbf{T}C, B) + \varepsilon.$$

By the triangle inequality applied to sets $A' = \mathbf{T}_2\mathbf{T}_1 A$, B and $C' = \mathbf{T}_2 C$, accounting the nonexpanding property of operator $\mathbf{T}_2 \in \mathfrak{I}$ (Assumption 3), we have

$$\begin{aligned}\Delta(\mathbf{T}_2\mathbf{T}_1 A, B) &\leq \Delta(\mathbf{T}_2\mathbf{T}_1 A, \mathbf{T}_2 C) + \Delta(\mathbf{T}_2 C, B)\\ &\leq \Delta(\mathbf{T}_1 A, C) + \Delta(\mathbf{T}_2 C, B)\\ &\leq \inf_{\mathbf{T}\in\mathfrak{I}} \Delta(\mathbf{T}A, C) + \inf_{\mathbf{T}\in\mathfrak{I}} \Delta(\mathbf{T}C, B) + 2\varepsilon.\end{aligned}$$

Since $\mathbf{T}_2\mathbf{T}_1 \in \mathfrak{I}$, then $\inf_{\mathbf{T}\in\mathfrak{I}} \Delta(\mathbf{T}A, B) \leq \Delta(\mathbf{T}_2\mathbf{T}_1 A, B)$. Due to arbitrariness of $\varepsilon > 0$, we obtain the first triangle inequality:

$$\inf_{\mathbf{T}\in\mathfrak{I}} \Delta(\mathbf{T}A, B) \leq \inf_{\mathbf{T}\in\mathfrak{I}} \Delta(\mathbf{T}A, C) + \inf_{\mathbf{T}\in\mathfrak{I}} \Delta(\mathbf{T}C, B).$$

Similar arguments show that

$$\inf_{\mathbf{T}\in\mathfrak{I}} \Delta(\mathbf{T}B, A) \leq \inf_{\mathbf{T}\in\mathfrak{I}} \Delta(\mathbf{T}B, C) + \inf_{\mathbf{T}\in\mathfrak{I}} \Delta(\mathbf{T}C, A).$$

As a result, we obtain the required triangle inequality,

$$\begin{aligned}&\max\left\{\inf_{\mathbf{T}\in\mathfrak{I}} \Delta(\mathbf{T}A, B), \inf_{\mathbf{T}\in\mathfrak{I}} \Delta(\mathbf{T}B, A)\right\}\\ &\quad\leq \max\left\{\inf_{\mathbf{T}\in\mathfrak{I}} \Delta(\mathbf{T}A, C), \inf_{\mathbf{T}\in\mathfrak{I}} \Delta(\mathbf{T}C, A)\right\}\\ &\quad+ \max\left\{\inf_{\mathbf{T}\in\mathfrak{I}} \Delta(\mathbf{T}C, B), \inf_{\mathbf{T}\in\mathfrak{I}} \Delta(\mathbf{T}B, C)\right\}\end{aligned}$$

i.e.,

$$SH(A, B) \leq SH(A, C) + SH(C, B).$$

The statement for SH-deviation is proved similarly.

Statements (iii) and (iv) follow from the triangle inequality (ii) as in the proof of Lemma 4.

□

We can define similarity relations between sets differently; A and A' are similar, $A \sim A'$, if there is a transformation $\mathbf{T} \in \mathfrak{I}$ such that $\mathbf{T}A = B$, where $\mathfrak{I}$ is a group or semigroup of transformations. The next two lemmas give sufficient conditions when two definitions of similarity imply each other.

Lemma 6

(i) If $A \sim B$, i.e., for some $\mathbf{T}_{AB} \in \mathfrak{I}$ holds $\mathbf{T}_{AB}A = B$, then $GH(A, B) = 0$.
(ii) Under Assumption 1 (attainability), if $GH(A, B) = 0$, then $A \sim B$.

Proof (i) is obvious. (ii) Under Assumption 1 for some $\mathbf{T}_{AB} \in \mathfrak{I}$, it holds

$$\begin{aligned} GH(A, B) &= \inf_{\mathbf{T} \in \mathfrak{I}} H(\mathbf{T}A, B) = H(\mathbf{T}_{AB}A, B) \\ &= \max\{\Delta(\mathbf{T}_{AB}A, B), \Delta(B, \mathbf{T}_{AB}A)\} = 0. \end{aligned}$$

Then $\Delta(\mathbf{T}_{AB}A, B) = \Delta(B, \mathbf{T}_{AB}A) = 0$ implies $\mathbf{T}_{AB}A \subseteq B$ and $B \subseteq \mathbf{T}_{AB}A$, hence $\mathbf{T}_{AB}A = B$, i.e., $A \sim B$. □

Lemma 7 *Under Assumptions 1, 2, 4,*

(i) If $A \sim B$, then $SH(A, B) = 0$;
(ii) If $SH(A, B) = 0$ for $A, B \in \mathfrak{D}$, then $A \sim B$.

Proof

(i) If $A \sim B$, then there is $\mathbf{T} \in \mathfrak{I}$ such that $\mathbf{T}A = B$. Since $\mathfrak{I}$ is an isometric group, then $A = \mathbf{T}^{-1}B$. So, $\inf_{\mathbf{T} \in \mathfrak{I}} \Delta(\mathbf{T}A, B) = 0$, $\inf_{\mathbf{T} \in \mathfrak{I}} \Delta(\mathbf{T}B, A) = 0$, and hence $SH(A, B) = 0$.
(ii) If $SH(A, B) = 0$, then by Assumption 1, there exist $\mathbf{T}_1, \mathbf{T}_2 \in \mathfrak{I}$ such that $\mathbf{T}_1 A \subseteq B$ and $\mathbf{T}_2 B \subseteq A$, i.e., $\mathbf{T}_2\mathbf{T}_1 A \subseteq \mathbf{T}_2 B \subseteq A$. Set $\mathbf{T} = \mathbf{T}_2\mathbf{T}_1 \in \mathfrak{I}$, then $\mathbf{T}A \subseteq A$. Suppose that $\mathbf{T}A \subset A$, then there is $x \in A$ such that $d(x, \mathbf{T}A) > 0$. Let O be an open set in $\{y \in A : d(y, x) < \varepsilon\}$ with Lebesgue measure $\lambda(O) = \eta > 0$. Then $\lambda(A) = \lambda(\mathbf{T}A) \leq \lambda(A) - \eta$. This contradiction implies that $\mathbf{T}A = A$. Since $\mathbf{T}A = \mathbf{T}_2\mathbf{T}_1 A \subseteq \mathbf{T}_2 B \subseteq A$, then $\mathbf{T}_2 B = A$, i.e., $A \sim B$.

□

Example 10 Let $\mathfrak{I}_1 = \{\mathbf{T}_\lambda, \lambda > 0\}$ be the scaling group and $\mathfrak{I}_2 = \{\mathbf{T}_\lambda, \lambda \in [0, 1]\}$ be the nonexpanding scaling semigroup of transformations in $\mathbb{R}^n$ such that $\mathbf{T}_\lambda A = \{\lambda x : x \in A\}$.

For two ball sets $A = \{x : ||x|| \leq 1\}$, $B = \{x : ||x|| \leq 2\}$ and $m \geq 2$, we have

$$\begin{aligned} \Delta_{SH}(A, B) &= \inf\nolimits_{\mathbf{T} \in \mathfrak{I}_2} \Delta(\mathbf{T}A, B) = \inf\nolimits_{\lambda \in [0,1]} \Delta(\lambda A, B) = 1, \\ \Delta_{SH}(B, A) &= \inf\nolimits_{\mathbf{T} \in \mathfrak{I}_2} \Delta(\mathbf{T}B, A) = \inf\nolimits_{\lambda \in [0,1]} \Delta(\lambda B, A) = 0, \\ SH(A, B) &= \max\{\Delta_{SH}(A, B), \Delta_{SH}(B, A)\} = 1, \\ SH(A, B) &= \min\{\Delta_{SH}(A, B), \Delta_{SH}(B, A)\} = 0, \\ \Delta_{SH}(mA, B) &= \inf\nolimits_{\mathbf{T} \in \mathfrak{I}_2} \Delta(\mathbf{T}mA, B) = \inf\nolimits_{\lambda \in [0,1]} \Delta(m\lambda A, B) = 0, \\ \Delta_{SH}(mB, A) &= \inf\nolimits_{\mathbf{T} \in \mathfrak{I}_2} \Delta(\mathbf{T}mB, A) = \inf\nolimits_{\lambda \in [0,1]} \Delta(m\lambda B, A) = 0. \end{aligned}$$

Remark that for any bounded sets $A, B \subset \mathbb{R}^n$

$$GH(A, B) = \inf_{\mathbf{T}_1, \mathbf{T}_2 \in \mathfrak{I}_1} H(\mathbf{T}_1 A, \mathbf{T}_2 B) = \inf_{\lambda_1 > 0, \lambda_2 > 0} H(\lambda_1 A, \lambda_2 B) = 0.$$

Lemma 8 *Under Assumptions 1, 2, property (ii) in Lemma 7 holds also for sets that consist of finite number of points and/or a finite number of continuous line intervals. The fulfillment of Assumption 4 (preservation of volumes) is not required.*

Proof If $SH(A, B) = 0$, then there exist $\mathbf{T}_1, \mathbf{T}_2 \in \mathfrak{I}$ such that $\mathbf{T}_1 A \subseteq B$ and $\mathbf{T}_2 B \subseteq A$. Disjoint compact sets remain disjoint under transformations $\mathbf{T} \in \mathfrak{I}$. Then the number of disjoint elements in A, B is equal, and for each line element $E_B \in B$, there is element $E_A \in A$ such that $\mathbf{T}_1 E_A \subseteq E_B$. Points are transformed to points under $\mathbf{T} \in \mathfrak{I}$, so numbers of point elements in A and B coincide, then the number of nonpoint elements in A and B also coincide. Nonpoint elements consist of bounded intervals, so we can calculate the total length of all intervals in any nonpoint element and thus can find the total lengths $\lambda(A)$ and $\lambda(B)$ of intervals in A and B. Since $\mathbf{T}_1 A \subseteq B$ and $\mathbf{T}_2 B \subseteq A$, then total lengths coincide, $\lambda(A) = \lambda(B)$. Let us prove that not only $\mathbf{T}_1 A \subseteq B$ but $\mathbf{T}_1 A = B$. Suppose the contrary, $\mathbf{T}_1 A \subset B$. Then there is a point $b \in B$ and $b \notin \mathbf{T}_1 A$. This point b belongs to some interval $I_B \subseteq B$. Since $\mathbf{T}_1 A$ is closed, then there is a subinterval $I'_B \subset I_B$, which is not contained in $\mathbf{T}_1 A$. This leads to the contradiction $\lambda(A) = \lambda(\mathbf{T}_1 A) < \lambda(B)$. Hence, $\mathbf{T}_1 A = B$ and thus $A \sim B$. □

Remark 3 (Sets of Transformations, Which Do Not Form a Semigroup) Isometric transformations, although having nice analytical properties, represent a rather limited class of transformations used in the problem of image matching. For example, a natural extension of the isometric class

$$\mathfrak{I} = \left\{\mathbf{T}_{z_0,z_1} : \mathbf{T}_{z_0,x_1} z = z_0 x + z_1, |z_0| = 1\right\}$$

is the class of transformations with limited scaling:

$$\mathfrak{I}' = \left\{\mathbf{T}_{z_0,z_1} : \mathbf{T}_{z_0,z_1} z = z_0 z + z_1,\ m \le |z_0| \le M\right\}, \quad m \le 1 \le M.$$

Of course, in this case, transformations $\mathfrak{I}'$ do not constitute a semigroup. Then a corresponding similarity measures are defined as in (4) and Definition 7. We obtain distances,

$$\begin{aligned} GH_{m,M}(A, B) := \inf_{m \le |z_0|, |z_0'| \le M; z_1, z_1'} \max &\left\{ \sup_{a\in A} \inf_{b\in B} \left|z_0 a + z_1 - z_0' b - z_1'\right|, \right. \\ &\left. \sup_{b\in B} \inf_{a\in A} \left|z_0 a + z_1 - z_0' b - z_1'\right| \right\} \le GH(A, B), \\ SH_{m,M}(A, B) := \max &\left\{ \inf_{m \le |z_0| \le M, z_1} \sup_{a\in A} \inf_{b\in B} |z_0 a + z_1 - b|, \right. \\ &\left. \inf_{m \le |z_0| \le M, z_1} \sup_{b\in B} \inf_{a\in A} |z_0 b + z_1 - a| \right\} \le SH(A, B). \end{aligned}$$

Remark 4 There are other transformation-based similarity measures for convex bodies [8, 25]. Let $\mathfrak{I}$ be a transformation group in $\mathbb{R}^n$. For convex bodies A, B in $\mathbb{R}^n$, the Banach-Mazur deviation Δ_{BM} and the Banach-Mazur metric ρ_{BM} are defined as follows:

$$\Delta_{BM}(A, B) = \inf_{\mathbf{T}\in\mathfrak{I}} \left\{ \frac{\lambda(\mathbf{T}A)}{\lambda(B)} : \mathbf{T}(A) \supseteq B \right\},$$

$$\rho_{BM}(A, B) = \ln \inf_{\mathbf{T}\in\mathfrak{I}} \left\{ \frac{\lambda(\mathbf{T}A)}{\lambda(B)} : \mathbf{T}(A) \supseteq B \right\} + \ln \inf_{\mathbf{T}\in\mathfrak{I}} \left\{ \frac{\lambda(\mathbf{T}B)}{\lambda(A)} : \mathbf{T}(B) \supseteq A \right\}.$$

Here $\lambda(A)$ denotes the Lebesgue area/volume of A.

8 The Existence of Optimal Transformations

Below we present some sufficient conditions for the fulfilment of Assumption 1.

Let A, B be some bounded sets of complex numbers and

$$\mathfrak{I} = \left\{ \mathbf{T}_{z_0,z_1} : \mathbf{T}_{z_0,z_1} x = z_0 x + z_1, 0 \le m \le |z_0| \le M < \infty \right\}.$$

Recall that we have introduced the notation

$$\|A\| = \sup_{a\in A} \|a\|;\ \|B\| = \sup_{b\in B} \|b\|.$$

Consider the functions $\Delta_{AB}(z_0, z_1) = \Delta(\mathbf{T}_{z_0,z_1} A, B)$, $\Delta_{BA}(z_0, z_1) = \Delta(\mathbf{T}_{z_0,z_1} B, A)$, $h_{AB}(z_0, z_1) = H(\mathbf{T}_{z_0,z_1} A, B)$ for fixed A, B.

Lemma 9 *The functions $\Delta_{AB}(z_0, z_1)$, $\Delta_{BA}(z_0, z_1)$, $h_{AB}(z_0, z_1)$ are continuous, and the following estimates hold true:*

$$\begin{aligned} \Delta_{AB}(z_0, z_1) &\ge |z_1| - M\|A\| - \|B\|, \\ \Delta_{BA}(z_0, z_1) &\ge |z_1| - \|A\| - M\|B\|, \\ h_{AB}(z_0, z_1) &\ge |z_1| - M\|A\| - \|B\|. \end{aligned}$$

Proof The function $|z_0 a + z_1 - b|$ is jointly continuous in all variables, so by ([5], Proposition 23), the functions $\Delta_{AB}(z_0, z_1) = \sup_{a\in A} \inf_{b\in B} |z_0 a + z_1 - b|$, $\Delta_{BA}(z_0, z_1)$ and, hence, $h_{AB}(z_0, z_1)$ are all continuous in z_0, z_1. It holds that

$$\begin{aligned} \Delta_{AB}(z_0, z_1) &= \Delta(\mathbf{T}_{z_0,z_1} A, B) \\ &= \sup_{a\in A} \inf_{b\in B} |z_0 a + z_1 - b| \ge \inf_{a\in A, b\in B} |z_0 a + z_1 - b| \\ &\ge \inf_{a\in A, b\in B} (|z_1| - |z_0 a - b|) \ge |z_1| - \sup_{a\in A, b\in B} |z_0 a - b| \\ &\ge |z_1| - M\|A\| - \|B\|. \end{aligned}$$

Changing the roles of A and B, we get that

$$\Delta_{BA}(z_0, z_1) \ge |z_1| - \|A\| - M\|B\|$$

Finally,

$$\begin{aligned}\Delta(B, \mathbf{T}_{z_0,z_1}A) &= \sup_{b\in B}\inf_{a\in A}|z_0a + z_1 - b| \geq \inf_{a\in A, b\in B}|z_0a + z_1 - b| \geq \\ &\geq \inf_{a\in A, b\in B}(|z_1| - |z_0a - b|) \geq |z_1| - \sup_{a\in A, b\in B}|z_0a - b| \\ &\geq |z_1| - M\|A\| - \|B\|,\end{aligned}$$

which gives also

$$\begin{aligned}h_{AB}(z_0, z_1) &= \max\left\{\Delta(\mathbf{T}_{z_0,z_1}A, B), \Delta(B, \mathbf{T}_{z_0,z_1}A)\right\} \\ &\geq |z_1| - M\|A\| - \|B\|.\end{aligned}$$

□

Lemma 10 *The following sets are nonempty and compact:*

$$\begin{aligned}Z_{AB} &= \{(z_0, z_1) : m \leq |z_0| \leq M, \quad \Delta_{AB}(z_0, z_1) \leq m\|A\| + \|B\|\|\}, \\ Z_{BA} &= \{(z_0, z_1) : m \leq |z_0| \leq M, \quad \Delta_{BA}(z_0, z_1) \leq \|A\| + m\|B\|\}, \\ Z &= \{(z_0, z_1) : m \leq |z_0| \leq M, \quad h_{AB}(z_0, z_1) \leq m\|A\| + \|B\|\}\end{aligned}$$

Proof Remark that point ($z_0' = m, z_1' = 0$) belongs to Z_{AB}, $Z_{B,A}$ and Z. Indeed, $\left|z_0'\right| = m$, and

$$\begin{aligned}\Delta_{AB}(m, 0) &= \sup_{a\in A}\inf_{b\in B}|ma - b| \leq \sup_{a\in A}\sup_{b\in B}|ma - b| \\ &\leq m\|A\| + \|B\|.\end{aligned}$$

Similarly, the point $(m, 0)$ also belongs to Z_{BA} and Z.

Let $(z_0, z_1) \in Z$, then by Lemma 9

$$m\|A\| + \|B\| \geq \Delta_{AB}(z_0, z_1) \geq |z_1| - M\|A\| - \|B\|$$

and, hence, $|z_1| \leq (m + M)\|A\| + 2\|B\|$, i.e., the set Z_{AB} is bounded. Since, $\Delta_{AB}(z_0, z_1)$ is continuous in z_0, z_1, the set Z is closed and hence compact. Similar statements hold for Z_{BA} and Z. □

Compactness and continuity imply the attainability of the infimum.

Corollary 1 *By Lemmas 9, 10, the infinum in*

$$GH(A, B) = \inf_{|z_0|=1, z_1} H(\mathbf{T}_{z_0,z_1}A, B) = \min_{(z_0,z_1)\in Z} H(\mathbf{T}_{z_0,z_1}A, B),$$

is attained and thus $GH(A, B) = \Delta(\mathbf{T}_{z_0,z_1}A, B)$ *for some* $\mathbf{T}_{z_0,z_1}$*; i.e., Assumption 1 is fulfilled.*

Similarly, infinums in

$$\inf_{|z_0|\le 1, z_1} \Delta\left(\mathbf{T}_{z_0,z_1}A, B\right) = \min_{(z_0,z_1)\in Z_{AB}} \Delta\left(\mathbf{T}_{z_0,z_1}A, B\right)$$

$$\inf_{|z_0|\le 1, z_1} \Delta\left(\mathbf{T}_{z_0,z_1}B, A\right) = \min_{(z_0,z_1)\in Z_{BA}} \Delta\left(\mathbf{T}_{z_0,z_1}B, A\right)$$

are attained; thus, there exist $\mathbf{T}_1, \mathbf{T}_2 \in \mathfrak{I}$, *such that*

$$SH(A, B) = \max\{\Delta(\mathbf{T}_1 A, B), \Delta(\mathbf{T}_2 B, A)\},$$

i.e., also here Assumption 1 is fulfilled.

9 Lipschitz Property of Distance Functions

In this section, we consider for simplicity the case $m = M = 1$, i.e., the group of isometric transformations. To calculate SH-deviations and distances between sets $A \subset R^2$ and $B \subset R^2$, one has to solve the following minimization problems:

$$\Delta_{AB}(z_0, z_1) = \min_{|z_0|=1, |z_1|\le\|A\|+\|B\|} \Delta\left(\mathbf{T}_{z_0,z_1}A, B\right); \tag{5}$$

$$\Delta_{BA}(z_0, z_1) = \min_{|z_0|=1, |z_1|\le\|A\|+\|B\|} \Delta\left(\mathbf{T}_{z_0,z_1}B, A\right); \tag{6}$$

$$h_{AB}(z_0, z_1) := \min_{|z_0|=1, |z_1|\le\|A\|+\|B\|} H\left(\mathbf{T}_{z_0,z_1}A, B\right). \tag{7}$$

Remark that since complex numbers z_0, z_1 can be represented as $z_0 = e^{i\varphi}$ and $z_1 = x + iy$, with $i = \sqrt{-1}$, then the functions $\Delta_{AB}, \Delta_{AB}, h_{AB}$ can be considered as functions of three variables (φ, x, y). We can also treat sets $A, B \subset \mathbb{R}^2$ as sets of complex numbers, depending on context.

The problems (5)–(7) are nonsmooth and multi-extremal (see, e.g., [12, 45, 57]), so they should be solved by global optimization algorithms (enumerative, annealing, genetic, branch and bound, etc.). Next, we prove that the objective functions of these problems are Lipschtz ones.

Lemma 11 *The functions* $\Delta_{AB}(z_0, z_1)$ *and* $\Delta_{BA}(z_0, z_1)$ *in (5), (6) are Lipschitz continuous in* (z_0, z_1), *namely,*

$$\left|\Delta_{AB}(z_0, z_1) - \Delta_{AB}(z_0', z_1')\right| \le \|A\| \left|z_0 - z_0'\right| + \left|z_1 - z_1'\right|,$$
$$\left|\Delta_{BA}(z_0, z_1) - \Delta_{BA}(z_0', z_1')\right| \le \|B\| \left|z_0 - z_0'\right| + \left|z_1 - z_1'\right|.$$

Proof We prove the first inequality, the second is obtained similarly. Consider the function

$$\varphi(z_0, z_1) = \sup_{a\in A} d(z_0 a + z_1, B) = d(z_0 a_{z_0,z_1} + z_1, B).$$

By the Lipschitz property of the distance function $d(\cdot, B)$, we have:

$$\begin{aligned}\varphi(z_0, z_1) - \varphi(z_0', z_1') &= d(z_0 a_{z_0,z_1} + z_1, B) - d(z_0' a_{z'_0,z'_1} + z_1', B) \\ &\le d(z_0 a_{z_0,z_1} + z_1, B) - d(z_0' a_{z_0,z_1} + z_1', B) \\ &\le \left| z_0 a_{z_0,z_1} + z_1 - (z_0' a_{z_0,z_1} + z_1')\right| \\ &\le \left|z_0 - z_0'\right| \left|a_{z_0,z_1}\right| + \left|z_1 - z_1'\right| \\ &\le \|A\| \left|z_0 - z_0'\right| + \left|z_1 - z_1'\right|.\end{aligned}$$

The right-hand side is invariant w.r.t. exchanging the roles of z and z', and therefore we get

$$\left|\varphi(z_0, z_1) - \varphi(z_0', z_1')\right| \le \|A\| \left|z_0 - z_0'\right| + \left|z_1 - z_1'\right|.$$

□

Remark 5 The complexity of problems (5)–(7) depends on parameters $\|A\|$ and $\|B\|$. They define a Lipschitz constant and the size of the feasible domains in (5)–(7). In many cases, these constants may be reduced. Define the vectors (Chebyshev centers)

$$a^* \in \arg\min_x \max_{a\in A} d(x, a),\ b^* \in \arg\min_x \max_{b\in B} d(x, b). \tag{8}$$

We show that we may solve equivalenty the problem

$$\Delta_{AB}(z_0, z_1) = \min_{|z_0|=1, |z_1|\le\|A-a^*\|+\|B-b^*\|} \Delta\left(\mathbf{T}_{z_0,z_1} A, B\right). \tag{9}$$

Unless $a^* = 0, b^* = 0$, the latter problem has smaller parameters $\|A - a^*\|$ and $\|B - b^*\|$.

We show that problems (5) and (9) are equivalent. Indeed, due to Corollary 1,

$$\begin{aligned}&\min_{|z_0|=1,\ |z_1|\le\|A-a^*\|+\|B-b^*\|} \Delta\left(z_0\left(A - a^*\right) + z_1, B - b^*\right) \\ &= \min_{|z_0|=1,\ z_1} \Delta\left(z_0 A - z_0 a^* + z_1, B - b^*\right) \\ &= \min_{|z_0|=1,\ z_1} \Delta\left(z_0 A - z_0 a^* + b^* + z_1, B\right) = \min_{|z_0|=1,\ z_1} \Delta\left(z_0 A + z_1, B\right) \\ &= \min_{|z_0|=1,\ |z_1|\le\|A\|+\|B\|} \Delta\left(z_0 A + z_1, B\right). \quad \square\end{aligned}$$

Remark 6 (Tangent Minorants) Let $z_0 = \cos\varphi + i \sin\varphi$, $z'_0 = \cos\varphi' + i\sin\varphi'$, $z_1 = x_1 + i y_1$, $z'_1 = x'_1 + i y'_1$ be complex numbers. By substituting these numbers into $\Delta_{A,B}(z_0, z_1)$ and $\Delta_{A,B}(z'_0, z'_1)$, we obtain the representation of these quantities as functions of real numbers, $\Delta_{A,B}(\varphi_0, x_1, y_1)$ and $\Delta_{A,B}(\varphi'_0, x'_1, y'_1)$. By Lemma 11

$$\begin{aligned}
\left|\Delta_{A,B}(\varphi_0, x_1, y_1) - \Delta_{A,B}(\varphi'_0, x'_1, y'_1)\right| &= \left|\Delta_{A,B}(z_0, z_1) - \Delta_{A,B}(z'_0, z'_1)\right| \\
&\le \|A\| \left|z_0 - z'_0\right| + \left|z_1 - z'_1\right| \\
&= 2\|A\| \sin \tfrac{|\varphi_0 - \varphi'_0|}{2} + \left(\left(x_1 - x'_1\right)^2 + \left(y_1 - y'_1\right)^2\right)^{1/2} \\
&\le \|A\| \left|\varphi_0 - \varphi'_0\right| + \left|x_1 - x'_1\right| + \left|y_1 - y'\right|.
\end{aligned}$$

Thus the functions

$$\begin{aligned}
\mu'(\varphi, x, y) &= \Delta_{A,B}(\varphi', x', y') - 2\|A\| \sin \tfrac{|\varphi_0 - \varphi'_0|}{2} \\
&\quad - \left(\left(x_1 - x'_1\right)^2 + \left(y_1 - y'_1\right)^2\right)^{1/2}, \\
\nu'(\varphi, x, y) &= \Delta_{A,B}(\varphi', x', y') - \|A\| \left|\varphi_0 - \varphi'_0\right| - \left|x_1 - x'_1\right| - \left|y_1 - y'\right|,
\end{aligned}$$

are tangent minorants of $\Delta_{A,B}(\cdot, \cdot, \cdot)$ at the point (φ', x', y'). Tangent minorants can be used to build methods for global optimization of functions [40, 41].

10 Calculation of Deviations and Distances Between Fixed Objects

Before solving the optimization problems (5)–(7), we discuss how to calculate objective function values in these problems in an efficient manner, i.e., how to calculate $\Delta(A, B)$ for $A \subset \mathbb{R}^2$ and $B \subset \mathbb{R}^2$. We will consider this task for some classes of sets.

The Case of Finite Sets If A and B are finite point sets, then the value $\Delta(A, B)$ is calculated by definition as $\max_{a \in A} \min_{b \in B} \|a - b\|$. If $|A| = n$ and $|B| = m$, then the complexity of this operation is $n \cdot m$. For very large sets, the calculation may be quite time-consuming [15, 26, 58, 65]. In this case, it can be useful the following observation.

Suppose for some $a \in A$, we have already calculated $d_B(a)$ and found $b \in B$ such that $d(a, b) = d_B(a)$. Then for another $a' \in A$ such that $d(a', b) < d_B(a)$, it holds $d_B(a') < d_B(a)$, and we need not actually calculate the value $d_B(a')$.

The Case of Infinite Sets For arbitrary families of line segments $C \subset \mathrm{R}^2$ and $D \subset \mathrm{R}^2$, the values of $\Delta(C, D)$ and $H(C, D)$ can be approximately calculated by approximating these families by finite sets of points. Let ℓ_C and ℓ_D be the total lengths of families C and D, and e be the accuracy of the approximation of segments. Then the complexity of the approximate calculation of quantities $\Delta(C, D)$ and $H(C, D)$ is proportional to the value $\ell_C \cdot \ell_D / e^2$. To approximate a set of length ℓ with accuracy e, one needs ℓ / e points, and this leads to complexity of $n \cdot m = \ell_C \cdot \ell_D / e^2$.

Lemma 12 *If C is a convex polygon, and D is a convex set, then $\Delta(C, D) = \Delta(\hat{C}, D)$, where $\hat{C}$ denotes the set of vertices of C.*

The lemma follows from the convexity of the function $c \mapsto \min_{d\in D} \|c - d\|$ $\square$.

If C and D are non-convex polygons, then the deviation $\Delta(C, D)$ and distance $H(C, D)$, obviously, are equal to the deviation/distance between the boundaries of the polygons.

Therefore, we consider next the case when the two sets are finite unions of line segments (or graphs consisting of line segments).

Let families A and B be given by sets of intervals on the plane,

$$A = \bigcup_{i=1}^{n} \alpha_i, \text{ where } \alpha_i = \left\{\lambda \underline{a}_i + (1-\lambda)\bar{a}_i : \lambda \in [0,1]\right\}, \quad i = 1, \ldots, n;$$
$$B = \bigcup_{j=1}^{m} \beta_j, \text{ where } \beta_j = \left\{\lambda \underline{b}_j + (1-\lambda)\bar{b}_j : \lambda \in [0,1]\right\}, \quad j = 1, \ldots, m.$$

We call such sets polylines.

Each interval α_i is defined by its endpoints $\{\underline{a}_i, \bar{a}_i\}$, and each interval β_j is defined by its endpoints $\{\underline{b}_j, \bar{b}_j\}$. Thus, each polyline consists of possibly unconnected intervals. In the case of a linked polyline A, one has $\bar{a}_i = \underline{a}_{i+1}$, $i = 1, \ldots, n-1$.

It is required to find methods for calculation of the Hausdorff deviation between two polylines $C = \mathbf{T}A$ and B, where C consists of intervals

$$C = \bigcup_{i=1}^{n} \sigma_i, \text{ where } \sigma_i = \left\{\lambda \underline{c}_i + (1-\lambda)\bar{c}_i : \lambda \in [0,1]\right\}, i = 1, \ldots, n,$$

with $\underline{c}_i = \mathbf{T}\underline{a}_i$ and $\bar{c}_i = \mathbf{T}\bar{a}_i$. Obviously, this subtask comes down to calculating the distance between intervals and a polyline, because

$$\Delta(C, B) = \max_{\sigma\in C} \Delta(\sigma, B), \quad \Delta(B, C) = \max_{\beta\in B} \Delta(\beta, C).$$

Moreover, it is enough to be able to calculate the distance from a point to a polyline, since

$$\Delta(\sigma, B) = \max_{c\in\sigma} \Delta(c, B) = \max_{c\in\sigma} \min_{\beta\in B} \Delta(c, \beta) = \max_{c\in\sigma} \min_{1\le j<m} \Delta(c, \beta_j),$$
$$\Delta(\beta, C) = \max_{b\in\beta} \Delta(\beta, C) = \max_{b\in\beta} \min_{\sigma\in C} \Delta(b, \sigma) = \max_{b\in\beta} \min_{1\le i<n} \Delta(b, \sigma_i).$$

The distance $\Delta(c, \beta_j)$ from point c to the interval $\beta_j = \left\{\lambda \underline{b}_j + (1-\lambda)\bar{b}_j : \lambda \in [0,1]\right\}$ is the solution of the problem:

$$\Delta(c, \beta_j) = \min_{\lambda\in[0,1]} \left\| c - (\lambda \underline{b}_j + (1-\lambda)\bar{b}_j) \right\|. \tag{10}$$

Obviuosly, the distance from point b to an interval σ_i is of similar structure.

The distance of a point to an interval is given by a formula; no optimization algorithm is needed:

$$\Delta(c, \beta_j) = \left\| c - (\lambda^* \underline{b}_j + (1 - \lambda^*)\bar{b}_j) \right\| ,$$

where

$$\lambda^* = \begin{cases} \underline{b}_j, & \lambda' \le 0, \\ \lambda' \underline{b}_j + (1 - \lambda')\bar{b}_j, & 0 < \lambda' < 1, \\ \bar{b}_j, & \lambda' \ge 1; \end{cases} \qquad \lambda' = \frac{\left\langle c - \underline{b}_j, \bar{b}_j - \underline{b}_j \right\rangle}{\left\| \bar{b}_j - \underline{b}_j \right\|^2};$$

Here $\langle \cdot, \cdot \rangle$ denotes the scalar product, $\|\cdot\|$ is the norm of a vector.

Calculation of the deviation of an interval σ from a polyline B.

Consider an interval with endpoints $\underline{c}_i, \bar{c}_i$, and an intermediate point $c^t = t\underline{c}_i + (1-t)\bar{c}$. Recall that the function $f(c) = \Delta(c, B)$ is Lipschitz with a constant 1 with respect to the argument c [51, Example 9.6].

Lemma 13 *The one-dimensional function* $\varphi(t) = \Delta\left(t\underline{c}_i + (1-t)\bar{c}_i, B\right)$ *is Lipschitz in* $t \in \mathbb{R}^1$ *with constant* $L_i = \left(\|\underline{c}_i\| + \|\bar{c}_i\|\right)$.

Proof By the Lipschitz property of $c \mapsto \Delta(c, B)$,

$$\begin{aligned} |\varphi(t_1) - \varphi(t_2)| &= \left|\Delta\left(t_1\underline{c}_i + (1-t_1)\bar{c}_i, B\right) - \Delta\left(t_2\underline{c}_i + (1-t_2)\bar{c}_i, B\right)\right| \\ &\le \left\|(t_1 - t_2)\underline{c}_i + (t_2 - t_1)\bar{c}_i\right\| \le \left(\|\underline{c}_i\| + \|\bar{c}_i\|\right) \cdot |t_1 - t_2| . \end{aligned}$$

Thus, the problem of finding the distance of the interval σ_i from the set B, $\Delta(\sigma_i, B) = \max_{c \in \sigma_i} \Delta(c, B) = \max_{0 \le t \le 1} \varphi(t)$, in the case of a general set (i.e., a polyline) B is a problem of multiextremal one-dimensional optimization and can be solved by Lipschitz global optimization methods [27, 47, 56]. □

11 The Branch-and-Bound Method for Calculation of the SH Deviation and SH Distance

In this section, we demonstrate how to calculate the SH distance and SH deviation between two figures (sets) A and B. Let a^* and b^* be some points related to figures A and B, respectively, like centers for point sets (8) or centers of endpoints for polyline sets.

Define $A^* = A - a^*$, $B^* = B - b^*$ and

$$\mathbf{T}_{z_0,z_1} A = \begin{pmatrix} \cos\varphi & \sin\varphi \\ -\sin\varphi & \cos\varphi \end{pmatrix} A^* + \left(b^* - a^*\right) + (x, y)^T,$$

with $z_0 = \cos\varphi + i\sin\varphi$, $z_1 = x + iy$.

The transformation $\mathbf{T}_{z_0,z_1}$ depends on three parameters: φ, x, y. Clearly, $\varphi \in [-\pi, +\pi]$. Evaluate $c^* := \|A - a*\| + \|B - b^*\|$. The range of possible values of the parameters is a compact rectangle $R = [-\pi, \pi] \times [-c^*, c^*] \times [-c^*, c^*]$.

We solve problems (5), (6) by a standard branch-and-bound (B&B) method applied to nonsmooth Lipschitz functions $(z_0, z_1) \mapsto \Delta\left(\mathbf{T}_{z_0,z_1}A^*, B^*\right)$ and $(z_0, z_1) \mapsto \Delta\left(\mathbf{T}_{z_0,z_1}B^*, A^*\right)$ defined on the rectangle R and with known Lipschitz constants as explained in Sect. 9, Lemma 12. The only peculiarity is that the calculation of functions $\Delta\left(\mathbf{T}_{z_0,z_1}A^*, B^*\right)$ and $(z_0, z_1) \mapsto \Delta\left(\mathbf{T}_{z_0,z_1}B^*, A^*\right)$ themselves at a given point (z_0, z_1) includes solution of a number of one-dimensional Lipschitz global optimization problems with known Lipschitz constant as described in Sect. 10, Lemma 13, if sets A and B both contain line segments.

In the standard B&B method for the global minimization of a Lipschitz function, the initial search box is subdivided into smaller subboxes, and thus the initial optimization problem is subdivided into a number of subproblems. For each subproblem, the minimum objective function value is estimated from above and from below. The initial upper bound is just the value of the objective function at any feasible point of the subproblem. The lower bound is obtained from the upper bound by subtracting the product of the Lipschitz constant and the size of the subproblem. The minimum of the upper bound of all subproblems considered so far is the achieved record value of the objective function for the whole problem. If the lower bound of some subproblem appears larger than the achieved record value, this subproblem can be safely deleted from further consideration. Otherwise, the initial upper bound of this subproblem can be improved by application of any effective local optimization method applied to this subproblem. Even if it does not converge, it can be used to improve the initial objective function value. The difference between the record achieved objective function value at some subproblem, and the lower bound for this subproblem give the achieved accuracy of the method. To get better accuracy, one has to refine the partition and reevaluate bounds for subproblems. To evaluate lower bounds one can use also tangent minorants of the objective function from Remark 6.

Figure 2 illustrates solving problems (5)–(7) by the B&B method for checking similarity of two rigid figures (black and gray ones), where distances between all elements of each figure are fixed (Fig. 2).

12 Calculation of the Gromov-Hausdorff Distance for Graphs

For two metric spaces A, B, the Gromov-Hausdorff distance is defined as a minimal Hausdorff distance between all possible isometric embeddings of A, B into all possible common metric spaces [11, 24, 59]. In our case, the closed subsets $A, B \in \mathbb{R}^2$ can be considered as metric spaces with induced Euclidean distance between

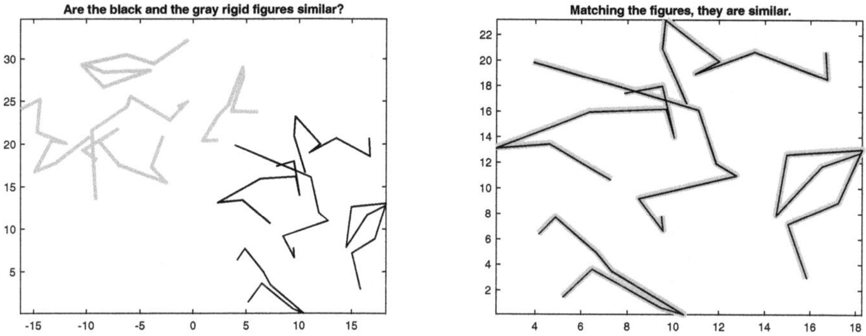

Fig. 2 Testing the B&B algorithm: checking if two rigid figures in the left box (black and gray, in fact, the latter is a blind transformation of the black one) are similar. They are similar indeed, since they match (as shown in the right box) after optimal transformation (by minimizing GH-distance/deviation between them, see (5)–(7))

points within A and B, and transformations $\mathbf{T} \in \mathfrak{T}$ can be considered as isometric embeddings of A and B into common space $\mathbb{R}^2$. In pattern recognition, the Gromov-Hausdorff distance is considered as a similarity measure between bendable objects, where not all distances between elements of the object are fixed.

Calculation of the Gromov-Hausdorff distance between general metric spaces is a hard task. We consider a simplified version of this task, namely, calculation of this distance between finite connected graphs in $\mathbb{R}^n$. In what follows, we reduce this problem to a nonconvex nonsmooth global optimization problem with smooth equality constraints.

Let $\Gamma_1 = (V_1, E_1)$ and $\Gamma_2 = (V_2, E_2)$ be two graphs with vertices in $\mathbb{R}^n$ ($n = 2$ below) $V_1 = \{v_1, \ldots, v_{n_1}\}$, $V_2 = \{\nu_1, \ldots, \nu_{n_2}\}$ and edges $E_1 = \{e_1, \ldots, e_{m_1}\}$, $E_2 = \{\epsilon_1, \ldots, \epsilon_{m_2}\}$. The lengths of the edges are considered fixed and known. Let us introduce variables $x = (x_1 \in \mathbb{E}^k, \ldots, x_{n_1} \in \mathbb{E}^k)$ and $y = (y_1 \in \mathbb{E}^k, \ldots, y_{n_2} \in \mathbb{E}^k)$ and consider the transformation of vertices $V_1(x) = \{v_1 + x_1, \ldots, v_{n_1} + x_{n_1}\}$ and $V_2(y) = \{\nu_1 + y_1, \ldots, \nu_{n_2} + y_{n_2}\}$. Consider Hausdorff distances $H(V_1(x), V_2)$ and $H(V_1, V_2(y))$ and the sets

$$X = \{x : \|v_i + x_i - (v_j + x_j)\| = \|v_i - v_j\|, (i, j) \in E_1\},$$

$$Y = \{y : \|\nu_i + y_i - (\nu_j + y_j)\| = \|\nu_i, \nu_j\|, (i, j) \in E_2\}.$$

The Gromov-Hausdorff distance $GrH(\Gamma_1, \Gamma_2)$ between Γ_1 and Γ_2 is defined as the minimal value of

$$GrH(\Gamma_1, \Gamma_2) = \min_{x \in X, y \in Y} \{H(V_1(x), V_2(y))\} \tag{11}$$

$$\leq \min\left\{\min_{x \in X} H(V_1(x), V_2), \min_{y \in Y} H(V_1, V_2(y))\right\}. \tag{12}$$

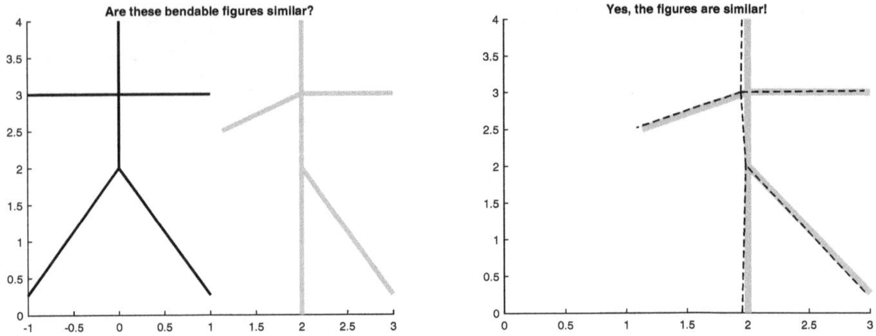

Fig. 3 Checking similarity between two bendable figures on the left (black and gray, the lengths of their line segments are fixed) by calculating GrH-distance between them, see (11), (12). The result (black-dashed and gray ones on the right) is visualized; GrH-distance between them is less than 0.05

Figure 3 illustrates the calculation of the Gromov-Hausdorff distance by solving minimization problems (11), (12) by the heuristic B&B method from [42, subsection 5.2], constraints can be taken into account, e.g., by the exact nonsmooth penalty method. The number of variables corresponds to the (doubled) number of vertexes, the number of equality constraints equals to the number of line segments in the transformed figure. Thus minimization subproblems (12) in this case have 14 variables and 6 nonlinear equality constraints.

13 Application to Pattern Classification/Recognition

The introduced similarity measures GH, Δ_{SH}, and SH can be used for complex figure location, recognition, or classification within the nearest or k-nearest neighbor classification methods. We treat the distance between a figure and elements of a class of figures as a random distance variable. As a robust measure of similarity of the figure and the class of figures, we use the average value at risk for this random distance variable. Calculation of this value assumes calculation of the similarity measure between the figure and all figures in the class. Each such calculation is a global optimization problem. We suggest a scheme how not to solve all these problems, at least not to solve them to the very end, by exploiting Lipschitz property and triangle inequality.

In pattern recognition, usually, a class (distribution) $\mathcal{S}$ of objects is represented by a discrete (e.g., empirical) distribution $\mathcal{S}^m$, given by sample objects $\{S_1, \ldots, S_m\}$ with probabilities $\{p_1, \ldots, p_m\}$. For a given object F, consider random variables $\xi = SH(F, \mathcal{S})$ and ξ^m taking values in the set $\{SH(F, S_i),\ i = 1, \ldots, m\}$ with corresponding probabilities $\{p_1, \ldots, p_m\}$. Denote cumulative distribution functions:

$$G(t) = \Pr\{\xi < t\} \quad \text{and} \quad G^m(t) = \Pr\left\{\xi^m < t\right\}.$$

The similarity measure between object F and class $\mathcal{S}$ can be defined in terms of SH-distance distribution $\xi = SH(F, S)$ by a variety of ways:

$$\begin{aligned} AV(F, \mathcal{S}) &= \textstyle\int t\,dG(t) = \mathbb{E}\left[\min H(\mathbf{T}_{\phi,x,y}F, \cdot)\right] \text{ (average distance)}, \\ Q_\alpha(F, \mathcal{S}) &= G^{-1}(\alpha) = \min_t \{t : G(t) \geq 1 - \alpha\} \text{ (quantile distance)}, \\ AVaR_\alpha(F, \mathcal{S}) &= \textstyle\int_0^\infty \max\{1 - G(t)/\alpha\,, 0\}\,dt \\ &= \max_x \left\{x - \tfrac{1}{\alpha}\mathbb{E}\min\{0, \xi - x\}\right\} \end{aligned}$$

(average value at risk α, where it is taken into account that $SH(F, \cdot) \geq 0$, see [46, subsection 2.2.3]).

The discrete counterparts of these similarity measures are as follows:

$$\begin{aligned} AV(F, \mathcal{S}^m) &= \tfrac{1}{m}\sum_{i=1}^{m} SH(F, S_i), \\ Q_\alpha(F, \mathcal{S}^m) &= G_m^{-1}(\alpha) = \min_t \left\{t : G_m^{-1}(t) \geq 1 - \alpha\right\} \text{ (quantile distance)}, \\ AVaR_\alpha(F, \mathcal{S}^m) &= \textstyle\int_0^\infty \max\{1 - G_m(t)/\alpha\,, 0\}\,dt \\ &= \max_u \left\{u - \tfrac{1}{\alpha m}\sum_{i=1}^{m} \min\{0, SH(F, S_i) - u\}\right\}. \end{aligned}$$

Quantile-Based Similarity Measures In case of empirical sample S^m, the quantile value $Q_\alpha(F, S^m)$ can be calculated by ordering the set $\{SH(F, S_i), i = 1, \ldots, m\}$ as $SH(F, S_{i_1}) \leq SH(F, S_{i_2}) \leq \ldots \leq SH(F, s_{i_m})$ and finding i_k such that $i_k(\alpha)/m \leq \alpha < i_{k+1}(\alpha)/m$. As a measure of similarity of object F and the set $\mathcal{S}^m$, one can take any quantity from the set

$$\left\{SH(F, S_{i_1}), SH(F, S_{i_2}), \ldots, SH(F, S_{i_k})\right\},$$

e.g., the first quantile $Q_{1/m}(F, \mathcal{S}^m) = SH(F, S_{i_1})$, or the k-st quantile $Q_{k/m}(F, \mathcal{S}^m) = SH(F, s_{i_k(\alpha)})$, or the average value

$$AVQ_\alpha(F, \mathcal{S}^m) = \frac{1}{k}\left(SH(F, S_{i_1}) + \ldots + SH(F, S_{i_k(\alpha)})\right).$$

In the nearest neighbor (NN-) classifier, any of these similarity measures $Q_{1/m}(F, \mathcal{S}^m)$, $Q_{k/m}(F, \mathcal{S}^m)$, $AVQ_\alpha(F, \mathcal{S}^m)$ can be used. The chosen level α may serve as a tuning parameter of the method and may be different for different classes. The optimal level α may be learned from additional training data.

Reducing the Complexity of a Classifier by Using the Triangle Inequality There is a complexity issue for the NN classifier connected with the necessity to calculate large number of distances $\{SH(F, s_i), i = 1, \ldots, m\}$ that may be time-consuming. The computational time can be reduced by the pre-calculation of within class distances $d_{ij} := SH(S_i, S_j)$ and using the triangle inequality

$$SH(F, S_j) + SH(F, S_i) \geq SH(S_i, S_j) = d_{ij}.$$

If we have already calculated quantity $SH(F, S_i)$ and want to calculate $SH(F, S_j)$ but we know that $d_{ij} \geq 2SH(F, S_i)$, then $SH(F, S_j) \geq d_{ij} - SH(F, S_i) \geq SH(F, S_i)$, and we could omit the calculation of $SH(F, S_j)$.

The same computation reduction trick can be applied for iterative calculation of the (k/m)-quantile of the set $\{SH(F, S_i), i = 1, \ldots, m\}$. First, we sequentially construct an initial ordered subset

$$SH(F, S_{i_1}) \leq SH(F, S_{i_2}) \leq \ldots \leq SH(F, S_{i_k}).$$

Then, we take object $S_j \notin \left\{S_{i_1}, \ldots, S_{i_k}\right\}$ and want to check if $SH(F, S_j) < SH(F, S_{i_k})$. But before actual calculation of the quantity $SH(F, S_j)$, we check the inequality $SH(S_{i_k}, S_j) \geq 2SH(F, S_{i_k})$. If so, then $SH(F, S_j) \geq SH(S_{i_k}, S_j) - SH(F, S_{i_k}) \geq SH(F, S_{i_k})$ and we need not actually calculate $SH(F, S_j)$. If $SH(S_{i_k}, S_j) < 2SH(F, S_{i_k})$, we calculate $SH(F, S_j)$ and check if $SH(F, S_j) < SH(F, S_{i_k})$. In the latter case, set $\left\{SH(F, S_{i_1}), \ldots, H(F, S_{i_{k-1}}, S_j)\right\}$ is reordered to give the new k smallest distance set. Then we take a new not considered yet object $S_{j'}$ and again try to update the k smallest numbers of the set $\{SH(F, S_i), i = 1, \ldots, m\}$, and so on.

The same trick can also be used within k nearest neighbors classification method. Suppose there are two training sets

$$\mathcal{S}^{m_1} = \{S_i, i = 1, \ldots, m_1\}$$

and $\mathcal{R}^{m_2} = \left\{R_j, j = 1, \ldots, m_2\right\}$ of objects, which represent the two corresponding classes $\mathcal{S}$ and $\mathcal{R}$. To classify a new object F, we have to calculate distances of the object F to all known class representatives $\mathcal{S}^{m_1}$ and $\mathcal{R}^{m_2}$ and to find k smallest numbers in the set

$$\left\{SH(F, S_i), i = 1, \ldots, m_1;\ SH(F, R_j), j = 1, \ldots, m_2\right\}.$$

This can again be done by using the pre-calculated distances $SH(S_i, S_j), i, j = 1, \ldots, m_1$; $SH(R_i, R_j), i, j = 1, \ldots, m_2$; $SH(S_i, R_j), i = 1, \ldots, m_1, j = 1, \ldots, m_2$; and by using the triangle inequality for excluding some calculations.

Reducing Computational Complexity Within a Branch-and-Bound Method Used for Calculation of Similarity Measures Quantile-based similarity measures between object F and class representatives $\mathcal{S}^m = \{S_1, \ldots, S_m\}$ are calculated iteratively by gradually ordering the set $\{SH(F, S_i), i = 1, \ldots, m\}$. At each iteration, there is a current estimate q of the similarity measure, and it is updated by a new quantity $SH(F, S_j)$ if $SH(F, S_j) < q$. The calculation of the quantities $SH(F, S)$ usually is a global optimization problem, which can be solved by the branch-and-bound method. If within iterations of the branch-and-bound method the current lower estimate of $SH(F, S_{j*})$ appears to be greater than q, then the branch can be excluded from further consideration, and the current object S_{j*} can be safely discarded, since this object cannot improve the similarity estimate q. Remark that within this procedure, the triangle inequality is not used.

Acknowledgments The first author thanks Vienna University for the hospitality during Spring 2022 that made it possible to work on an early version of this paper.

References

1. Agarwal, P.K., Fox, K., Nath, A., Sidiropoulos, A., Wang, Y.: Computing the Gromov-Hausdorff distance for metric trees. ACM Trans. Algorithms **14**(2), Article 24 (2018)
2. Ahn, H.-K., Knauer, C., Scherfenberg, M., Schlipf, L., Vigneron, A.: Computing the Discrete Fréchet distance with imprecise input. Int. J. Comput. Geom. Appl. **22**(01), 27–44 (2012). https://doi.org/10.1142/s0218195912600023
3. Alt, H., Mehlhorn, K., Wagener, H., Welzl, E.: Congruence, similarity, and symmetries of geometric objects. Discret. Comput. Geom. **3**(3), 237–256 (1988). https://doi.org/10.1007/bf02187910
4. Andoni, A., Indyk, P.: Nearest neighbors in high-dimensional spaces. In: Goodman, J.E., O'Rourke, J., Tóth, C.D. (eds.) Handbook of Discrete and Computational Geometry, 3rd edn., pp. 1135–1155. CRC Press, Taylor & Francis Group, Boca Raton (2018)
5. Aubin, J-P., Ekeland, I.: Applied Nonlinear Analysis. John Wiley & Sons, Hoboken (1984)
6. Birsan, T., Tiba, D.: One hundred years since the introduction of the set distance by Dimitrie Pompeiu. In: Ceragioli, F., Dontchev, A., Furuta, H., Marti, K., Pandolfi, L. (eds.) IFIP International Federation for Information Processing, vol. 199. System Modeling and Optimization, pp. 35–39. Springer, Boston (2006). https://doi.org/10.1007/0-387-33006-2_4
7. Braß, P.: Combinatorial geometry problems in pattern recognition. Discret. Comput. Geom. **28**(4), 495–510 (2002). https://doi.org/10.1007/s00454-002-2884-3
8. Bronstein, E.M.: Approximation of convex sets by polytopes. J. Math. Sci. **153**(6), 727–762 (2008). https://doi.org/10.1007/s10958-008-9144-x
9. Bronstein, A., Bronstein, M., Kimmel, R.: Efficient computation of isometry-invariant distances between surfaces. SIAM J. Sci. Comput. **28**(5), 1812–1836 (2006)
10. Bronstein, A., Bronstein, M.M., Kimmel, R., Mahmoudi, M., Sapiro, G.: A Gromov-Hausdorff framework with diffusion geometry for topologically-robust nonrigid shape matching. Int. J. Comput. Vis. **89**(2–3), 266–286 (2010)
11. Burago, D., Burago, Yu., Ivanov, Yu.: A Course in Metric Geometry. Graduate Studies in Mathematics 33. AMS, Providence (2001)
12. Cenker, Ch., Pflug, G., Mayer, M.: Stochastic shape theory. In: Kropatsch, W.G., Bischof, H. (eds.) Digital Image Analysis: Selected Techniques and Applications, pp. 49–80. Springer-Verlag, New York (2001)
13. Chazal, F.: High-dimensional topological data analysis. In: Goodman, J.E., O'Rourke, J., Tóth, C.D. (eds.) Handbook of Discrete and Computational Geometry, 3rd edn., pp. 663–683. CRC Press, Taylor & Francis Group, Boca Raton (2018)
14. Chen, J., Leung, M.K., Gao, Y.: Noisy logo recognition using line segment Hausdorff distance. Pattern Recogn. **36**, 943–955 (2003)
15. Chew, L.P., Goodrich, M.T., Huttenlocher, D.P., Kedem, K., Kleinberg, J.K., Kravets, D.: Geometric pattern matching under Euclidean motion. Comput. Geom. **7**, 113–124 (1997). https://doi.org/10.1016/0925-7721(95)00047-X

16. Cohen, T.S., Geiger M., Weiler M.: A general theory of equivariant CNNs on homogeneous spaces. Adv. Neural Inf. Proces. Syst. **32**, 9142–9153 (2019)
17. Conci, A., Kubrusly, C.: Distances between sets – a survey. Adv. Math. Sci. Appl. **26**, 1–18 (2017). http://www.ele.puc-rio.br/~carlos/84-msa17.pdf
18. De, O., Aeberhard, S.: Line-based face recognition under varying pose. IEEE Trans. Pattern Anal. Mach. Intell. **21**, 1081–1088 (1999)
19. Deza, M.M., Deza, E.: Encyclopedia of Distances. Springer-Verlag, Berlin (2009)
20. Distante, A., Distante, C.: Geometric transformations. In: Handbook of Image Processing and Computer Vision, pp. 149–208. Springer, Cham (2020). https://doi.org/10.1007/978-3-030-42374-2_3
21. Dudar, V.V., Semenov, V.V.: Construction of neural architectures with desired behavior under geometric transformations of the input. J, Numer, Appl, Math. **133**(1), 31–49 (2020)
22. Fehri, A., Velasco-Forero, S., Meyer, F.: Characterizing images by the GromovHausdorff distances between derived hierarchies. International Conference on Image Processing (ICIP) (2018), Athens. HAL Id: hal-01823866ff (2018). https://hal.archives-ouvertes.fr/hal-01823866
23. Fujita, O.: Metrics based on average distance between sets. Japan J. Indust. Appl. Math. **30**, 1–19 (2013). https://doi.org/10.1007/s13160-012-0089-6
24. Gromov, M.: Metric Structures for Riemannian and Non-Riemannian Spaces. Birkhauser, Boston (2006)
25. Gruber, P.M.: The space of convex bodies. In: Handbook of Convex Geometry, pp. 301–317. Elsevier, Amsterdam (1993)
26. Gualtieri, J.A., Moigne, J.L., Packer, C.V.: Distance between images. Fourth Symp. Front. Massively Parallel Comput. **10**, 216–223 (1992). https://doi.org/10.1109/FMPC.1992.234956
27. Horst, R., Pardalos, P.M. (eds.): Handbook of Global Optimization, vol. 1. Kluwer, Dordrecht (1995)
28. Huet, B., Hancock, E.R.: Line pattern retrieval using relational histograms. IEEE Trans. Pattern Anal. Mach. Intell. **21**(12), 1369–1370 (1999)
29. Huttenlocher, D.P., Klanderman, G.A., Rucklidge, W.J.: Comparing images using the Hausdorff distance. IEEE Trans. Pattern Anal. Mach. Intell. **15**(9), 850–863 (1993). https://doi.org/10.1109/34.232073
30. Javaheri, A., Brites, C., Pereira, F., Ascenso, J.: A generalized Hausdorff distance based quality metric for point cloud geometry. In: 2000 Twelfth International Conference on Quality of Multimedia Experience (QoMEX). IEEE, Piscataway (2000). https://doi.org/10.1109/QoMEX48832.2020.9123087
31. Khiyari, H.E., Wechsler, H.: Age invariant face recognition using convolutional neural networks and set distances. J. Inf. Secur. **8**(3), 174–185 (2017). https://doi.org/10.4236/jis.2017.83012
32. Knauer, C., Luffler, M., Scherfenberg, M., Wollec, T.: The directed Hausdorff distance between imprecise point sets. Theor. Comput. Sci. **412**, 4173–4186 (2011). https://doi.org/10.1016/j.tcs.2011.01.039
33. Kovalevsky, V.A.: Image Pattern Recognition. Springer-Verlag, New York (1980)
34. Kumar, K.S., Manigandan, T., Chitra, D., Murali, L.: Object recognition using Hausdorff distance for multimedia applications. Multimed Tools Appl. **79**, 4099–4114 (2020). https://doi.org/10.1007/s11042-019-07774-z
35. Lampert, C.H., Blaschko, M.B., Hofmann, T.: Efficient subwindow search: a branch and bound framework for object localization. IEEE Trans. Pattern Anal. Mach. Intell. **31**(12), 2129–2142 (2009). https://doi.org/10.1109/tpami.2009.144
36. Lin, K.-H., Lam, K.-M., Siu, W.-C.: Spatially eigen-weighted Hausdorff distances for human face recognition. Pattern Recognit. **36**(8), 1827–1834 (2003). https://doi.org/10.1016/s0031-3203(03)00011-6
37. Lu, Y., Tan, C.L., Huang, W., Fan, L.: An approach to word image matching based on weighted Hausdorff distance. In: Proceedings of Sixth International Conference on Document Analysis and Recognition, pp. 921–925 (2001). https://doi.org/10.1109/icdar.2001.953920

38. Mémoli, F.: On the use of Gromov-Hausdorff distances for shape comparison. In: Botsch, M., Pajarola, R. (eds.). Eurographics Symposium on Point-Based Graphics (2007)
39. Mémoli, F., Sapiro, G.: A theoretical and computational framework for isometry invariant recognition of point cloud data. Found. Comput. Math. **5**(3), 313–347 (2005)
40. Norkin, V.I.: Pijavski's method for solving general global optimization problem. Comp. Maths and Math. Phys. **32**(7), 873–886 (1992)
41. Norkin, V.I., Onischenko, B.O.: Minorant methods of stochastic global optimization. Cybernet. Syst. Anal. **41**(2), 203–214 (2005). https://doi.org/10.1007/s10559-005-0053-4
42. Norkin V.I., Pichler A.: Portfolio reshaping under 1st order stochastic dominance constraints by the exact penalty function methods. Optimization 1–34 (2024). Published online 09 Feb 2024. https://doi.org/10.1080/02331934.2024.2313086
43. O'Rourke, J., Toussaint, G.T.: Pattern recognition. In: Goodman, J.E., O'Rourke, J., Tóth, C.D. (eds.) Handbook of Discrete and Computational Geometry, 3rd edn., pp. 1421–1450. CRC Press, Taylor & Francis Group, Boca Raton (2018)
44. Paumard, J.: Robust comparison of binary images. Pattern Recognit. Lett. **18**, 1057–1063 (1997). https://doi.org/10.1016/S0167-8655(97)80002-5
45. Pflug, G.Ch.: Random planar shapes and their statistical recognition. Ann. Math. Artif. Intell. **13**, 267–279 (1995)
46. Pflug, G.Ch., Römisch, W.: Modeling, Measuring and Managing Risk. World Scientific, River Edge (2007)
47. Pintér, J.D.: Global optimization in action (continuous and Lipschitz optimization: algorithms, implementations and applications). Kluwer Academic Publishers, Dordrecht (1996)
48. Pompeiu, D.: Sur la continuite' desfonctions de variables complexes (These). Gauthier-Villars, Paris, 1905; Ann. Fac. Sci. de Toulouse. **7**, 264–315 (1905). https://doi.org/10.5802/afst.226
49. Rachkovsky, D.A.: An Introduction to Quick Similarity Search. Interservice, Kyiv (2019). In Russian
50. Ribera, J., Guera, D., Chen, Y., Delp, E.J.: Locating objects without bounding boxes. In: 2019 IEEE/CVF Conference on Computer Vision and Pattern Recognition (CVPR), pp. 6479–6489 (2019). https://doi.org/10.1109/cvpr.2019.00664
51. Rockafellar, R.T., Wets, R.J-B.: Variational Analysis. Springer-Verlag, Berlin (1998)
52. Rodrıguez, M., Berdugo, G., Jabba, D., et al.: Neural networks and Hausdorff distance applied to number recognition in electrical meters. Appl. Artif. Intell. **26**, 921–940 (2012). https://doi.org/10.1080/08839514.2012.731344
53. Rucklidge, W.: Efficient visual recognition using the Hausdorff distance. Lecture Notes in Computer Science, vol. 1173. Springer, Berlin (1996)
54. Rucklidge, W.: Efficiently locating objects using the Hausdorff distance. Int. J. Comput. Vis. **24**(3), 251–270 (1997). https://doi.org/10.1023/A:1007975324482
55. Schlesinger, M.I., Vodolazskii, Y.V., Yakovenko, V.M.: Recognizing the similarity of polygons in a strengthened Hausdorff Metric. Cybern. Syst. Anal. **50**, 476–486 (2014). https://doi.org/10.1007/s10559-014-9636-2
56. Sergeyev, Y.D., Kvasov, D.E.: Deterministic Global Optimization. An Introduction to the Diagonal Approach. Springer, New York (2017)
57. Shapiro, M.D., Blaschko, M.B.: On Hausdorff distance measures. Technical Report UM-CS-2004-071. University of Massachusetts, Department of Computer Science (2004). https://web.cs.umass.edu/publication/docs/2004/UM-CS-2004-071.pdf
58. Taha, AA, Hanbury, A.: An efficient algorithm for calculating the exact Hausdorff distance. IEEE Trans. Pattern. Anal. **37**(11), 2153–2163 (2015). https://doi.org/10.1109/TPAMI.2015.2408351
59. Tuzhilin, A.A.: Who Invented the Gromov-Hausdorff Distance? ArXiv e-prints, arXiv:1612.00728 (2016)
60. Ushakov, V.N., Lebedev, P.D., Tarasyev, A.M., Ushakov, A.V.: Optimization of the Hausdorff distance between convex polyhedrons in R^3. IFAC-PapersOnLine **48**(25), 197–201 (2015). https://doi.org/10.1016/j.ifacol.2015.11.084

61. Vivek, E.P., Sudha, N.: Robust Hausdorff distance measure for face recognition. Pattern Recognit. **40**(2), 431–442 (2007). https://doi.org/10.1016/j.patcog.2006.04.019
62. Yang, C-H.T., Lai, S-H., Chang, L-W.: Hybrid image matching combining Hausdorff distance with normalized gradient matching. Pattern Recognit. **40**(4), 1173–1181 (2007). https://doi.org/10.1016/j.patcog.2006.09.014
63. Yi, X., Camps, O.C.: Line-based recognition using a multidimensional Hausdorff distance. IEEE Trans. Pattern Anal. Mach. Intell. **21**(9), 901–916 (1999)
64. Zhang, D.P., Shu, W.: Two novel characteristics in palmprint verification: datum point invariance and line feature matching. Pattern Recognit. **32**, 691–702 (1999)
65. Zhang, D., He, F., Han, S., Zou, L., Wu, Y., Chen, Y.: An efficient approach to directly compute the exact Hausdorff distance for 3D point sets. Integ. Comput.-Aided Eng. **24**, 261–277 (2017). https://doi.org/10.3233/ICA-170544
66. Zhao, C., Shi, W., Deng, Y.: A new Hausdorff distance for image matching. Pattern Recognit. Lett. **26**, 581–586 (2005). https://doi.org/10.1016/j.patrec.2004.09.022

Approximating Multicriteria Matrix Game Solution with Finite Sets

Natalia M. Novikova and Irina I. Pospelova

1 Introduction

Games with multiple criteria payoffs, further called *multicriteria (MC) games*, have finite number of players; each player's payoff is a vector function; and players want to optimize their vector payoffs. Such games were firstly introduced by Shapley in [1] for two players and zero-sum matrix payoffs. In this case, both players have a finite number of pure strategies and the same payoff, but the first player maximizes it while the second one minimizes. In the general case, the solution and the value of an MC game in mixed strategies are infinite sets (*solution or equilibrium and value sets*). MC games arise in decision-making and may be used, for example, for modeling negotiation process (see problems described in [2]). We believe that, in such cases, it is easier for players to use points from the finite sets rather than from the whole equilibrium or value set. Besides, proper visualization may provide decision-makers and negotiators with better understanding of the decision situation. Therefore, the paper is aimed at approximating solution and value sets by finite ones in order to represent them visually and provide decision-making support.

Our approach is based on Shapley's idea: He used linear combinations of the matrices for characterizing the equilibrium in the MC matrix game. It is a particular case of well-known method of scalarization [3, 4], when the vector function is substituted with a parametric scalar one using a special scalarizing function. Linear

N. M. Novikova
Federal Research Center "Computer Science and Control" of the Russian Academy of Sciences, Moscow, Russia
e-mail: novikova@gse.cs.msu.ru

I. I. Pospelova (✉)
CMC Faculty, Lomonosov Moscow State University, Moscow, Russia
e-mail: ipospelova05@yandex.ru

B. Goldengorin (ed.), *Theory, Algorithms, and Experiments in Applied Optimization*, Springer Optimization and Its Applications 226,
https://doi.org/10.1007/978-3-031-91357-0_14

scalarizing function leads to the linear combination of the partial criteria called *linear scalarization* (LS) [5] and is widely used in real-world applications of multicriteria optimization problems and decision-making. The method of scalarization applied to the vector payoffs in the MC game reduces it to a set of parametric scalar games (each vector function is substituted with a parametric scalar one). Their equilibria and values allow, in some cases, to represent equilibrium and value sets of the MC game in a parametric form.

Earlier in [6], the authors studied application of Germeier's scalarization (GS) that is based on so-called logical scalarization [3] or Rawlsian social welfare function [7] to MC games. This type of scalarization has some advantages [8]. In particular, it was shown [6] that approximation of the MC matrix game solution and value with finite sets can be obtained when the payoffs are parameterized with GS. Although the solution and the value of the MC game can be parameterized with LS as well, the authors did not find any studies devoted to the analysis of approximation properties for LS. Approximation properties of other methods used for representing MC equilibria such as, for example, reference points method [9, 10] or successive concessions [4], are also not studied. As for possible types of scalarizing functions besides LS and GS, the question of their application in MC games is not raised yet, while a lot of such functions are proposed for multicriteria optimization problems (see [4, 11–14] for other scalarizing functions and a combination of GS and LS). This question is still waiting to be answered.

An essential requirement for the finite approximation is its representativeness which means that any optimal point has a close point in the approximating set. This problem differs from the problem of finding approximate equilibria [15], because each point of the constructed finite approximation is exact equilibrium. The difficulty of the stated problem is explained by the fact that solutions of parametric games, in contrast to parametric optimization problems, are often not continuous in the parameters [15]. The property of Hausdorff approximation with the finite net on the set of the parameters is weaker [6] than continuity; therefore, it can be fulfilled in more cases.

The paper is organized as follows. In Sect. 2, we introduce definitions of the solution and the value of an MC matrix game, indicate the difficulties of their approximation using LS, and suggest the ways to overcome the difficulties in the case of MC matrix games with 2×2 matrices. In Sect. 3, we find general solution for such games. In Sect. 4, we form so-called basic values of LS parameters for obtaining approximation. In Sect. 5, we describe the method for constructing a finite approximation of the equilibrium set with the help of LS. Section 6 gives some examples and shows difference between LS and GS in approximating MC matrix games. In Sect. 7, possible directions for further research are pointed out.

2 Formal Statement of the Problem

Following [1], two-person multicriteria matrix game is defined by the set of $m \times n$ matrices $A^1, \ldots, A^k$ which determine gains of Player 1 and losses of Player 2 for each of k criteria, respectively, when Player 1 chooses a row $i \in \{1, \ldots, m\}$ and Player 2 chooses a column $j \in \{1, \ldots, n\}$. The solution of this game in mixed strategies x, y is a set, and it is defined by Shapley in the form

$$E \stackrel{\text{def}}{=} \{(x^0, y^0) |\, x^0 \in \operatorname{Arg} \operatorname*{Max}_{x \in S^m} F(x, y^0),\ y^0 \in \operatorname{Arg} \operatorname*{Min}_{y \in S^n} F(x^0, y)\}, \tag{1}$$

where

$$F(x, y) = (f^1(x, y), \ldots, f^k(x, y)),\quad f^h(x, y) \stackrel{\text{def}}{=} xA^h y,\quad h = 1, \ldots, k,$$

and $S^l \stackrel{\text{def}}{=} \{v \in \mathbb{R}^l_+ |\sum_{h=1}^{l} v_h = 1\}$, $l = m, n$. Here and further we deal with weakly Pareto efficient Max and Min of a criterion vector [14], i.e., for an arbitrary set $W \subset \mathbb{R}^k_+$ (in the criterion space)

$$\operatorname{Max} W \stackrel{\text{def}}{=} \{v^0 \in W | \forall v \in W\ \exists i :\ v_i \le v_i^0\},$$

$$\operatorname{Min} W \stackrel{\text{def}}{=} \{v^0 \in W | \forall v \in W\ \exists i :\ v_i \ge v_i^0\}.$$

Thus, we consider the widest of possible sets as a solution [16] keeping in mind the idea that these sets are usually used for finding a compromise. Shapley's solution (1) of the MC game generalizes the concept of Nash equilibrium [17] to the multicriteria case. We also refer to it as the equilibrium set meaning MC equilibrium. Other approaches to defining the equilibrium in MC games can be found in [2, 18, 19]. We do not dwell on them, as well as on various methods for selecting a specific point from the equilibrium set.

In the paper, we limit ourselves with the simplest case of MC matrix games with $m = n = 2$. We denote them $k \times 2 \times 2$ for short. In such games, vectors of mixed strategies have the form $x = (p, 1 - p)$ and $y = (q, 1 - q)$, and the equilibrium set E can be redefined as

$$R \stackrel{\text{def}}{=} \{(p^0, q^0) |\, p^0 \in \operatorname{Arg} \operatorname*{Max}_{p \in [0,1]} \Phi(p, q^0),\ q^0 \in \operatorname{Arg} \operatorname*{Min}_{q \in [0,1]} \Phi(p^0, q)\},$$

where Φ has the components

$$\varphi^h(p, q) \stackrel{\text{def}}{=} pqa^h_{11} + p(1 - q)a^h_{12} + (1 - p)qa^h_{21} + (1 - p)(1 - q)a^h_{22} \tag{2}$$

and a_{ij}^h, $i = 1, 2$, $j = 1, 2$, are elements of the matrix A^h, $h = 1, \dots, k$. Further without loss of generality, we assume all elements of A^h, $h = 1, \dots, k$, to be positive integers. This assumption allows us not to use rounding, because there are only rational points in approximating sets we construct. All matrices with rational points satisfy this condition, since adding the same constant to all elements and multiplication by the same positive constant does not affect the set R.

Now we introduce formally the value of the MC game as the set of averaged criterion vectors for mixed strategies corresponding to the equilibrium outcomes:

$$V \overset{\text{def}}{=} \{\Phi(p^0, q^0) | \, (p^0, q^0) \in R\} = \{(x^0 A^1 y^0, \dots, x^0 A^k y^0) | \, (x^0, y^0) \in E\}.$$

Elements of V (points in the criterion space) are vector evaluations of the players' payoffs when they choose one or another equilibrium solution. The set V may have a very bizarre structure, including curved pieces of the boundary, even for the simplest $2 \times 2 \times 2$ games (see, e.g., Sect. 6). As a rule, V and R are non-convex and contain parts with empty interior that makes difficult their finite approximation. We consider closeness of the approximating sets to the initial ones in terms of Hausdorff metric. It means that for any point of the initial set, there should exist a close point in the finite set composed of points from the initial set, representing (in total) its approximation. It is proved in [6] for $k = 2$ that the required approximation can be constructed with GS; however, our aim is to propose approximating method based on LS, because it is more popular and suitable for many applications [5].

Shapley stated that if LS is applied to the matrices of the MC game, i.e., if $A(\nu) \overset{\text{def}}{=} \sum_{h=1}^{k} \nu_h A^h$, $\nu = \lambda, \eta$, are considered, then the set E coincides with the union of the solutions in mixed strategies of scalar bimatrix games with the matrices $A(\lambda)$ and $-A(\eta)$ when LS parameters λ and η independently take values from the simplex S^k. Let $\Gamma(\lambda, \eta) \overset{\text{def}}{=} \langle S^m, S^n, A(\lambda), -A(\eta)\rangle$ be the parametric game obtained from the initial MC matrix game with the aid of LS.

Even for $k = 2$ (when each player has only one scalarization parameter, because $\nu_1 + \nu_2 = 1$) and for an arbitrary or random (very fine) finite net on the set $[0, 1] \times [0, 1]$ of LS parameters, the union of all corresponding equilibrium outcomes of the games $\Gamma(\lambda, \eta)$ does not have to be an approximation of R in Hausdorff metric. (Similar property is also typical for LS-based approximation of a weakly Pareto efficient set in MC optimization problems.) The aim of the paper is to point out such a finite net (depending on elements of the matrices A^h), for which LS approximation properties are satisfied, and to find corresponding approximating sets for V and R.

Let us emphasize that in the considered case of 2×2 criterion matrices, the equilibrium set for the obtained 2×2 bimatrix games $\Gamma(\lambda, \eta)$ are known in explicit form for almost all (excluding degenerate cases) bimatrices [20]. Unfortunately, considering only those pairs (λ, η), which lead to nondegenerate bimatrix games $\Gamma(\lambda, \eta)$, we usually cannot construct approximation of R (and V). Therefore, the solution sets for 2×2 bimatrix games should be obtained in degenerate cases (see

Sect. 3) as well. Further (see Sects. 4 and 5), this helps us to find finite approximation of infinite solution sets.

3 Solution of an Arbitrary Bimatrix Game $\Gamma(\lambda, \eta)$

Now obtain the solution of a parametric bimatrix game in explicit form in dependence of LS parameters. The gain of Player 1 and the loss of Player 2 in mixed strategies are equal to

$$x \sum_{h=1}^{k} \nu_h A^h y =$$

$$pq \sum_{h=1}^{k} \nu_h a_{11}^h + p(1-q) \sum_{h=1}^{k} \nu_h a_{12}^h + (1-p)q \sum_{h=1}^{k} \nu_h a_{21}^h + (1-p)(1-q) \sum_{h=1}^{k} \nu_h a_{22}^h =$$

$$pq \sum_{h=1}^{k} \nu_h (a_{11}^h - a_{12}^h - a_{21}^h + a_{22}^h) + p \sum_{h=1}^{k} \nu_h (a_{12}^h - a_{22}^h) + q \sum_{h=1}^{k} \nu_h (a_{21}^h - a_{22}^h) + \sum_{h=1}^{k} \nu_h a_{22}^h$$

when the players choose the vector of parameters ν independently, with $\nu = \lambda$ for Player 1 and $\nu = \eta$ for Player 2 in the game $\Gamma(\lambda, \eta)$.

To shorten the writing, we introduce the following notation:

$$A_h \stackrel{\text{def}}{=} a_{11}^h - a_{12}^h - a_{21}^h + a_{22}^h,\ a_1^h \stackrel{\text{def}}{=} a_{12}^h - a_{22}^h,\ a_2^h \stackrel{\text{def}}{=} a_{21}^h - a_{22}^h,\ h = 1, \dots, k,$$

represent aggregates formed from elements of the matrices and

$$A \stackrel{\text{def}}{=} (A_1, \dots, A_k),\ a_1 \stackrel{\text{def}}{=} (a_1^1, \dots, a_1^k),\ a_2 \stackrel{\text{def}}{=} (a_2^1, \dots, a_2^k),\ a_{22} \stackrel{\text{def}}{=} (a_{22}^1, \dots, a_{22}^k)$$

are vectors consisting of the aggregates. Further $\langle \cdot, \cdot \rangle$ denotes the dot product in $\mathbb{R}^k$.

In the introduced notation, in the game $\Gamma(\lambda, \eta)$, Player 1 tries to maximize the function

$$p\langle \lambda, Aq + a_1 \rangle + \langle \lambda, qa_2 + a_{22} \rangle \tag{3}$$

over $p \in [0, 1]$, and Player 2 wants to minimize

$$q\langle \eta, Ap + a_2 \rangle + \langle \eta, pa_1 + a_{22} \rangle \tag{4}$$

over $q \in [0, 1]$.

For the fixed q, function (3) is linear in p; therefore, it takes maximum either in the endpoints of the segment [0, 1] or in its arbitrary point when coefficient of p is equal to 0. Minimal points of (4) in q for fixed p are determined analogously. Optimal points depend on the signs of coefficients in p for Player 1 in (3) and in q for Player 2 in (4); therefore, we can find all pairs of equilibrium strategies by considering all variants of the signs and zero for the dot products

$$\langle \lambda, Aq + a_1 \rangle, \quad \langle \eta, Ap + a_2 \rangle. \tag{5}$$

I. At first, let us assume that both expressions in (5) are not equal to 0 for a fixed pair of parameters (λ, η). Then the equilibrium pair of strategies is to be pure. If, for example, $\langle \lambda, Aq^0 + a_1 \rangle > 0$ for some q^0, then $p^0 = 1$ is optimal strategy for Player 1. In this case, if $\langle \eta, Ap^0 + a_2 \rangle = \langle \eta, A + a_2 \rangle \neq 0$, then q^0 can only be equal to 0 or 1. The same reasoning for all possible pairs of signs of expressions in (5) leads to the following conditions for the equilibrium (p^0, q^0) in the considered case:

$$\begin{cases} \langle \lambda, a_1 \rangle < 0, \\ \langle \eta, a_2 \rangle > 0 \end{cases} \Rightarrow (0, 0), \quad \begin{cases} \langle \lambda, a_1 \rangle > 0, \\ \langle \eta, A + a_2 \rangle > 0 \end{cases} \Rightarrow (1, 0),$$

$$\begin{cases} \langle \lambda, A + a_1 \rangle < 0, \\ \langle \eta, a_2 \rangle < 0 \end{cases} \Rightarrow (0, 1), \quad \begin{cases} \langle \lambda, A + a_1 \rangle > 0, \\ \langle \eta, A + a_2 \rangle < 0 \end{cases} \Rightarrow (1, 1).$$

Note that the obtained conditions are sufficient for the existence of a pure strategy equilibrium, and a pure equilibrium may also appear in other situations.

II. Now consider the case when one of the expressions in (5) is equal to zero. The component of the equilibrium that has zero coefficient (it is denoted with upper index 0) is any from [0, 1], and the second component is equal to 0 or 1 according to the corresponding inequality condition for the coefficient in it. In particular,

$$\begin{cases} \langle \lambda, Aq^0 + a_1 \rangle < 0, \\ \langle \eta, a_2 \rangle = 0 \end{cases} \Rightarrow (0, q^0).$$

The inequality results in the restriction on the strategy q^0:

$$q^0 \langle \lambda, A \rangle + \langle \lambda, a_1 \rangle < 0 \Rightarrow \left[\begin{array}{ll} q^0 < -\dfrac{\langle \lambda, a_1 \rangle}{\langle \lambda, A \rangle} & \text{if } \langle \lambda, A \rangle > 0, \\ q^0 > -\dfrac{\langle \lambda, a_1 \rangle}{\langle \lambda, A \rangle} & \text{if } \langle \lambda, A \rangle < 0. \end{array} \right.$$

If $\langle \lambda, A \rangle = 0$, then q^0 is any from [0, 1] when $\langle \lambda, a_1 \rangle < 0$. Subject to the condition $q^0 \in [0, 1]$, finally we come to the first variant of the equilibrium outcomes (p^0, q^0) for the considered case II: $p^0 = 0$ and

$$\left[\begin{array}{l} q^0 \in \left[0, \min\left(-\frac{\langle \lambda, a_1\rangle}{\langle \lambda, A\rangle}, 1\right)\right) \quad \text{if } \langle \lambda, A\rangle > 0, \\ q^0 \in \left(\max\left(-\frac{\langle \lambda, a_1\rangle}{\langle \lambda, A\rangle}, 0\right), 1\right] \quad \text{if } \langle \lambda, A\rangle < 0, \\ q^0 \in [0, 1] \;\; \text{if } \langle \lambda, A\rangle = 0 \text{ and } \langle \lambda, a_1\rangle < 0. \end{array}\right. \tag{6}$$

Other combinations of signs of expressions in (5) lead to three more variants:

$$\left\{\begin{array}{l} \langle \lambda, q^0 A + a_1\rangle > 0, \\ \langle \eta, A + a_2\rangle = 0 \end{array}\right. \Rightarrow (1, q^0),$$

$$\left[\begin{array}{l} q^0 \in \left(\max\left(-\frac{\langle \lambda, a_1\rangle}{\langle \lambda, A\rangle}, 0\right), 1\right] \quad \text{if } \langle \lambda, A\rangle > 0, \\ q^0 \in \left[0, \min\left(-\frac{\langle \lambda, a_1\rangle}{\langle \lambda, A\rangle}, 1\right)\right) \quad \text{if } \langle \lambda, A\rangle < 0, \\ q^0 \in [0, 1] \;\; \text{if } \langle \lambda, A\rangle = 0 \text{ and } \langle \lambda, a_1\rangle > 0; \end{array}\right. \tag{7}$$

$$\left\{\begin{array}{l} \langle \lambda, a_1\rangle = 0, \\ \langle \eta, A p^0 + a_2\rangle > 0 \end{array}\right. \Rightarrow (p^0, 0),$$

$$\left[\begin{array}{l} p^0 \in \left(\max\left(-\frac{\langle \eta, a_2\rangle}{\langle \eta, A\rangle}, 0\right), 1\right] \quad \text{if } \langle \eta, A\rangle > 0, \\ p^0 \in \left[0, \min\left(-\frac{\langle \eta, a_2\rangle}{\langle \eta, A\rangle}, 1\right)\right) \quad \text{if } \langle \eta, A\rangle < 0, \\ p^0 \in [0, 1] \;\; \text{if } \langle \eta, A\rangle = 0 \text{ and } \langle \eta, a_2\rangle > 0; \end{array}\right. \tag{8}$$

$$\left\{\begin{array}{l} \langle \lambda, A + a_1\rangle = 0, \\ \langle \eta, A p^0 + a_2\rangle < 0 \end{array}\right. \Rightarrow (p^0, 1),$$

$$\left[\begin{array}{l} p^0 \in \left[0, \min\left(-\frac{\langle \eta, a_2\rangle}{\langle \eta, A\rangle}, 1\right)\right) \quad \text{if } \langle \eta, A\rangle > 0, \\ p^0 \in \left(\max\left(-\frac{\langle \eta, a_2\rangle}{\langle \eta, A\rangle}, 0\right), 1\right] \quad \text{if } \langle \eta, A\rangle < 0, \\ p^0 \in [0, 1] \;\; \text{if } \langle \eta, A\rangle = 0 \text{ and } \langle \eta, a_2\rangle < 0. \end{array}\right. \tag{9}$$

III. Finally, consider the case when both expressions in (5) are equal to 0. We obtain the system of equations

$$\left\{\begin{array}{l} \langle \lambda, A q^0 + a_1\rangle = 0, \\ \langle \eta, A p^0 + a_2\rangle = 0, \end{array}\right.$$

which leads to the pair (p^0, q^0):

$$\begin{cases} p^0 = -\dfrac{\langle \eta, a_2 \rangle}{\langle \eta, A \rangle} & \text{if } \langle \eta, A \rangle \neq 0, \\ q^0 = -\dfrac{\langle \lambda, a_1 \rangle}{\langle \lambda, A \rangle} & \text{if } \langle \lambda, A \rangle \neq 0, \end{cases} \tag{10}$$

that is equilibrium when $p^0, q^0 \in [0, 1]$. Such an equilibrium is completely mixed, as a rule.

If $\langle \lambda, A \rangle = 0$, then $\langle \lambda, a_1 \rangle = 0$ should be for the case III. Hence, the equilibrium pairs are composed of all $q^0 \in [0, 1]$ and p^0 from (10) for $\langle \eta, A \rangle \neq 0$. Similarly, when $\langle \eta, A \rangle = 0$, the components of the equilibrium are all $p^0 \in [0, 1]$ and q^0 from (10) for $\langle \lambda, A \rangle \neq 0$. If

$$\begin{cases} \langle \lambda, A \rangle = 0, \\ \langle \eta, A \rangle = 0, \end{cases} \tag{11}$$

then all pairs $(p^0, q^0) \in [0, 1] \times [0, 1]$ are equilibria in $\Gamma(\lambda, \eta)$.

The bimatrix game $\Gamma(\lambda, \eta)$ does not have any other equilibrium, because one of the conditions I–III is always satisfied. Thus, the equilibrium set constructed with the help of I–III is always not empty (in accordance with bimatrix game theory from [20]). Uniting the found equilibria over $(\lambda, \eta) \in S^k \times S^k$, we obtain MC equilibrium R, that is the set defined as the solution of $k \times 2 \times 2$ MC matrix game.

Note that conditions from the case I and (10) in III lead to the formulas from [20] obtained by direct solving the bimatrix game. Nevertheless, the case II does not fit the known formulas, since it is degenerate from the viewpoint of [20]. At the same time, when it is realized, it usually gives an infinite set of equilibrium pairs, i.e., makes a significant contribution to the construction of R. Besides, if there exist $(\lambda, \eta) \in S^k \times S^k$ such that (11) in III is satisfied, then $R = [0, 1] \times [0, 1]$, i.e., the equilibrium set coincides with the set of all outcomes of the MC game. However, it is true not for all MC game matrices. Indeed, (11) is satisfied for $\lambda \neq \eta$ only if the vector A has zero components (parameter vectors are not collinear when not equal, because they belong to the simplex S^k). Moreover, in the case III, condition (11) assumes that

$$\begin{cases} \langle \lambda, a_1 \rangle = 0, \\ \langle \eta, a_2 \rangle = 0. \end{cases} \tag{12}$$

Thus, for $\lambda = \eta$, aggregates a_1 and a_2 obtained from the game matrices should be proportional. This property is satisfied for specific matrices only; therefore, an arbitrary MC game usually has nontrivial solution (see, e.g., Sect. 6).

Note 1 From the bimatrix game theory, any 2×2 game with the matrices B^1, B^2 can be represented in the form of $\Gamma(\lambda^0, \eta^0)$ for two-criteria matrix game with $A^1 = B^1$, $A^2 = -B^2$ and $\lambda^0 = (1, 0)$, $\eta^0 = (0, 1)$. The solution of the scalarized MC game represented in I–III gives complete set of equilibria for any bimatrix game including degenerate cases. (The initial condition for the elements of matrices A^h to be positive is provided by adding a large constant to all elements.)

4 Basic Values for δ-Net on the Set of LS Parameters When Approximating R

In spite of the analytically obtained equilibrium set for $k \times 2 \times 2$ MC matrix game, its practical application in the process of finding a compromise, for example, during negotiations, may be difficult. According to the MC optimization terminology [13, 14, 21], MC game players model decision-makers (DMs). And DMs do not always consider the whole set of feasible values of their objective functions, limiting themselves to understandable, justified, and interpretable weights of the criteria. In this case, the search for a compromise, which should be one of the points of the equilibrium set of the MC game, becomes a finite iterative procedure. Thus, in order to support decision-making, it is necessary to provide a finite enumeration of the weights, which allows to find an approximation to the equilibrium set with a predetermined accuracy. However, simply by setting a net on the sets of scalarization parameters (even one parameter with fixed another), close enough points may not be obtained for some of the equilibrium values. This is explained by the presence of equality conditions for the parameters in the case II (Sect. 3). If a value of the scalarization parameter providing the equality is not chosen, the corresponding pair of equilibrium strategies may not be obtained. Moreover, even for $R = [0, 1] \times [0, 1]$ and an arbitrary net of $(\lambda, \eta) \in S^k \times S^k$, the most part of potentially compromise equilibria may not be found if none of points of the net satisfies (11), (12).

Further we use complete solution from Sect. 3 for constructing for any $\varepsilon > 0$ a finite δ-net on the set of scalarization parameters which, in turn, allows to construct a finite ε-net on the set R, approximating the whole set of equilibrium outcomes, i.e., MC equilibrium, in Hausdorff metric.

Note that according to I–III, there are areas in the set of LS parameters, in which possible variants of equilibrium strategies are continuous in parameters. Indeed, pure strategy pairs are provided by strict inequalities in I. These inequalities are fulfilled for sufficiently small variations of the parameters; therefore, the corresponding strategies remain unchanged. The pairs of the form $(0, q^0)$, $(p^0, 0)$, $(1, q^0)$, $(p^0, 1)$, (p^0, q^0) have components p^0, q^0 which are continuous over η and λ, respectively, (under the condition $\langle \nu, A \rangle \neq 0$, $\nu = \lambda, \eta$). It is possible $\forall \varepsilon > 0$ to choose a finite δ-net in each area, where components of the equilibrium strategy are continuous. This provides ε-approximation in the part of the equilibrium set related to such areas.

However, continuity of the equilibrium outcomes over η, λ may fail in transition from one area to another. This happens when one of the inequalities in the case I turns into equality, and vice versa, when the equalities in II turn into inequalities. In addition, it is necessary to take into account points of discontinuity that occur when $\langle \nu, A\rangle = 0$, $\nu = \lambda, \eta$.

In Sect. 3, inequality conditions describe the equilibrium strategies not only in case I but also in cases II and III. Indeed, p^0, q^0 depend on the ratios

$$-\frac{\langle \lambda, a_1\rangle}{\langle \lambda, A\rangle}, -\frac{\langle \eta, a_2\rangle}{\langle \eta, A\rangle},$$

which are considered for those values of parameters, for which these expressions belong to [0, 1]. Let $\langle \nu, A\rangle > 0$, $\nu = \lambda, \eta$, for certainty. Then

$$0 < -\frac{\langle \lambda, a_1\rangle}{\langle \lambda, A\rangle} < 1 \ \Leftrightarrow \begin{cases} \langle \lambda, A + a_1\rangle > 0, \\ \langle \lambda, a_1\rangle < 0, \end{cases}$$
$$0 < -\frac{\langle \eta, a_2\rangle}{\langle \eta, A\rangle} < 1 \ \Leftrightarrow \begin{cases} \langle \eta, A + a_2\rangle > 0, \\ \langle \eta, a_2\rangle < 0. \end{cases}$$

Other variants of inequalities are studied similarly. Thus, the following dot products are indicative for all conditions I–III:

$$\langle \lambda, a_1\rangle,\ \langle \lambda, A + a_1\rangle,\ \langle \eta, a_2\rangle,\ \langle \eta, A + a_2\rangle,\ \langle \lambda, A\rangle,\ \langle \eta, A\rangle. \tag{13}$$

To take into account the peculiarities arising from the equality to zero of any of these expressions, it is necessary to include the corresponding values of parameters in the δ-net.

Consider a case of MC matrix game for $k = 2$. Then $\lambda = (\lambda_1, \lambda_2)$, $\eta = (\eta_1, \eta_2)$, $\lambda, \eta \in S^2$, i.e., each vector of scalarization parameters is only determined by its first component: $\lambda = (\lambda_1, 1 - \lambda_1)$, $\eta = (\eta_1, 1 - \eta_1)$, $\lambda_1, \eta_1 \in [0, 1]$. In this case, the δ-net should be constructed on $[0, 1] \times [0, 1]$ in the coordinate plane (λ_1, η_1).

Conditions of the dot products (13) equal 0 take the following form:

$$\left[\begin{array}{l} \lambda_1 a_1^1 + (1 - \lambda_1)a_1^2 = \lambda_1(a_1^1 - a_1^2) + a_1^2 = 0, \\ \lambda_1(A_1 + a_1^1) + (1 - \lambda_1)(A_2 + a_1^2) = \lambda_1(A_1 - A_2 + a_1^1 - a_1^2) + A_2 + a_1^2 = 0, \\ \eta_1 a_2^1 + (1 - \eta_1)a_2^2 = \eta_1(a_2^1 - a_2^2) + a_2^2 = 0, \\ \eta_1(A_1 + a_2^1) + (1 - \eta_1)(A_2 + a_2^2) = \eta_1(A_1 - A_2 + a_2^1 - a_2^2) + A_2 + a_2^2 = 0, \\ \lambda_1 A_1 + (1 - \lambda_1)A_2 = \lambda_1(A_1 - A_2) + A_2 = 0, \\ \eta_1 A_1 + (1 - \eta_1)A_2 = \eta_1(A_1 - A_2) + A_2 = 0. \end{array}\right.$$

Hence we obtain the set of *basic values* for the parameters λ_1, η_1:

$$\left[\begin{aligned} \lambda_1 &= -\frac{a_1^2}{a_1^1 - a_1^2}, \\ \lambda_1 &= -\frac{A_2 + a_1^2}{A_1 - A_2 + a_1^1 - a_1^2}, \\ \eta_1 &= -\frac{a_2^2}{a_2^1 - a_2^2}, \\ \eta_1 &= -\frac{A_2 + a_2^2}{A_1 - A_2 + a_2^1 - a_2^2}, \\ \lambda_1 &= -\frac{A_2}{A_1 - A_2}, \\ \eta_1 &= -\frac{A_2}{A_1 - A_2}. \end{aligned}\right. \tag{14}$$

All those λ_1, η_1 from (14), for which $\lambda_1, \eta_1 \in [0, 1]$, are included in the δ-net: Possible λ_1 are combined with possible η_1 and with boundary values (that are 0 and 1), and all such combinations (λ_1, η_1) give *basic nodes* on $[0, 1] \times [0, 1]$. The remaining points of the δ-net are situated uniformly on the intersections of the lines parallel to the coordinate axes. The intersections of these lines with similar lines at basic values are also added into the δ-net.

In the case of an arbitrary k, only first $k-1$ components of the parameter vectors should be considered. And instead of the δ-net on the product of simplexes $S^k \times S^k$, a δ-net D_δ on the set S that is the product of two $(k-1)$-dimensional unit semicubes (without parts above the main diagonal because of the condition $\sum_{h=1}^{k-1} \nu_h \leq 1 \; \forall \nu = \lambda, \eta$) should be constructed. These diagonals determine boundary relations for D_δ points. For the kth components of the parameter vectors, we put $\nu_k = 1 - \sum_{h=1}^{k-1} \nu_h \leq 1 \; \forall \nu = \lambda, \eta$. Equating each dot product from (13) to zero, we obtain *basic relations* between LS parameters. Intersections of the corresponding hyperplanes with each other and with the boundary hyperplanes determine a partition of S into the areas, where equilibrium sets of the scalar games $\Gamma(\lambda, \eta)$ are continuous over parameters. Note that these areas may be of any dimension from 1 to $k-1$.

If the intersection of the hyperplanes is a point, then we choose it as a node for D_δ. For a line obtained as the intersection, we take points on this line having distance not greater than δ and also choose them as nodes for D_δ. When there is a hyperplane in the intersection, a δ-net is constructed on it and included into D_δ (for example, δ-net on a hyperplane can be constructed by intersecting this hyperplane with a regular grid, see below for details). Similar δ-nets on basic and boundary hyperplanes determine other basic nodes for D_δ. Finally, nodes of regular δ-grid on S are added. This grid is formed of the hyperplanes parallel to the coordinate hyperplanes of $(2(k-1)-1)$-dimension, constructed with a step less than δ (on each coordinate axis), in the set S. As a result, the distance between any two neighboring nodes in D_δ is not greater than δ in maximum norm. (Here and further, we consider the norm $||\nu|| = \max_h \nu_h$.) Then the distance from the other values of the LS parameters to the nearest nodes is less than δ due to the chosen norm.

The finite net constructed on the set of LS parameters according to the described method provides approximation of the MC game solution in Hausdorff metric, when the distance between neighboring nodes is not greater than sufficiently small δ. Formally, the set R_δ that is the union of equilibria of the games $\Gamma(\lambda^\delta, \eta^\delta)$ over all $(\lambda^\delta, \eta^\delta)$ from the constructed δ-net on $S^k \times S^k$ tends in Hausdorff metric to the set R that is the union of equilibria of the games $\Gamma(\lambda, \eta)$ over all $(\lambda, \eta) \in S^k \times S^k$.

Indeed, take an arbitrary point (p^0, q^0) from the set R representing MC equilibrium. As it was proved in [1], it belongs to the equilibrium set of the game $\Gamma(\lambda^0, \eta^0)$ for some (λ^0, η^0) from $S^k \times S^k$. If (λ^0, η^0) is included into δ-net on $S^k \times S^k$, then $(p^0, q^0) \in R_\delta$. Otherwise, in the simplest case, λ^0, η^0 do not belong to the boundary values of the parameters, and none of the expressions in (13) is equal to 0 for them. Hence for sufficiently small δ, variations of λ^0, η^0 by δ do not change the signs of the dot products in (13). Thus, the equilibrium remains the same for the nodes of the δ-net close to λ^0, η^0.

In the general case, the pair (λ^0, η^0) not included in the δ-net may correspond to the basic or boundary relations for either λ^0 or η^0. For $k = 2$, assuming λ_1^0 to be equal to 0, 1, or one of the variants in (14), we obtain the corresponding line, and we have points from δ-net constructed for η_1 on this line. Let us choose one of them nearest to η_1^0. Then for sufficiently small δ, inequalities for the second component of the pair remain unchanged, and the equilibrium is the same.

Consider $k > 2$. Let for certainty $\langle \lambda^0, a_1 \rangle = 0$ and the other dot products in (13) be positive. Then (λ^0, η^0) satisfies case II from Sect. 3 (see first line in (8)). According to the construction, in the δ-neighborhood of (λ^0, η^0), there exists a δ-net node (λ', η) on the hyperplane $\langle \lambda', a_1 \rangle = 0$ such that $||\lambda' - \lambda^0|| \le \delta$ and $||\eta - \eta^0|| \le \delta$. Thus, equilibria of the game $\Gamma(\lambda', \eta)$ have the same form:

$$(p, 0), \quad p \in \left(\max\left(-\frac{\langle \eta, a_2 \rangle}{\langle \eta, A \rangle}, 0 \right), 1 \right]. \tag{15}$$

Since the numerator and the denominator of the fraction

$$-\frac{\langle \eta, a_2 \rangle}{\langle \eta, A \rangle} \tag{16}$$

are strictly positive under the assumption, the set of equilibrium p in (15) is equal to $(0, 1]$ for sufficiently small δ (the same is satisfied for $\Gamma(\lambda^0, \eta^0)$).

More complicated reasonings are needed for this variant when $\langle \eta^0, a_2 \rangle < 0$ and positiveness of other nonzero expressions in (13), i.e., formula (15) is fulfilled but (16) is now greater than 0. Then the left part of the half-interval in (15) for the equilibrium $(p, 0)$ is greater than 0, i.e., it depends on η. However, when max in (15) is not equal to 1, it is continuous in η, i.e., the half-interval in (15) is continuous in Hausdorff metric. Otherwise,

$$-\frac{\langle \eta, a_2 \rangle}{\langle \eta, A \rangle} = 1$$

implies

$$-\langle \eta, a_2 \rangle = \langle \eta, A \rangle \ \Rightarrow \langle \eta, a_2 + A \rangle = 0.$$

If the latter equality is satisfied for any sufficiently small δ, then it is satisfied for η^0, i.e., $\langle \eta^0, a_2 + A \rangle = 0$. Thus, the pair (λ^0, η^0) is situated in the intersection of the hyperplanes corresponding to two basic relations and has a node (λ^*, η^*) from the δ-net in its δ-neighborhood, for which both relations are satisfied (i.e., $\langle \lambda^*, a_1 \rangle = 0$ and $\langle \eta^*, a_2 + A \rangle = 0$). The single equilibrium here is $(p^*, q^*) = (1, 0)$, which coincides with (p^0, q^0).

Other combinations of the basic relations are considered similarly, and we do not dwell on them further.

5 Approximating R with Finite Set with ε-Accuracy

Note that the scalar game $\Gamma(\lambda^\delta, \eta^\delta)$ for a pair of parameters may have infinite number of equilibria, in particular, for the basic values $(\lambda^\delta, \eta^\delta)$ obtained in the cases II and III (except (10)). Therefore, R_δ composed of the equilibria at the δ-net nodes (see Sect. 4) is infinite for most of MC game matrices in spite of the finite number of parameters in the net. For constructing a finite ε-net on R, it is sufficient to construct it on R_δ, i.e., on those its components (that are equilibria of the games $\Gamma(\lambda^\delta, \eta^\delta)$), for which there exist equilibrium segments or half-intervals for one or both players. Remind that such bimatrix games are degenerate, but their LS parameters are included as nodes into the δ-net.

In case III, infinite sets of equilibria in $\Gamma(\lambda^\delta, \eta^\delta)$ are segments [0, 1] for one or both players and we choose an ε-net on these segments. The ε-net is constructed similarly when third lines in (6), (7), (8), or (9) are realized for $(\lambda^\delta, \eta^\delta)$ in case II. For first and second lines in (6)–(9), we include boundary values in the ε-net if they are included in the corresponding half-intervals. And for the endpoint not included in the half-interval (as left endpoint in (15)), we take a point in the half-interval and at the distance of $\varepsilon/2$ from the endpoint or, if the length of the half-interval is less than ε, we take its midpoint. We choose other points in the half-interval uniformly with rational nodes, for example, of the form K/N, where $0 \leq K \leq N$, $N \geq 1/\varepsilon$, K, N are integers. Thus, we limit ourselves to a finite number of equilibria in the game $\Gamma(\lambda^\delta, \eta^\delta)$.

When the game $\Gamma(\lambda^\delta, \eta^\delta)$ has finite number of equilibria (see case I, conditions (10), and some situations in case II) for $(\lambda^\delta, \eta^\delta)$ from the δ-net constructed in Sect. 4, all these equilibria are included into the final finite approximation R_δ^ε for the equilibrium set R of the MC matrix game $k \times 2 \times 2$. For δ and ε tending to zero, the set R_δ^ε tends to R in Hausdorff metric.

To complete the picture, we point out the existence of MC-degenerate games, for which it does not matter what net to choose on the set of LS parameters, since all $\Gamma(\lambda, \eta)$ have the same solution. Sometimes it is not necessary to approximate R, because it is a singleton.

Indeed, cases I–III assume that the expressions obtained from (5) depend on the parameters λ and η, and the systems of equalities and inequalities allow to set different values of parameter pairs at which one or another case is realized (and the corresponding variants of equilibrium outcomes are obtained). However, with a certain combination of elements of criterion matrices, it may turn out that there is actually no such dependence. We call these cases *MC-degenerate*. They are characterized by the fact that the condition determining, which outcome is equilibrium, does not change, but it is fulfilled for all values of parameters.

Since the dot product of any vector ν of LS parameters and an arbitrary vector $a = (a^1, \dots, a^k)$ can be represented in the form

$$\langle \nu, a \rangle = \sum_{h=1}^{k-1} \nu_h a^h + \left(1 - \sum_{h=1}^{k-1} \nu_h\right) a^k = \sum_{h=1}^{k-1} \nu_h (a^h - a^k) + a^k,$$

the condition of this expression does not depend on ν means that $a^1 = \dots = a^k$. Thus, MC-degeneracy for the inequalities from I and II appears when several from the following equalities are satisfied:

$$\begin{aligned}
a_1^1 &= \dots = a_1^k & &\Rightarrow a_{12}^1 - a_{22}^1 = \dots = a_{12}^k - a_{22}^k,\\
a_2^1 &= \dots = a_2^k & &\Rightarrow a_{21}^1 - a_{22}^1 = \dots = a_{21}^k - a_{22}^k,\\
A_1 + a_1^1 &= \dots = A_k + a_1^k & &\Rightarrow a_{11}^1 - a_{21}^1 = \dots = a_{11}^k - a_{21}^k,\\
A_1 + a_2^1 &= \dots = A_k + a_2^k & &\Rightarrow a_{11}^1 - a_{12}^1 = \dots = a_{11}^k - a_{12}^k.
\end{aligned}$$

MC-degeneracy for the equalities from II and III means that the relation determining equilibrium strategy in III or the boundary of the half-interval similar to (15) for equilibrium strategies in II does not depend on scalarization parameters; i.e., one or both of the following equalities are satisfied:

$$-\frac{\langle \eta, a_2 \rangle}{\langle \eta, A \rangle} = \alpha \; \forall \eta \in S^k, \quad -\frac{\langle \lambda, a_1 \rangle}{\langle \lambda, A \rangle} = \beta \; \forall \lambda \in S^k, \quad \alpha, \beta \in [0, 1]. \tag{17}$$

If both equalities in (17) are satisfied, the pair (α, β) is the unique equilibrium in the MC matrix game. If the only one equality is satisfied, then equilibrium pairs are (α, q^0) or (p^0, β), where p^0 and q^0 are determined by the conditions from I–III and may form a segment. In the last case, for finite ε-approximation of R, the abovementioned approximation is needed, whereas arbitrary LS parameters may be chosen for λ or for η, respectively.

Note that all MC-degenerate cases are only determined by the relations between the elements of the initial matrices of the game. This means that their presence can be found a priori, before starting the search for equilibrium. But it is not necessary to find out, because the results of the MC approximation are not affected by the degeneracy of the MC game (any values of the LS parameters can be chosen).

6 Approximation of V and Examples

The finite approximation of the set V, which is the value of the MC matrix game in mixed strategies, is based on the formula $V_\delta^\varepsilon = \Phi(R_\delta^\varepsilon)$ due to the continuity of the vector function Φ composed of the mathematical expectations of the payoff components over mixed strategies (2). From the following examples (Figs. 1, 2), it can be seen that the structure of V is, generally speaking, not simpler than R. Therefore, it does not make sense to approximate V apart from R.

Further examples are considered for the $2 \times 2 \times 2$ games (two-criterion 2×2 matrix games), taking into account the clarity of the representation of V in two-dimensional criterion space. The equilibrium set R is also represented in two-dimensional space of first coordinates (p, q) of the mixed strategies corresponding to the equilibrium outcomes. Note that a classification of the feasible sets $\Phi([0, 1] \times [0, 1])$ for such games is proposed in [22]; however, from the point of view of the types of MC equilibria, this classification is incomplete. We do not repeat here the figures given in [2, 6], for which the approximation of MC equilibria is constructed using GS. It is not possible to obtain an approximation for them based on LS with an arbitrary uniform net on a set of parameters, but the approach described in Sect. 4, 5 allows this to be done. Instead, we discuss other examples from the same classes "butterfly" and "envelope" according to the classification of [22], but with different forms of equilibrium sets (comparison of Figs. 1, 2 with Figs. 5, 3 from [6] confirm the incompleteness of the classification).

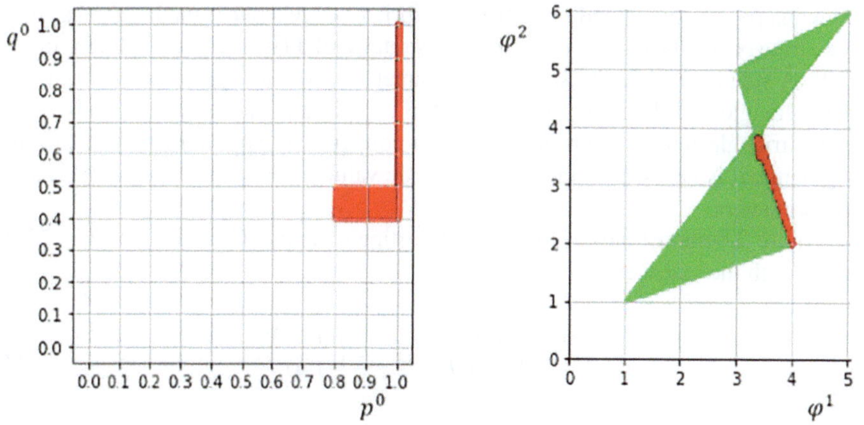

Fig. 1 Solution of $2 \times 2 \times 2$ game (18) of "butterfly" type

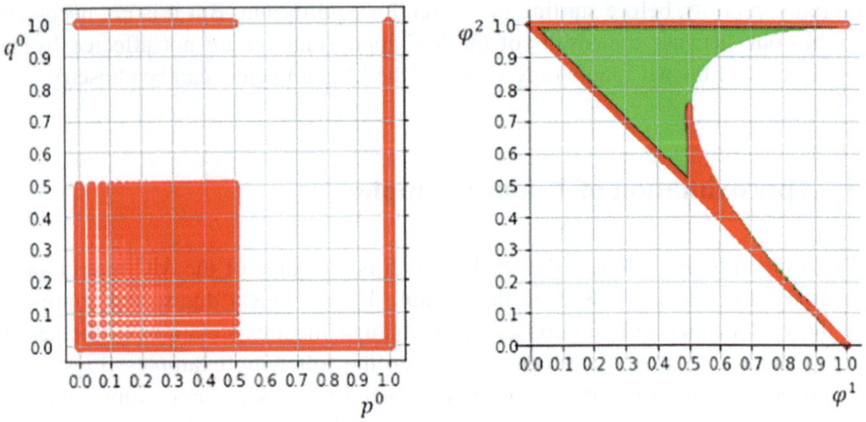

Fig. 2 Solution of $2 \times 2 \times 2$ game (19) of "envelope" type

Figure 1 shows equilibrium values and outcomes for $2 \times 2 \times 2$ MC game with Player 1's payoff

$$A^1 = \begin{pmatrix} 4 & 3 \\ 1 & 5 \end{pmatrix}, \quad A^2 = \begin{pmatrix} 2 & 5 \\ 1 & 6 \end{pmatrix}. \tag{18}$$

The equilibrium set R in the coordinates of p (horizontal) and q (vertical) is on the left and the value V (on the right) is highlighted red against the green background of the feasible set in the plane of partial criteria φ^1, φ^2.

The feasible set for the game (on the right in Fig. 1) is of the "butterfly" type [22]; the set V is a straight line along one side of its "wing" to the junction point of wings with a small triangular descent from it (looks like a half of an arrow). Looking at the equilibrium set R (on the left in Fig. 1), note that the vertical line ($p^0 = 1$) contains points corresponding to the equilibria of degenerate bimatrix games, since the strategy of Player 1 is pure and the strategy of Player 2 is completely mixed [20]. A regular net on the set of LS parameters gives an approximation only for the rectangle and the point (1, 1) from the left picture, and only the points of the upper triangle remain from the set V. To construct an approximation of the line segments, it is necessary to include the basic value $\eta_1 = 0.75$ in the δ-net on a set of parameters.

Figure 2 demonstrates even more interesting example. Consider $2 \times 2 \times 2$ MC game with the matrices

$$A^1 = \begin{pmatrix} 1 & 0 \\ 0 & 1 \end{pmatrix}, \quad A^2 = \begin{pmatrix} 1 & 1 \\ 1 & 0 \end{pmatrix}. \tag{19}$$

As Figs. 1 and 2 shows the equilibrium set R on the left and the value V (on the right) against the background of the feasible set of the type named "envelope" in

[22] in the plane of values of partial criteria

$$\varphi^1(p, q) = 2pq - p - q + 1, \varphi^2(p, q) = -pq + p + q.$$

In correspondence with the quadratic boundary of the feasible set, the value V has a curved part of the boundary but also contains one-dimensional components: the horizontal $\varphi^2 = 1$ and half of the diagonal $\varphi^1 + \varphi^2 = 1$. The latter maps the segment $q^0 = 0, p^0 \in [0.5, 1]$ into the criterion space from the left picture of Fig. 2; i.e., it corresponds to degenerate bimatrix games (excluding the point (0, 1)). Two other segments in R also correspond to degenerate bimatrix games. To find these equilibrium outcomes using LS, it is necessary to include the basic values of $\lambda_1 = 0.5, \eta_1 = 0.4$ in the δ-net on the set of the parameters.

Now we compare approximation of the equilibrium with LS for the game (19) with what is obtained for approximation with GS. Germeier's scalarizing function is introduced in [3, 23] for multicriteria maximization problem in the form

$$\min_{h \in I(\mu)} \frac{f^h}{\mu_h}, \ \mu \in S^k, \ I(\mu) \stackrel{\text{def}}{=} \{h = 1, \dots, k | \mu_h > 0\},$$

where μ is the vector of parameters and S^k is standard simplex in $\mathbb{R}^k$. When this function is used for approximation of R and V [6], it does not require preliminary calculations of the basic parameter values. An arbitrary δ-net (for sufficiently small δ) on the set of GS parameters may be chosen, it should only include the boundary values of μ ($\mu_h = 1$ for some h). The results of calculations using GS, for example, (19), are shown in Fig. 3, where δ-net with $\delta = 0.001$ for μ_1 and ε-net with $\varepsilon = 0.001$ for the values of p and q are used.

From the mathematical point of view, the solution and the value of the MC game can be successfully represented both with GS and LS, but it may be difficult to interpret the scalar parametric representation when it does not meet the sense of the considered problem. Players (and decision-makers) usually choose a particular scalarizing function for the criterion vectors based on their understanding of the type of scalarization. GS is more preferable when it is important to increase minimum level of normalized vector components. LS is used when components of the criterion vector are compared with respect to their coefficients of relative importance. Therefore, the question of approximating equilibrium sets with the help of LS is of great interest despite of non-convexity of the considered sets.

7 Conclusion

An important task lies in the visual representation of the solution set for the problems with infinite set of solutions. Such problems include, among others, MC optimization [11, 21] and MC games. However, it is equally important for MC

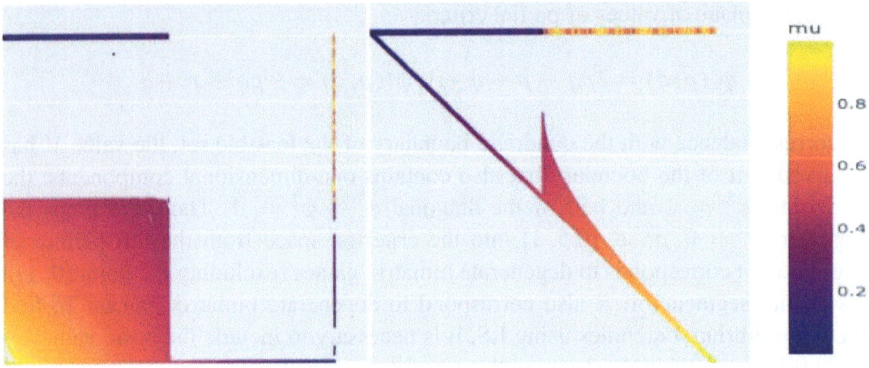

Fig. 3 Results of approximation of game (19) with GS (the axes are labeled similar to Fig. 2)

problems not to get just a picture with a view of Pareto frontier or equilibrium but to find the connection of the points in the picture with the scalarization parameters, so that the DMs understand which strategies and which optimal values correspond to which relations between partial criteria. The use of GS is most convenient for this purpose, since it gives a one-parameter description of R and V [6] and allows to show in color the dependence on the value of the parameter (see Fig. 3). Nevertheless, if the DM argues in terms of LS, then she/he is interested in the dependence on the coefficients of LS. The question of convenient visual representation of this dependence that arises here will be the subject of our further research.

Competing Interests The authors have no conflicts of interest to declare that are relevant to the content of this paper.

References

1. Shapley, L.S.: Equilibrium points in games with vector payoffs. Nav. Res. Logist. Q. **6**(1), 57–61 (1959)
2. Novikova, N.M., Pospelova, I.I.: A game-theoretic approach to two-person negotiation under multiple criteria. Group Decis. Negot. (2024). https://doi.org/10.1007/s10726-023-09859-5
3. Germeier, Y.B.: Introduction to the Theory of Operations Research. Nauka, Moscow (1971) [in Russian]
4. Steuer, R.E.: Multiple Criteria Optimization: Theory, Computation, and Application. Wiley, Hoboken (1986)
5. Noghin, V.D.: Reduction of the Pareto Set. An Axiomatic Approach. Springer, Cham (2018)
6. Novikova, N.M., Pospelova, I.I.: Germeier's scalarization for approximating solution of multicriteria matrix games. Mathematics (2023). https://doi.org/10.3390/math11010133
7. Rawls, J.: A Theory of Justice. Harvard University Press, Cambridge (1971)
8. Zapata, A., Mármol, A.M., Monroy, L., Caraballo, M.A.: A maxmin approach for the equilibria of vector-valued games. Group Decis. Negot. (2019). https://doi.org/10.1007/s10726-018-9608-4

9. Wierzbicki, A.P.: Reference point approaches. In: International Series in Operations Research and Management Science, pp. 237–275. Springer US, Berlin (1999)
10. Kruś, L.: On computer-based support in noncooperative multicriteria games. In: Atanassov, K., et al. Uncertainty and Imprecision in Decision Making and Decision Support: Cross-Fertilization, New Models and Applications (2017). https://doi.org/10.1007/978-3-319-65545-1_12
11. Pardalos, P.M., Žilinskas, A., Žilinskas, J.: Non-convex Multi-objective Optimization. Springer, Berlin (2018)
12. Krasnoshchekov, P.S., Morozov, V.V., Popov, N.M., Fedorov, V.V.: Hierarchical design schemes and decompositional numerical methods. J. Comput. Sist. Sci. Int. **40**, 754–763 (2001)
13. Miettinen, K.: Nonlinear Multiobjective Optimization. Springer, New York (1998)
14. Ehrgott, M.: Multicriteria Optimization, 2nd ed,. Springer, Heidelberg (2005)
15. Feinstein, Z.: Continuity and sensitivity analysis of parameterized Nash games. Econ. Theory Bull. **10**, 233–249 (2022)
16. Voorneveld, M., Vermeulen, D., Borm, P.: Axiomatizations of pareto equilibria in multicriteria games. Games Econ. Behav. (1999). https://doi.org/10.1006/game.1998.0680
17. Nash, J.F.: Non cooperative games. Ann. Math. **54**(2), 286–295 (1951)
18. Rettieva, A.N.: Equilibria in dynamic multicriteria games. Int. Game Theory Rev. **19**, 1750002 (2017)
19. Rettieva, A.N.: Solutions of dynamic multicriteria games: classical and untraditional approaches. Autom. Remote Control. **82**, 902–910 (2021)
20. Moulin, H.: Théorie des jeux pour l'économie et la politique. Hermann, Paris (1981)
21. Lotov, A.V., Bushenkov, V.A., Kamenev, G.K.: Interactive Decision Maps: Approximation and Visualization of Pareto Frontier. Springer, New York (2004)
22. Tanaka, T., Higuchi, M.: Classification of matrix types for multicriteria two-person zero-sum matrix games. IFAC Proc. Vol. **33**(16), 659–668 (2000)
23. Smirnov, M.M.: On the logical scalarization of criteria in the problem of approximation of the pareto set. Comput. Math. Math. Phys. **36**(3), 605–614 (1996)

Optimization of an E-Commerce Supply Chain: Middle Mile and Last Mile

Maria Luíza Teixeira Santos, Franklina Maria Bragion Toledo, and Mariá Cristina Vasconcelos Nascimento

1 Introduction

The e-commerce sector in Brazil has experienced remarkable growth in recent years. According to the Brazilian Electronic Commerce Association (ABCOMM), the sector's turnover reached BRL 187 million in 2023, representing an increase of nearly 10% compared to 2022. Additionally, the number of buyers rose to approximately 87 million [1].

Apart from low prices, one of the most attractive factors for a consumer is the delivery time of the goods [2]. Often, a customer prefers to pay a higher price to deliver the product more quickly [3]. Therefore, in addition to trying to satisfy consumers with attractive prices, companies are increasingly concerned about managing their logistics activities to enable quick delivery of their orders and, consequently, high customer satisfaction.

The location of companies' distribution centers directly impacts transport times and costs, which are fundamental to the efficiency of the logistics process [4]. Furthermore, meeting delivery deadlines requires stock control and distribution center management policies. It is important to note that some orders must be dispatched as quickly as possible, typically by the day following the customer's order placement [5].

M. L. T. Santos (✉) · F. M. B. Toledo
University of São Paulo, Avenida Trabalhador São-carlense, São Carlos, SP, Brazil
e-mail: marialuizats@usp.br; fran@icmc.usp.br

M. C. V. Nascimento
Aeronautics Institute of Technology, Praça Marechal Eduardo Gomes, São José dos Campos, SP, Brazil
e-mail: maria.nascimento@gp.ita.br

B. Goldengorin (ed.), *Theory, Algorithms, and Experiments in Applied Optimization*, Springer Optimization and Its Applications 226,
https://doi.org/10.1007/978-3-031-91357-0_15

One way to improve these decisions is by applying optimization techniques to solve the problems of managing companies' logistics processes. In this way, high-quality, nontrivial solutions can be found. In addition, different scenarios can be simulated. As highlighted in the literature, optimization as a decision support tool has proved effective in different contexts [6], including reduction of times and costs in logistics processes.

The research presented in this chapter aims to develop solutions for a middle mile and last mile delivery problem, considering four layers of goods transport: fulfillment centers, cross-docking, service centers, and customers. The goal is to minimize the transportation costs of products, which are part of customers' orders served by each service center in the network.

The rest of this chapter is structured as follows. Section 2 presents a literature review on delivery systems, focusing on middle and last mile. Section 3 describes the studied problem and the developed mathematical model for the investigated problem. Section 4 shows the computational experiments, presenting the proposed scenarios for evaluation. Finally, Sect. 5 summarizes the chapter by presenting the final considerations of this research.

2 Literature Review

This section presents an overview of the literature related to the topic of delivery systems. The keywords used for the literature review were "cross-docking," "e-commerce," "logistics," "middle-mile," "last-mile," and "transportation." The studies found went through a selection filter, where content such as drones, robots, uncertainty, and nonlinear were discarded. The databases used were Web of Science and Scopus. Therefore, in this chapter, we first show the types of systems according to their most striking characteristics. Then, we briefly review the middle mile and last mile delivery problems.

2.1 Classifications of Delivery Systems

There are various classifications of goods delivery systems in the literature, including Buijs et al. [7] and ePowerTrucks [8].

Buijs et al. [7] point to four main strategies for delivering goods. The first is direct delivery, i.e., the cargo is transported directly from the supplier to the customer. This type of delivery is common for brands that do not have a physical shop or warehouse. The strategy known as milk run aims to define a sequence of collections and deliveries on a single route, where a driver is responsible for picking up products and taking them to customers or warehouses. The warehousing and cross-docking strategies use warehouses from which deliveries are carried out.

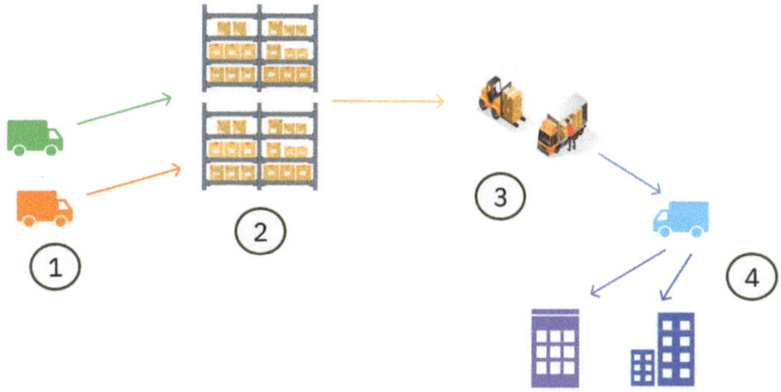

Source: Adapted from Flaticon ([14]), Freepik ([15]), Vecteezy ([16]) and VectorStock ([17]).

Fig. 1 Illustrative example of the warehousing and cross-docking strategies

In warehousing, goods are stored and delivered to customers. As well as considering delivery processes, the distribution of items in a warehouse is subject to structural and organizational studies and analyses to make the most of space and allocate goods, as highlighted by Boysen et al. [9], Ashayeri and Gelders [10], and Fan and Wang [11]. In cross-docking, goods are received from suppliers and consolidated and sent to customers in the shortest possible storage time [12]. Figure 1 illustrates the distribution of goods using the warehousing and cross-docking strategies. The warehousing strategy encompasses the entire process presented, made up of four stages: In Stage 1, vehicles take goods to a warehouse, where those goods remain stored (Stage 2) until they are moved (Stage 3) and then transferred to the next stage of transport (Stage 4). The cross-docking strategy ignores Stage 2, i.e., there is no prolonged storage of goods. These two strategies are often used to handle loads with stable and constant demand, as in the supermarket industry (e.g., case study of the retail chain Walmart [13]).

In addition to distribution strategies, according to ePowerTrucks [8], parcel/order distribution can be divided into three stages of planning: first mile, middle mile, and last mile. Figure 2 illustrates these distribution stages. The first mile refers to the logistics of delivering the product from the factory to a warehousing or distribution center. In the middle mile process, the product leaves the warehousing/distribution center and heads to a distribution hub. In the last mile process, which is the final stage of the journey, the product leaves this last hub and makes its way to the customer.

The structures and characteristics of each stage can differ according to the business and institutional organizations. Therefore, different storage structures or logistics strategies can be considered.

According to the classification of ePowerTrucks [8], the problem investigated in this chapter is classified as a logistics distribution problem of middle mile and last mile delivery. Therefore, the bibliographical review presented in this section addresses these delivery systems.

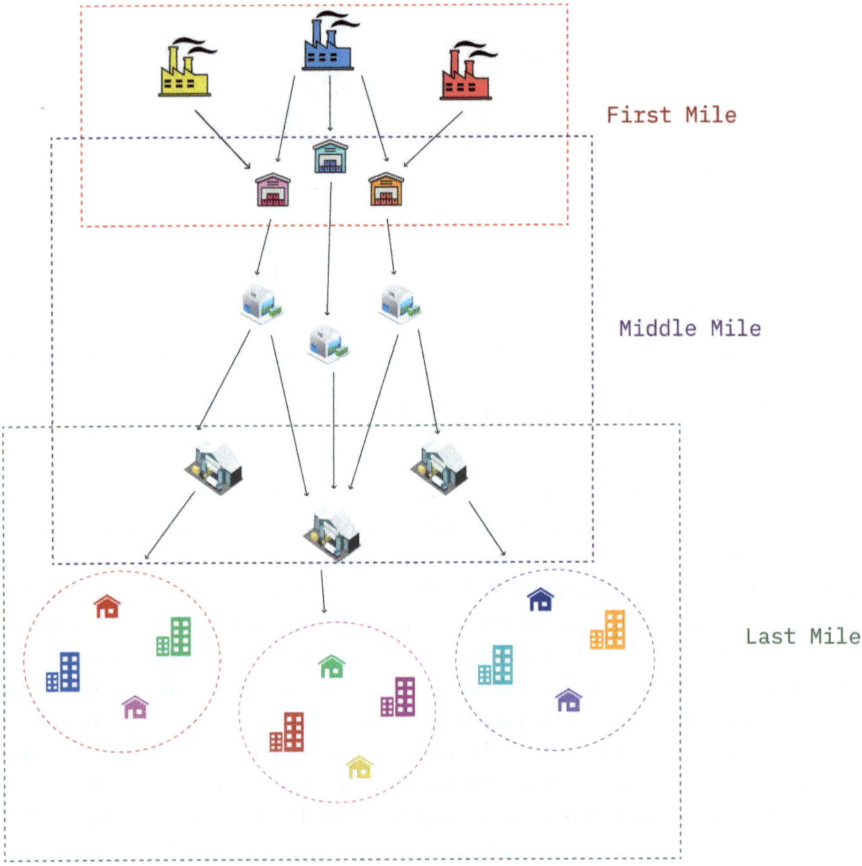

Fig. 2 Distribution stages according to ePowerTrucks [8]

2.2 *Middle Mile and Last Mile Delivery*

The middle mile and last mile problems have specific characteristics in the delivery process, such as speed and efficiency. These are the main foundations for greater customer satisfaction and their likely return to the shops, according to Gaggero-Vásquez et al. [18] and Ma et al. [19]. In addition, Vakulenko et al. [20] emphasize that free delivery, same-day delivery, delivery at a set time, and picking up the order at preestablished locations are equally important issues for consumers.

Considering the papers analyzed, two forms of last mile delivery addressed by the authors are highlighted: direct delivery (Fig. 3) and routing delivery (Fig. 4). The following section describes the approaches proposed in the literature for these two problems.

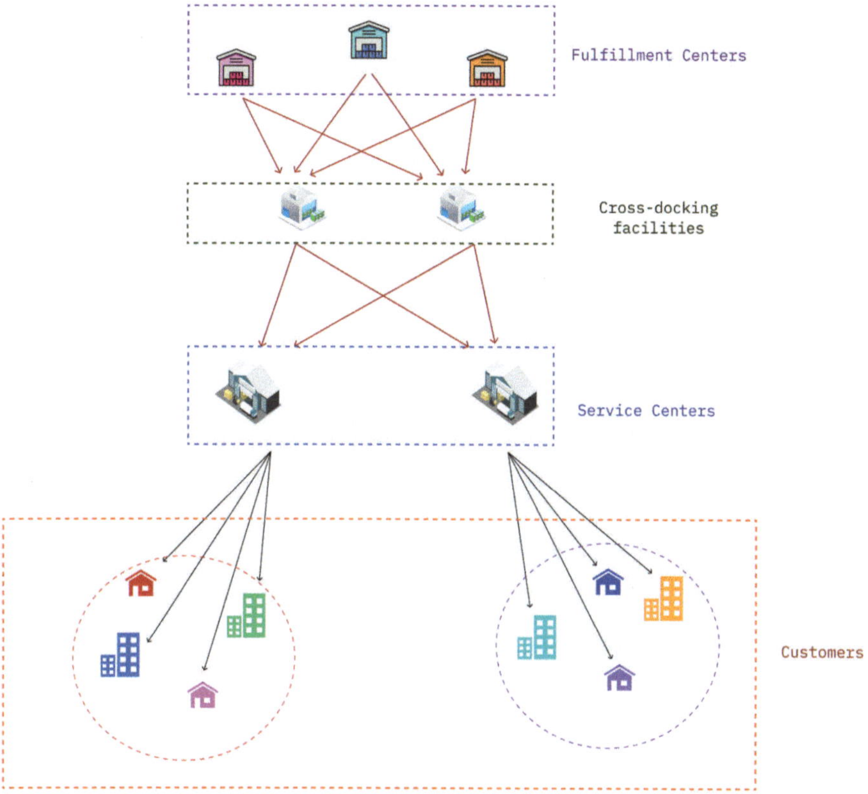

Source: Adapted from Flaticon ([14]), Freepik ([15]), Vecteezy ([16]) and VectorStock ([17]).

Fig. 3 Distribution process by direct delivery

2.2.1 Direct Delivery

Akyüz et al. [21] and Zhang et al. [22] approach direct deliveries in online retailers where the fleet is homogeneous. Akyüz et al. [21] consider consolidating the products of an order in a single shop and sending the order from this shop directly to each customer. The authors study the problem in real time, i.e., planning is carried out for each placed order. Mathematical models and heuristics are proposed to deal with the problem. The heuristics were used to improve the results found by solved models using Gurobi 8.1.1 with a CPU time limit of 1800s. Final results showed that heuristic solutions are most viable for online problems, reaching a better accuracy of 1.39% compared to solving mathematical models with no further strategies.

Zhang et al. [22] investigate the problem of direct delivery to customers respecting a deadline and analyzing two strategies: (i) FIFO (first in first out), in which all arrived packages are delivered from the station at the next time period available, and (ii) order consolidation, where the dispatch of some packages may be

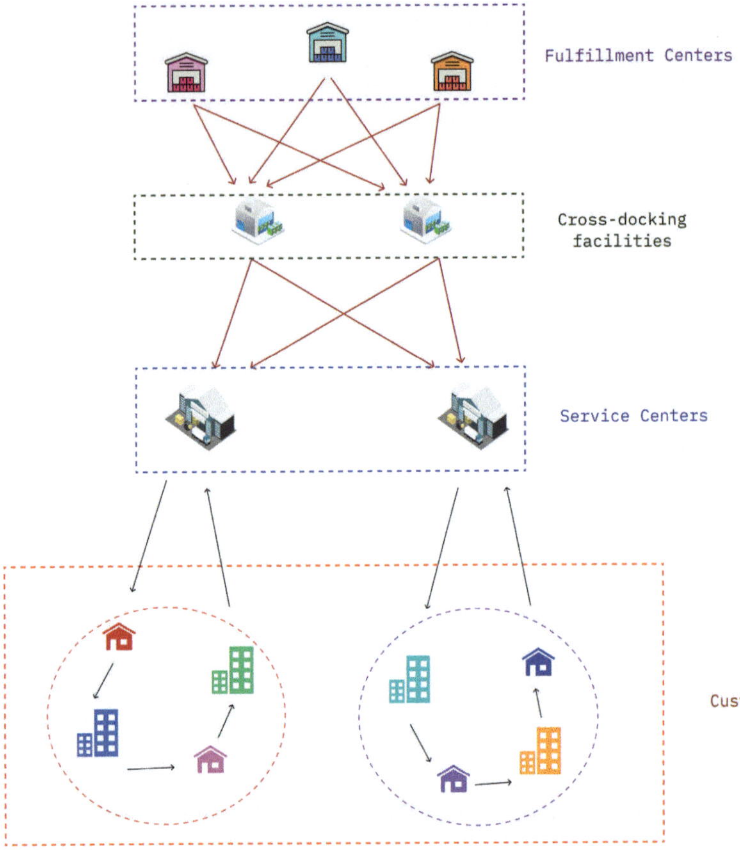

Source: Adapted from Flaticon ([14]), Freepik ([15]), Vecteezy ([16]) and VectorStock ([17]).

Fig. 4 Distribution process by routing delivery

postponed for order consolidation. The arrival of products at distribution centers is also analyzed. Results showed that low capacity in distribution centers can lead to order consolidation problems and that distribution centers located on the outskirts of the city are more beneficial for order consolidation than those located in city centers. Moreover, the order consolidation strategy can reduce an average of 5% of total costs when compared to FIFO strategy.

2.2.2 Delivery with Vehicle Routing

Pichka et al. [23], Zhou et al. [24], Rautela et al. [25], Li et al. [26], Hong et al. [27], and Jiang et al. [28] approached the delivery problem in the e-commerce context considering routing in the last mile level. In particular, [23–26] considered heterogeneous fleets and two-echelon distribution problem (from warehouses/distribution centers and from the distribution centers to customers). Furthermore, Hong et al. [27] approach a two-echelon problem in rural areas, using a homogeneous fleet. Jiang et al. [28] deal with a single echelon, delivering from distribution centers to customers or parcel lockers on the same route.

Pichka et al. [23] studied a two-echelon vehicle routing problem addressing the distribution of a single product. Mathematical models and a hybrid heuristic have been developed to deal with the problem. For the majority of instances, the heuristic obtained better solutions in terms of quality and time when compared to the results obtained using solver IBM CPLEX 12.5 with a time limit of two hours (7200 seconds).

Zhou et al. [24] dealt with a distribution system where each customer can choose how to receive the product, i.e., in their house, or pick up at one of the available pickup points. A hybrid genetic algorithm was proposed to solve the problem. The results showed a reduction of around 16% in transport costs compared to not allowing delivery options (home delivery and customer pickup) and with independent distribution system. Rautela et al. [25] analyzed different scenarios, including and excluding pickup and delivery points. As a result, for an e-commerce company in São Paulo, the authors showed that the scenario with the inclusion of hubs as delivery and pickup points saves the cost of the process by around 8%, diminishes the number of missed deliveries, and reduces the vehicle congestion on the established routes.

Li et al. [26] developed a mathematical model and a metaheuristic ALNS (adaptive large neighborhood search) for order delivery of customers (home delivery). A time limit of eight hours was imposed for the solver IBM CPLEX 12.5, whereas the maximum execution time for the metaheuristic was two hours. In this setting, the metaheuristic obtained better solutions than the solver's, reaching an improvement of around 21% in reducing costs in some cases.

Hong et al. [27] study a two-echelon format in a rural area comprising a logistics center, transit stations, and farmers. A mathematical model and the IACO (Improved adaptive Ant Colony Optimization) heuristic were proposed. When evaluating the methods developed, the heuristics were more efficient when finding satisfactory solutions in lower times than IBM CPLEX 12.6 results within a running time limit of 300s. Moreover, the authors carried out single- and two-echelon experiments on real case instances for the province of Yunnan-China, which used the single echelon (farmers served directly by logistics centers). The results showed a reduction of around 45% in the distance travelled and 24% in vehicle costs when the two-echelon strategy was implemented.

Jiang et al. [28] developed a mathematical model based on the travelling salesman problem (TSP) aiming at reducing transport costs and carbon emissions. In the good delivery process through a single warehouse, there can be delivery directly to the customer and/or to some parcel lockers for the respective customers to pick up their parcels. The authors also proposed an ILS (iterated local search) metaheuristic. The results showed that when parcel lockers are used, total cost can be reduced by between 4.5% and 51.2%, while the reduction in carbon emissions varies between 18.7% and 51.2%.

Table 1 Main characteristics of related work

Papers	Fleet	Echelons	Exchange B/W depots	Delivery strategy		Solution approach	
				Direct	Routing	Math. model	Heuristic
Akyüz et al. [21]	HO	1	✓	✓		✓	✓
Zhang et al. [22]	HO	1		✓		✓	✓
Pichka et al. [23]	HE	2			✓	✓	✓
Zhou et al. [24]	HE	2			✓	✓	✓
Rautela et al. [25]	HE	2			✓	✓	
Li et al. [26]	HE	2			✓	✓	✓
Hong et al. [27]	HO	2			✓	✓	✓
Jiang et al. [28]	–	1			✓	✓	✓
This research	HE	3		✓	✓	✓	

Note: HO = Homogeneous fleet; HE = Heterogeneous fleet

2.3 Considerations

Different logistics organizations are becoming increasingly common. Each organization encompasses particular features of the companies in which they occur. In the context of e-commerce, as well as aiming to reduce shipping costs, companies are also concerned with increasing customer satisfaction, especially with the rapid dispatch of goods.

The literature suggests that more studies on integrated middle mile and last mile delivery must be carried out, especially regarding multiple warehouses and route generation. Based on the strategies used in the middle mile and last mile delivery processes, the investigation presented in this chapter proposes mathematical models to solve a problem in the segment inspired by a company in the e-commerce sector. The investigated problem considers heterogeneous vehicles and the delivery of orders directly and through routing.

In Table 1, we summarize the related literature briefly discussed earlier in this section. The first column presents the reference, and the following four columns show some of the main problem features investigated in the reference. Finally, the last column points to whether the corresponding reference introduces mathematical models and/or heuristic solutions.

3 Problem Definition and Mathematical Modeling

The problem addressed in this chapter was inspired by the order delivery process of an e-commerce company in Latin America. The challenge herein investigated is determining a plan for the merchandise distribution system where operating costs and the speed of the distribution of orders are prioritized.

In general, the problem consists of deciding when, how, and through which warehouse the products will be sent until they reach the service centers, which are responsible for distributing the orders to customers. These decisions must be made focusing on minimizing distribution costs and speeding up the distribution network, directly affecting customer satisfaction.

The studied problem deals with the following assumptions for distribution planning:

- The transport network has four distribution layers, illustrated in Fig. 5, composed by:
  - Fulfillment center, warehouses where products are stored until a purchase
  - Cross-docking facilities, where occurs consolidation of customer orders to be sent to service centers
  - Service center, where sorting orders occur for delivery, usually in urban vehicles
  - Customers, the last layer of the network, where each customer is linked to an order (purchase)
- Distribution is carried out on a multi-echelon supply chain: fulfillment centers (warehouses) to cross-docking facilities (where orders are consolidated); from cross-docking facilities to service centers; and from service centers to customers.
- The capacity of the vehicles transporting from fulfillment centers (warehouses) to cross-docking facility transportation is sufficient to complete the requested route.
- Delivery planning is carried out daily.
- Transport costs between locations are known and defined for each vehicle type.
- Orders can be shipped from the service centers to customers directly or in a route.

Generally, when the customer places an order, the products go through a transport process until the customer receives the order. First, the products in each order are transported from one or more fulfillment centers to a cross-docking facility. From there, the customer's order is consolidated and sent to a service center which is the order integration point for a given region. Finally, the orders are delivered to the customers.

The research presented in this chapter analyzes two different strategies: direct and routing deliveries. Figure 3 illustrates the direct delivery system, where the customer receives their order directly from the service center, while Fig. 4 displays the delivery with the routing strategy, where there may be delivery routes from the service centers to groups of customers. In both strategies, the exchange of products between locations in the same layer, such as fulfillment centers $\rightarrow$ fulfillment centers, cross-docking facility $\rightarrow$ cross-docking facility and service centers $\rightarrow$ service centers, is not allowed.

Another critical point in a distribution system with the possibility of storage is the definition of inventory levels according to the demands. In this research, it is assumed that stocks are sufficient to fulfill orders.

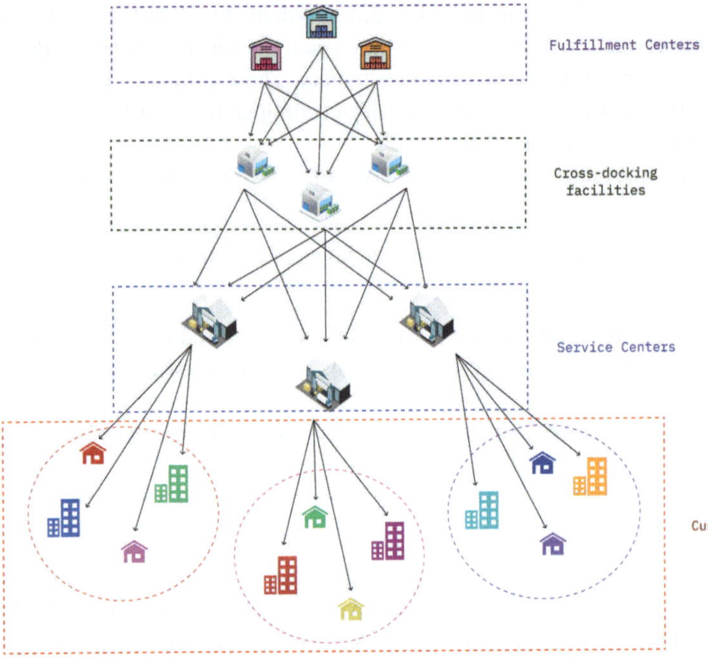

Source: Adapted from Flaticon ([14]), Freepik ([15]), Vecteezy ([16]) and VectorStock ([17]).

Fig. 5 Layers of distribution planning

3.1 Mathematical Model

The problem under investigation encompasses two strategies for delivering orders to the customers: (i) directly or (ii) using delivery routes. In both cases, the cross-docking facilities receive the items from the fulfillment centers, where the orders are consolidated and forwarded to the service centers and then to the final customers. The fleet is privately owned, and each cross-docking facility can dispatch only one vehicle to each service center. All vehicles are available for transfers from cross-docking facilities to service centers. However, only a single type of vehicle is considered for transfers from service centers to customers and between customers.

This section introduces two models: The first focuses on direct delivery, detailed in Sect. 3.1.1, while the second incorporates vehicle routing for order delivery, described in Sect. 3.1.2.

3.1.1 Direct Delivery Model

The direct delivery model is described by (1)–(14). The sets, parameters, and variables that define the model are given next.

Sets	
B	Set of products, $B = \{1, 2, \ldots, \|B\|\}$, where $\|B\|$ is the cardinality of B
FC	The set of nodes that represents fulfillment centers, $FC = \{1, 2, \ldots, \|FC\|\}$, where $\|FC\|$ is the cardinality of FC
CD	The set of nodes that represents cross-dock facilities, $CD = \{1, 2, \ldots, \|CD\|\}$, where $\|CD\|$ is the cardinality of CD
CF	The set of nodes that represents customers, $CF = \{1, \ldots, \|CF\|\}$, where $\|CF\|$ is the cardinality of CF
SC	The set of nodes that represents service centers, $SC = \{\|CF\| + 1, 2, \ldots, \|CF\| + \|SC\|\}$, where $\|SC\|$ is the cardinality of SC
V	The set of available vehicles, $V = \{1, 2, \ldots, \|V\|\}$, where $\|V\|$ is the cardinality of V, and it contains all available vehicles
V_ℓ	The set of vehicles available in cross-dock facility ℓ, $V_\ell \subset V$, $V_{\ell_1} \cap V_{\ell_2} = \emptyset, \forall \ell_1, \ell_2 \in CD$, where $\ell_1 \neq \ell_2$
VS_k	The set of vehicles available at service center k, $VS_k \subset V$, $VS_{k_1} \cap VS_{k_2} = \emptyset$, $\forall k_1, k_2 \in SC$, where $k_1 \neq k_2$
Parameters	
$cr_{h\ell}$	Cost of transportation from fulfillment center h to cross-dock facility ℓ
$cd^v_{\ell k}$	Cost of transportation from cross-dock facility ℓ to service center k using vehicle v
cf^v	Fixed cost of using vehicle v
c_{ki}	Cost of transportation from service center k to customer i
I^b_h	Initial stock of product b at fulfillment center h
q^b_i	Quantity of product b demanded by customer i
p_i	Weight of customer order i, which is the sum of the weights of all products requested by customer i
Cap^v	Vehicle type capacity v
Decision variables	
$\alpha_{h\ell}$	Binary variable that receives value 1 if there is a vehicle leaving the fulfillment center h to the cross-dock facility ℓ, and 0 otherwise
$\beta^v_{\ell k}$	Binary variable that receives value 1 if vehicle v leaves cross-dock facility ℓ to service center k, and 0 otherwise
x_{ki}	Binary variable that receives value 1 if a vehicle goes from service center k to customer i, and 0 otherwise
$z^b_{h\ell}$	Integer variable that receives the quantity of item b delivered from fulfillment center h to cross-dock facility ℓ
$y_{i\ell k}$	Binary variable that receives value 1 if customer order i is sent from cross-dock facility ℓ to service center k, and 0 otherwise

$$\begin{aligned} \min \quad & \sum_{h \in FC} \sum_{\ell \in CD} cr_{h\ell}\, \alpha_{h\ell} \\ & + \sum_{\ell \in CD} \sum_{v \in V_\ell} \sum_{k \in SC} cd^v_{\ell k}\, \beta^v_{\ell k} + \\ & + \sum_{k \in SC} \sum_{v \in VS_k} \sum_{i \in CF} (cf^v + 2\, c_{ki})\, x_{ki} \end{aligned} \tag{1}$$

subject to

 # Product Transportation Constraints

$$\sum_{\ell \in CD} z_{h\ell}^b \leq I_h^b \qquad \forall b \in B, \forall h \in FC \tag{2}$$

$$z_{h\ell}^b \leq I_h^b \, \alpha_{h\ell} \qquad \forall b \in B, \forall h \in FC, \forall \ell \in CD \tag{3}$$

$$\sum_{h \in FC} z_{h\ell}^b \geq \sum_{i \in CF} \sum_{k \in SC} q_i^b \, y_{i\ell k} \qquad \forall b \in B, \forall \ell \in CD \tag{4}$$

$$\sum_{i \in CF} p_i \, y_{i\ell k} \leq \sum_{v \in V_\ell} Cap^v \, \beta_{\ell k}^v \qquad \forall \ell \in CD, \forall k \in SC \tag{5}$$

$$\sum_{v \in V_\ell} \beta_{\ell k}^v \leq 1 \qquad \forall \ell \in CD, \forall k \in SC \tag{6}$$

$$\sum_{\ell \in CD} \sum_{k \in SC} y_{i\ell k} = 1 \qquad \forall i \in CF \tag{7}$$

 # Direct Delivery Constraints

$$\sum_{i \in CF} x_{ki} \leq |VS_k| \qquad \forall k \in SC \tag{8}$$

$$\sum_{\ell \in CD} y_{i\ell k} = x_{ki} \qquad \forall k \in SC, \forall i \in CF \tag{9}$$

 # Domain of Variables

$$\alpha_{h\ell} \in \{0, 1\} \qquad \forall h \in FC, \forall \ell \in CD \tag{10}$$

$$z_{h\ell}^b \in \mathbb{Z}^+ \qquad \forall b \in B, \forall h \in FC, \forall \ell \in CD \tag{11}$$

$$y_{i\ell k} \in \{0, 1\} \qquad \forall i \in CF, \forall \ell \in CD, \forall k \in SC \tag{12}$$

$$\beta_{\ell k}^v \in \{0, 1\} \qquad \forall \ell \in CD, \forall k \in SC, \forall v \in V_\ell \tag{13}$$

$$x_{ki} \in \{0, 1\} \qquad \forall k \in SC, \forall i \in CF \tag{14}$$

The objective function (1) aims to minimize the total transportation costs, including those between fulfillment centers and cross-dock facilities, cross-dock facilities and service centers, and service centers and customers. Additionally, it accounts for the costs associated with the vehicles selected for deliveries. Constraints (2) ensure that the quantity of products dispatched from each fulfillment center does not exceed the available stock. The set of constraints (3) guarantees that if transportation occurs between a fulfillment center h and a cross-docking facility ℓ, the corresponding transport cost is accounted for. Constraints (4) require that for orders to be consolidated at a given cross-docking facility ℓ, the facility must first receive all the items demanded in those orders. Constraints (5) enforce that the

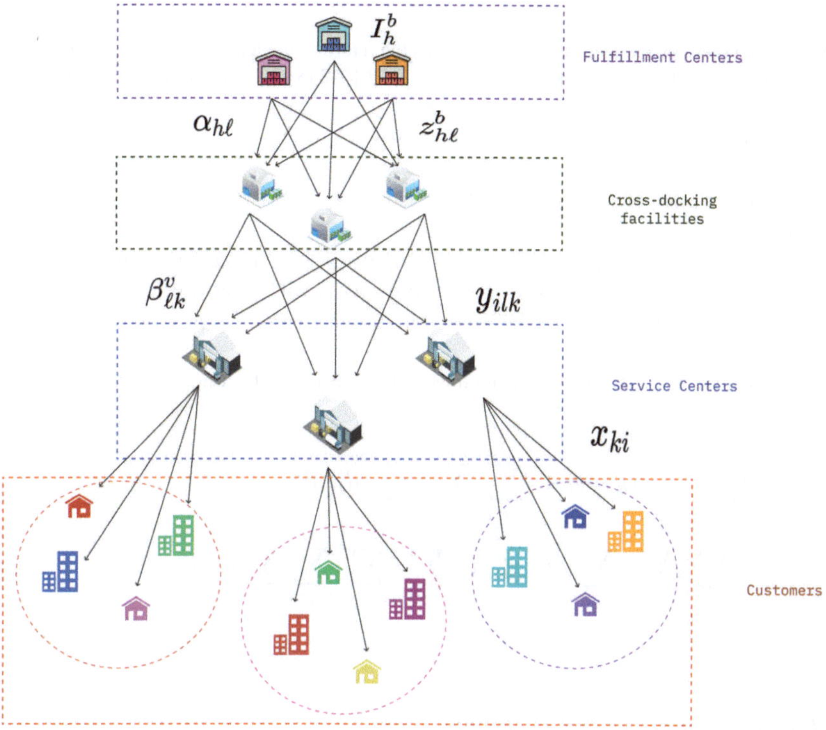

Source: Adapted from Flaticon ([14]), Freepik ([15]), Vecteezy ([16]) and VectorStock ([17]).

Fig. 6 The model transport flow with its variables—direct delivery model

number of orders sent from cross-docking facility ℓ to service center k must respect the capacity of the vehicle v used. The set of constraints (6) ensures that only one vehicle can depart from cross-docking facility ℓ to service center k. Constraints (7) ensure that an order i can only be transported from one cross-docking facility ℓ to a single service center k. The set of constraints (8) limits the number of deliveries made by a service center k to the number of vehicles it has available. Constraints (9) ensure that the service center k receiving an order for customer i is responsible for its delivery. The domain of the variables is defined in (10)–(14). Figure 6 illustrates the model transport flow and its associated variables.

3.1.2 Routing Delivery Model

In this strategy, orders are distributed to customers using routes. The model is described by (15)–(27). The x variables from the previous model have been redefined to represent routes, and new variables, δ and w, have been introduced. The additional sets, parameters, and variables are detailed next.

Sets	
GSC	The set of artificial nodes that represents the return of vehicles to the service centers, $GSC = \{\lvert CF\rvert + \lvert SC\rvert + 1, \lvert SC\rvert + 2, \ldots, \lvert CF\rvert + 2\,\lvert SC\rvert\}$, where GSC_k corresponds to the artificial node to service center k
N	The set of service centers and customer nodes ($N = SC \cup CF \cup GSC$)
Parameters	
c_{ij}	Cost of transportation between nodes i and j $(i,\ j \in N, i \neq j)$
Decision variables	
x^v_{ij}	Binary variable that receives value 1 if the vehicle v goes from the service center/customer i to the service center/customer j, and 0 otherwise
δ^v_i	Binary variable that receives value 1 if customer order i is in vehicle v, and 0 otherwise
w^v_i	Real variable that receives the total weight of orders sent to customer i on vehicle v

$$\min \sum_{h\in FC}\sum_{\ell\in CD} cr_{h\ell}\,\alpha_{h\ell} + \sum_{\ell\in CD}\sum_{v\in V_\ell}\sum_{k\in SC} cd^v_{\ell k}\,\beta^v_{\ell k} + \sum_{k\in SC}\sum_{v\in VS_k}\sum_{i\in CF} cf^v\,x^v_{ki} + \sum_{k\in SC}\sum_{v\in VS_k}\sum_{i\in CF}\sum_{\substack{j\in N\\ i\neq j}} c_{ij}\,x^v_{ij} \tag{15}$$

subject to

 # Product Transportation Constraints

(2)–(7)

 # Routing Delivery Constraints

$$\sum_{v\in VS_k} \delta^v_i = \sum_{\ell\in CD} y_{i\ell k} \qquad \forall i \in CF, \forall k \in SC \tag{16}$$

$$\sum_{\substack{j\in CF\cup\{GSC_k\}\\ i\neq j}} x^v_{ij} = \delta^v_i \qquad \forall i \in CF, \forall k \in SC, \forall v \in VS_k \tag{17}$$

$$\sum_{j\in CF\cup\{k\}} x^v_{j,GSC_k} = 1 \qquad \forall k \in SC, \forall v \in VS_k \tag{18}$$

$$\sum_{\substack{j\in CF\cup\{GSC_k\}\\ i\neq j}} x^v_{ij} - \sum_{\substack{j\in CF\cup\{k\}\\ j\neq i}} x^v_{ji} = 0 \qquad \forall i \in CF, \forall k \in SC, \forall v \in VS_k \tag{19}$$

$$\sum_{\substack{j\in CF\cup\{GSC_i\}\\ i\neq j}} x^v_{ij} = 1 \qquad \forall i \in SC, \forall v \in VS_i \tag{20}$$

$$w^v_i \geq w^v_j + p_i - Cap^v\,(1 - x^v_{ij}) \qquad \forall k \in SC, \forall v \in VS_k, \forall i \in CF, \forall j \in CF \cup \{GSC_k\},\ i \neq j \tag{21}$$

$$w_i^v \leq Cap^v\ \delta_i^v \qquad \forall k \in SC, \forall v \in VS_k, \forall i \in CF \tag{22}$$

$$w_{GSC_k}^v = 0 \qquad \forall k \in SC, \forall v \in VS_k \tag{23}$$

$$x_{ij}^v = 0 \qquad \forall i, j \in SC, i \neq j, \forall v \in VS_i \tag{24}$$

 # Domain of Variables

(10)–(13)

$$\delta_i^v \in \{0, 1\} \qquad \forall v \in V, \forall i \in N \tag{25}$$

$$x_{ij}^v \in \{0, 1\} \qquad \forall v \in V, \forall i, j \in N,\ i \neq j \tag{26}$$

$$w_i^v \geq 0 \qquad \forall i \in CF \cup GSC,\ \forall k \in SC,\ \forall v \in VS_k \tag{27}$$

The objective function (15) is similar to (1). However, in addition to minimizing the transport costs between fulfillment centers and cross-docking facilities, cross-docking facilities and service centers, and service centers and customers, it also minimizes the transport costs between customers.

Constraints (2)–(7) describe the transportation of products. Constraints (16) ensure that each customer's order i is assigned to a vehicle v from the service center k responsible for its delivery. Constraints (17) ensure that vehicle v departs from customer i to another customer or an artificial node representing its service center. The set of constraints (18) guarantees that vehicle v returns to its service center of origin. Constraints (19) establish the flow consistency by ensuring that for each customer i, a vehicle v has both an entry and an exit. Constraints (20) specify that each vehicle can leave the service center only once. The set of constraints (21) ensures that vehicle v carries all the orders assigned to its route. Constraints (22) ensure that the capacity of vehicle v is not exceeded. The set of constraints (23) requires vehicles to return empty to their respective service centers. Constraints (24) ensure that no transfers occur between service centers. The domain of the variables is defined in Constraints (10)–(13), (25), (26), and (27).

Note: The value of $BigM$ can be restricted to $2{\cdot}Cap^v$. To enhance understanding, Fig. 7 illustrates the transport flow, highlighting the variables used in the model described in this section.

4 Computational Experiments

The direct delivery and routing delivery approaches proposed in this chapter for addressing e-commerce delivery systems were evaluated by analyzing the solutions obtained using an optimization solver on artificially generated instances based on real-world data. Additionally, three strategies for assigning customers to service centers were examined.

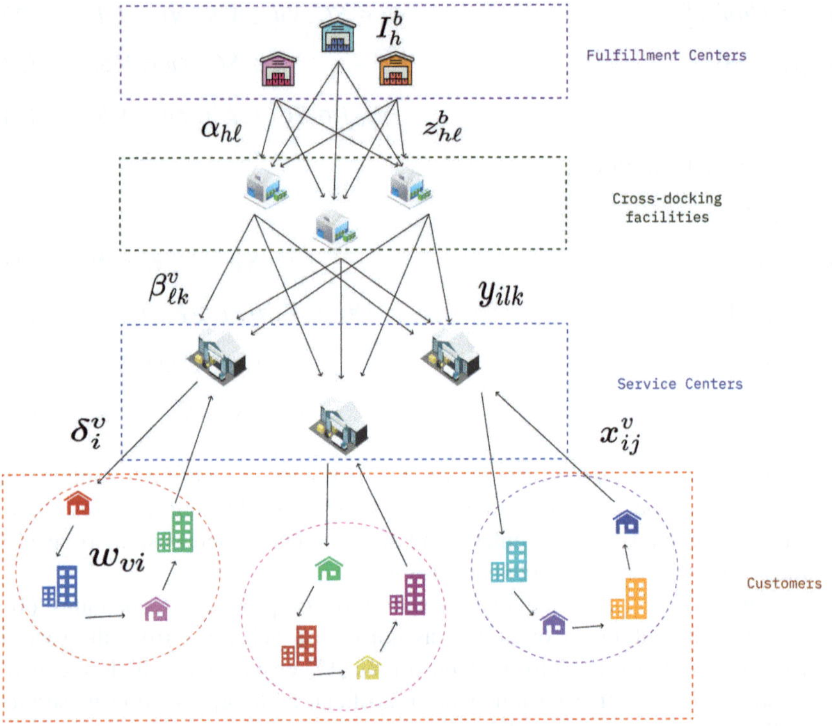

Source: Adapted from Flaticon ([14]), Freepik ([15]), Vecteezy ([16]) and VectorStock ([17]).

Fig. 7 Routing delivery model illustration

In the first strategy, customers are assigned a priori to the closest service centers. In the second, customer-to-service center assignments are treated as optimization decisions based on delivery approaches. Finally, the third strategy represents a middle ground between the two: Some customers are assigned a priori to the nearest service centers, while others remain as optimization decisions.

The following section describes the instances generated, and in Sect. 4.2, we summarize the computational results.

4.1 Instance Generation

The instances used in the experiments were generated from addresses in the municipality of São Paulo-SP (Brazil) and the metropolitan region extracted from the Olist [29] database. The distance between addresses (in kilometers) was calculated using the Haversine formula [30].

The database includes customer and seller addresses, and information about the requested orders (items and their quantities). Five instances were generated, each containing 1,000 customers. From these, smaller instances were derived with 15, 50, 75, 100, 250, and 500 customers. All instances share the same configuration, featuring two fulfillment centers, four cross-docking facilities, and two service centers.

The selection of customers per instance is based on the proximity to service centers. For each service center, customers located within a maximum distance of 11 km are chosen from the pool of available customers. These customers are then filtered for each service center, forming two separate sets. A third set of customers, for instance, combining some customers from the intersection of these two sets. Then, customers of these three sets are randomly chosen to obtain each instance of 1,000 customers. As explained before, five instances of 1,000 customers were generated. For the smaller instances (15, 50, 75, 100, 250, and 500 customers), the customers were selected according to the order of the clients in the largest instance (1,000 customers).

The locations of the fulfillment centers, cross-docking facilities, and service centers remain fixed across all instances. Twenty products were selected to define the orders, with weights ranging from 1 to 5 kg. According to Hübner et al. [31], most e-commerce purchases consist of two to four items. Following this observation, each order consists of exactly two items, with the quantity for each item varying between 1 and 3 units.

Figure 8 illustrates the selection process for generating the instances. Light blue markers represent end customers located within 11 km of the service centers (red markers) (Fig. 8a, b). Four cross-docking facilities (brown markers) and two fulfillment centers (navy blue markers) were strategically positioned (Fig. 8a). These locations were chosen based on an analysis of the regional map to approximate the locations of cross-docking facilities and fulfillment centers operated by Brazilian companies.

To define the stock levels of each item in the fulfillment centers, an occurrence matrix is generated, and an item can only be in one fulfillment center. The stock level of each item for each instance is equal to the total customer demand.

Two types of vehicles are considered: medium and large. Medium vehicles have a capacity of 12,000 kg, while large vehicles can carry up to 750,000 kg. Large vehicles are used for transfers between fulfillment centers $\rightarrow$ cross-docking facilities. For medium vehicles, the quantity is the total number of orders plus one medium vehicle for each cross-docking facility $\rightarrow$ service center relationship. Only one large vehicle is needed for each relationship between a fulfillment center and cross-docking facility, as well as between cross-docking facilities and service centers.

Shipping costs for deliveries from service centers to customers, as well as from customer to customer, include the cost of the medium-sized vehicle if that route is selected.

The instances generated are available at the GitHub[1] repository with public access. This repository describes the available data and instructions on how to use it.

[1] https://github.com/marialuizats/e-commerce-book-chapter

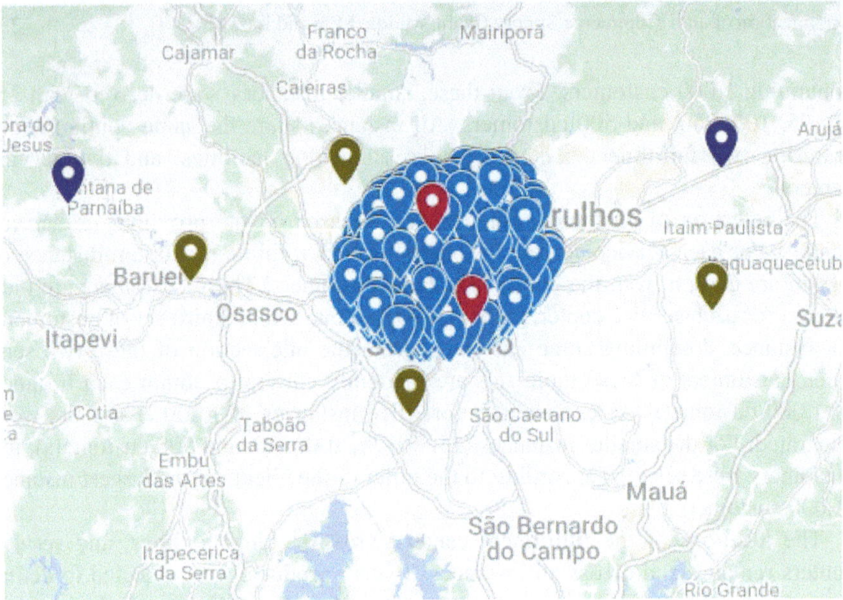

(a)

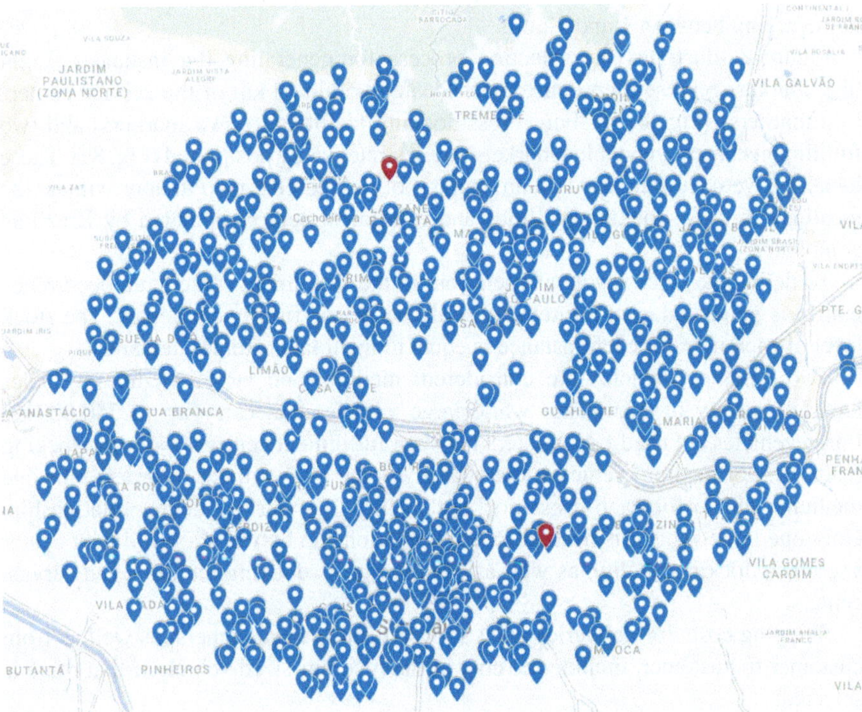

(b)

Source: Adapted from Google Maps ([32])

Fig. 8 Instance example. (**a**) Markers of FCs, CDs, SCs, and customers. (**b**) Customers' locations markers

4.2 Computational Tests

This section presents an experiment to evaluate the delivery strategies employing scenarios produced from the set of instances, generated based on the characteristics outlined in Sect. 4.1.

The computational experiments were carried out on a cluster with a parallel processing of 104 IP113 computational blades, each with 2 Intel Xeon E5-2680v2 2.8 GHz processors with ten cores and 128 GB DDR3 1866MHz memory, on a CentOS Linux 7 system. The models developed were described using the Python 3.10 [33] programming language. The instances were solved using the solver optimization tool Gurobi 11.0.3 in its default configuration via Gurobi API.

On the one hand, the experiments on the direct delivery model were carried out considering all discussed instances. On the other, the experiments reported for the routing delivery model considered only the subgroups with 15, 50, 75, and 100 customers, as the problem complexity prevented the representation and solution of larger instances on the used computer.

The instances were analyzed under three distinct planning scenarios:

- **Scenario 1—Fixed**

  This scenario was inspired by the context of an e-commerce company, where each customer is assigned to the nearest service center beforehand, i.e., customers must receive their order from the nearest service center. Consequently, for Models (1)–(14), the x_{ki} variables have their values fixed a priori, while for Models (15)–(27), we need to define a set of customers served by each service center k ($CS_k \subset CF$, $CS_{k_1} \cap CS_{k_2} = \emptyset$, $\forall k_1, k_2 \in SC$, where $k_1 \neq k_2$). Furthermore, we assign customers to service centers a priori by

$$\sum_{v \in VS_k} \delta_i^v = 1 \qquad \forall k \in SC, \forall i \in CS_k. \tag{28}$$

- **Scenario 2—Partially flexible**

  Given the sets of customers for each specific service center (Scenario 1), each customer must receive their order from the established service center. There is no assignment for the intersection set of customers (customers from both service centers), so these customers will be served by the service center chosen when the instance is solved.

- **Scenario 3—Flexible**

  No customer assignment to the service center is defined in advance. Therefore, defining which service center will deliver to each customer is a decision variable in the problem.

4.2.1 Direct Delivery Experiments

This section presents the results of the direct delivery problem. When analyzing the results for each instance, the values were the same in all scenarios. These same values were expected since the direct delivery model focuses on the direct distance between customers and service centers. Therefore, using different scenarios did not interfere with the optimal values.

Table 2 reports the results obtained considering the direct delivery strategy. The number of customers of each instance is indicated in column "Customers," and column "Instance" corresponds to the label of the corresponding instance. The value of the best solution obtained is reported in the third column, referred to as "D.D.

Table 2 Computational results obtained using the direct delivery strategy

Customers	Instance	D.D. solution
15	1	4,074.67
	2	3,817.47
	3	4,085.07
	4	4,921.47
	5	3,937.07

Customers	Instance	D.D. solution
100	1	25,813.07
	2	26,982.67
	3	25,643.07
	4	26,461.87
	5	26,349.87

Customers	Instance	D.D. solution
50	1	12,809.47
	2	13,938.27
	3	13,037.47
	4	13,993.87
	5	13,227.87

Customers	Instance	D.D. solution
250	1	63,770.67
	2	64,938.27
	3	65,073.87
	4	64,831.47
	5	67,333.07

Customers	Instance	D.D. solution
75	1	18,818.67
	2	20,410.27
	3	19,749.07
	4	20,356.67
	5	20,673.47

Customers	Instance	D.D. solution
500	1	132,501.47
	2	129,575.07
	3	129,585.87
	4	133,977.47
	5	131,672.67

Customers	Instance	D.D. solution
1000	1	265,227.87
	2	262,873.07
	3	263,979.47
	4	267,247.07
	5	265,889.07

Solution." The solver proved the optimality of all instances within one second in computational time.

4.2.2 Routing Delivery Experiments

This section shows the results obtained by the solver on the problem where delivery to customers is allowed through routes (routing delivery strategy).

Tables 3, 4, 5, and 6 present the results of instances with 15, 50, 75, and 100 customers, respectively. The tables show, for each instance, the value of the best solution obtained considering the time limit of 3,600 seconds (RT. Solution), the Gap, the time in seconds, and the reduction in relation to the solution obtained for the problem considering direct delivery (ROF). Equations (29) and (30) show the formulas to calculate Gap and ROF, respectively.

$$\text{Gap} = \frac{|\text{Primal Objective Bound} - \text{Dual Objective Bound}|}{|\text{Primal Objective Bound}|} \tag{29}$$

$$\text{ROF} = \frac{\text{D.D. Solution} - \text{RT. Solution}}{\text{D.D. solution}} \times 100\%. \tag{30}$$

A provably optimal solution was obtained for instances with 15 customers (Table 3) except for Instance 5 in Scenario 3. For all instances, there is a reduction in the value of the objective function concerning direct delivery. This reduction is at least 49% for Scenarios 1 and 2, and 58% to Scenario 3. As a result, we obtained an improvement in cost reduction with routing delivery. In addition, one may observe

Table 3 Results—routing delivery—15 customers

Scenario	Instance	RT. solution	Gap (%)	Time (s)	ROF (%)
1	1	2,068.47	0.00	3.52	49.24
	2	1,917.27	0.00	4.42	49.78
	3	2,017.87	0.00	4.78	50.60
	4	2,356.47	0.00	110.89	52.12
	5	1,908.47	0.00	5.31	51.53
2	1	2,068.47	0.00	42.00	49.24
	2	1,742.67	0.00	189.07	54.35
	3	1,781.87	0.00	201.10	56.38
	4	2,128.07	0.00	391.80	56.76
	5	1,821.67	0.00	69.17	53.73
3	1	1,693.27	0.00	59.22	58.44
	2	1,571.67	0.00	19.70	58.83
	3	1,584.27	0.00	7.63	61.22
	4	1,706.67	0.00	209.77	65.32
	5	1,463.67	6.81	3600.03	62.82

Table 4 Results—routing delivery—50 customers

Scenario	Instance	RT. solution	Gap (%)	Time (s)	ROF (%)
1	1	3,076.07	12.59	3601.35	75.99
	2	3,107.87	8.89	3600.77	77.70
	3	3,145.67	14.85	3601.34	75.87
	4	2,965.67	14.08	3600.96	78.81
	5	3,178.47	5.46	3600.86	75.97
2	1	3,083.67	18.82	3601.02	75.93
	2	3,039.87	15.64	3600.49	78.19
	3	3,151.67	18.44	3600.51	75.83
	4	2,985.47	17.83	3600.44	78.67
	5	3,246.67	16.36	3600.50	75.46
3	1	2,634.87	19.42	3600.45	79.43
	2	2,557.27	15.49	3600.16	81.65
	3	2,463.87	12.18	3600.13	81.10
	4	2,463.07	18.20	3600.08	82.40
	5	2,704.67	14.44	3600.47	79.55

Table 5 Results—routing delivery—75 customers

Scenario	Instance	RT. solution	Gap (%)	Time (s)	ROF (%)
1	1	4,432.27	39.40	3600.33	76.45
	2	3,560.07	19.76	3600.47	82.56
	3	3,514.87	17.76	3601.95	82.20
	4	3,659.27	27.84	3600.47	82.02
	5	3,665.07	21.89	3600.65	82.27
2	1	3,619.47	26.16	3600.85	80.77
	2	3,551.67	23.49	3601.12	82.60
	3	3,468.27	22.01	3601.01	82.44
	4	3,468.87	23.48	3600.63	82.96
	5	3,661.07	22.60	3601.33	82.29
3	1	5,536.07	58.86	3600.85	70.58
	2	5,734.87	58.68	3601.40	71.90
	3	5,025.03	51.96	3602.13	74.56
	4	5,531.27	58.15	3601.17	72.83
	5	5,577.47	55.36	3601.24	73.02

that Scenario 1 always presents optimal solutions with higher values than those obtained by Scenario 3, since each customer is assigned a priori to the nearest service center. As expected, Scenario 3 showed the greatest cost reduction, since there is no prior assignment of customers to service centers in this scenario.

Considering the instances with 50 customers (Table 4), with a time limit of one hour, no proven optimal solution was obtained, with an average gap of around 15%.

Table 6 Results—routing delivery—100 customers

Scenario	Instance	RT. solution	Gap (%)	Time (s)	ROF (%)
1	1	4,005.47	19.94	3601.29	84.48
	2	4,452.67	26.16	3600.44	83.50
	3	4,167.87	25.58	3600.45	83.75
	4	5,088.27	38.12	3600.49	80.77
	5	4,241.47	24.98	3600.48	83.90
2	1	5,726.83	45.72	3601.69	77.81
	2	5,616.87	43.78	3601.33	79.18
	3	6,352.83	52.27	3600.34	75.23
	4	5,262.47	41.61	3601.91	80.11
	5	5,493.67	40.99	3601.35	79.15
3	1	11,009.27	74.98	3601.94	57.35
	2	13,560.07	79.30	3602.59	49.75
	3	9,240.67	71.18	3603.20	63.96
	4	9,299.23	70.39	3602.77	64.86
	5	9,759.47	70.63	3602.82	62.96

It can also be observed that for all instances, the cost reduction was above 75%. In relation to the scenarios, Scenario 2 shows higher values for most instances.

For the instances with 75 customers (Table 5), similar to the instances with 50 customers, no proven optimal solution was obtained within the established time limit. The average gap was around 35%, while the cost reduction was at least 70%. It is worth emphasizing that in none of the instances did Scenario 3 obtain better solutions within the established time limit. For all instances, Scenario 2 obtained better results, followed by Scenario 1. These results are due to Scenario 3 having more flexibility than the others and being more complex to solver.

Finally, for instances with 100 customers (Table 6), with a time limit of one hour, as well as instances with 50 and 75 customers, no proven optimal solution was obtained, with an average gap of around 48%. The cost reduction was above 49% for all instances. Regarding the scenarios, Scenario 1 shows better values for all instances within the established time limit.

4.2.3 Delivery Times Analysis

In addition to the costs associated with sending goods, we analyzed delivery times. Taking into account the Brazilian speed limits for local roads (30 km/h) [34], and that one kilometer can be covered in two minutes, we carried out a comparative analysis of delivery times for the instances of 100 customers (Table 7). For all instances, the average (in minutes) for delivery of goods analyzing the direct delivery model is around 12 minutes for all scenarios, while for the routing model, for Scenario 1, the delivery time is at least 80 minutes; for Scenario 2, we have at

Table 7 Smallest, largest, and mean values of time delivery

Scenario	Instance	Direct delivery			Routing delivery		
		S.T.	L.T.	A.T.	S.T.	L.T.	A.T.
1	1	2.00	21.96	11.66	2.00	217.36	95.35
	2	2.00	21.74	12.24	5.42	234.60	100.95
	3	2.12	21.66	11.55	2.50	212.66	80.78
	4	2.00	21.96	11.98	2.00	243.12	97.23
	5	2.00	21.80	11.91	2.00	244.16	99.42
2	1	2.00	21.96	11.66	2.00	108.64	45.04
	2	2.00	21.74	12.24	5.24	187.42	73.79
	3	2.12	21.66	11.55	3.10	143.30	46.60
	4	2.00	21.96	11.98	2.00	154.68	50.34
	5	2.00	21.80	11.91	2.00	191.84	56.98
3	1	2.00	21.96	11.66	2.00	90.62	29.80
	2	2.00	21.74	12.24	2.00	99.20	28.17
	3	2.12	21.66	11.55	3.96	103.84	35.67
	4	2.00	21.96	11.98	2.50	75.80	26.95
	5	2.00	21.80	11.91	2.00	92.90	34.88

Obs.: S.T. = shortest time; L.T. = longest time; A.V. = average time

least 45 minutes, and finally, for Scenario 3, at least 26 minutes. As expected, the average time tends to be lower for Scenario 3, where there is no prior assignment of service center → customer. When comparing the two models, the average time for delivery via routing can be up to 8 times longer than the average time for direct delivery (instance 2—Scenario 1).

Regarding the shortest delivery times, for direct delivery, the shortest time for each instance will always be the shortest time needed to deliver to the customer closest to one of the service centers. For delivery with routing, the shortest time is no more than 6 minutes. Concerning the longest delivery times, the results from direct delivery modeling show 22 minutes as the maximum time. About the possibility of routing, the longest time value increases to a maximum of 245 minutes (about 4.1 hours). Given the deliveries are made in a single day, 4 hours can still be considered a relatively short time.

We also carried out an analysis of the number of customers by looking at the distribution of delivery times for the solutions with the possibility of routing (Table 8), using the following intervals: deliveries with a time of less than 10 minutes, between 10 and 20 minutes, between 20 and 30 minutes, between 30 and 40 minutes, between 40 and 50 minutes, and a time of more than 50 minutes. As expected, the prior designation of service center → customer (Scenario 1) makes the last interval (more than 50 minutes) to have more deliveries (Fig. 9), since priority will be given to solutions with the smallest possible number of vehicles used. The results for Scenario 2 are slightly better when compared to Scenario 1 (Fig. 10); we can see a slight reduction in the differences between the number of

Table 8 Number of customers—time intervals

Scenario	Instance	<10 (Min)	10–20 (Min)	20–30 (Min)	30–40 (Min)	40–50 (Min)	>50 (Min)
1	1	16	5	4	5	4	66
	2	5	8	6	6	3	72
	3	18	10	6	7	7	52
	4	16	14	5	5	8	52
	5	14	12	9	4	5	56
2	1	25	13	13	9	11	29
	2	8	7	8	9	12	56
	3	22	21	11	11	9	26
	4	24	26	11	6	6	27
	5	33	16	10	9	5	27
3	1	42	17	12	16	6	7
	2	25	29	21	7	2	16
	3	22	20	21	11	11	15
	4	41	39	11	6	2	1
	5	43	21	6	8	12	10

Obs.: Min. = Minutes

Fig. 9 Number of customers per time interval—scenario 1

Fig. 10 Number of customers per time interval—scenario 2

Fig. 11 Number of customers per time interval—scenario 3

customers for each interval. Finally, when we look at the results for Scenario 3 (Fig. 11), there is a more appropriate distribution, with fewer customers in the time interval of more than 50 minutes. It is due to the fact the number of vehicles used is greater, since there is no prior assignment of service center $\rightarrow$ customer, compared to the other scenarios (Fig. 12).

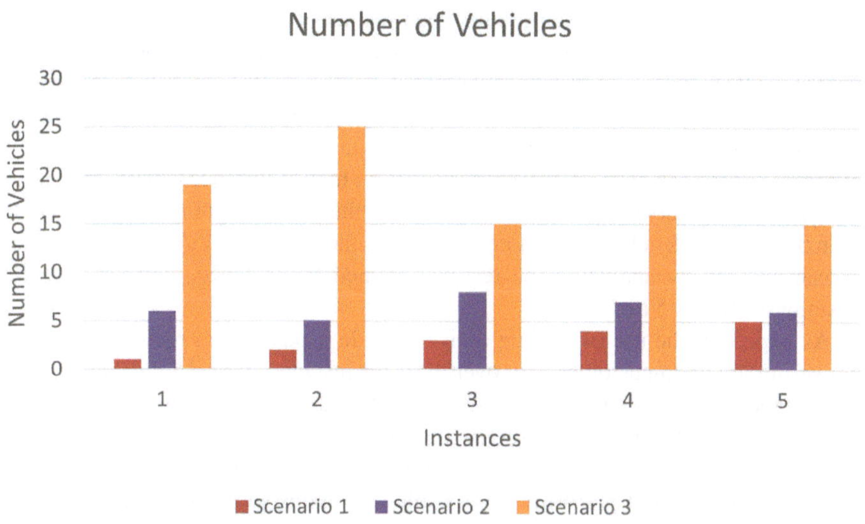

Fig. 12 Number of vehicles per scenario—100 customers—routing delivery

5 Conclusion and Further Researches

This research investigated the problem of distributing goods and delivering orders in the e-commerce sector, focusing on middle mile and last mile delivery. Multiple layers of distribution logistics were analyzed based on the structure of an e-commerce company in Latin America. The aim was to solve the problem using mathematical models and resolution methods and generate instances similar to orders placed in reality.

In the first stage of this research, mathematical models were developed to represent the problem of delivering orders to customers, a frequent problem in the area of last mile delivery in the e-commerce sector. The first model seeks to carry out direct deliveries between the four layers of order distribution logistics: fulfillment center $\rightarrow$ cross-docking facilities $\rightarrow$ service center $\rightarrow$ customers. In contrast, the second model presents all the characteristics of the previous model with the addition of allowing routing delivery in the last echelon (service center $\rightarrow$ customers), seeking a reduction in order delivery costs.

To generate instances similar to reality, data from a Brazilian e-commerce [29] service was used. Stock availability was considered through the probability of an item being contained or not in the fulfillment centers, with the total quantity made available being the total demand for the items. Customer order items were generated following the Hübner et al. [31] proposal, and the vehicles selected had a capacity of 12,000 kg and 750,000 kg.

The different strategies considered sought to assess the compromise between the flexibility of the problem and its resolution. The solutions obtained by the direct delivery model confirmed there is no other option but to use one vehicle per customer. Therefore, the distances between the service centers and customers are the only costs associated with delivering products in the last layer. As a result, the values in all the scenarios remained the same.

The direct impact of the different scenarios can be seen in the performance of the instances where delivery routes are allowed and where the solutions found have different values for each of the scenarios. For smaller instances, with 15 and 50 customers, Scenario 3 shows better solutions in the established time limit. It can be seen that the problem becomes more complex as the number of customers increases, as can be seen by analyzing the results obtained for instances with 75 and 100 customers (Tables 5 and 6).

As expected, the use of routing delivery made it possible to reduce the cost of delivering goods to customers, by over 49% for instances with 15 customers, 75% for instances with 50 customers, 71% for instances with 75 customers, and 49% for instances with 75 customers. The high difference between the gaps between Scenarios 2 and 3 and the brief distinction in the solutions' values indicate that Scenario 2 has a better performance in exploiting nodes with better solutions when compared to Scenario 3. In this way, choosing Scenario 2 to use a resolution method could be a good alternative for obtaining good solutions more quickly.

Analyzing the delivery time between the two scenarios (direct delivery and routing delivery) showed the average time to deliver an order to a customer can vary considerably according to each Scenario studied, ranging from 26 to 100 minutes for routing and 11 to 12 minutes for direct delivery. For the shortest and longest times, the difference between values for different scenarios in routing delivery is about 4 minutes (instance 2) and 170 minutes (instance 4). Through internal analyses within the companies, we believe that delivery time for routing could be reduced by making a trade-off between the cost of delivery and an increase in the number of vehicles used, prioritizing fast delivery to the customers.

When it comes to the number of customers at the distribution of delivery times, Scenario 1 takes more time to deliver orders to most customers (a minimum of 50 customers receiving orders in more than 50 minutes); it is due to the fact a smaller number of vehicles are used compared to Scenarios 2 and 3. A significant reduction in delivery time can be seen in Scenario 3 (no more than 16 customers receiving orders in more than 50 minutes), where more vehicles are used.

Even though good solutions were obtained for instances over 50 customers, the computational experiments show that the model developed is highly complex to solve, especially for instances larger than those used in this research. Therefore, the computer used becomes a very relevant factor when solving instances, as well as when developing mathematical models. Thus, computational machines with low configurations tend to be very limited in obtaining good solutions in relatively low time limits.

Therefore, for future research, we suggest: i) incorporating route time limit constraints (one-hour limit), ii) carrying out new computational tests on larger instances, and iii) developing other resolution methods to compare the solutions obtained.

Among the methods to be investigated, it is suggested to evaluate the use of heuristics and/or metaheuristics (e.g., [21, 22, 35–37]) that have been successfully used to solve routing problems.

Acknowledgments We gratefully acknowledge the São Paulo Research Foundation (FAPESP) for providing research funding (master's scholarship) for this project (Process Number #2022/01022-7). This study was financed in part by the Coordenação de Aperfeiçoamento de Pessoal de Nível Superior - Brasil (CAPES) - Finance Code 001, and by the Conselho Nacional de Desenvolvimento Científico e Tecnológico (CNPq) (Grant No. 403735/2021-1 and 309161/2022-3). The research was carried out using the computational resources of the Center for Mathematical Sciences Applied to Industry (CeMEAI), funded by FAPESP (grant 2013/07375-0).

References

1. ABCOMM. Principais Indicadores do e-Commerce: 2018–2023 (2024). Accessed 18 Oct 2024. https://dados.abcomm.org/crescimento-do-ecommerce-brasileiro
2. ABCOMM. Fulfillment: entenda a importância desta estratégia (2019). Accessed Oct 2024. https://abcomm.org/noticias/fulfillment-entenda-a-importancia-desta-estrategia/
3. Starlinks. How Delivery Lead Time Impacts Online Shopping Decisions (2024). Accessed 25 Oct 2024. https://www.starlinks-me.com/blog-post/how-delivery-lead-timeimpacts-online-shopping-decisions
4. Perl, J., Daskin, M.S.: A warehouse location-routing problem. Transp. Res. B Methodol. **19**, 381–396 (1985)
5. Yaman, H., Karasan, O.E., Kara, B.Y.: Release time scheduling and hub location for next-day delivery. Oper. Res. **60**, 906–917 (2012)
6. Petropoulos, F. et al.: Operational Research: methods and applications. J. Oper. Res. Soc. **75**, 423–617 (2024)
7. Buijs, P., Vis, I.F., Carlo, H.J.: Synchronization in cross-docking networks: a research classification and framework. Eur. J. Oper. Res. **239**, 593–608 (2014)
8. ePowerTrucks. What Does First, Middle, & Last Mile Delivery Mean? (2021). Accessed 20 Oct 2024. https://www.epowertrucks.co.uk/news/what-does-first-middle-last-mile-delivery-mean/
9. Boysen, N., de Koster, R., Weidinger, F.: Warehousing in the e-commerce era: a survey. Eur. J. Oper. Res. **277**, 396–411 (2019)
10. Ashayeri, J., Gelders, L.F.: Warehouse design optimization. Eur. J. Oper. Res. **21**, 285–294 (1985)
11. Fan, J., Wang, G.: Joint optimization of dynamic lot and warehouse sizing problems. Eur. J. Oper. Res. **267**, 849–854 (2018)
12. Van Belle, J., Valckenaers, P., Cattrysse, D.: Cross-docking: state of the art. Omega **40**, 827–846 (2012)
13. Mohtashami, A.: Scheduling trucks in cross docking systems with temporary storage and repetitive pattern for shipping trucks. Appl. Soft Comput. **36**, 468–486 (2015)
14. Freepik-Company-S.L.: Flaticon (2024). Accessed 20 Oct 2024. https://www.flaticon.com/
15. Freepik-Company.: Freepik (2024). Accessed 20 Oct 2024. https://www.freepik.com/
16. Eezy-Inc.: Vecteezy (2024). Accessed 20 Oct 2024. https://www.vecteezy.com/
17. VectorStock-Media. VectorStock (2024). Image ID: 1253866. Accessed 20 Oct 2024. https://www.vectorstock.com/
18. Gaggero-Vásquez, M., Mercado-Daneri, V., Chavez-Ugaz, R.: The impact of the global context on last-mile delivery and the relationship with the e-commerce channel. In: 2022 Congreso Internacional de Innovación y Tendencias en Ingeniería (CONIITI), pp. 1–4 (2022)

19. Ma, H., Kuang, H., Huang, M.: A last-mile delivery problem with alternative delivery options based on prospect theory. In: 2021 33rd Chinese Control and Decision Conference (CCDC), pp. 4202–4207 (2021)
20. Vakulenko, Y., Shams, P., Hellström, D., Hjort, K.: Online retail experience and customer satisfaction: the mediating role of last mile delivery. Int. Rev. Retail, Distrib. Consum. Res. **29**, 306–320 (2019)
21. Akyüz, M.H., Muter, İ., Erdoğan, G., Laporte, G.: Minimum cost delivery of multi-item orders in e-commerce logistics. Comput. Oper. Res. **138**, 105613 (2022)
22. Zhang, Y., Sun, L., Hu, X., Zhao, C.: Order consolidation for the last-mile split delivery in online retailing. Transp. Res. Part E Logist. Transp. Rev. **122**, 309–327 (2019)
23. Pichka, K., Bajgiran, A.H., Petering, M.E.H., Jang, J., Yue, X.: The two echelon open location routing problem: mathematical model and hybrid heuristic. Comput. Ind. Eng. **121**, 97–112 (2018)
24. Zhou, L., Baldacci, R., Vigo, D., Wang, X.: A multi-depot two-echelon vehicle routing problem with delivery options arising in the last mile distribution. Euro. J. Oper. Res. **265**, 765–778 (2018)
25. Rautela, H., Janjevic, M., Winkenbach, M.: Investigating the financial impact of collection-and-delivery points in last-mile E-commerce distribution. Res. Transport. Bus. Manage. **45**, 100681 (2022)
26. Li, X., Li, J., Aneja, Y., Guo, Z., Tian, P.: Integrated order allocation and order routing problem for e-order fulfillment. IISE Trans. **51**, 1128–1150 (2019)
27. Hong, W., Du, C., Xu, S., Pu, X.: Routing optimization of two-echelon two-way logistics in rural areas based on E-commerce environment. Transport. Res. Record **2679**, 564–574 (2024)
28. Jiang, L., Chang, H., Zhao, S., Dong, J., Lu, W.: A travelling salesman problem with carbon emission reduction in the last mile delivery. IEEE Access **7**, 61620–61627 (2019)
29. Olist: Brazilian E-Commerce Public Dataset by Olist (2022). Accessed 20 Oct 2024. https://www.kaggle.com/datasets/olistbr/brazilian-ecommerce?resource=download&select=olist_products_dataset.csv
30. Jimoh, O.D., Ajao, L.A., Adeleke, O.O., Kolo, S.S.: A vehicle tracking system using greedy forwarding algorithms for public transportation in urban arterial. IEEE Access **8**, 191706–191725 (2020)
31. Hübner, A., Holzapfel, A., Kuhn, H.: Operations management in multichannel retailing: an exploratory study. Oper. Manage. Res. **8**, 84–100 (2015)
32. LLC., G.: Google My Maps (2024). Accessed 20 Oct 2024. https://www.google.com/maps/d/
33. Foundation, P.S.: Python Language Reference, version 3.10 (2021). www.python.org
34. DETRANMS. Respeite os limites de velocidade das vias (2017). Accessed 20 Oct 2024. www.detran.ms.gov.br/respeite-os-limites-de-velocidade-das-vias-alerta-detran-ms/
35. Zhen, L., Baldacci, R., Tan, Z., Wang, S., Lyu, J.: Scheduling heterogeneous delivery tasks on a mixed logistics platform. Euro. J. Oper. Res. **298**, 680–698 (2022)
36. Giménez-Palacios, I., et al.: First-mile logistics parcel pickup: Vehicle routing with packing constraints under disruption. Transp. Res. Part E Logist. Transp. Rev. **164**, 102812 (2022)
37. Fachini, R.F., Armentano, V.A., Toledo, F.M.B.: A granular local search matheuristic for a heterogeneous fleet vehicle routing problem with stochastic travel times. Netw. Spat. Econ. **22**, 33–64 (2022)

Unified Presentation of the Generalized Ellipsoid Method

Petro Stetsyuk, Andreas Fischer, and Olha Khomiak

1 Introduction

The classical ellipsoid method was published by Yudin and Nemirovsky [8] in 1976. They derived this method from the successive cutoff scheme and called it the modified method of centered sections. Shor invented the ellipsoid method independently; see [14] from 1977. In this study, the method is presented as a special case of subgradient methods with space dilation proposed by Shor in the late 1960s [13, 18]. Note that he was one step away from discovering the classical ellipsoid method in 1970. This becomes obvious if we compare the proof of the theorem on the convergence of the ellipsoid method in [14] with the proof of Theorem 2 in [13].

Based on subgradient methods with space dilation [16], the generalized ellipsoid method (GEM) with space dilation was suggested; see [24, 25]. A parametric version of the ellipsoid method with space scaling by a scalar parameter $\lambda > 0$ is proposed in [26, 27]. In this paper, the same parametrization is applied to the GEM and yields the parametrized general ellipsoid method (PGEM).

The general problem that should be solved by the PGEM is given at the beginning of Sect. 2, whereas the PGEM is described thereafter by two algorithmic versions. The first algorithm is based on updating a nonsingular matrix B, as in the ellipsoid method by Shor, and the second one updates a positive definite symmetric matrix $H = BB^\top$, as proposed by Skokov [21] for subgradient methods with space

P. Stetsyuk (✉) · O. Khomiak
V. M. Glushkov Institute of Cybernetics, National Academy of Sciences of Ukraine, Kyiv, Ukraine
e-mail: stetsyuk@incyb.kiev.ua; homyak@nas.gov.ua

A. Fischer
Faculty of Mathematics, Technische Universität Dresden, Dresden, Germany
e-mail: Andreas.Fischer@tu-dresden.de

B. Goldengorin (ed.), *Theory, Algorithms, and Experiments in Applied Optimization*, Springer Optimization and Its Applications 226,
https://doi.org/10.1007/978-3-031-91357-0_16

dilation. Theorem 1 provides the convergence results for the PGEM and shows that they are the same as known for the GEM. Numerical issues of the two versions are discussed as well. In Sect. 2.2, a detailed proof of Theorem 1 is given. Moreover, in Sect. 2.3, we describe how the PGEM can be applied to particular problem classes like unconstrained and constrained convex optimization or saddle point problems for convex-concave functions. Obtaining Lagrangian upper bounds for Boolean optimization problems is discussed as well, particularly for computing an upper bound for the independence number of an undirected graph.

Special cases of the PGEM are reviewed in detail in Sect. 3. We start with the GEM. Thereafter, in Sect. 3.2, it is shown that, for certain values of $\lambda > 0$, well-known versions of the ellipsoid method by Shor [14], Nemirovsky and Yudin [9], and Khachiyan [4] are obtained. Moreover, Sect. 3.3 reviews two approximate ellipsoid methods, where a step-by-step volume reduction of the ellipsoids occurs that is only close to the optimal reduction.

The last section is devoted to an implementation and applications of the ellipsoid method by Shor. Based on this method, an implementation as Octave program is shown (Emshor program). This program is then applied to the minimization of a convex ravine function and to a problem by Sylvester, where the minimal hypersphere containing a given set of points has to be determined. The dependence of several parameters, like the dimension n of variables and an accuracy parameter, is shown for several examples.

For further reading, we recommend [31], where the ellipsoid method and variants are studied for either solving a system of linear inequalities or to prove that no solution exists. We would also mention the paper [11], where a combination of an ellipsoid and a subgradient method is suggested. Last but not least, both papers contain also a number of helpful references.

Our notation is rather standard. We just mention that $\langle \cdot, \cdot \rangle$ denotes the Euclidean scalar product and $\| \cdot \|$ the Euclidean vector norm.

2 The Parameterized Generalized Ellipsoid Method (PGEM) and Its Properties

Consider the following problem. Given a mapping $g : \mathbb{R}^n \to \mathbb{R}^n$, assume that there is a point $x^* \in \mathbb{R}^n$ such that

$$\langle x - x^*, g(x) \rangle \geq 0 \quad \text{for all } x \in \mathbb{R}^n \tag{1}$$

and $g(x) \neq 0$ for all $x \neq x^*$. The PGEM is now used to approximately determine x^*. Without loss of generality, we assume that the point x^* is unique. To find an approximation to the point x^*, we can use the ellipsoid method and its parameterized versions.

The generalized ellipsoid method [24, 25] is an algorithm with dilation of the n-dimensional space, where the space dilation coefficient $\alpha > 1$ satisfies the inequality

$$\alpha + \frac{1}{\alpha} < 2\sqrt[n]{\alpha}. \tag{2}$$

2.1 B-Form and H-Form of the PGEM

Using the additional space scaling parameter $\lambda > 0$, the PGEM (Algorithm 1) is described below. For $\lambda = 1$, a geometric explanation of the GEM can be found in Sect. 3.1.

Algorithm 1 ***B*-form of the PGEM**

Step 0. Select $\lambda > 0$ and α according to (2). Choose $x_0 \in \mathbb{R}^n$, a nonsingular matrix $B_0 \in \mathbb{R}^{n\times n}$ and r_0 so that

$$\left\| B_0^{-1}(x_0 - x^*) \right\| \leq r_0.$$

Moreover, set $k := 0$.
Step 1. Set $g_k := g(x_k)$. If $g_k = 0$, then set $k^* := k$, $x^* := x_{k^*}$, and STOP.
Step 2. Calculate

$$x_{k+1} := x_k - h_k B_k \xi_k, \quad \text{where} \quad \xi_k := \frac{B_k^\top g_k}{\left\| B_k^\top g_k \right\|}, \quad h_k := \frac{1}{2}\left(1 - \frac{1}{\alpha^2}\right) r_k.$$

Step 3. Update

$$B_{k+1} := \lambda \left(B_k + \left(\frac{1}{\alpha} - 1\right) (B_k \xi_k)\, \xi_k^\top \right) \quad \text{and} \quad r_{k+1} := \frac{1}{2\lambda}\left(\alpha + \frac{1}{\alpha}\right) r_k.$$

Step 4. Set $k := k + 1$ and go to Step 1.

At each iteration of the above algorithm, the matrix B_k is updated. Setting $A_k := B_k^{-1}$, we have that x is transformed into $y := A_k x$, i.e., $x = B_k y$. Updating the B-matrix (see Step 3) requires $O(n^2)$ operations for dense B_k. This is due to the use of the parameter $\lambda > 0$ and the space dilation operator $R_\alpha(\xi) : \mathbb{R}^n \to \mathbb{R}^n$ defined as

$$R_\alpha(\xi) := I_n + (\alpha - 1)\xi\xi^\top, \tag{3}$$

where $\xi \in \mathbb{R}^n$, $\|\xi\| = 1$ is the direction of dilation and $I_n \in \mathbb{R}^{n\times n}$ is the identity matrix. The properties of the space dilation operator are analyzed in detail in [16].

For $\beta := 1/\alpha$, and denoting the inverse dilation operator by $R_\alpha^{-1}(\xi)$, we obtain $R_\alpha^{-1}(\xi) = R_\beta(\xi)$ and $B_{k+1} = \lambda B_k R_\beta(\xi_k)$. The last expression shows the space dilation operator's role in updating the B-matrix in Step 3. To update A-matrices, we have

$$A_{k+1} = \frac{1}{\lambda} R_\alpha(\xi_k) A_k, \tag{4}$$

which follows from

$$A_{k+1}=B_{k+1}^{-1}=(\lambda B_k R_\beta(\xi_k))^{-1} = \frac{1}{\lambda} R_\beta^{-1}(\xi_k) B_k^{-1} = \frac{1}{\lambda} R_{1/\beta}(\xi_k) A_k = \frac{1}{\lambda} R_\alpha(\xi_k) A_k.$$

To formulate the main convergence theorem for Algorithm 1, let us define the ellipsoid

$$\mathcal{E}_k := \{x \mid \|A_k(x_k - x)\| \le r_k\},$$

where x_k is generated by the algorithm. Moreover, let $vol(\mathcal{E}_k)$ denote this ellipsoid's volume.

Theorem 1 *Let the sequence* $\{x_k\}$ *be generated by Algorithm 1. Then*

$$x_k \in \mathcal{E}_k \tag{5}$$

is satisfied for all $k \in \mathbb{N}$. *The ratio of the volumes of two subsequent ellipsoids is a constant and equals*

$$q_n(\alpha) := \frac{vol(\mathcal{E}_{k+1})}{vol(\mathcal{E}_k)} = \frac{1}{\alpha}\left(\frac{1}{2}\left(\alpha + \frac{1}{\alpha}\right)\right)^n < 1 \tag{6}$$

for all $k \in \mathbb{N}$.

For a detailed proof of Theorem 1, see Sect. 2.2.

The PGEM can also be written in H-form by means of a positive definite symmetric matrix H_k, which replaces $B_k B_k^\top$ in Algorithm 1.

Algorithm 2 ***H*-form of the PGEM**

Step 0. Select $\lambda > 0$ and α according to (2). Choose $x_0 \in \mathbb{R}^n$, a positive definite symmetric matrix $H_0 \in \mathbb{R}^{n\times n}$, and r_0 so that

$$(x_0 - x^*)^\top H_0^{-1}(x_0 - x^*) \le r_0^2.$$

Moreover, set $k := 0$.

Step 1. Set $g_k := g(x_k)$. If $g_k = 0$, then set $k^* := k$, $x^* := x_{k^*}$, and STOP.

Step 2. Calculate

$$x_{k+1} := x_k - h_k \frac{H_k g_k}{\sqrt{g_k^\top H_k g_k}}, \quad \text{where } h_k := \frac{1}{2}\left(1 - \frac{1}{\alpha^2}\right) r_k.$$

Step 3. Update

$$H_{k+1} := \lambda^2 \left(H_k + \left(\frac{1}{\alpha^2} - 1 \right) \frac{H_k g_k g_k^\top H_k}{g_k^\top H_k g_k} \right) \quad \text{and} \quad r_{k+1} := \frac{1}{2\lambda}\left(\alpha + \frac{1}{\alpha}\right) r_k.$$

Step 4. Set $k := k + 1$, and go to Step 1.

The sequences $\{x_k\}$ and $\{\mathcal{E}_k\}$ generated by the PGEM in H-form are the same (for exact arithmetic) as those generated by the PGEM in B-form if, in Step 0 of the latter, the same values for x_0, r_0, and α are chosen as for the PGEM in H-form and B_0 is chosen so that $H_0 = B_0 B_0^\top$. Hence, Theorem 1 is valid for the PGEM in H-form as well.

To see that the H-form indeed produces the same sequences $\{x_k\}$ and $\{\mathcal{E}_k\}$, one inductively shows by simple calculations that

$$x_{k+1} = x_k - h_k B_k \xi_k = x_k - h_k \frac{H_k g_k}{\sqrt{g_k^\top H_k g_k}},$$

$$B_{k+1} B_{k+1}^\top = \lambda^2 \left(H_k + \left(\frac{1}{\alpha^2} - 1 \right) \frac{H_k g_k g_k^\top H_k}{g_k^\top H_k g_k} \right)$$

holds for all $k \in \mathbb{N}$, implying

$$H_{k+1} = B_{k+1} B_{k+1}^\top.$$

Finally note that $\mathcal{E}_k$ has an equivalent representation by means of H_k-matrices:

$$\mathcal{E}_k = \{x \mid \|B_k^{-1}(x_k - x)\| \le r_k\} = \{x \mid (x_k - x) H_k^{-1}(x_k - x) \le r_k^2\}.$$

On the one hand, the PGEM in H-form requires just a half of operations than the PGEM in B-form does. On the other hand, the PGEM in H-form is computationally less stable, since the matrices H_k may become unsymmetric and indefinite. Thus, the H-form requires to monitor these properties of the matrices H_k. We will demonstrate this numerical instability in Sect. 4.2 by means of an example on minimizing a ravine function.

2.2 Proof of Theorem 1 on Algorithm 1

We will prove the theorem by analogy with the proof developed by Shor for the ellipsoid method [14]. To this end, we use the expression

$$R_\alpha^\top(\xi)R_\alpha(\xi) = R_{\alpha^2}(\xi), \tag{7}$$

which easily follows from the definition (3) of the space dilation operator (also see [16, pp. 68–69]).

Proof We will prove Theorem 1 by induction with respect to k. For $k = 0$, the inequality in (5) becomes $\|B_0^{-1}(x_0-x^*)\| \le r_0$ and is true by Step 0 of Algorithm 1. Let us now assume that the inequality in (5) holds for some k. Then, this inequality will be proved to hold for $k = k + 1$.

From (7) and the fact that $A_{k+1} = \frac{1}{\lambda} R_\alpha(\xi_k)A_k$ follows from (4), we get

$$\begin{aligned}
\|A_{k+1}(x_{k+1} - x^*)\|^2 &= \langle A_{k+1}(x_{k+1} - x^*),\, A_{k+1}(x_{k+1} - x^*)\rangle \\
&= \frac{1}{\lambda^2}\langle R_\alpha(\xi_k)A_k(x_{k+1} - x^*),\, R_\alpha(\xi_k)A_k(x_{k+1} - x^*)\rangle \\
&= \frac{1}{\lambda^2}\left\langle A_k(x_{k+1} - x^*),\, R_\alpha^\top(\xi_k)R_\alpha(\xi_k)A_k(x_{k+1} - x^*)\right\rangle \\
&= \frac{1}{\lambda^2}\langle A_k(x_{k+1} - x^*),\, R_{\alpha^2}(\xi_k)A_k(x_{k+1} - x^*)\rangle \\
&= \frac{1}{\lambda^2}\left\langle A_k(x_{k+1} - x^*),\, (I + (\alpha^2 - 1)\xi_k\xi_k^\top)A_k(x_{k+1} - x^*)\right\rangle \\
&= \frac{1}{\lambda^2}\langle A_k(x_{k+1} - x^*),\, A_k(x_{k+1} - x^*)\rangle + \\
&\quad \frac{1}{\lambda^2}(\alpha^2 - 1)\langle \xi_k,\, A_k(x_{k+1} - x^*)\rangle^2 \\
&= \frac{1}{\lambda^2}\|A_k(x_{k+1} - x^*)\|^2 + \frac{1}{\lambda^2}(\alpha^2 - 1)\langle \xi_k,\, A_k(x_{k+1} - x^*)\rangle^2 .
\end{aligned}$$

This leads

$$\lambda^2\|A_{k+1}(x_{k+1} - x^*)\|^2 = \|A_k(x_{k+1}-x^*)\|^2 + (\alpha^2-1)\langle \xi_k,\, A_k(x_{k+1}-x^*)\rangle^2 . \tag{8}$$

To specify both terms on the right-hand side of (8), we use the equation

$$A_k(x_{k+1} - x^*) = A_k(x_k - x^*) - h_k\xi_k, \tag{9}$$

which follows from the definition of x_{k+1} in Step 2 of Algorithm 1 and $A_k = B_k^{-1}$. Thanks to (9) and $\|\xi_k\| = 1$, the first term on the right-hand side of (8) can be written as

$$\|A_k(x_{k+1} - x^*)\|^2 = \|A_k(x_k - x^*)\|^2 - 2h_k \left\langle A_k(x_k - x^*), \xi_k \right\rangle + h_k^2 \tag{10}$$

and, with regard to the second term on the right-hand side of (8), we further get

$$\begin{aligned}\langle A_k(x_{k+1} - x^*), \xi_k\rangle^2 &= \langle A_k(x_k - x^*) - h_k\xi_k, \xi_k\rangle^2 \\ &= (\langle A_k(x_k - x^*), \xi_k\rangle - h_k \langle \xi_k, \xi_k\rangle)^2 \\ &= (\langle A_k(x_k - x^*), \xi_k\rangle - h_k)^2 \\ &= \langle A_k(x_k - x^*), \xi_k\rangle^2 - 2h_k \langle A_k(x_k - x^*), \xi_k\rangle + h_k^2.\end{aligned} \tag{11}$$

Substituting (10) and (11) into (8), we have

$$\begin{aligned}&\lambda^2 \|A_{k+1}(x_{k+1} - x^*)\|^2 \\ &= \|A_k(x_{k+1} - x^*)\|^2 + (\alpha^2 - 1) \langle \xi_k, A_k(x_{k+1} - x^*)\rangle^2 \\ &= \|A_k(x_k - x^*)\|^2 - 2h_k \langle A_k(x_k - x^*), \xi_k\rangle + h_k^2 \\ &\quad + (\alpha^2 - 1) \left(\langle A_k(x_k - x^*), \xi_k\rangle^2 - 2h_k \langle A_k(x_k - x^*), \xi_k\rangle + h_k^2\right) \\ &= \|A_k(x_k - x^*)\|^2 - 2\alpha^2 h_k \langle A_k(x_k - x^*), \xi_k\rangle \\ &\quad + (\alpha^2 - 1) \langle A_k(x_k - x^*), \xi_k\rangle^2 + \alpha^2 h_k^2 \\ &= \|A_k(x_k - x^*)\|^2 \\ &\quad - \langle A_k(x_k - x^*), \xi_k\rangle \left(2\alpha^2 h_k - (\alpha^2 - 1) \langle A_k(x_k - x^*), \xi_k\rangle\right) + \alpha^2 h_k^2.\end{aligned}$$

Hence, taking into account $h_k = \frac{1}{2}\left(1 - \frac{1}{\alpha^2}\right) r_k$ from Step 2 in Algorithm 1, it follows that

$$\begin{aligned}&\lambda^2 \|A_{k+1}(x_{k+1} - x^*)\|^2 \\ &= \|A_k(x_k - x^*)\|^2 \\ &\quad - (\alpha^2 - 1) \langle A_k(x_k - x^*), \xi_k\rangle (r_k - \langle A_k(x_k - x^*), \xi_k\rangle) \\ &\quad + \frac{(\alpha^2-1)^2}{4\alpha^2} r_k^2.\end{aligned} \tag{12}$$

Now, we investigate the sign of the product

$$\left\langle A_k(x_k - x^*), \xi_k\right\rangle \left(r_k - \left\langle A_k(x_k - x^*), \xi_k\right\rangle\right), \tag{13}$$

which appears within (12). The first multiplier in this product is nonnegative. Since $\langle x - x^*, g(x)\rangle \geq 0$ for all $x \in \mathbb{R}^n$, this can be easily seen as follows:

$$\begin{aligned}\langle A_k(x_k - x^*), \xi_k\rangle &= \left\langle A_k(x_k - x^*), \frac{B_k^\top g_k}{\|B_k^\top g_k\|}\right\rangle \\ &= \|B_k^\top g_k\|^{-1} \left\langle A_k(x_k - x^*), B_k^\top g_k\right\rangle \\ &= \|B_k^\top g_k\|^{-1} \langle B_k A_k(x_k - x^*), g_k\rangle \\ &= \|B_k^\top g_k\|^{-1} \langle x_k - x^*, g_k\rangle \\ &\geq 0.\end{aligned} \tag{14}$$

According to Step 0 of Algorithm 1 and $B_k^{-1} = A_k$, we have

$$0 \le \langle A_k(x_k - x^*), \xi_k \rangle \le \|A_k(x_k - x^*)\| \le r_k,$$

and, thus,

$$r_k - \langle A_k(x_k - x^*), \xi_k \rangle \ge 0. \tag{15}$$

Hence, (14) and (15) imply that (13) is nonnegative. From (12) and Step 0 of Algorithm 1, we therefore obtain

$$\begin{aligned} \lambda^2 \|A_{k+1}(x_{k+1} - x^*)\|^2 &\le \|A_k(x_k - x^*)\|^2 + \tfrac{(\alpha^2-1)^2}{4\alpha^2} r_k^2 \\ &\le r_k^2 + \tfrac{(\alpha^2-1)^2}{4\alpha^2} r_k^2 \\ &= \left(\tfrac{\alpha^4+2\alpha^2+1}{4\alpha^2}\right) r_k^2 \\ &= \left(\tfrac{1}{2}\left(\alpha + \tfrac{1}{\alpha}\right)\right)^2 r_k^2. \end{aligned}$$

This yields

$$\|A_{k+1}(x_{k+1} - x^*)\|^2 \le \left(\frac{1}{2\lambda}\left(\alpha + \frac{1}{\alpha}\right)\right)^2 r_k^2 = r_{k+1}^2,$$

i.e., inequality (5) is valid for $k = k + 1$.

The set of points which satisfies the inequality $\|A_k(x_k - x)\| \le r_k$ is the ellipsoid $\mathcal{E}_k$ containing the point x^*. The ellipsoid $\mathcal{E}_k$ has the volume

$$vol(\mathcal{E}_k) = \frac{v_0 r_k^n}{\det A_k}, \tag{16}$$

where v_0 is the volume of the n-dimensional unit ball. Since $A_{k+1} = \frac{1}{\lambda} R_\alpha(\xi_k) A_k$ (see (4)), we have

$$\det A_{k+1} = \left(\frac{1}{\lambda}\right)^n \det R_\alpha(\xi_k) \det A_k. \tag{17}$$

Based on (16) and (17), we get the volume reduction factor

$$q_n(\alpha) = \frac{vol(\mathcal{E}_{k+1})}{vol(\mathcal{E}_k)} = \frac{v_0 r_{k+1}^n \det A_k}{v_0 r_k^n \det A_{k+1}} = \left(\frac{r_{k+1}}{r_k}\right)^n \frac{\lambda^n \det A_k}{\det R_\alpha(\xi_k) \det A_k},$$

and further, with the definition of r_{k+1} in Step 3 of Algorithm 1,

$$q_n(\alpha) = \left(\frac{r_{k+1}}{r_k}\right)^n \frac{\lambda^n}{\alpha} = \left(\frac{1}{2\lambda}\left(\alpha + \frac{1}{\alpha}\right)\right)^n \frac{\lambda^n}{\alpha} = \frac{1}{\alpha}\left(\frac{1}{2}\left(\alpha + \frac{1}{\alpha}\right)\right)^n.$$

Since, by (2), the space dilation coefficient α satisfies the inequality $\alpha + \frac{1}{\alpha} < 2\sqrt[n]{\alpha}$, it follows that

$$q_n(\alpha) = \frac{1}{\alpha}\left(\frac{1}{2}\left(\alpha + \frac{1}{\alpha}\right)\right)^n < 1.$$

This completes the proof of Theorem 1. □

2.3 *Problem Classes for the Application of the PGEM*

Any PGEM can be employed for solving instances of the problem classes mentioned below. To see this, we show for each class how to construct a function $g : \mathbb{R}^n \to \mathbb{R}^n$ satisfying (1), i.e., that $\langle x - x^*, g(x)\rangle \geq 0$ holds for all $x \in \mathbb{R}^n$.

2.3.1 Minimization of Convex Functions

Let $f : \mathbb{R}^n \to \mathbb{R}$ be a convex function with minimizer x^*. The minimum value of f is denoted by $f^* := f(x^*)$. For simplicity, we assume that x^* is the only minimizer of f. To apply a PGEM to approximate x^*, let the mapping g used in Sect. 2.1 be specified by $g_f : \mathbb{R}^n \to \mathbb{R}^n$ with $g_f(x)$ being a subgradient of f at x, i.e., g_f satisfies

$$\langle x - x^*, g_f(x)\rangle \geq f(x) - f(x^*) = f(x) - f^* \geq 0 \quad \text{for all } x \in \mathbb{R}^n. \tag{18}$$

A convex function f can be either differentiable or nondifferentiable. Examples of two nondifferentiable functions $f_1, f_2 : \mathbb{R}^2 \to \mathbb{R}$ of two variables are given in Fig. 1. They are defined by

$$f_1(x_1, x_2) := |x_1|+10|x_2| \text{ and } f_2(x_1, x_2) := \max\{x_1^2+(2x_2-2)^2-3, x_1^2+(x_2+1)^2\}.$$

The function f_1 is a ravine piecewise linear function and has elongated level curves, whereas function f_2 is essentially ravine and piecewise quadratic. Results of applying an ellipsoid method to a ravine piecewise linear function are given in Sect. 4.2.

Let us finally mention that Sect. 4.3 presents the application of the ellipsoid method to the problem of minimizing a convex piecewise quadratic function, which is related to the well-known Sylvester problem.

2.3.2 Constrained Convex Programming

Let us consider the constrained nonlinear program

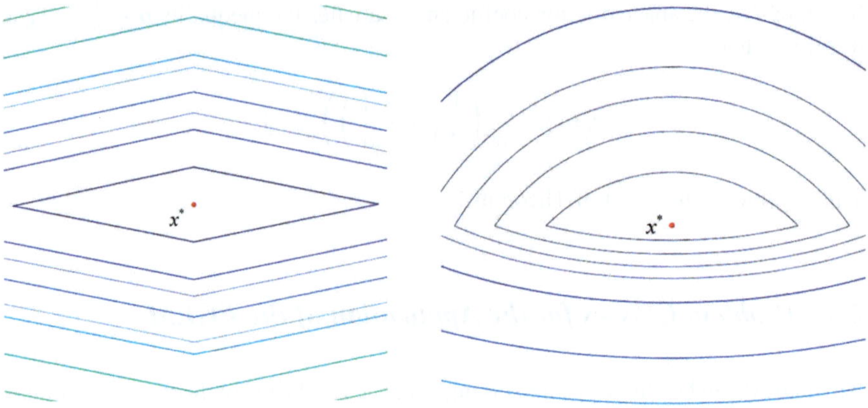

Fig. 1 Level curves of functions f_1 (left) and f_2 (right)

$$\min_{x \in \mathbb{R}^n} \quad f_0(x) \quad \text{subject to} \quad f_i(x) \le 0 \quad \text{for} \quad i = 1, 2, \ldots, m, \tag{19}$$

where the functions $f_i : \mathbb{R}^n \to \mathbb{R}$ are convex for $i = 0, 1, \ldots, m$. Further, for the same indices i, let $g_i : \mathbb{R}^n \to \mathbb{R}^n$ denote a mapping, so that $g_i(x)$ is a subgradient of f_i at x. We assume that problem (19) has the unique solution x^* and that the Slater condition is satisfied. Moreover, let the mapping $g : \mathbb{R}^n \to \mathbb{R}^n$ be defined by

$$g(x) := g_{i(x)}(x),$$

where $x \mapsto i(x)$ satisfies the conditions

$$\begin{aligned} i(x) &= 0, && \text{if } f_1(x) \le 0, \ldots, f_m(x) \le 0, \\ i(x) &\in \{i \mid f_i(x) > 0\}, && \text{if } \max\{f_1(x), \ldots, f_m(x)\} > 0. \end{aligned} \tag{20}$$

It then holds that $\langle x - x^*, g(x) \rangle \ge 0$ for all $x \in \mathbb{R}^n$. To see this, let us first consider the case that x satisfies $\max\{f_1(x), \ldots, f_m(x)\} \le 0$. According to (20), we have $g(x) = g_0(x)$. Moreover, by $f_0(x) \ge f_0(x^*)$ and the convexity of f, it follows that

$$\langle x - x^*, g(x) \rangle = \langle x - x^*, g_0(x) \rangle \ge f_0(x) - f_0(x^*) \ge 0.$$

If, otherwise, $\max\{f_1(x), \ldots, f_m(x)\} > 0$, then (20) implies $f_j(x) > 0$ for $j = i(x)$. By the convexity of f_j and by $f_j(x^*) \le 0$, we get

$$\langle x - x^*, g(x) \rangle = \langle x - x^*, g_j(x) \rangle \ge f_j(x) - f_j(x^*) \ge 0.$$

Thus, $\langle x - x^*, g(x) \rangle \ge 0$ holds for all $x \in \mathbb{R}^n$.

As a special case of problem (19), we consider the problem of minimizing a convex function $f : \mathbb{R}^n \to \mathbb{R}$ on a ball. Then a minimizer x^* of f on the ball $S(x_0, R) = \{x : \|x - x_0\| \leq R\}$ is characterized by

$$\langle g_f(x), x - x^* \rangle \geq f(x) - f(x^*) \geq 0 \quad \text{for all } x \in S(x_0, R).$$

For any $x \notin S(x_0, R)$, we have

$$\begin{aligned} \langle x - x_0, x - x^* \rangle &= \|x - x_0\|^2 - \langle x - x_0, x^* - x_0 \rangle \\ &\geq \|x - x_0\| \left(\|x - x_0\| - \|x^* - x_0\| \right) \geq 0. \end{aligned}$$

Thus, a mapping $g : \mathbb{R}^n \to \mathbb{R}^n$ satisfying $\langle g(x), x - x^* \rangle \geq 0$ for all $x \in \mathbb{R}^n$ can be defined by

$$g(x) := \begin{cases} g_f(x), & x \in S(x_0, R), \\ \frac{x - x_0}{\|x - x_0\|}, & x \notin S(x_0, R). \end{cases}$$

2.3.3 Saddle Point Problems for Convex-Concave Functions

Let $f : \mathbb{R}^n \times \mathbb{R}^m \to \mathbb{R}$ be a convex-concave function, i.e., $f(\cdot, y)$ is convex for any $y \in \mathbb{R}^m$ and $f(x, \cdot)$ is concave for any $x \in \mathbb{R}^n$. A pair $z^* := (x^*, y^*) \in \mathbb{R}^n \times \mathbb{R}^m$ is called saddle point of f if

$$f(x^*, y) \leq f(x^*, y^*) \leq f(x, y^*) \quad \text{for all } (x, y) \in \mathbb{R}^n \times \mathbb{R}^m.$$

Moreover, with $z := (x, y)$, let the set $G_x(z) \subset \mathbb{R}^n$ contain all partial subgradients of $f(\cdot, y)$ with y fixed, whereas $G_y(z) \subset \mathbb{R}^m$ contains all partial supergradients of $f(x, \cdot)$ with x fixed. Based on this, let a mapping $g : \mathbb{R}^n \times \mathbb{R}^m \to \mathbb{R}^n \times \mathbb{R}^m$ be defined by

$$g(z) := \begin{pmatrix} g_x(z) \\ -g_y(z) \end{pmatrix} \quad \text{with} \quad (g_x(z), g_y(z)) \in G_x(z) \times G_y(z).$$

Taking into account the notion of a saddle point and the convexity of the functions $f(\cdot, y)$ and $-f(x, \cdot)$ for arbitrarily but fixed y and x, respectively, it follows that

$$\begin{aligned} 0 &\leq f(x, y^*) - f(x^*, y) \\ &= f(x, y^*) - f(x, y) + f(x, y) - f(x^*, y) \\ &\leq \langle y - y^*, -g_y(z) \rangle + \langle x - x^*, g_x(z) \rangle \\ &= \langle z - z^*, g(z) \rangle. \end{aligned}$$

Thus, $\langle z - z^*, g(z) \rangle \geq 0$ holds for all $z \in \mathbb{R}^n \times \mathbb{R}^m$.

2.3.4 Lagrangian Bounds for Nonconvex Optimization Problems

Consider the problem of nonlinear programming in the following general form

$$f^* := \max_{x \in X} f_0(x) \quad \text{subject to} \quad f_i(x) = 0, \quad i = 1, \ldots, m, \tag{21}$$

where $X \subseteq \mathbb{R}^n$ is a closed set and $f_0, f_1, \ldots, f_m : \mathbb{R}^n \to \mathbb{R}$ are differentiable functions. Let us further assume that the problem has a feasible solution. As usual, the Lagrange function $L : \mathbb{R}^n \times \mathbb{R}^m$ is given by

$$L(x, u) = f_0(x) + \sum_{i=1}^{m} u_i f_i(x),$$

where $u = (u_1, \ldots, u_m)$ denotes the vector of Lagrange multipliers. For each $u \in \mathbb{R}^m$, we define

$$\psi(u) := \max_{x \in X} L(x, u). \tag{22}$$

The resulting function $\psi : \mathbb{R}^m \to \mathbb{R}$ is convex but nondifferentiable in general. It is easy to verify that $\psi(u) \geq f^*$ for an arbitrary $u \in \text{dom}(\psi)$. Thus, one is interested in finding the best (least) upper bound for f^* in this class of Lagrangian bounds, i.e., the minimization problem

$$\psi^* := \min_{u \in \mathbb{R}^m} \psi(u) \tag{23}$$

arises. Note that finding the best Lagrangian bound ψ^* is a main source of nonsmooth optimization problems.

Many combinatorial optimization problems can be formulated as nonlinear programming problems [17, 20], and the corresponding Lagrangian bounds may be obtained by a PGEM. Such problems include problems about the maximum clique [10] and the maximum independent set of graph vertices.

We now consider a class of quadratic optimization problems that can be regarded as special case of problem (21). To this end, let $X = \mathbb{R}^n$ and the functions $f_i : \mathbb{R}^n \to \mathbb{R}$ be defined by

$$f_i(x) := x^\top K_i x + x^\top b_i + c_i$$

with given symmetric matrices $K_i \in \mathbb{R}^{n \times n}$, $b_i \in \mathbb{R}^n$, and $c_i \in \mathbb{R}$, for $i = 0, 1, \ldots, m$. Then the Lagrange function $L : \mathbb{R}^n \times \mathbb{R}^m \to \mathbb{R}$ reads as

$$L(x, u) = f_0(x) + \sum_{i=1}^{m} u_i f_i(x) = x^\top K(u) x + x^\top b(u) + c(u),$$

where

$$K(u) := K_0 + \sum_{i=1}^{m} u_i K_i, \quad b(u) := b_0 + \sum_{i=1}^{m} u_i b_i, \quad c(u) := c_0 + \sum_{i=1}^{m} u_i c_i.$$

The Lagrangian bound ψ^* according to (23) is given by (see [17]):

$$\psi^* = \min_{u \in \mathbb{R}^m, K(u) \preceq 0} \max_{x \in \mathbb{R}^n} \left\{ x^\top K(u) x + x^\top b(u) + c(u) \right\}.$$

A symmetric matrix $K \in \mathbb{R}^{n \times n}$ is called negative definite ($K \prec 0$ for short) if all of its eigenvalues are negative. K is said to be negative semidefinite ($K \preceq 0$) if its eigenvalues are negative or zero. For the function ψ, we obtain

$$\psi(u) = \max_{x \in \mathbb{R}^n} L(x, u) = \begin{cases} L(x(u), u), & \text{if } K(u) \preceq 0 \text{ and } \mathrm{Proj}_{\ker K(u)}(b(u)) = 0, \\ +\infty, & \text{otherwise.} \end{cases}$$

If $K(u) \prec 0$, the kernel of the matrix $K(u)$ contains the zero vector only, so that $\mathrm{Proj}_{\ker K(u)}(b(u)) = 0$ follows and the necessary optimality condition $L_x(x, u) = 0$ for maximizing $L(\cdot, u)$ yields

$$x(u) := -\frac{1}{2} (K(u))^{-1} b(u).$$

If $K(u) \preceq 0$ with $\ker K(u) \supsetneq \{0\}$, then $\mathrm{Proj}_{\ker K(u)}(b(u)) = 0$ guarantees that $L_x(x, u) = 0$ has a unique solution $x(u)$ that can be determined as a linear combination of an orthogonal system of eigenvectors of $K(u)$ which do not belong to vanishing eigenvalues of this matrix.

For the use of an ellipsoid method, it is sufficient to define $g(u)$ in (1) as follows

$$g(u) := \begin{cases} (f_1(x(u)), \ldots, f_m(x(u)))^\top, & \text{if } K(u) \preceq 0 \text{ and } \mathrm{Proj}_{\ker K(u)}(b(u)) = 0, \\ (\xi^\top K_1 \xi, \ldots, \xi^\top K_m \xi)^\top & \text{otherwise,} \end{cases}$$

where $\xi = \xi(u)$ is any normalized eigenvector of the largest eigenvalue of $K(u)$. With the help of an ellipsoid method, one can specify an upper bound on the number of iterations required to find ψ^* with a given accuracy. But, depending on m and the predefined accuracy, this number can become very large. Therefore, the program DSQTPr was developed to find ψ^* based on the r-algorithm [19]. On the one hand, it was not possible to specify an upper bound on the number of iterations for this algorithm. On the other hand, the program was successfully used to find Lagrangian bounds for several quadratic test problems [19, 23]. In the latter paper, the program was applied to globally solve small balanced packing problems.

To complete this section, we consider an example of a discrete optimization problem to which the previous solution approach can be applied. Let $G = (V, E)$

be an undirected graph (without loops), where V denotes the set of vertices, and $\emptyset \neq E \subseteq \{\{i, j\} \mid i, j \in V, i \neq j\}$ the set of edges. Note that $\{i, j\}$ is the same edge as $\{j, i\}$. A subset of vertices $S \subseteq V$ is called independent set of graph G if, for any two vertices in S, there is no edge connecting these vertices.

The independence number $\alpha(G)$ of a graph G denotes the maximum number of vertices that provide an independent set. In general, the problem of determining $\alpha(G)$ is an NP-hard problem [1, 3]. Moreover, this problem can be equivalently formulated by means of quadratic Boolean optimization [17]:

$$\begin{aligned} &\alpha(G) = \max \sum_{i \in V} x_i \\ &\text{subject to} \quad x_i x_j = 0 \;\text{ for all } \{i, j\} \in E, \quad x_i^2 - x_i = 0 \;\text{ for all } i \in V. \end{aligned} \tag{24}$$

The constraint $x_i^2 - x_i = 0$ for $i \in V$ ensures that the variable x_i is Boolean. This variable is equal to 1 if the vertex $i \in V$ is included in an independent set, and 0 otherwise. The quadratic constraints for the edges of the graph G mean that two vertices connected by an edge in the graph G do not simultaneously belong to an independent set.

Problem (24) can be regarded as special case of (21). If we then derive the nonsmooth minimization problem (23) for obtaining a Lagrangian upper bound $\psi^*(G)$ for $\alpha(G)$, it has $|E| + |V|$ variables. Shor proved in [17] that $\psi^*(G)$ equals the well-known Lovász number [7] of graph G. In other words, the Shor bound $\psi^*(G)$ has the same properties as the Lovász number. For example, if the graph G belongs to the family of perfect graphs, then $\psi^*(G) = \alpha(G)$, i.e., the bound $\psi^*(G)$ is already the desired independence number. Another way to obtain an upper bound for the independence number is given in [28].

Let us finally mention that the way of using Lagrangian bounds for nonconvex optimization (including discrete optimization) as described in this subsection is not widely known. More research exists on applying Lagrangian bounds for models of linear programming with binary variables; for example, see [12].

3 Special Cases of PGEM

In this section, we will discuss and illustrate partial cases of the PGEM:

- the GEM, i.e., the PGEM with $\lambda = 1$ and α according to (2),
- a unified representation of the classical ellipsoid method with $\lambda > 0$, $\alpha = \sqrt{\frac{n+1}{n-1}}$,
- two approximate ellipsoid methods with $\lambda = 1$ and $\alpha_1 = \sqrt{1 + \frac{1}{n^2}} + \frac{1}{n}$ or $\alpha_2 = \sqrt{\sqrt{1 + \frac{4}{n^2}} + \frac{2}{n}}$.

3.1 The Generalized Ellipsoid Method (GEM)

The following algorithm is just Algorithm 1 with $\lambda = 1$.

Algorithm 3 B-form of the GEM

Step 0. Select α according to (2). Choose $x_0 \in \mathbb{R}^n$, a nonsingular matrix $B_0 \in \mathbb{R}^{n\times n}$ and $r_0 > 0$ so that

$$\left\| B_0^{-1}(x_0 - x^*) \right\| \leq r_0.$$

Moreover, set $k := 0$.
Step 1. Set $g_k := g(x_k)$. If $g_k = 0$, then set $k^* := k$, $x^* := x_{k^*}$ and STOP.
Step 2. Calculate

$$x_{k+1} := x_k - h_k B_k \xi_k, \quad \text{where} \quad \xi_k := \frac{B_k^\top g_k}{\|B_k^\top g_k\|}, \quad h_k := \frac{1}{2}\left(1 - \frac{1}{\alpha^2}\right) r_k.$$

Step 3. Update

$$B_{k+1} := B_k + \left(\frac{1}{\alpha} - 1\right)(B_k\xi_k)\,\xi_k^\top \quad \text{and} \quad r_{k+1} := \frac{1}{2}\left(\alpha + \frac{1}{\alpha}\right) r_k.$$

Step 4. Set $k := k + 1$ and go to Step 1.

Recall that, for any x_k generated by the GEM in B-form, the ellipsoid $\mathcal{E}_k$ is given by $\{x \mid \|B_k^{-1}(x_k - x)\| \leq r_k\}$. The following theorem can be derived from Theorem 1.

Theorem 2 ([24]) *Let x_k and x_{k+1} be generated by the GEM in B-form (Algorithm 3). Then,*

$$x^* \in \mathcal{E}_k \tag{25}$$

is satisfied. Moreover, the ratio of volumes of the ellipsoids $\mathcal{E}_{k+1}$ and $\mathcal{E}_k$ does not depend on k and is equal to

$$q_n(\alpha) = \frac{vol(\mathcal{E}_{k+1})}{vol(\mathcal{E}_k)} = \frac{1}{\alpha}\left(\frac{1}{2}\left(\alpha + \frac{1}{\alpha}\right)\right)^n < 1. \tag{26}$$

The GEM uses an ellipsoid described around a half-ball of radius r in $\mathbb{R}^n$ ($n \geq 2$), the volume of this ellipsoid is less than the volume of the full ball. Such an ellipsoid has an oblate shape in the direction that is normal to the hyperplane defining the half-ball. The parameters of the ellipsoid are shown in Fig. 2, where a is the length of the minor semiaxis of the ellipsoid, b is the length of the major semiaxes of the

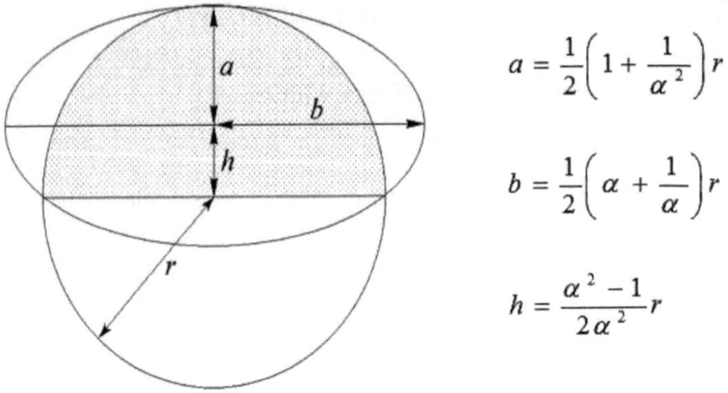

Fig. 2 The parameters of an ellipsoid containing a half-ball in $\mathbb{R}^n$

ellipsoid (the number of such semiaxes is equal to $n - 1$), and h is the distance from the center of the ball to the center of the ellipsoid in the direction of its minor semiaxis.

The volume of this ellipsoid is $v_e = v_0 a b^{n-1}$, and the volume of the ball is $v_b = v_0 r^n$, where v_0 denotes the volume of the unit ball in $\mathbb{R}^n$. Therefore, the volume reduction factor is equal to

$$\frac{v_e}{v_b} = \left(\frac{a}{r}\right)\left(\frac{b}{r}\right)^{n-1} = \left(\frac{a}{b}\right)\left(\frac{b}{r}\right)^{n} = \frac{1}{\alpha}\left(\frac{1}{2}\left(\alpha + \frac{1}{\alpha}\right)\right)^{n} = q_n(\alpha).$$

If $\alpha = \frac{b}{a}$ satisfies condition (2), then $q_n(\alpha) < 1$ can be guaranteed. To transform the ellipsoid, containing a half-ball, into a new ball, it is sufficient to dilate the space of variables in the direction of the minor semiaxis with the coefficient $\alpha = \frac{b}{a}$. This can be done using the space dilation operator $R_\alpha(\xi)$, where the direction ξ coincides with the direction of the minor semiaxis of the ellipsoid.

If $X = \mathbb{R}^n$ is the original space of variables, then in the transformed space of variables $Y = R_\alpha(\xi)X$, we get a new ball of radius b, which contains the solution of our problem. Repeating the same procedure, but for the new ball in the transformed space, we obtain the GEM. Here, in Step 2, the direction of the minor semiaxis of the ellipsoid in the transformed space $Y_k = B_k^{-1}X$ is calculated, and the transition to its center is performed. The calculated direction is used for the next space dilation, which is implemented in Step 3 by determining the matrix B_{k+1}. In the next transformed space $Y_{k+1} = B_{k+1}^{-1}X$, we get a ball of radius r_{k+1}.

The ellipsoids in the classical method of Nemirovsky and Yudin [8] can be obtained in Algorithm 3 for the dilation coefficient $\alpha^* = \sqrt{\frac{n+1}{n-1}}$. This value corresponds to the ellipsoid of minimal volume with respect to α satisfying (2). To see this, consider

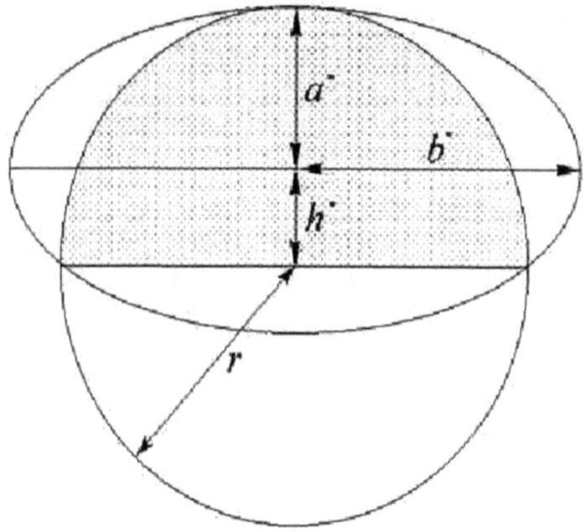

$$a^{*} = r\,\frac{n}{n+1}$$

$$b^{*} = r\,\frac{n}{\sqrt{n^2-1}}$$

$$h^{*} = r\,\frac{1}{n+1}$$

Fig. 3 The parameters of an ellipsoid of minimum volume containing a half-ball

$$\begin{aligned} q_n'(\alpha) &= \left(\frac{1}{\alpha}\left(\frac{1}{2}\left(\alpha+\frac{1}{\alpha}\right)\right)^n\right)' \\ &= -\frac{1}{\alpha^2}\left(\frac{1}{2}\left(\alpha+\frac{1}{\alpha}\right)\right)^n + \frac{1}{\alpha}\left(\frac{n}{2}\left(1-\frac{1}{\alpha^2}\right)\right)\left(\frac{1}{2}\left(\alpha+\frac{1}{\alpha}\right)\right)^{n-1} \\ &= \left(\frac{1}{2}\left(\alpha+\frac{1}{\alpha}\right)\right)^{n-1}\left(-\frac{1}{\alpha^2}\left(\frac{1}{2}\left(\alpha+\frac{1}{\alpha}\right)\right)+\frac{1}{\alpha}\left(\frac{n}{2}\left(1-\frac{1}{\alpha^2}\right)\right)\right) \\ &= \frac{1}{2\alpha^3}\left(\frac{1}{2}\left(\alpha+\frac{1}{\alpha}\right)\right)^{n-1}\left(-\left(\alpha^2+1\right)+n\left(\alpha^2-1\right)\right), \end{aligned}$$

where $q_n'(\alpha) = 0$ is equivalent to $-\left(\alpha^2+1\right)+n\left(\alpha^2-1\right) = 0$. Thus, the only solution of $q_n(\alpha) = 0$ is $\alpha^* = \sqrt{\frac{n+1}{n-1}}$, which also satisfies condition (2).

The parameters of the ellipsoid of minimum volume are shown in Fig. 3, where a^* is the length of its minor semi-axis, b^* is the length of the major semi-axes (their number is equal to $n-1$), h^* is the distance from the center of the ball to the center of the ellipsoid in the direction of its minor semi-axis.

The volume reduction factor $q_n(\alpha^*)$ (see (26)) can be estimated by

$$\begin{aligned} q_n(\alpha^*) &= \frac{1}{\alpha^*}\left(\frac{1}{2}\left(\alpha^*+\frac{1}{\alpha^*}\right)\right)^n \\ &= \frac{n}{n+1}\left(\frac{n}{\sqrt{n^2-1}}\right)^{n-1} \end{aligned}$$

$$
\begin{aligned}
&= \frac{n}{n+1}\left(\frac{n^2}{n^2-1}\right)^{\frac{n-1}{2}} \\
&= \left(1-\frac{1}{n+1}\right)\left(1+\frac{1}{n^2-1}\right)^{\frac{n-1}{2}} \\
&< \exp\left(-\frac{1}{n+1}\right)\exp\left(\frac{n-1}{2\left(n^2-1\right)}\right) \\
&= \exp\left(-\frac{1}{n+1}+\frac{n-1}{2\left(n^2-1\right)}\right) \\
&= \exp\left(-\frac{1}{2(n+1)}\right) \\
&< 1.
\end{aligned}
$$

A more accurate estimate of $q_n(\alpha^*)$ is given in [3, p. 77] by

$$
q_n(\alpha^*) < \exp\left(-\frac{1}{2n}\right) < 1.
$$

The length of the major semiaxis of the ellipsoid of minimum volume (Fig. 3) is determined by the formula $b^* = r\frac{n}{\sqrt{n^2-1}}$ for $n > 1$. If $n = 1$, then [9, p. 74] yields a volume reduction factor of 0.5. In Sect. 3.3, two approximate ellipsoids are considered, which are close to the ellipsoid of minimum volume.

3.2 *Unified Representation of the Ellipsoid Method*

Let us consider the problem of minimizing a convex function $f : \mathbb{R}^n \to \mathbb{R}$ with f^* denoting the minimal value of f and $g_f : \mathbb{R}^n \to \mathbb{R}^n$ being a subgradient mapping; see Sect. 2.3.1 for details. Moreover, we set the space dilation coefficient α equal to α^* (as derived in Sect. 3.1). However, as suggested in [26], we include the space scaling parameter $\lambda > 0$ in Algorithm 4. Adapting this parameter allows us to obtain different versions of the ellipsoid method as will be discussed in this subsection. Based on g_f, Algorithm 4 presents a more sophisticated stopping rule, which enables us to control the achieved distance of $f(x^k)$ from f^*.

Algorithm 4 Parameterized ellipsoid method

Step 0. Select $\lambda > 0$ and $\varepsilon > 0$. Choose $x_0 \in \mathbb{R}^n$ and $r_0 > 0$ such that

$$
\left\|x_0 - x^*\right\| \le r_0.
$$

Moreover, set $B_0 := I_n$ and $k := 0$.
Step 1. If $\left\| B_k^\top g_f(x_k) \right\| r_k \leq \varepsilon$, then set $k_\varepsilon := k$, $x_\varepsilon := x_{k_\varepsilon}$, and STOP.
Step 2. Calculate

$$x_{k+1} := x_k - h_k B_k \xi_k, \quad \text{where} \quad \xi_k := \frac{B_k^\top g_f(x_k)}{\left\| B_k^\top g_f(x_k) \right\|}, \quad h_k := \frac{1}{n+1} r_k.$$

Step 3. Update

$$B_{k+1} := \lambda \left(B_k + \left(\sqrt{\frac{n-1}{n+1}} - 1 \right) (B_k \xi_k)\, \xi_k^\top \right) \quad \text{and} \quad r_{k+1} := \frac{1}{\lambda} \frac{n}{\sqrt{n^2-1}} r_k.$$

Step 4. Set $k := k+1$, and go to Step 1.

Theorem 3 *Let the sequence* $\{x_k\}_{k=0}^{k_\varepsilon}$ *be generated by Algorithm 4. Then*

$$x_k \in \mathcal{E}_k \tag{27}$$

is satisfied for all $k = 0, 1, \ldots, k_\varepsilon$. *The ratio of the volumes of the ellipsoid* $\mathcal{E}_{k+1}$ *and* $\mathcal{E}_k$ *(for* $k = 0, \ldots, k_\varepsilon - 1$*) does not depend on k and is equal to*

$$q_n := \frac{vol(\mathcal{E}_{k+1})}{vol(\mathcal{E}_k)} = \sqrt{\frac{n-1}{n+1}} \left(\frac{n}{\sqrt{n^2+1}} \right)^n < \exp\left\{ -\frac{1}{2n} \right\} < 1. \tag{28}$$

Moreover, $f(x_{k_\varepsilon}) - f^* \leq \varepsilon$ *is fulfilled.*

For any $\lambda > 0$, the proof of this theorem can be carried out similar to that of Theorem 1 and [25, Theorem 2] or (30). For certain values of λ, existing variants of the ellipsoid method are obtained, namely, those by Shor [14], Nemirovsky and Yudin [9, page 76]), and Khachiyan [4, Lemma 4]. A fourth almost unknown variant is given in [2, 27] and denoted by [2] in Tables 1 and 2. Details of these variants can be found in Table 1. All variants (with different λ) are theoretically equivalent, though slight differences could be observed due to the accumulation of numerical errors. For example, using an Octave code for Algorithm 4 for the minimization of the nonsmooth function $f(x) = \sum_{i=1}^{n} 2^{i-1} |x_i - 1|$ with $x_0 = 0 \in \mathbb{R}^{10}$, $r_0 = 5$, and three different values of ε, the results in Table 2 are obtained.

From Step 3 of Algorithm 2, one can easily obtain the well-known versions of the GEM in H-form by inserting the corresponding value of λ from Table 1.

Table 1 Characteristics of four variants of the ellipsoid method

λ	Multiplications	Updated radius	Reference
1	$3n^2+n$	$r_{k+1}=\frac{n}{\sqrt{n^2-1}}r_k$	[14]
$\frac{n}{\sqrt{n^2-1}}$	$4n^2+n$	$r_{k+1}=r_k$	[4]
$\sqrt[n]{\frac{n+1}{n-1}}$	$4n^2+n$	$r_{k+1}=\left(\frac{n-1}{n+1}\right)^{\frac{1}{2n}}r_k$	[9]
$\left(\frac{n}{\sqrt{n^2-1}}\right)^{\frac{3}{2}}$	$4n^2+n$	$r_{k+1}=\left(\frac{\sqrt{n^2-1}}{n}\right)^{\frac{1}{2}}r_k$	[2]

Table 2 Results of Algorithm 4 applied to $f(x)=\sum_{i=1}^{n}2^{i-1}\,|x_i-1|$ for different values of λ

ε	λ according to	k^*	$f(x_\varepsilon^*)$	$\|x_\varepsilon^*-x^*\|$	$\|B_{k^*}\|$	r_{k^*}
1.0e−07	Shor [14]	4351	9.2e−09	9.2e−10	2.6e−18	1.6e+10
1.0e−07	Khachiyan [4]	4351	9.2e−09	9.2e−10	8.2e−09	5.0e+00
1.0e−07	Nemirovsky and Yudin [9]	4351	9.2e−09	9.2e−10	2.4e+01	1.7e−09
1.0e−07	Fischer et al. [2]	4351	9.2e−09	9.2e−10	4.6e−04	8.9e−05
1.0e−08	Shor [14]	4821	9.0e−10	1.1e−10	2.3e−20	1.7e+11
1.0e−08	Khachiyan [4]	4807	9.2e−10	1.8e−10	7.9e−10	5.0e+00
1.0e−08	Nemirovsky and Yudin [9]	4811	9.3e−10	9.0e−11	2.5e+01	1.7e−10
1.0e−08	Fischer et al. [2]	4819	9.0e−10	1.6e−10	1.4e−04	2.8e−05
1.0e−14	Shor [14]	6716	0.0e+00	0.0e+00	2.0e−29	2.3e+15
1.0e−14	Khachiyan [4]	6724	2.2e−16	2.2e−16	2.4e−14	5.0e+00
1.0e−14	Nemirovsky and Yudin [9]	6741	1.1e−16	1.1e−16	4.0e+00	1.1e−14
1.0e−14	Fischer et al. [2]	6738	6.7e−16	6.7e−16	3.1e−07	2.2e−07

3.3 *Two Approximate Ellipsoid Methods*

Section 3.1 shows how the ellipsoid of minimal volume is obtained for the space dilation coefficient α^* among all α satisfying (2). In contrast to this, it is suggested in [22] to replace the half-balls in $\mathbb{R}^n$ not by ellipsoids of minimum volume but by ellipsoids whose volume reduction coefficient $Q_n(\alpha^{**})$ is an approximation of $q_n(\alpha^*)$. More in detail, it holds

$$
\begin{aligned}
q_n(\alpha) &= \frac{1}{\alpha}\left(\frac{1}{2}\left(\alpha+\frac{1}{\alpha}\right)\right)^n = \frac{1}{\alpha}\left(1+\frac{1}{2}\left(\alpha+\frac{1}{\alpha}-2\right)\right)^n \\
&< \frac{1}{\alpha}\exp\left(\frac{n}{2}\left(\alpha+\frac{1}{\alpha}-2\right)\right) =: Q_n(\alpha).
\end{aligned}
$$

To find a space dilation coefficient α^{**} at which $Q_n(\cdot)$ reaches a minimum for $\alpha > 0$, we consider the derivative

$$
Q_n'(\alpha) = \left(\frac{1}{\alpha}\exp\left(\frac{n}{2}\left(\alpha+\frac{1}{\alpha}-2\right)\right)\right)'
$$

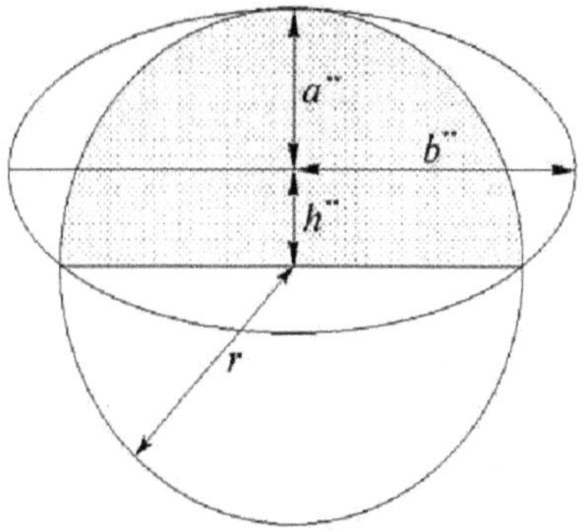

$$a^{**} = r\left(1 - \frac{1}{n}\left(\sqrt{1+\frac{1}{n^2}} - \frac{1}{n}\right)\right)$$

$$b^{**} = r\sqrt{1+\frac{1}{n^2}}$$

$$h^{**} = \frac{r}{n}\left(\sqrt{1+\frac{1}{n^2}} - \frac{1}{n}\right)$$

Fig. 4 The parameters of an approximate ellipsoid containing a half-ball

$$\begin{aligned}
&= -\frac{1}{\alpha^2}\exp\left(\frac{n}{2}\left(\alpha + \frac{1}{\alpha} - 2\right)\right) \\
&\quad + \frac{1}{\alpha}\left(\frac{n}{2}\left(1 - \frac{1}{\alpha^2}\right)\right)\exp\left(\frac{n}{2}\left(\alpha + \frac{1}{\alpha} - 2\right)\right) \\
&= \frac{1}{\alpha}\exp\left(\frac{n}{2}\left(\alpha + \frac{1}{\alpha} - 2\right)\right)\left(-\frac{1}{\alpha} + \frac{n}{2}\left(1 - \frac{1}{\alpha^2}\right)\right) \\
&= \frac{1}{2\alpha^3}\exp\left(\frac{n}{2}\left(\alpha + \frac{1}{\alpha} - 2\right)\right)\left(-2\alpha + n(\alpha^2 - 1)\right).
\end{aligned}$$

Then $Q_n'(\alpha) = 0$ is equivalent to $(\alpha)^2 - \frac{2}{n}\alpha - 1 = 0$ and has just one positive solution

$$\alpha^{**} = \frac{1}{n} + \sqrt{1 + \frac{1}{n^2}}.$$

The parameters of the approximate ellipsoid are shown in Fig. 4, where a^{**} is the length of its minor semiaxis, b^{**} is the length of the major semiaxes (their number is equal to $n-1$), and h^{**} is the distance from the center of the ball to the center of the ellipsoid in the direction of its minor semiaxis. The volume of the approximate ellipsoid is smaller than the volume of the sphere, and the volume reduction factor is

$$q_n(\alpha^{**}) = \left(\frac{a^{**}}{r}\right)\left(\frac{b^{**}}{r}\right)^{n-1} = \left(\sqrt{1+\frac{1}{n^2}} - \frac{1}{n}\right)\left(1 + \frac{1}{n^2}\right)^{n/2} < 1.$$

For $n \geq 2$, the volume reduction factor $q_n(\alpha^{**})$ can be estimated as follows

$$
\begin{aligned}
q_n(\alpha^{**}) &= \left(\sqrt{1+\frac{1}{n^2}}-\frac{1}{n}\right)\left(1+\frac{1}{n^2}\right)^{n/2} \\
&\le \left(\sqrt{1+\frac{1}{n^2}+\frac{1}{4n^4}}-\frac{1}{n}\right)\exp\left(\frac{1}{2n}\right) \\
&= \left(1+\frac{1}{2n^2}-\frac{1}{n}\right)\exp\left(\frac{1}{2n}\right) \\
&\le \exp\left(\frac{1}{2n^2}-\frac{1}{n}\right)\exp\left(\frac{1}{2n}\right) \\
&= \exp\left(-\frac{1}{2n}+\frac{1}{2n^2}\right).
\end{aligned}
$$

For large values n, the reduction factor $q_n(\alpha^{**})$ can be approximated by $1-\frac{1}{2n}$. The same holds if we consider $q_n(\alpha^*)$ for the ellipsoid of minimum volume. In the case $n = 1$, the reduction coefficient of the segment length is

$$
q_1(\alpha^{**}) = q_1(\sqrt{2}+1) = 2-\sqrt{2} \approx 0.5858.
$$

The construction of another ellipsoid in [5] is based on the approximation of the reduction coefficient $q_n(\alpha)$ by $Q_n(\alpha)$ that comes from

$$
q_n(\alpha) = \tfrac{1}{\alpha}\left(\tfrac{1}{2}\left(\alpha+\tfrac{1}{\alpha}\right)\right)^n = \tfrac{1}{\alpha}\left(1+\tfrac{(\alpha^2-1)^2}{4\alpha^2}\right)^{n/2} \le \tfrac{1}{\alpha}\exp\left(\tfrac{n(\alpha^2-1)^2}{8\alpha^2}\right) =: Q_n(\alpha).
$$

To find a dilation coefficient α^{***} at which $Q_n(\cdot)$ reaches a minimum, a similar technique as before exploits $Q_n'(\alpha) = 0$ and yields a polynomial equation of fourth order, namely, $n\alpha^4 - 4\alpha^2 - n = 0$. It follows that the only positive solution is

$$
\alpha^{***} = \sqrt{\sqrt{1+\frac{4}{n^2}}+\frac{2}{n}}.
$$

The parameters of the approximate ellipsoid are shown in Fig. 5, where a^{***} is the length of its minor semiaxis, b^{***} is the length of the major semiaxes (their number is equal to $n-1$), and h^{***} is the distance from the center of the ball to the center of the ellipsoid in the direction of its minor semiaxis. The volume of the approximate ellipsoid is smaller than the volume of the sphere with the volume reduction factor

$$
q_n(\alpha^{***}) = \left(\frac{a^{***}}{r}\right)\left(\frac{b^{***}}{r}\right)^{n-1} = \sqrt{\sqrt{1+\frac{4}{n^2}}-\frac{2}{n}}\left(\frac{1}{2}\left(1+\sqrt{1+\frac{4}{n^2}}\right)\right)^{n/2} < 1.
$$

For $n > 2$, the volume reduction factor $q_n(\alpha^{***})$ can be estimated as follows

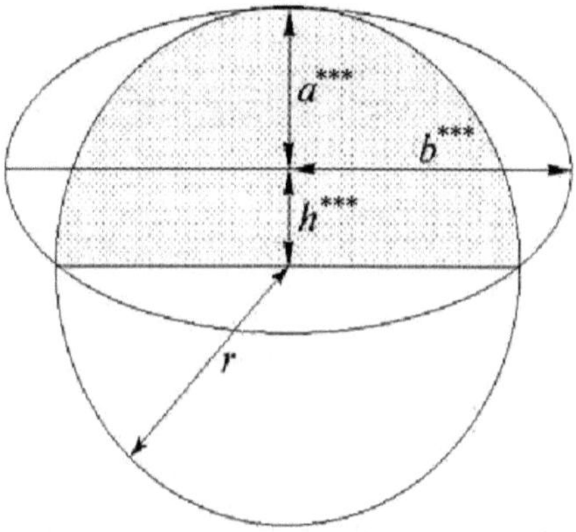

$$a^{***} = \frac{1}{2}\left(1+\sqrt{1+\frac{4}{n^2}}-\frac{2}{n}\right)r$$

$$b^{***} = \frac{1}{\sqrt{2}}\sqrt{1+\sqrt{1+\frac{4}{n^2}}}\,r$$

$$h^{***} = \frac{1}{2}\left(1+\frac{2}{n}-\sqrt{1+\frac{4}{n^2}}\right)r$$

Fig. 5 The parameters of the new approximate ellipsoid containing a half-ball

$$\begin{aligned}
q_n(\alpha^{***}) &= \sqrt{\sqrt{1+\frac{4}{n^2}}-\frac{2}{n}}\left(\frac{1}{2}\left(1+\sqrt{1+\frac{4}{n^2}}\right)\right)^{n/2} \\
&\le \sqrt{\sqrt{\left(1+\frac{2}{n^2}\right)^2}-\frac{2}{n}}\left(\frac{1}{2}\left(1+1+\frac{2}{n^2}\right)\right)^{n/2} \\
&= \left(1+\frac{2}{n^2}-\frac{2}{n}\right)^{1/2}\left(1+\frac{1}{n^2}\right)^{n/2} \\
&\le \exp\left(\frac{1}{n^2}-\frac{1}{n}\right)\exp\left(\frac{1}{2n}\right) \\
&= \exp\left(-\frac{1}{2n}+\frac{1}{n^2}\right).
\end{aligned}$$

For large values n, the reduction factor $q_n(\alpha^{***})$ can again be approximated by the expression $1-\frac{1}{2n}$ like $q_n(\alpha^*)$ for the ellipsoid of minimum volume. For $n=1$ and $n=2$, the reduction factor is

$$q_1(\alpha^{***}) = q_1\left(\sqrt{\sqrt{5}+2}\right) = \sqrt{\sqrt{5}-2}\sqrt{\frac{1+\sqrt{5}}{2}} = \frac{\sqrt{5}-1}{2} \approx 0.6180$$

and

$$q_2(\alpha^{***}) = q_2\left(\sqrt{\sqrt{2}+1}\right) = \sqrt{\sqrt{2}-1}\frac{1+\sqrt{2}}{2} \approx 0.7769,$$

respectively.

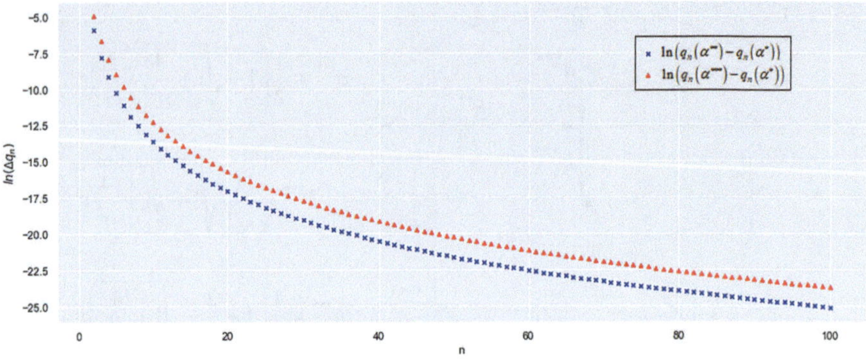

Fig. 6 The difference of reduction factors of approximate ellipsoids and the minimum ellipsoid in logarithmic scale

Table 3 Number of iterations for solving problems with relative accuracy 10^{-10}

n	10	20	30	40	50
$K(\alpha^*)$	4598	18,414	41,439	73,676	115,122

The use of the first or second approximate ellipsoid for large n with reduction factor $q_n(\alpha^{**})$ or $q_n(\alpha^{***})$ will be close to the classical ellipsoid method with $q_n(\alpha^*)$ in terms of convergence speed. This corresponds to the results in Fig. 6, where

$$\Delta q_n := \ln(q_n(\alpha^{**}) - q_n(\alpha^*)) \quad \text{or} \quad \Delta q_n := \ln(q_n(\alpha^{***}) - q_n(\alpha^*))$$

are shown for increasing n. For example, if $n = 100$, the difference between the reduction factors of the approximate ellipsoids and the minimum ellipsoid is between 10^{-25} and 10^{-23}.

For $n \in \{10, 20, 30, 40, 50\}$, Table 3 shows the number of iterations for the classical ellipsoid method to guarantee that the volume reduction is at least 10^{-10n}. Here, $K(\alpha^*)$ denotes the smallest integer for which $(q_n(\alpha^*))^{K(\alpha^*)} \leq 10^{-10n}$ holds. Accordingly, $K(\alpha^{**})$ and $K(\alpha^{***})$ can be defined for the two approximate ellipsoid methods and their reduction factors $q_n(\alpha^{**})$ and $q_n(\alpha^{***})$. It turns out that in most cases $K(\alpha^*) = K(\alpha^{**}) = K(\alpha^{***})$, sometimes with a difference of 1 or 2 iterations.

4 The Algorithm Emshor for Minimizing Convex Functions

This section shows an implementation of the ellipsoid method by Shor (Emshor) in the next subsection. Moreover, it provides numerical results if this implementation is applied to a ravine function in Sect. 4.2 and to a problem raised by Sylvester in Sect. 4.3.

4.1 The Algorithm Emshor and Its Implementation

To apply a GEM to the minimization of a convex function $f : \mathbb{R}^n \to \mathbb{R}$, we assume for simplicity that x^* is the only minimizer of f. Then, as in Sect. 2.3.1, let us specialize the mapping g from Sect. 2 by $g_f : \mathbb{R}^n \to \mathbb{R}^n$ with $g_f(x)$ being a subgradient of f at x, i.e., g_f satisfies

$$(x - x^*)^\top g_f(x) \geq f(x) - f(x^*) = f(x) - f^* \geq 0 \quad \text{for all } x \in \mathbb{R}^n. \tag{29}$$

The Emshor algorithm in B-form is nothing else than Algorithm 4 in Sect. 3.2 with $\lambda = 1$. According to this, the convergence results of Theorem 3 are true for the Emshor algorithm. Recall that the stopping criterion used in Step 1 of Algorithm 4 guarantees that the Emshor algorithm finds x_k with $f(x_k) \leq f^* + \varepsilon$ for some predefined $\varepsilon > 0$. This criterion follows for $x_k \in \mathbb{R}^n$, $B_k \in \mathbb{R}^{n \times n}$, and r_k generated by the Emshor algorithm because

$$\begin{aligned} f(x_k) - f^* &\leq \langle x_k - x^*, g_f(x_k) \rangle \\ &= \langle B_k^{-1}(x_k - x^*), B_k^\top g_f(x_k) \rangle \\ &\leq \| B_k^{-1}(x_k - x^*) \| \| B_k^\top g_f(x_k) \| \\ &\leq r_k \| B_k^\top g_f(x_k) \|, \end{aligned} \tag{30}$$

where the first inequality follows from (29) and the last is a consequence of (27). Hence, if $r_k \| B_k^\top g_f(x_k) \| \leq \varepsilon$ holds for some iterate x_k, then $f(x_k) - f^* \leq \varepsilon$ follows.

The Emshor algorithm is implemented in Octave and then called Emshor program. To calculate the value $f(x)$ and a subgradient $g_f(x)$ at x, the user has to provide an Octave function `function[f,g]=calcfg(x)`. The code of the Emshor program and some short comments are given below.

```
# Input parameters:
# calcfg - name of the function for calculation of f and g
# x0 - starting point, x0(1:n)
# rad - the radius of the ball localizing the minimum point
# epsf - parameter for stopping (accuracy)
# maxitn - parameter for stopping (maximal no. of iterations)
# intp - printing interval (after each intp iterations)
# Output parameters:
# x - approximation of the minimum point, x(1:n)
# f - value of function f at the point x
# itn - number of iterations performed
# ist - exit code (1 = epsf, 4 = maxitn)
function[x,f,itn,ist] =
emshor(calcfg,x0,rad,epsf,maxitn,intp);                      #row01
dn=double(length(x0)); beta=sqrt((dn-1.d0)/(dn+1.d0)); #row02
x=x0; radn=rad; B=eye(length(x));                            #row03
```

```
for (itn = 0:maxitn)                                                    #row04
   [f,g1] = calcfg(x); g=B'*g1; dg=norm(g);                             #row05
   if(radn*dg < epsf) ist = 1;  return; endif                           #row06
   xi=(1.d0/dg)*g; dx = B * xi;                                         #row07
   hs=radn/(dn+1.d0); x -= hs * dx;                                     #row08
   B += (beta - 1) * (B * xi) * xi';                                    #row09
   radn=radn/sqrt(1.d0-1.d0/dn)/sqrt(1.d0+1.d0/dn);                     #row10
   if(mod(itn,intp)==0)                                                 #row11
     printf("itn %4d  f %14.6e\n",itn,f);                               #row12
   endif                                                                #row13
endfor                                                                  #row14
ist = 4;                                                                #row15
endfunction                                                             #row16
```

The iterative process is executed in a `for`-loop (`#row04`-`#row14`), where `#row05` and `#row06` implement Step 1, `#row07` and `#row08` implement Step 2, and `#row09` and `#row10` implement Step 3 of Emshor (Algorithm 4). After every `intp` iterations in the `for`-loop intermediate results are printed (see `#row11`-`#row13`). The Emshor program is stopped if either a vector $\mathbf{x}$ with $f(\mathbf{x}) \leq f^* + \varepsilon$ is found (`ist` $= 1$ in `#row06`) or the maximal number of iterations `maxitn` is reached (see `#row04` and `#row15`).

4.2 *Computational Experiments for a Ravine Function*

We will demonstrate the work of the Emshor program for the minimization of a ravine piecewise linear convex function $f : \mathbb{R}^n \to \mathbb{R}$ with

$$f(x) = \sum_{i=1}^{n} 2^{i-1} |x_i - 1| \tag{31}$$

and refer the reader to [25, Sect. 3.3]. Obviously, the unique minimizer of f is $x^* = (1, 1, \dots, 1)^\top \in \mathbb{R}^n$ with optimal value $f^* = f(x^*) = 0$.

The degree of elongation of the level sets of the function f in (31) is determined by the ratio of the maximum coefficient of $|x_i - 1|$ to the minimal coefficient. The latter is always equal to 1. For $n = 20$, the maximal coefficient equals $2^{19} \approx 5.2\text{e}+05$.

Table 4 provides results obtained by the Emshor program for the starting values $x_0 := 0 \in \mathbb{R}^n$, $r_0 \in \{5, 500, 50000\}$, and $\varepsilon \in \{10^{-3}, 10^{-6}, 10^{-9}\}$. The calculations were performed for $n \in \{5, 10, 15, 20\}$ on a Pentium 2.5 GHz computer using GNU Octave version 3.0.0. The table shows the number of iterations (*itn*), the function value $f(x_{itn})$ at the last iteration, and the computing time (*time*) in seconds. It should be noted that all execution times can be significantly reduced when one makes use of compiled commands for matrix times vector operations.

Table 4 Results for applying the Emshor program to the minimization of f in (31)

$r_0 = 5$									
	$\varepsilon = 10^{-3}$			$\varepsilon = 10^{-6}$			$\varepsilon = 10^{-9}$		
n	itn	$f(x_{itn})$	$time$	itn	$f(x_{itn})$	$time$	itn	$f(x_{itn})$	$time$
5	519	6.1e−06	0.07	873	1.1e−07	0.08	1201	1.2e−10	0.11
10	2484	8.7e−05	0.23	3829	7.2e−08	0.34	5253	7.4e−11	0.47
15	6541	6.5e−06	0.59	9594	1.0e−08	0.85	12,757	5.4e−11	1.12
20	13,126	5.7e−05	1.19	18,703	4.2e−08	1.66	23,300	1.7e−11	2.01
$r_0 = 500$									
	$\varepsilon = 10^{-3}$			$\varepsilon = 10^{-6}$			$\varepsilon = 10^{-9}$		
n	itn	$f(x_{itn})$	$time$	itn	$f(x_{itn})$	$time$	itn	$f(x_{itn})$	$time$
5	747	1.1e−05	0.10	1080	1.6e−07	0.13	1392	1.6e−10	0.19
10	3429	9.0e−05	0.42	4810	9.3e−08	0.60	6165	8.8e−11	0.73
15	8615	5.6e−05	1.07	11,671	6.1e−08	1.48	14,774	2.3e−11	1.32
20	16,760	4.7e−05	2.09	22,416	4.1e−08	2.82	27,072	2.6e−11	2.29
$r_0 = 50{,}000$									
	$\varepsilon = 10^{-3}$			$\varepsilon = 10^{-6}$			$\varepsilon = 10^{-9}$		
n	itn	$f(x_{itn})$	$time$	itn	$f(x_{itn})$	$time$	itn	$f(x_{itn})$	$time$
5	951	1.7e−04	0.09	1328	9.1e−08	0.13	1651	1.3e−11	0.24
10	4323	8.4e−05	0.37	5726	7.7e−09	0.51	7106	3.0e−12	0.71
15	10,707	6.3e−05	0.97	13,797	1.1e−08	1.25	16,914	4.4e−11	1.48
20	20,459	4.6e−05	1.88	26,095	4.0e−08	2.40	30,819	2.4e−11	1.90

From Table 4, it can be seen that the Emshor program finds very precise approximations to the minimizer of the ravine convex function. The number of iterations grows slightly faster than n^2 does. For details on complexity, see [15].

We have already discussed possible numerical instabilities of the ellipsoid method in H-form (see Algorithm 2); see Sect. 2.1. It is now demonstrated that the Emshor algorithm in H-form might not be able to approximate the minimum point of the ravine function with high accuracy. To this end, in the Emshor program, the operations with the matrix B in `#row05`, `#row07`, and `#row09` are replaced by the corresponding operations with the symmetric matrix $H = BB^\top$. In Table 5, the results obtained by applying the modified Emshor program (in H-form) are shown, where $\varepsilon = 10^{-3}$, $x_0 := 0 \in \mathbb{R}^n$, and $r_0 \in \{5, 500\}$ are used for partial comparability with Table 4. In Table 5, the number ind says that the matrix H_{ind+1} became indefinite, the iteration number $itr \le ind$ is the iteration, where the smallest value $f(x_{itr})$ of f is attained during the whole iteration process.

It is easy to see that for all $n \in \{5, 10, 15, 20\}$, the accuracies in Table 5 are insufficient compared with those from Table 4, i.e., the modified Emshor program stops long before an acceptable accuracy is achieved.

Table 5 Results for applying the Emshor program in H-form to the minimization of f in (31)

$\varepsilon = 10^{-3}$								
	$r_0 = 5$				$r_0 = 500$			
n	ind	$f(x_{ind})$	itr	$f(x_{itr})$	ind	$f(x_{ind})$	itr	$f(x_{itr})$
5	461	0.001310	446	0.00001033	453	0.208985	443	0.00201819
10	1664	0.030775	1467	0.00008307	1767	2.357996	1690	0.00195698
15	6541	0.000065	6528	0.00000029	8615	0.000056	7804	0.00000031
20	5627	5.700439	5356	0.00240983	5434	1142.453	4980	0.20944151

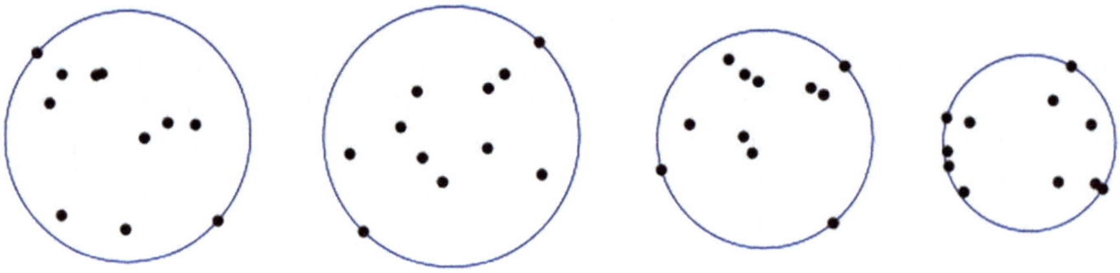

Fig. 7 The smallest bounding circles containing four sets of points in $\mathbb{R}^2$

4.3 Applying Emshor to a Problem by Sylvester

Let a set of points $A_m = \{a_j \in \mathbb{R}^n \mid j = 1, \ldots, m\} \subset \mathbb{R}^n$ be given. We are looking for a hypersphere in $\mathbb{R}^n$ of minimal radius that contains all points of the set A_m. In 1857, this problem was formulated by the English mathematician James Joseph Sylvester for the case $n = 2$; see [30]. Therefore, for any integer $n \geq 2$, we name the problem after Sylvester.

For $n = 2$, Fig. 7 shows point sets and a corresponding minimal circle. The smallest bounding circle for a set of points in $\mathbb{R}^2$ can be determined by at most three points from this set, which lie on the boundary of the circle. If a circle is defined by only two points, then the straight line connecting these points is a diameter of the smallest bounding circle. If a circle is defined by three points, then a triangle with vertices at these points cannot be obtuse.

The problem of finding a hypersphere of minimal radius that contains $A_m \subset \mathbb{R}^n$ for any integer $n \geq 2$ can be formulated as a convex programming problem [6, 29] with a piecewise quadratic function $f : \mathbb{R}^n \to \mathbb{R}$ so that

$$f(x) := \max_{j=1,\ldots,m} \|x - a_j\|^2.$$

Then the problem by Sylvester for $n \geq 2$ is given by

$$\min_{x \in \mathbb{R}^n} f(x). \tag{32}$$

It has a unique solution x^*, which denotes the center of the minimal hypersphere that contains the set A_m. The radius of this hypersphere is equal to

$$\sqrt{f(x^*)} = \sqrt{\max_{j=1,\dots,m} \|x^* - a_j\|^2} = \max_{j=1,\dots,m} \|x^* - a_j\|.$$

To solve problem (32), the Emshor algorithm (Algorithm 4 for $\lambda = 1$) can be employed. In Step 0, we use

$$x_0 = \frac{1}{m}\sum_{j=1}^{m} a_j \quad \text{and} \quad r_0 = \max_{j=1,\dots,m} \|x_0 - a_j\|,$$

since $\|x^* - x_0\| \le r_0$ is fulfilled as needed to start the algorithm. The input parameter $\varepsilon > 0$ is used to stop the Emshor algorithm according to (30) so that

$$f(x_{k_\varepsilon}) - f(x^*) = \max_{j=1,\dots,m} \|x_{k_\varepsilon} - x^*\|^2 \le f(x^*) + \varepsilon.$$

Finally note that the convergence results in Theorem (3) are valid for the Emshor algorithm (i.e., for Algorithm 4 with $\lambda = 1$).

The Octave implementation of the Emshor algorithm is given as Emshor program in Sect. 4.1. To calculate function values and subgradients of f at x, the following Octave function was implemented for Sylvester's problem.

```
function [f,g] = fg(x)
global A m n
temp = ones(m,1)*x' - A;
temp1 = sum(temp'.*temp');
[tmax imax] = max(temp1);
f = tmax;
g = 2.0*temp(imax,:)';
end
```

The set of points $A_m = \{a_j \in \mathbb{R}^n \mid j = 1, \dots, m\}$ is passed to this function using global parameters `A`, `m`, and `n`, where `A` is an $m \times n$-matrix.

To verify whether the Emshor program works well for Sylvester's problem, we use a test example with $m = n + 1$ points in $\mathbb{R}^n$, where the n points are vertices of the n-dimensional simplex $\left\{\sum_{i=1}^n x_i = 1 \mid x_i \in \mathbb{R}^n,\ x_i \ge 0,\ i = 1, \dots, n\right\}$, and the $(n+1)$-point coincides with the origin. For this test example, the matrix A, the optimal center x^*, and the optimal radius $R_* := \sqrt{f(x^*)}$ are as follows:

$$A = \begin{pmatrix} 1\,0 \cdots 0\,0 \\ 0\,1 \cdots 0\,0 \\ \vdots\,\vdots \cdots \vdots\,\vdots \\ 0\,0 \cdots 1\,0 \\ 0\,0 \cdots 0\,1 \\ 0\,0 \cdots 0\,0 \end{pmatrix} \in \mathbb{R}^{(n+1)\times n}, \quad x^* = \frac{1}{n}\begin{pmatrix} 1 \\ 1 \\ \vdots \\ 1 \\ 1 \end{pmatrix} \in \mathbb{R}^n, \quad R_* = \sqrt{1 - \frac{1}{n}}.$$

The computational results for our test example are shown in Table 6 for $n = 30$, $\varepsilon \in \{10^{-2}, 10^{-4}, \ldots, 10^{-28}, 10^{-30}\}$, where the column k_ε provides the number of iterations needed to stop the according to Step 1 of Algorithm 4. The column *time* shows the computing time until stopping.

Table 6 demonstrates that the Emshor program fairly quickly finds rather accurate approximations to x^*. This is evidenced by the results from the last column $\|x_{k_\varepsilon} - x^*\|$. For example, to find an approximation to the point x^* with accuracy 10^{-8} sufficient number of iterations to make is 41800, which takes only 2.5 seconds. It can be seen from the second last column of the table that the objective function value of problem (32) at x_{k_ε} has a high accuracy characterized by up to 14 significant digits. Table 6 also shows that, for increasing values of the accuracy ε, the achieved values of the objective function do not decrease monotonically. This is typical for the ellipsoid method, which provides a monotonous decrease of the volume of the ellipsoids containing the solution x^* but does not ensure a monotonic decrease of the values of the function being minimized.

Table 6 Results of the Emshor program for the Sylvester test example with $n = 30$

ε	k_ε	*time*	$f(x_{k_\varepsilon})$	$\|x_{k_\varepsilon} - x^*\|$
1.0e−02	9248	0.6	0.96691514094205810	1.419648e−03
1.0e−04	17,344	1.1	0.96666675076144826	4.038524e−05
1.0e−06	25,522	1.6	0.96666669634550528	8.408335e−06
1.0e−08	33,675	2.1	0.96666666694991410	1.177884e−06
1.0e−10	41,800	2.5	0.96666666666926282	5.148086e−08
1.0e−12	49,954	3.1	0.96666666666666778	4.682847e−09
1.0e−14	58,115	3.5	0.96666666666666790	1.519273e−09
1.0e−16	65,514	4.2	0.96666666666666712	1.009356e−11
1.0e−18	67,685	4.4	0.96666666666666712	3.935437e−12
1.0e−20	69,468	4.5	0.96666666666666701	7.548093e−12
1.0e−22	78,832	5.1	0.96666666666666712	5.063893e−13
1.0e−24	83,439	5.4	0.96666666666666690	4.869797e−13
1.0e−26	87,876	5.8	0.96666666666666690	5.051377e−13
1.0e−28	95,098	6.2	0.96666666666666701	5.061942e−13
1.0e−30	95,921	6.3	0.96666666666666712	5.064286e−13

Acknowledgments We gratefully acknowledge funding by the Volkswagen Foundation under grant No. 97 775 and by the National Academy of Sciences of Ukraine under grant No. 2.3/25-P.

References

1. Butenko, S, Pardalos, P.M., Sergienko, I., Shylo, V., Stetsyuk P.: Finding maximum independent sets in graphs arising from coding theory. In: Proceedings of the 2002 ACM Symposium on Applied Computing, pp. 542–546. ACM Digital Library, New York (2002)
2. Fischer, A., Khomyak, O., Stetsyuk, P.: The ellipsoid method and computational aspects. Commun. Optim. Theory **21**, 1–14 (2023)
3. Grötschel, M., Lovász, L., Schrijver, A.: Geometric Algorithms and Combinatorial Optimization. Springer, Berlin (1988)
4. Khachiyan, L.G.: Polynomial algorithms in linear programming. USSR Comput. Math. Math. Phys. **20**(1), 53–72 (1980)
5. Khomiak, O.: Three ellipsoids for external approximation of the half-ball (in Ukrainian). Cybernet. Comput. Technol. **3**, 48–59 (2024)
6. Khomiak, O.M., Davydov, O.O.: Ellipsoid method for a hyperball of a minimal radius. In: Proceedings of VII International Scientific Conference Mathematical Modeling, Optimization and Information Technologies, Chisinau-Kyiv-Batumi, pp. 340–342 (2021)
7. Lovász, L.: On the Shannon-capacity of a graph. IEEE Trans Inform. Theory. **25**(1), 1–7 (1979)
8. Nemirovsky, A.S., Yudin, D.B.: Information complexity and efficient methods of solving convex extremal problems. Ekonomika i Mat. Metody **2**, 357–369 (1976)
9. Nemirovsky, A.S., Yudin, D.B.: Problem Complexity and Method Efficiency in Optimization. John Wiley, New York (1983), translated from book published by Nauka, Moscow (1979)
10. Pardalos, P., Xue, J.: The maximum clique problem. J. Glob. Optim. **4**, 301–328 (1994)
11. Rodomanov, A., Nesterov, Y.: Subgradient ellipsoid method for nonsmooth convex problems. Math. Program. **199**, 305–341 (2023)
12. Shibasaki, R.S., Baiou, M., Barahona, F., Mahey, P., de Souza, M.C.: Lagrangian bounds for large multicommodity network design: a comparison between volume and bundle methods. Int. Trans. Oper. Res. **28**, 296–326 (2021)
13. Shor, N.Z.: Convergence rate of the gradient descent method with dilatation of the space. Cybernetics **6**, 102–108 (1970)
14. Shor, N.Z.: Cut-off method with space extension in convex programming problems. Cybernetics **13**, 94–96 (1977)
15. Shor, N.Z.: New development trends in nondifferentiable optimization. Cybernetics **13**, 881–886 (1977)
16. Shor, N.Z.: Methods of Minimization of Non-Differentiable Functions and their Applications (in Russian). Naukova Dumka, Kyiv (1979)
17. Shor, N.Z.: Nondifferentiable Optimization and Polynomial Problems. Kluwer Academic Publishers, Dordrecht (1998)
18. Shor, N.Z., Biletskii, V.I.: The method of space dilation for accelerating convergence in ravine-type problems. In: Proc. Sem. Sci. Council of the Academy of Sciences of the Ukrainian SSR on Cybernetics: Optimal Decision Theory, vol. 2, pp. 3–18, Kyiv (1969)
19. Shor, N.Z., Stetsyuk, P.I.: The use of modification of r-algorithm for finding a global minimum of polynomial functions. Cybern. Syst. Anal. **33**, 482–497 (1997)
20. Shor, N.Z., Stetsyuk, P.I.: Lagrangian bounds in multiextremal polynomial and discrete optimization problems. J. Glob. Optim. **23**, 1–41 (2002)
21. Skokov, V.A.: Note on minimization methods employing space stretching. Cybernetics **10**, 689–692 (1974)
22. Stetsyuk, P.I.: An approximate method of ellipsoids. Cybern. Syst. Anal. **39**(3), 435–439 (2003)

23. Stetsyuk, P., Romanova, T., Scheithauer, G.: On the global minimum in a balanced circular packing problem. Optim. Lett. **10**, 1347–1360 (2016)
24. Stetsyuk, P.I., Fesiuk, O.V., Khomyak, O.N.: The generalized ellipsoid method. Cybern. Syst. Anal. **54**, 576–584 (2018)
25. Stetsyuk, P.I., Fischer, A., Khomyak, O.: The generalized ellipsoid method and its implementation. In: Jaćimović, M., Khachay, M., Malkova, V., Posypkin, M. (eds.) Optimization and Applications. OPTIMA 2019. Communications in Computer and Information Science, vol. 1145, pp. 355–370. Springer, Cham (2020)
26. Stetsyuk, P.I., Fischer, A., Khomiak, O.M.: Ellipsoid methods with space scaling. In: Aliev, F., Başar, T. (eds.) Proceedings of the 8th International Conference on Control and Optimization with Industrial Applications, vol. I, pp. 402–404, Baku (2022)
27. Stetsyuk, P.I., Fischer, A., Khomiak, O.M.: Unified representation of the classical ellipsoid method. Cybern. Syst. Anal. **59**, 784–793 (2023)
28. Stetsyuk, P., Fischer, A., Khomiak, O.: An exact nonsmooth penalty approach for a special class of linear programs. In: Proceedings of the 5th International Conference on Problems of Cybernetics and Informatics, Baku, Azerbaijan, 3 p. IEEE Explore, Piscataway (2023)
29. Stetsyuk, P., Khomiak, O., Davydov, O.: Using the ellipsoid method for Sylvester's problem and its generalization (in Ukrainian). Cybernet. Comput. Technol. **1**, 27–46 (2024)
30. Sylvester, J.J.: A question in the geometry of situation. Quarter. J. Pure Appl. Math. **1**, 79 (1857)
31. Todd, M.J. The ellipsoid method redux. Preprint, available from http://dx.doi.org/10.48550/arXiv.2301.07632 (2023)

Non-submodular Optimization and Non-convex Relaxation

Weili Wu, Zhao Zhang, Wei Li, and Ding-Zhu Du

1 Introduction

Consider a set function $f : 2^X \to \mathbb{R}$. f is called a *submodular function* if for any $A, B \in 2^X$,

$$f(A) + f(B) \geq f(A \cup B) + f(A \cap B).$$

f is called a *supermodular function* if $-f$ is submodular. f is said to be *monotone nondecreasing* if for any $A, B \in 2^X$,

$$A \subseteq B \Rightarrow f(A) \leq f(B).$$

We emphasize the distinction between maximization (minimization) of submodular (supermodular) functions which are computationally intractable (NP-hard) [17] and their polynomially (efficiently) solvable counterparts, namely, minimization (maximization) of submodular (supermodular) functions [36]. The unconstrained submodular (supermodular) function is quasiconcave (quasiconvex) on any chain of the corresponding Boolean (Hasse diagram) [20] A *nonsubmodular optimization* is a

W. Wu · D.-Z. Du (✉)
Department of Computer Science, University of Texas at Dallas, Richardson, TX, USA
e-mail: weiliwu@utdallas.edu; dzdu@utdallas.edu

Z. Zhang
School of Mathematics, Zhejiang Normal University, Jinhua, Zhejiang, China
e-mail: hxhzz@sina.com

W. Li
Department of Computer Science, Texas Southern University, Houston, TX, USA
e-mail: liw@tsu.edu

B. Goldengorin (ed.), *Theory, Algorithms, and Experiments in Applied Optimization*, Springer Optimization and Its Applications 226,
https://doi.org/10.1007/978-3-031-91357-0_17

maximization or minimization problem with either objective function or a constraint which is neither submodular nor supermodular.

The submodular function is usually considered as a sort of discrete convex function, since its unconstrained minimization can be solved in polynomial-time [21, 34, 36]. In the set function optimization, the analysis on submodular functions or monotone nondecreasing submodular functions has been very well studied [30]. Especially, many theoretical guaranteed approximation solutions have been found for them [14, 16, 28, 32, 39, 44].

In recent developments of computer science and technology, one encounters many nonsubmodular optimizations. The following is an example.

Problem 1.1 (Interactive-Aware Viral Marketing [19]) *Consider a social network $G = (V, E)$ with independent cascade model. The problem can be reduced to the following set function maximization:*

$$\max \quad f(S)$$
$$\textit{subject to} \quad |S| \le k, S \subseteq V,$$

where S is called the seed set. The objective function $f(S)$ can be represented as

$$f(S) = \mathbb{E}_{g\sim D}\left[\alpha \cdot |I_g(S)| + \beta \cdot \sum_{\{u,v\}\subseteq I_g(S)} b(u, v)\right]$$

where g is the subgraph of G consisting of all alive edges at end of cascade process, D is the distribution of g, $I_g(S)$ is the set of active nodes, and $b(u, v)$ is benefit generated by interaction between nodes u and v. α and β are used to balance two types of profits. Moreover, $\mathbb{E}_{g\sim D}$ means that the expectation is taken for g over distribution D.

Lemma 1.2 ([19]) *$f(S)$ is neither submodular nor supermodular.*

Proof Consider a counterexample in Fig. 1. Each edge (u, v) is associated with a pair of labels, $(p_{uv}, b(u, v))$, i.e., the first label is the propagation probability, and

Fig. 1 A counterexample

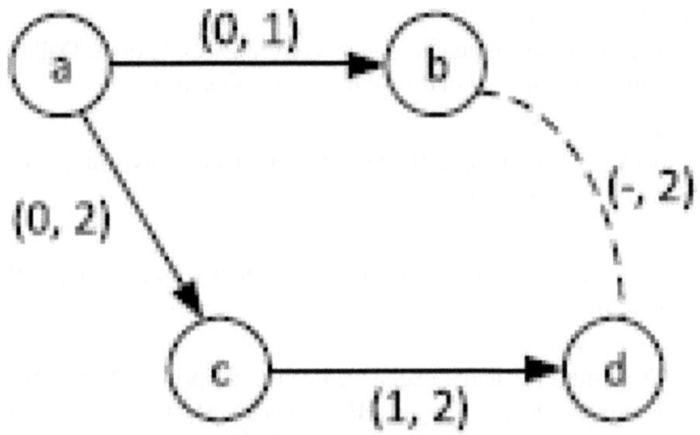

the second label is the activity benefit between two nodes. Set $\alpha > 0$ and $\beta > 0$. Then,

$$\begin{aligned} f(\{a,d\}) - f(\{a\}) &= (2\alpha + \beta) - \alpha \\ &< f(\{a,b,d\}) - f(\{a,b\}) \\ &= (3\alpha + 3\beta) - 2\alpha, \end{aligned}$$

Which means that f is not submodular. Moreover,

$$\begin{aligned} f(\{c\}) - f(\emptyset) &= 2\alpha + 2\beta \\ &> f(\{c,d\}) - f\{d\}) \\ &= (2\alpha + 2\beta) - \alpha, \end{aligned}$$

which means that f is not supermodular. □

More examples are listed as follows:

Problem 1.3 (Viral Marketing for Complementary Products [22]) *Consider a social network and a set of products with complementary relationships, i.e., buying a group of products may influence to buy a certain product. Given a budget B, find a set of customers for giving discounts (e.g., free samples) within the budget B to maximize the expected total sales of the product.*

Problem 1.4 (Content Spread [48]) *Sharing and spreading various contents is growing importance of information diffusion in online social networks. To develop effective and efficient techniques for content processing, a content spread maximization problem is studied in [48], which is a nonsubmodular optimization.*

Problem 1.5 (Target Influence [51]) *This problem aims at influence toward certain target user, that is, given a social network $G = (V, E)$ with an information diffusion model, a target set $T \subseteq V$, and a positive integer k, the problem is to find k locations for seeds to maximize the influence spread within T.*

Problem 1.6 (Rumor Blocking) *The rumor blocking is an important research problem in the study of social networks. There are two popular approaches to deal with the rumor blocking. The first one is to cut connections at nodes or links (e.g., see [25, 27]). In [46], consider link deletion to minimize the rumor spread, which is a nonsubmodular minimization. In [18], consider node cuts to maximize the number of protected nodes, which is a nonsubmodular maximization. The second approach is to place protectors in the network. Each protector would spread the truth to clarify the rumor [33, 41]. There are many formulation on how to allocate protectors, which gives more nonsbmodular optimizations [23, 24, 40, 47, 54].*

Problem 1.7 (Corpora Extraction [8]) *The sentiment analysis has been widely used in mining information from online social networks for election prediction and decision-making in business management [1, 3, 37]. The language corpora plays an important role in the sentiment analysis [7]. The size of corpora effects the power and accuracy of analysis. However, it also brings redundancy. Consider such a tradeoff. In [8], a nonsubmodular optimization is formulated about corpora extraction.*

Currently, no satisfied solution has been found for nonsubmodular optimizations. In this article, we give an extensive survey to study approximations of nonsubmodular optimizations and an important related technique, nonconvex relaxation. Meanwhile, we explore a few open problems on those subjects.

2 Parameterized Method

To deal with nonsubmodular optimization, it is a naive idea to introduce some measures to evaluate how far the function differs from the submodularity. With this idea, several parameters are introduced, and theoretical results for submodular optimization are extended to nonsubmodular optimization. For example, the supermodular degree $\mathcal{D}_f^+$ for set function f is introduced in [13] with property that f is submodular if and only if $\mathcal{D}_f^+ = 0$.

Definition 2.1 (Supermodular Degree [13]) Consider a set function $f : 2^X \to R$. The *supermodular degree of an element* $u \in X$ by a function f is defined to be $|\mathcal{D}^+(u)|$ where

$$\mathcal{D}_f^+(u) = \{v \in X \mid \exists A \subseteq X : \Delta_u f(A \cup \{v\}) > \Delta_u f(A)\}.$$

The *supermodular degree of function* f is defined by

$$\mathcal{D}_f^+ = \max_{u \in X} |\mathcal{D}^+(u)|.$$

When only function f is studied on supmodular degree, we may simply write $\mathcal{D}^+ = \mathcal{D}_f^+$.

With this parameter, a result for monotone nondecreasing submodular function maximization with k matroid constraints is generalized to the following.

Theorem 2.2 ([15]) *There exists a greedy algorithm for monotone nondecreasing nonsubmodular maximization with k matroid constraints, having approximation performance ratio* $\frac{1}{k(\mathcal{D}_f^+ + 1) + 1}$.

The following is another example, which is a nice result with two parameters.

Definition 2.3 (Submodularity Ratio [38]) The submodularity ratio of a nonnegative set function f on 2^V is the largest scalar γ such that for any $\Omega, S \subset V$,

$$\sum_{omega \in \Omega \setminus S} \Delta_\omega f(S) \geq \gamma \cdot \Delta_\Omega f(S).$$

Definition 2.4 (Curvature [6]) The curvature of a nonnegative set function f on 2^V is the smallest scalar α such that for any $\Omega, S \subset V$ and $i \in S \setminus \Omega$,

$$\Delta_i f((S \setminus \{i\}) \cup \Omega) \geq \alpha \cdot \Delta_i f(S \setminus \{i\}).$$

Theorem 2.5 ([2]) *There exists a greedy algorithm for monotone nondecreasing nonsubmodular maximization with size constraint, having approximation performance ratio* $\frac{1}{\alpha}(1 - e^{-\alpha\gamma})$ *where* α *is the curvature of objective set function* f *and* γ *is the submodularity ratio.*

Above two results have been utilized in [49, 54], respectively. However, parameters in those applications are not clearly determined. In fact, one found that in each specific nonsubmodular optimization problem in the real-world problem, the parameter value often tends to give a performance ratio 0 for maximization and ∞ for minimization. This means that the parameter method tends to give nice theorems, but not useless.

3 Sandwich Method

After too many unsuccessful efforts have been made on establishing the approximation performance ratio for algorithms in solving nonsubmodular maximization problems, a new method, called the sandwich method, was initiated in [29] in the study of data science. Later, this method was used extensively in many research works [4, 18, 22, 24, 40, 43, 46, 47, 51–53]. To explain this method, suppose we face a problem $\max_{A \in \Omega} f(A)$ where Ω is a collection of subsets of 2^X and X is a finite set. Then the sandwich method works in the following way.

- **Input** a set function $f : 2^X \to R$.
- Initially, find two submodular functions u and l such that $u(A) \geq f(A) \geq l(A)$ for $A \in \Omega$. Then, carry out following
  1. Compute an α-approximation solution S_u for $\max_{A \in \Omega} u(A)$ and a β-approximation solution S_l for $\max_{A \in \Omega} l(A)$.
  2. Compute a feasible solution S_o for $\max_{A \in \Omega} f(A)$.
- **Output** $S = \text{argmax}(f(S_u), f(S_o), f(S_l))$.

The performance ratio of this algorithm is as follows:

Theorem 3.1 ([29]) *The solution S produced by the sandwich method satisfies the following:*

$$f(S) \geq \max\left\{\frac{f(S_u)}{u(S_u)} \cdot \alpha, \frac{opt_l}{opt_f} \cdot \beta\right\} \cdot opt_f,$$

where opt_f *(*opt_l*) is the objective function value of the optimal solution for* $\max_{A \in \Omega} f(A)$ *(*$\max_{A \in \Omega} l(A)$*).*

For a specific example, let us consider the interaction-aware viral marketing problem (Problem 1.1).

Define two set functions $B_1(S)$ and $B_2(S)$ on 2^X as follows:

$$B_1(S) = \mathbb{E}_{g \sim D}\left[\alpha \cdot |I_g(S)| + \beta \cdot \sum_{u \in I_g(S), v \in V} b(u, v)\right],$$

$$B_2(S) = \mathbb{E}_{g \sim D}[\alpha \cdot |I_g(S)|].$$

Lemma 3.2 ([19]) *Both* $B_1(S)$ *and* $B_2(S)$ *are monotone nondecreasing and submodular such that for any* $S \in 2^X$*,*

$$B_1(S) \geq f(S) \geq B_2(S).$$

Therefore, there is a $(1 - 1/e)$-approximation solution S_1 for $\max\{B_1(S) \mid |S| \leq k\}$, and there is a $(1 - 1/e)$-approximation solution S_2 for $\max\{B_2(S) \mid |S| \leq k\}$. Choose

$$S_3 = \text{argmax}(f(S_1), f(S_2)).$$

Theorem 3.3 ([19]) *The solution* S_3 *produced by above sandwich method satisfies the following:*

$$f(S_3) \geq (1 - e^{-1}) \cdot \max\left\{\frac{f(S_1)}{u(S_1)}, \frac{opt_{B_2}}{opt_f}\right\} \cdot opt_f,$$

where opt_f *(*opt_{B_2}*) is the objective function value of the optimal solution for* $\max_{A \in \Omega} f(A)$ *(*$\max_{A \in \Omega} B_2(A)$*).*

Note that in this example, we did not compute S_o. Actually, S_o is usually used to improve the performance of implementation. Here, we did not consider implementation and, hence, ignored it.

A theoretical foundation can be established from the following fundamental theorem on set functions.

Theorem 3.4 (DS (Difference of Submodular) Decomposition [26, 31]) *Every set function f can be decomposed into the difference of two monotone nondecreasing submodular functions g and h, i.e., $f = g - h$.*

Corollary 3.5 *For any set function $f : 2^X \rightarrow \mathbb{R}$, there exist two monotone nondecreasing submodular functions μ and ℓ on 2^X such that $\mu \geq f \geq \ell$.*

Proof Let $f = g - h$ where g and h are monotone nondecreasing submodular. For any set $S \in 2^X$, set $\mu(S) = g(S) - h(\emptyset)$ and $\ell(S) = g(S) - h(X)$. Then, for any set $S \in 2^X$,

$$\mu(S) \geq f(S) \geq \ell(S)$$

and μ and ℓ are monotone nondecreasing submodular. □

Although above theorem and corollary have indicated the existence of DS decomposition and the existence of monotone nondecreasing submodular upper and lower bounds, no efficient method has been found to compute them.

Open Problem 1 *Is there an efficient algorithm to produce a DS decomposition for any given set function?*

Open Problem 2 *Is there an efficient algorithm to produce a pair of monotone nondecreasing submodular upper and lower bounds for any given set function?*

Actually, even for some specific set functions, finding nontrivial such bounds is still not an easy task. Let us present an example.

Consider a directed graph $G = (V, E)$. The positive influence process is defined as follows: Every node has two states, active and in active. Before process, every node is inactive. Initially, select at most k nodes, which form a seed set A, and activate them. In each subsequent step, every inactive node v checks whether the number of active incoming neighbors is not less than the number of inactive incoming neighbors. If yes, then v becomes active. The process will end if no new inactive node becomes active. The influence spread $\sigma(A)$ is the number of active nodes at the end of process.

For this influence spread, a trivial lower bound is the zero function $\theta(A) = 0$. However, no nontrivial lower bound has been found.

Open Problem 3 *Please find a nontrivial monotone nondecreasing submodular function for the influence spread of the positive influence.*

Clearly, the disadvantage of sandwich method is that the performance ratio is really ugly, which cannot give a clean indication for the quality of approximation.

4 Stop at Local Optimal Solution

While sandwich method is used extensively in the study of data science, a class of algorithms are designed in the study of machine learning, with the following properties.

- The algorithm will terminate within finitely many iterations.
- Each iteration runs in a polynomial time.
- The algorithm will stop at a local optimal solution.

What is a local optimal solution? The definition is as follows:

Definition 4.1 (Local Optimal Solution) Consider a set function minimization problem $\min_{A \in \Omega} f(A)$ where Ω is the feasible domain. $A \in \Omega$ is called a *local minimal solution* if $A \in \Omega$ and for any A' obtained from A by deleting an element of adding an element, $f(A') \geq f(A)$. Similarly, we can define a local maximal solution.

Submodular-supermodular algorithm [31], modular-modular algorithm [26], and iterated sandwich algorithm [19, 45] are examples in this class. Those algorithms are designed based on DS decomposition of set functions or sandwich theorem as follows:

Theorem 4.2 (Sandwich Theorem [45]) *Consider a set function $f : 2^X \to R$. For any $A \subseteq X$, there exist two modular functions m_u and m_l such that $f(A) = m_u(A) = m_l(A)$ and $m_u(Y) \geq f(Y) \geq m_l(Y)$ for any $Y \in 2^X$. (A set function f is modular if $f(A) + f(B) = f(A \cup B) + f(A \cap B)$.)*

To get a little detail picture, let us consider the submodular-supermodular algorithm for unconstrained minimization $\min f(A)$ for $A \in 2^X$ is as follows.

Submodular-Supermodular Algorithm for Minimization.
Input: a set function $f : 2^X \to R$ and its DS decomposition
$f = g - h$ where g and h are submodular functions.
Output: A subset A of X.
$A \leftarrow \emptyset$;
choose a nonempty set $A^+ \subseteq X$;
while $A^+ \neq A$ **do**
$A \leftarrow A^+$;
compute a lower bound modular function m_{hl} for h such that $m_{hl}(A) = h(A$
compute a minimum solution A^+ for $\min_{Y \in 2^X} |g(Y) - m_{hl}(Y)|$;
end-while
return A.

Lemma 4.3 *Let $h : 2^X \to R$ be a submodular function. Let σ is a permutation of all elements in X such that $A = \{\sigma(1), \ldots, \sigma(|A|)\}$. Define the modular function m^{σ}_{hl} by*

$$m^{\sigma}_{hl}(Y) = h(\emptyset) + \sum_{\sigma(i)\in Y} \{h(S^{\sigma}_i) - h(S^{\sigma}_{i-1})\}$$

where $S^{\sigma}_i = \{\sigma(1), \sigma(2), \ldots, \sigma(i)\}$. *Then* m^{σ}_{hl} *is a lower bound of* h *such that for any* i, $h(S^{\sigma}_i) = m^{\sigma}_{hl}(S^{\sigma}_i)$, *and especially,* $h(A) = m^{\sigma}_{hl}(A)$.

Let Σ_A be a collection of at most $\max(|A|, |X \setminus A|)$ permutations σ of X such that $A = \{\sigma(1), \ldots, \sigma(|A|)\}$, $\sigma(|A|)$ goes over all elements of A, and $\sigma(|A| + 1)$ goes over all elements of $X \setminus A$. In the submodular-supermodular algorithm, let m_{hl} go over all m^{σ}_{hl} for $\sigma \in \Sigma_A$. Then, the algorithm will stop at a local minimal solution. Therefore, we can state as follows:

Theorem 4.4 ([31]) *The submodular-supermodular algorithm for unconstrained set function minimization stops at a local minimal solution if run each iteration in* $O(n)$ *time,*

For this class of algorithms, a serious problem is about the number of iterations. In [19, 45], it indicates that for some optimization problems, computing a local optimal solution is PLS-complete [35], that is, it is unlikely to have a polynomial-time algorithm to compute a local optimal solution. It follows that for submodular-supermodular algorithm, the number of iterations is unlikely to be bounded by a polynomial, since this algorithm is working generally for a large family of problems, containing those optimization problems for which computing a local optimal solution is PLS-complete. Therefore, it is unlikely that this class of algorithms runs in polynomial time, and hence they may not be efficient.

5 Global Convergence

We already see that all previous known algorithms and methods do not meet our expectation. Therefore, one proposed a new idea to evaluate the quality of approximation algorithms, motivated from convergence theory for nonlinear continuous optimization. Before introducing this new idea, let us recall some knowledge in nonlinear continuous optimization [9].

For convex function minimization with or without constraints, a satisfied algorithm is usually expected to generate a sequence of points which is convergent to an optimal solution. However, such an expectation cannot be realized for non-convex optimizations. Then instead, one established the following criterion.

Definition 5.1 (Global Convergence [9–11]) An algorithm for continuous optimization usually generates a sequence of infinitely many points. This sequence is said to be global convergent if every convergent subsequence converges to a Kuhn-Tucker point. In mathematical optimization, a Kuhn-Tucker (KT) point or Karush-Kuhn-Tucker (KKT) point is feasible solution which satisfies first-derivative necessary conditions for optimality.

This concept suggests a new measure for approximation performance.

6 Global Approximation of Local Optimality

This new measure is defined as follows:

Definition 6.1 (GL-Approximation) An algorithm is called a *global approximation of local optimality*, or *GL-approximation* for a brief name, if it can always produce an approximation solution within a guaranteed factor from some local optimal solution.

The following is an example of GL-approximation, obtained from modification of submodular-supermodular algorithm.

Modified Submodular-Supermodular Algorithm for Minimization.
Input: a set function $f : 2^X \to R$ and its DS decomposition
$f = g - h$ where g and h are submodular functions.
Output: A subset A of X.
choose a set $A \subseteq X$; set $ok \leftarrow 1$;
while $ok = 1$ **do**
for $\sigma \in \Sigma_A$ **do**
compute a lower bound modular function m^{σ}_{hl} for h;
compute a minimum solution A^+_σ for $\min_{Y \in 2^X}[g(Y) - m^{\sigma}_{hl}(Y)]$;
end-for
$\sigma \leftarrow \text{argmin}_{\sigma \in \Sigma_A} f(A^+_\sigma)$;
$A^+ \leftarrow A^+_\sigma$
if $f(A^+) \cdot (1 + \varepsilon) \geq f(A)$ **then** $ok \leftarrow 0$;
$A \leftarrow A^+$;
end-while
return A.

In the algorithm, Σ_A is a collection of at most $\max(|A|, |X \setminus A|)$ permutations σ of X such that $A = \{\sigma(1), \ldots, \sigma(|A|)\}$, $\sigma(|A|)$ goes over all elements of A, and $\sigma(|A| + 1)$ goes over all elements of $X \setminus A$, and m^{σ}_{hl} is a modular lower bound of h, constructed by using permutation σ (see [26] for detail.)

For this algorithm, we can prove the following result.

Theorem 6.2 *Let f be a nonnegative function. Then above algorithm runs within $O(\frac{1}{\varepsilon} \ln \zeta)$ iterations and always ends at an approximation solution with objective function value within a factor of $(1+\varepsilon)$ from a local optimal solution where ζ is the ratio of the maximum value and the minimum value of f.*

7 Non-convex Relaxation

Relaxation is a popular technique for design of approximation algorithms. Many traditional approximation algorithms are designed with LP relaxation, semi-definite relaxation, and convex relaxation. Now, non-convex relaxation may become a new development for GL approximation.

Let us consider a specific problem as follows:

Problem 7.1 (Maximum Group Set Coverage with Composed Targets [12]) *Consider m groups $\mathcal{G}_1, \mathcal{G}_2, \ldots, \mathcal{G}_m$ of subsets of a finite set X. All elements in X are partitioned into subsets $X_1, X_2, \ldots X_r$, each called a* composed target. *Suppose that for every composed target X_t, each subset in any group $\mathcal{G}_i$ covers at most one elements in X_t. The problem is to select one subset from each group to cover the maximum total number of composed targets, where a composed target is said to be covered if all elements in the composed target are covered.*

This problem can be formulated as the following nonsubmodular maximization with linear constraints (INLP).

$$
\begin{aligned}
\max\ & \sum_{t=1}^{\tau} \prod_{j\in X_t} y_j \\
\text{s.t.}\ & y_j \le \sum_{i=1}^{m} \sum_{S: j\in S\in\mathcal{G}_i} x_{iS} \quad \forall j = 1, \ldots, n, \\
& \sum_{S: S\in\mathcal{G}_i} x_{iS} \le 1 \quad \forall i = 1, \ldots, m, \\
& y_j \in \{0, 1\} \quad \forall j = 1, \ldots, n, \\
& x_{iS} \in \{0, 1\} \quad \forall S \in \mathcal{G}_i \text{ and } i = 1, 2, \ldots, m.
\end{aligned}
$$

(Note that the objective function of INLP is supermodular; see, e.g., [5].) Its relaxation is a linear constrained nonlinear programming (NLP) as follows:

$$
\begin{aligned}
\max\ & \sum_{t=1}^{\tau} \prod_{j\in X_t} y_j \\
\text{s.t.}\ & y_j \le \sum_{i=1}^{m} \sum_{S: j\in S\in\mathcal{G}_i} x_{iS} \quad \forall j = 1, \ldots, n, \\
& \sum_{S: S\in\mathcal{G}_i} x_{iS} \le 1 \quad \forall i = 1, \ldots, m, \\
& 0 \le y_j \le 1 \quad \forall j = 1, \ldots, n,
\end{aligned}
$$

$$0 \leq x_{iS} \leq 1 \quad \forall S \in \mathcal{G}_i \text{ and } i = 1, 2, \ldots, m.$$

Let (y_j^*, x_{iS}^*) be a $(1-\varepsilon)$-approximate for a local optimal solution (KKT-point) of (NLP). Let (x_{iS}, y_j) be a solution obtained from the following randomized rounding.

Randomized Rounding For each group $\mathcal{G}_i$, select one subset S (i.e., set $x_{iS} = 1$ and $x_{iS'} = 0$ for $S' \neq S$) with probability x_{iS}^* and no subset of group $\mathcal{G}_i$ is selected with probability $1 - \sum_{S \in \mathcal{G}_i} x_{iS}^*$. Set $y_j = 1$ if element j appears in a selected subset, and $y_j = 0$ otherwise.

Du et al. [12] showed the following:

Theorem 7.2 *Let (y_j, x_{iS}) be an approximate solution obtained by the above randomized rounding. Then*

$$E\left[\sum_{t=1}^{\tau} \prod_{j \in X_t} y_j\right] \geq (1 - e^{-1})^{\alpha}(1-\varepsilon)lopt$$

where lopt is the objective value of a local optimal solution of (NLP) and $\alpha = \max_{1 \leq t \leq \tau} |X_t|$.

There are several research directions raised by this result.

Open Problem 4 *What is the complexity of computing a Kuhn-Tucker point for a linear-constrained nonlinear optimization?*

Open Problem 5 *What is the complexity of computing a Kuhn-Tucker point for a linear-constrained multi-linear function maximization?*

Vavasis [42] gave an algorithm for computing an ε-KKT of a box-constrained quadratic programming with $O(n^3(M/\varepsilon)^2)$ arithmetic operations. Ye [50] presented an algorithm for computing an ε-KKT of a linearly constrained quadratic programming with $O((n^3/\varepsilon)\log(1/\varepsilon) + n \log n)$ iterations. Can those works be extended to other objective functions, especially multilinear functions? It would be an interesting effort. In fact, this is a research direction with a largely unexplored field. Research outcomes will become foundation for non-convex relaxation.

Open Problem 6 *What is the relationship between local optimal solutions of an integer programming and local optimal solutions of its non-convex relaxation?*

8 Conclusion

Design and analysis of G-L approximation (Global approximation of Local optimum) are an important new research direction in the study of nonsubmodular optimizations, with a lot of interesting open problems. In this direction, non-convex relaxation may attract attentions from not only discrete optimization but also continuous optimization. Surprising outcomes may grow from their combination.

Acknowledgments This research is supported in part by National Natural Science Foundation of China (61772154, 11771013, 11531011), Zhejiang Provincial Natural Science Foundation of China (LD19A010001). This work is also supported in part by National Science Foundation of USA under Grants 1747818, 2302469 and 2318662, alongside support from NASA under Grant 80NSSC22KM0052.

References

1. Aral, S., Walker, D.: Creating social contagion through viral product design: a randomized trial of peer influence in networks. Manag. Sci. **57**(9), 1623–1639 (2011)
2. Bian, A.A., Buhmann, J.M., Krause, A., Tschiatschek, S.: Guarantees for greedy maximization of non-submodular functions with applications. arXiv:1703.02100 (2017)
3. Bonchi, F., Castillo, C., Gionis, A., Jaimes, A.: Social network analysis and mining for business applications. ACM Trans. Intell. Syst. Technol. **2**(3), 22 (2011)
4. Chen, W., Lin, T., Tan, Z., Zhao, M., Zhou, X.: Robus Influence Maximization, KDD'16, San Francisco, CA (2016)
5. Crama, Y., Hammer, P.L.: Boolean Functions. Cambridge University Press, Cambridge (2011)
6. Das, A., Kempe, D.: Submodular meets spectral: Greedy algorithms for subset selection, sparse approximation and dictionary selection. arXiv:1102.3975 (2011)
7. Dong, L., Guo, Q., Wu, W.: Speech corpora subset selection based on time-continuous utterances features. J. Comb. Optim. **37**(4), 1237–1248 (2019)
8. Dong, L., Guo, Q., Wu, W., Satpute, M.N.: A semantic relatedness preserved subset extraction method for language corpora based on pseudo-Boolean optimization. Theor. Comput. Sci. **836**, 65–75 (2020)
9. Du, D.-Z., Pardalos, P.M., Wu, W.: Mathematical Theory of Optimization. Kluwer Academic Publishers, Dordrecht (2001)
10. Du, D.-Z., Zhang, X.-S.: A convergence theorem of Rosen's gradient projection method. Math. Program. **36**, 135–144 (1986)
11. Du, D.-Z., Zhang, X.-S.: Global convergence of Rosen's gradient projection method. Math. Program. **44**, 357–366 (1989)
12. Du, H., Zhang, Z., Duan, Z., Tian, C., Du, D.-Z.: Full view camera sensor coverage and group set coverage. In: Proceedings of 15th EAI International Wireless Internet Conference (2022)
13. Feige, U., Izsak, R.: Welfare maximization and the supermodular degree. In: ACM ITCS, pp. 247–256 (2013)
14. Feige, U., Mirrokni, V., Vondrák, J.: Maximizing nonmonotone submodular functions. In: Proceedings of the IEEE Foundations of Computer Science, pp. 461–471 (2007)
15. Feldman, M., Izsak, R.: Constrained monotone function maximization and the supermodular degree. In: ACM-SIAM SODA (2014)
16. Feldman, M., Naor, J., Schwartz, R.: A unified continuous greedy algorithm for submodular maximization. In: IEEE FOCS, pp. 570–579 (2011)
17. Fujishige, S.: Submodular Functions and Optimization. Annals of Discrete Mathematics, vol. 58. Elsevier Science, Amsterdam (2005)
18. Gai, L., Du, H., Wu, L., Zhu, J., Bu, Y.: Blocking rumor. Cut. J. Comb. Optim. **36**(2), 392–399 (2018)
19. Gao, C., Gu, S., Yang, R., Yu, J., Wu, W., Xu, D.: Interaction-aware influence maximization and iterated sandwich method. Theor. Comput. Sci. **821**, 23–33 (2020)

20. Goldengorin, B.: Maximization of submodular functions: theory and enumeration algorithms. Eur. J. Oper. Res. **198**, 102–112 (2009)
21. Grötschel, M., Lovász, L., Schrijver, A.: Geometric Algorithms and Combinatorial Optimization, 2nd edn. Springer-Verlag, Berlin (1988)
22. Guo, J., Wu, W.: A novel scene of viral marketing for complementary products. IEEE Trans. Comput. Soc. Syst. **6**(4), 797–808 (2019)
23. Guo, J., Li, Y., Wu, W.: Targeted protection maximization in social networks. IEEE Trans. Netw. Sci. Eng. **7**(3), 1645–1655 (2020)
24. Guo, J., Chen, T., Wu, W.: A multi-feature diffusion model: rumor blocking in social networks. IEEE/ACM Trans. Netw. **29**(1), 386–397 (2021)
25. Habiba, H., Yu, Y., Berger-Wolf, T.Y., Saia, J.: Finding spread blockers in dynamic networks, ser. SNAKDD 2008, pp. 55–76 (2010)
26. Iyer, R., Bilmes, J.: Algorithms for approximate minimization of the difference between submodular functions. In: Proceedings of the UAI (2012)
27. Kimura, M., Saito, K., Motoda, H.: Blocking links to minimize contamination spread in a social network. ACM Trans. Knowl. Discov. Data **3**(2), 9:1–9:23 (2009)
28. Lee, J., Mirrokni, V., Nagarajan, V., Sviridenko, M.: Nonmonotone submodular maximization under matroid and knapsack constraints. In: Proceedings of the ACM Symposium on Theory of Computing, pp. 323–332 (2009)
29. Lu, W., Chen, W., Lakshmanan, L.V.S.: From competition to complementarity: comparative influence diffusion and maximization. Proc. VLDB Endow. **9**(2), 60–71 (2015)
30. Murota, K.: Discrete Convex Analysis. SIAM, Philadelphia (2003)
31. Narasimhan, M., Bilmes, J.: A submodular-supermodular procedure with applications to discriminative structure learning. In: Proceedings of the UAI (2005)
32. Nemhauser, G.L., Wolsey, L.A., Fisher, M.L.: An analysis of approximations for maximizing submodular set functions - I. Math. Program. **14**(1), 265–294 (1978)
33. Nguyen, N.P., Yan, G., Thai, M.T., Eidenbenz, S.: Containment of misinformation spread in online social networks, ser. WebSci 2012, pp. 213–222. ACM, New York (2012)
34. Orlin, J.B.: A faster strongly polynomial time algorithm for submodular function minimization. Math. Program. **118**, 237–251 (2009)
35. Schäfer, A.A., Yannakakis, M.: Simple local search problems that are hard to solve. SIAM J. Comput. **20**, 56–87 (1991)
36. Schrijver, A.: A combinatorial algorithm minimizing submodular functions in strong polynomial time. J. Comb. Theory **80**, 346–355 (2000)
37. Stibe, A.: Exploring social influence and incremental online Persuasion on Twitter: a longitudinal study. In: Mobile Web Information Systems, pp. 286–300. Springer International Publishing, Cham (2014)
38. Sviridenko, W.J., Vondrk, M.J.: Optimal approximation for submodular and supermodular optimization with bounded curvature. Math. Oper. Res. **42**(2), 1197–1218 (2017)
39. Svitkina, Z., Fleischer, L.: Submodular approximation: sampling-based algorithms and lower bounds. SIAM J. Comput. **40**, 1715–1737 (2011)
40. Tong, G.A., Du, D.-Z., Wu, W.: On misinformation containment in online social networks. NeurIPS, 339–349 (2018)
41. Tripathy, R.M., Bagchi, A., Mehta, S.: A study of rumor control strategies on social networks, ser. CIKM 2010. ACM, New York (2010)
42. Vavasis, S.A.: Nonlinear Optimization: Complexity Issues. Oxford Science, New York (1991)
43. Wang, Z., Yang, Y., Pei, J., Chen, E.: Activity maximization by effective information diffusion in social networks. arXiv: 1610.07754v1 [cs.SI] (2016)
44. Wolsey, L.A.: An analysis of the greedy algorithm for the submodular set covering problem. Combinatorica **2**(4), 385–393 (1982)
45. Wu, W., Zhang, Z., Du, D.-Z.: Set function optimization. J. Oper. Res. Soc. China **7**(2), 183–193 (2019)
46. Yan, R., Li, Y., Wu, W., Li, D., Wang, Y.: Rumor blocking through online link deletion on social networks. ACM Trans. Knowl. Discov. Data **13**(2), 16:1–16:26 (2019)

47. Yan, R., Li, D., Wu, W., Du, D.-Z., Wang, Y.: Minimizing influence of rumors by blockers on social networks: algorithms and analysis. IEEE Trans. Netw. Sci. Eng. **7**(3), 1067–1078 (2020)
48. Yang, W., Ma, J., Li, Y., Yan, R., Yuan, J., Wu, W., Li, D.: Marginal gains to maximize content spread in social networks. IEEE Trans. Comput. Soc. Syst. **6**(3), 479–490 (2019)
49. Yang, W., Yuan, J., Wu, W., Ma, J., Du, D.-Z.: Maximizing activity profit in social networks. IEEE Trans. Comput. Soc. Syst. **6**(1), 117–126 (2019)
50. Ye, Y.: On the complexity of approximating a KKT point of quadratic programming. Math. Program. **80**, 195–211 (1998)
51. Zhang, Y., Guo, J., Yang, W., Wu, W.: Targeted activation probability maximization problem in online social networks. IEEE Trans. Netw. Sci. Eng. **8**(1), 294–304 (2021)
52. Zhu, J., Zhu, J., Ghosh, S., Wu, W., Yuan, J.: Social influence maximization in hypergraph in social networks. IEEE Trans. Netw. Sci. Eng. **6**(4), 801–811 (2019)
53. Zhu, J., Ghosh, S., Wu, W.: Group influence maximization problem in social networks. IEEE Trans. Comput. Soc. Syst. **6**(6), 1156–1164 (2019)
54. Zhu, J., Ghosh, S., Wu, W.: Robust rumor blocking problem with uncertain rumor sources in social networks. World Wide Web **24**(1), 229–247 (2021)

DR-Submodular Maximization and Its Application

Wenguo Yang, Shengminjie Chen, and Xiaoming Sun

1 Introduction

The DR-submodular function, characterized by diminishing returns property, has naturally emerged in numerous theoretical studies and practical applications. While DR-submodularity is equivalent to submodularity in the context of set functions, it generally represents a stronger condition for continuously differentiable functions. Not strictly speaking, DR-submodularity signifies diminishing marginal returns for every pair with ordered relations, whereas submodularity pertains to diminishing marginal returns in some special pairs with ordered relations. Conventionally, there is no distinction between submodular functions and DR-submodular functions when discussing set functions.

While minimizing unconstraints submodular set function is polynomially solvable [43], maximizing submodular set function is NP-hard. Seminal works were proposed by Nemhauser et al. [56] and Fisher et al. [35], who proved the conventional greedy algorithm is a $1 - e^{-1}$ approximation algorithm for monotone submodular set functions under cardinality constraint and a $\frac{1}{1+p}$ approximation algorithm for monotone submodular set functions under the intersection of p-matroids, respectively. Building on this foundation, [28] demonstrated that the

W. Yang (✉)
School of Mathematical Sciences, University of Chinese Academy of Sciences, Beijing, China
e-mail: yangwg@ucas.ac.cn

S. Chen · X. Sun
State Key Lab of Processors, Institute of Computing Technology, Chinese Academy of Sciences, Beijing, China

School of Computer Science and Technology, University of Chinese Academy of Sciences, Beijing, China
e-mail: csmj@ict.ac.cn; sunxiaoming@ict.ac.cn

B. Goldengorin (ed.), *Theory, Algorithms, and Experiments in Applied Optimization*, Springer Optimization and Its Applications 226,
https://doi.org/10.1007/978-3-031-91357-0_18

approximation ratio is impossible greater than $1 - e^{-1}$ for monotone submodular set function maximization subject to cardinality constraints under the assumption NP $\neq$ P. Filmus and Ward [34] embedded the non-oblivious local search within the greedy algorithm, achieving a $1 - e^{-1}$ approximation with $O(nr^3\epsilon^{-3}\log r)$ complexity for maximizing monotone submodular set function under matroid constraints. Although the greedy algorithm provides the tightest theoretical guarantee for maximizing monotone submodular set functions subject to cardinality constraints, it cannot be adopted directly for knapsack constraints. Sviridenko [63] combined the partial enumeration technique with the greedy algorithm, resulting in a $1-e^{-1}$ approximation algorithm with $O(n^5)$ complexity for monotone submodular set functions under knapsack constraints. Following these breakthrough works, the submodular set function maximization problem has become a hot issue in operation research and theoretical computer sciences, leading to further results, such as monotone submodular set functions under multiple knapsack constraints [61], k-exchange system constraints [32, 68], etc.

Although the aforementioned algorithms achieve the tight approximation ratios for maximizing submodular set functions, their complexities make them impractical for large-scale problems. Minoux [52] proposed a lazy greedy algorithm to accelerate the conventional greedy algorithm. Badanidiyuru and Vondrák [3] designed a rough version of the greedy algorithm, namely, threshold decrease algorithm, which is a $1 - e^{-1} - \epsilon$ approximation algorithm with $O(\frac{n}{\epsilon}\log\frac{n}{\epsilon})$ complexity for maximizing submodular set functions under cardinality constraints. Furthermore, [53] developed a much lazier greedy algorithm that maintains the same approximation ratio while achieving nearly linear complexity of $O(n\log\frac{1}{\epsilon})$. Additionally, [12] presented a deterministic non-oblivious local search algorithm for monotone submodular set functions under matroid constraints, keeping the $1 - e^{-1} - \epsilon$ approximation ratio while reducing the complexity to $\tilde{O}_{\epsilon}(nr)$. For knapsack constraints, [47] and [33] reduced the size of partially enumerated solution from 3 to 2 resulting in the complexity reduction to $O(n^4)$.

When the submodular set function lacks monotonicity, [13] introduced a random greedy algorithm with the $1/e$ approximation ratio under cardinality constraints. Chen et al. [23] and Tukan et al. [65] improved the approximation ratio of combinatorial algorithms from $1/e$ to 0.385 for cardinality constraints. Additionally, [62] proposed a deterministic algorithm with 0.283 and $1/4$ approximation ratios for matroid constraints and knapsack constraints, respectively. Chen et al. [23] also designed a random 0.305 approximation algorithm for matroid constraints. Feige et al. [29] proposed a $1/3$ approximation algorithm and a $1/4$ approximation algorithm for maximizing unconstrained non-monotone submodular set functions, demonstrating that the hardness results are $1/2$. On this basis, [14] adopted a double greedy algorithm to maximize unconstrained non-monotone submodular set functions, where the random version achieves a $1/2$ approximation ratio and the deterministic version has the $1/3$ approximation algorithm ratio. On the other hand, [38, 40] investigated the relationship between the local maxima and global maxima and proposed a preliminary preservation algorithm (PPA) to obtain the smallest subinterval $[S, T]$ of $[\emptyset, N]$ containing a global maximum.

However, the aforementioned algorithms are combinatorial algorithms. Over the last two decades, continuous optimization tools have played an important role in maximizing submodular set functions. Lovasz [49] first proposed the Lovasz extension, proving that the Lovasz extension of submodular set functions is a convex function. Unfortunately, the Lovasz extension is non-smooth. Vondrak [66] introduced a smooth extension, namely, the multilinear extension, and proposed a $1 - e^{-1}$ approximation algorithm for social welfare problem. Building on the multilinear extension, [16] developed a continuous greedy algorithm that is the $1 - e^{-1}$ approximation algorithm for the multilinear extension of monotone submodular set functions subject to down-closed constraints. Combining the pipage rounding [1], the nonintegral solution in $[0, 1]^n$ can be rounded to a $\{0, 1\}^n$-binary solution without loss in the expected sense under matroid constraints. Inspired by these breakthrough works, numerous studies on maximizing non-monotone submodular set functions has been emerged. Feldman et al. [31] designed a unified continuous algorithm, namely, measured continuous greedy, for maximizing submodular set functions subject to down-closed constraints. This algorithm achieves an approximation ratio is $1/e$ for non-monotone cases and $1 - 1/e$ for monotone cases. Because cardinality constraints, matroid constraints, and knapsack constraints are down-closed, the measured continuous greedy yields a non-integer approximation solution under these three constraints. Using the pipage rounding [1], swap rounding [17], and contention resolution schemes [18], these non-integer solutions can be rounded to integer solutions with strong theoretical guarantees. Furthermore, [14] introduced a continuous double greedy to maximizing the multilinear extension of non-monotone submodular set function, which is $1/2$ approximation algorithm for unconstrained cases. It is important to note that the solution returned by these algorithms is not always stationary points. Chekuri et al. [18] proved that the best stationary point in a two-stage local search and stationary points in the reduced feasible domain are $1/4$ and 0.309 approximation solutions, respectively.

These algorithms are independently searching for a stationary point, the implementation of continuous greedy algorithm, and the implementation of double greedy algorithm. Buchbinder et al. [13] first introduced the combination of continuous double greedy and random greedy, achieving a $e^{-1}+0.004$ approximation algorithm for maximizing submodular set functions under cardinality constraints. Building on this idea of composition, [27] adopted the convex combination of the solution returned by the double greedy and the measured continuous greedy, resulting in a 0.372 approximation ratio for down-closed constraints. Buchbinder and Feldman [10] proposed a 0.385 approximation algorithm for down-closed constraints that combines the local search with the measured continuous greedy.

From the upper bound of the approximation ratio, a significant theoretical gap remains in the problem of maximizing non-monotone submodular set functions. Gharan and Vondrák [36] first illustrated that no algorithm can achieve an approximation better than 0.478 under the partition matroid constraints. Qi [60] extended this inapproximability result to cardinality constraints. Collectively, these hardness results imply that there cannot be an approximation algorithm for maximizing non-monotone submodular set functions under down-closed constraints under the assumption P $\neq$ NP, whose approximation ratio is greater than 0.478.

All the aforementioned works consider submodular function maximization from the perspective of discrete optimization. Although the multilinear extension of submodular set functions is a continuous differentiable DR-submodular functions defined in $[0, 1]^n$, it represents a specific type of DR-submodular functions characterized by having zero-diagonal elements in its Hessian and nonpositive off-diagonal elements. In contrast, for general DR-submodular functions, all elements of the Hessian are nonpositive. Consequently, it is important to consider whether algorithms based on the multilinear extension of submodular set functions can be directly adopted for general DR-submodular functions.

Recently, DR-submodular optimization has also become a hot topic in continuous optimization. On one hand, [2] proved that minimizing DR-submodular functions is polynomially solvable. On the other hand, the NP-hardness of maximizing submodular set functions implies that maximizing DR-submodular functions is also NP-hard. Leveraging the nonpositive Hessian elements, [6] proposed a variant of Frank-Wolfe with the $1 - 1/e - \epsilon$ approximation ratio and $O(\frac{1}{\epsilon})$ iterations for maximizing monotone DR-submodular functions under down-closed and convex constraints. They also introduced a continuous double greedy with $1/3$ approximation ratio and linear computational complexity for maximizing unconstraints non-monotone DR-submodular functions. On this basis, [57] improved the approximation ratio of maximizing unconstrained non-monotone DR-submodular functions from $1/3$ to $1/2$, aligning with the hardness results for submodular set functions. As for maximizing non-monotone DR-submodular functions under down-closed convex constraints, [5] extended the $1/4$ and $1/e$ approximation algorithms from the multilinear extension of submodular set functions to general DR-submodular functions. Chen et al. [20] proposed a unified analysis about the 0.385 approximation algorithm, demonstrating its applicability to general DR-submodular functions. Furthermore, [11] improved the approximation ratio to 0.401 by combining local search, double greedy, and a variant of Frank-Wolfe while introducing recursion. When combined with a theoretically guaranteed rounding method, this algorithm also serves as a 0.401 approximation algorithm for maximizing non-monotone submodular set functions under matroid constraints and knapsack constraints. When the feasible domain is convex but releases the down-closeness, [26] established that Frank-Wolfe is $\frac{1}{3\sqrt{3}}$ approximation algorithm for maximizing non-monotone DR-submodular functions. Du et al. [25] offered a tight analysis for Frank-Wolfe, enhancing the approximation ratio to $1/4$. In addition, [58] proposed a unified approach for maximizing DR-submodular functions, which recover the $1/e$ and $1/4$ approximation algorithms. Du [24] introduced a framework to design and analyze approximation algorithms for maximizing DR-submodular functions, namely, Lyapunov framework. On the inapproximability side, [55] showed that no polynomial algorithm can beat the approximation ratio $1/4$ under the assumption P $\neq$ NP.

From the perspective of stationary points, [42] demonstrated that stationary points for monotone DR-submodular functions yield a $1/2$ approximation solution with this ratio being tight. To beat the $1/2$ theoretical guarantee associated with stationary points, [69] constructed a non-oblivious function of monotone DR-submodular functions, where stationary points provide a $1 - e^{-1}$ approximation solution for the original DR-submodular functions. In contrast, [20] illustrated that stationary points can exhibit arbitrarily poor approximation ratios in non-monotone cases by a counterexample. Although stationary points cannot provide theoretical guarantees for non-monotone cases, [70] devised a non-oblivious function of DR-submodular functions for which stationary points achieve a $\frac{1-\|\mathbf{x}_0\|_\infty}{4}$ approximation solution for the original DR-submodular functions.

Leveraging to the aforementioned approximation algorithms, various application fields have flourished, including influence maximization in social networks [45], quadratic cost partition problem [39], sensor placement [46], document summarization [48], determinantal point processes [37], among others. These compelling applications have, in turn, spurred advancements in the design of approximation algorithms for DR-submodular maximization. Notable developments include algorithms for submodular set functions minus modular set functions [41, 50], the difference between submodular set functions [19, 22, 44], the sum of DR-submodular functions and linear functions [8, 33], the sum of DR-submodular functions and concave functions [54], the ratio between two DR-submodular functions [4, 21, 59], etc.

Although [9] has summarized some approximation algorithms for submodular set functions, they did not take into account continuous DR-submodular functions. In addition, it has been more than several years. Numerous important works have been published. Consequently, in this work, our main contribution is to retrospect a spectrum of algorithms tailored for maximizing DR-submodular functions, covering both discrete and continuous domains, while providing proof sketches to illustrate their approximation ratios. In addition, we briefly outline some potential directions and applications for the DR-submodular maximization problem.

The remainder of this work is organized as follows. Section 2 introduces some essential definitions and lemmas and describes the formal statement of DR-submodular maximization problem. Section 3 discusses algorithms, and their proof sketches for maximizing submodular set functions. In Sect. 4, we review the approximation results for maximizing continuous differentiable DR-submodular functions. Finally, we outline some future directions and applications, albeit not exhaustively.

2 Preliminary

In this section, we provide a detailed explanation of notations, operation symbols, definitions, and lemmas that are relevant to the analysis of algorithms.

Consider a ground set N with n elements and a real-valued function f defined over 2^N, i.e., $f : 2^N \rightarrow \mathbb{R}$, if f is monotone, then for $\forall X \subseteq Y \subseteq N$, it holds that $f(X) \leq f(Y)$. Additionally, if f is submodular, then it satisfies the following definition:

Definition 1 (Submodular Set Function) $f : 2^N \rightarrow \mathbb{R}$ is a submodular set function, if for $\forall X, Y \subseteq N$, the following inequality holds:

$$f(X) + f(Y) \geq f(X \cap Y) + f(X \cup Y) \tag{1}$$

For set functions, there is an equivalent definition for submodular set functions known as the diminishing return property. Briefly, this property describes that the marginal increment from adding an element to a subset of the ground set is nonincreasing with the size of subset increasing. The formal statement is as follows:

Definition 2 (Diminishing Return Property) $f : 2^N \rightarrow \mathbb{R}$ is a submodular set function, if for $\forall X \subseteq Y \subseteq N$ and $e \in N \setminus Y$, the following inequality holds:

$$f(X \cup \{e\}) - f(X) \geq f(Y \cup \{e\}) - f(Y) \tag{2}$$

For convenience, let $f(e|X)$ and $f(Y|X)$ be the marginal gain of adding element e and set Y to the current set X, i.e., $f(e|X) = f(X \cup \{e\}) - f(X)$ and $f(Y|X) = f(X \cup Y) - f(X)$. Therefore, the diminishing return property and the monotonicity can be expressed by $f(e|X) \geq f(e|Y), \forall X \subseteq Y \subseteq N, \forall e \in N \setminus Y$ and $f(e|X) \geq 0$, respectively. Next, we show that the diminishing return property is equivalent with submodulartiy for set functions.

Lemma 1 *Definition 1* $\Leftrightarrow$ *Definition 2*

Proof

(1) Definition 1 $\Rightarrow$ Definition 2: $\forall A \subseteq B \subseteq N$ and $\forall e \in N \setminus B$, denote $X = A \cup \{e\}$ and $Y = B$.
(2) Definition 2 $\Rightarrow$ Definition 1: $\forall A, B \subseteq N$, denote $L = A \setminus A \cap B = \{e_1, \ldots, e_m\}$, then $f(A) - f(A \cap B) = \sum_{i=1}^{m} f((A \cap B) \cup \{e_1, \ldots, e_i\}) - f((A \cap B) \cup \{e_1, \ldots, e_{i-1}\}) \geq \sum_{i=1}^{m} f(B \cup \{e_1, \ldots, e_i\}) - f(B \cup \{e_1, \ldots, e_{i-1}\}) = f(A \cup B) - f(B)$.

□

To extend the submodular set function from $\{0, 1\}^n$ to $[0, 1]^n$, we first introduce the definition of threshold set and random set. On this basis, the Lovasz extension and the multilinear extension are as follows:

Definition 3 (Threshold Set) $T_\lambda(\mathbf{x})$ is a threshold subset of N depending on a vector $\mathbf{x} \in [0, 1]^n$. For the threshold $\lambda \in [0, 1]$, if the element e in the ground set N has the component $x_e \geq \lambda$ of the vector $\mathbf{x}$, then the element e is in $T_\lambda(\mathbf{x})$.

$$T_\lambda(\mathbf{x}) = \{e \in N | x_e \geq \lambda\} \tag{3}$$

Definition 4 (Random Set) $R(\mathbf{x})$ is a random subset of N depending on a vector $\mathbf{x} \in [0, 1]^n$. For the element e in the ground set N, it is in $R(\mathbf{x})$ with the probability x_e. The elements are independent of each other. Then, the probability of $R(\mathbf{x})$ is as follows:

$$Pr[R(\mathbf{x})] = \prod_{i \in R(\mathbf{x})} x_i \prod_{j \notin R(\mathbf{x})} (1 - x_j) \tag{4}$$

Definition 5 (Lovasz Extension) For a given function defined over the ground set N with n elements, i.e., $f : \{0, 1\}^n \to \mathbb{R}$, its Lovasz extension $F_L : [0, 1]^n \to \mathbb{R}$ is as follows:

$$F_L(\mathbf{x}) = \int_0^1 F(T_\lambda(\mathbf{x}))d\lambda \tag{5}$$

Definition 6 (Multilinear Extension) For a given function defined over the ground set N with n elements, i.e., $f : \{0, 1\}^n \to \mathbb{R}$, its multilinear extension $F_{ML} : [0, 1]^n \to \mathbb{R}$ is as follows:

$$F_{ML}(\mathbf{x}) = \mathbb{E}_{R(\mathbf{x})}[f(R(\mathbf{x}))] = \sum_{S \subseteq N} f(S) \prod_{i \in S} x_i \prod_{j \notin S} (1 - x_j) \tag{6}$$

It is straightforward to see that the multilinear extension is smooth, and its partial derivatives have the following closed-form expression:

$$\begin{aligned}
\frac{\partial F_{ML}(\mathbf{x})}{\partial x_e} &= \sum_{S \subseteq N \setminus \{e\}} f(e|S) \prod_{i \in S} x_i \prod_{j \notin S, j \neq e} 1 - x_j = \frac{F_{ML}(\mathbf{x} \vee \mathbf{1}_e) - F_{ML}(\mathbf{x})}{1 - x_e} \\
\frac{\partial F_{ML}^2(\mathbf{x})}{\partial x_e \partial x_m} &= \sum_{S \subseteq N \setminus \{e, m\}} \Big[f(m|S \cup \{e\}) - f(m|S) \Big] \prod_{i \in S} x_i \prod_{j \notin S, j \neq e, j \neq m} 1 - x_j \leq 0 \\
\frac{\partial F_{ML}^2(\mathbf{x})}{\partial^2 x_e} &= 0
\end{aligned} \tag{7}$$

where $(\mathbf{x} \vee \mathbf{y})_i = \max\{x_i, y_i\}$ and $(\mathbf{x} \wedge \mathbf{y})_i = \min\{x_i, y_i\}$ for $\forall i$. $\mathbf{1}_e$ is a unit vector that the component of e is 1 and otherwise 0. In addition, there is a famous inequality between the Lovasz extension and the multilinear extension.

Lemma 2 (Lemma A.4 [67]) *For a submodular set functions $f : \{0, 1\}^n \to \mathbb{R}$, the following inequality holds between its Lovasz extension and its multilinear extension.*

$$F_{ML}(\mathbf{x}) \geq F_L(\mathbf{x}) \tag{8}$$

Proof Assuming the decreasing order $1 = x_0 \geq x_1 \geq x_2 \geq \ldots \geq x_n \geq 0$, it is easy to see that the Lovasz extension can be rewritten as follows:

$$\begin{aligned}
F_L(\mathbf{x}) &= \sum_{i=0}^{n}(x_i - x_{i+1}) f([i]) = f(\emptyset) + \sum_{i=1}^{n} x_i (f([i]) - f([i-1])) \\
&= f(\emptyset) + \sum_{i=1}^{n} x_i f(i|[i-1]) = f(\emptyset) + \sum_{i=1}^{n} \mathbb{E}\left[f(R(\mathbf{x}) \cap \{i\}|[i-1])\right] \\
&\leq \mathbb{E}_{R(\mathbf{x})}\left[f(R(\mathbf{x}))\right] = F_{ML}(\mathbf{x})
\end{aligned}$$

where $[i] = \{1, 2, \ldots, i\}$ □

For convenience, let F be the multilinear extension of the submodular set function of f for the following description. Next, we introduce some lemmas that are widely used in approximation algorithm design for submodular set functions.

Lemma 3 ([30]) *Considering two vectors* $\mathbf{x}, \mathbf{x}' \in [0, 1]^n$ *such that for every* $e \in N$, $|x_e - x'_e| \leq \delta$, *then*

$$F(\mathbf{x}') - F(\mathbf{x}) \geq \sum_{e \in N}(x'_e - x_e) \cdot \partial_e F(\mathbf{x}) - O(n^3\delta^2) \cdot F(\mathbf{x}^*) \tag{9}$$

Lemma 4 ([29]) *Let* $f : 2^N \to \mathbb{R}$ *be a submodular set function. Denote by* $A(p)$ *a random subset of* A, *where each element appears with probability* p. *Then,*

$$\mathbb{E}[f(A(p))] \geq (1-p) f(\emptyset) + p f(A) \tag{10}$$

Proof Using the inductions: first, $A = \emptyset$, the lemma is trivial. Suppose $A = A' \cup \{e\}$, then $A'(p) = A(p) \cap A'$ and

$$\begin{aligned}
\mathbb{E}[f(A(p))] &= \mathbb{E}[f(A'(p))] + \mathbb{E}[f(A(p)) - f(A'(p))] \\
&= \mathbb{E}[f(A'(p))] + \sum_{A'(p)} Pr[A'(p)] p f(e|A'(p)) \\
&\geq \mathbb{E}[f(A'(p))] + p f(e|A') \\
&\geq (1-p) f(\emptyset) + p f(A') + p(f(A) - f(A'))
\end{aligned}$$

□

Lemma 5 ([13]) *Let* $f : 2^N \to \mathbb{R}$ *be a submodular set function. Denote by* $A(p)$ *a random subset of* A, *where each element appears with probability at most* p. *Then,*

$$\mathbb{E}[f(A(p))] \geq (1-p) f(\emptyset) \tag{11}$$

Proof Let $A = \{e_1, \ldots, e_m\}$ be a set with the descent order by the probability with elements, i.e., $Pr[e_i \in A(p)] = p_i \geq Pr[e_{i+1} \in A(p)] = p_{i+1}$. Denote $A_k = \{e_1, e_2, \ldots, e_k\}$. Then

$$\begin{aligned}
\mathbb{E}[f(A(p))] &= \mathbb{E}\left[f(\emptyset) + \sum_{i=1}^{m} f(e_i | A_{i-1} \cap A(p))\right] \\
&\geq f(\emptyset) + \mathbb{E}_{e_i}\left[\sum_{i=1}^{m} f(e_i | A_{i-1})\right] \\
&= f(\emptyset) + \sum_{i=1}^{m} p_i f(e_i | A_{i-1}) \\
&= (1 - p_1) f(\emptyset) + \sum_{i=1}^{m-1} [p_{i-1} - p_i] f(A_i) + p_m f(A) \geq (1 - p) f(\emptyset)
\end{aligned}$$

□

Since the previous description of DR-submodular function pertains only to set function, next, we introduce the definition of DR-submodular function in the n-dimensional real vector space, along with related lemmas.

Definition 7 (Submodular Function [64]) For a given function F defined over the subset of $\mathbb{R}^n$, i.e., $F : \mathcal{X} \subseteq \mathbb{R}^n \to \mathbb{R}$, if F is submodular, then for $\forall \mathbf{x}, \mathbf{y} \in \mathcal{X}$, the following inequality holds:

$$F(\mathbf{x}) + F(\mathbf{y}) \geq F(\mathbf{x} \vee \mathbf{y}) + F(\mathbf{x} \wedge \mathbf{y}) \tag{12}$$

where $(\mathbf{x} \vee \mathbf{y})_i = \max\{x_i, y_i\}$ and $(\mathbf{x} \wedge \mathbf{y})_i = \min\{x_i, y_i\}$ for $\forall i$.

While submodularity is equivalent to the diminishing return property for set functions, generally speaking, the diminishing return property is stronger than submodularity on n-dimensional real vector space. Consequently, the diminishing return property is also referred to as DR-submodularity, and submodularity can be equivalently defined by a weaker form of diminishing marginal returns. The formal description is as follows:

Definition 8 (Weak DR-Submodular Function [6]) The weak DR-submodular function F is defined over $\mathcal{X}$, i.e., $F : \mathcal{X} \to \mathbb{R}$. For $\forall \mathbf{x} \leq \mathbf{y} \in \mathcal{X}$ and $i \in [n]$ with $x_i = y_i$, if $\forall k \in \mathbb{R}_+$ such that $\mathbf{x} + k\mathbf{1}_i, \mathbf{y} + k\mathbf{1}_i \in \mathcal{X}$, then the following inequality holds:

$$F(\mathbf{x} + k\mathbf{1}_i) - F(\mathbf{x}) \geq F(\mathbf{y} + k\mathbf{1}_i) - F(\mathbf{y}) \tag{13}$$

If F is twice-differentiable, the off-diagonal elements of its Hessian are nonpositive.

$$\frac{\partial^2 F(\mathbf{x})}{\partial x_i \partial x_j} \leq 0 \quad \forall i, j \in [n], i \neq j \tag{14}$$

Definition 9 (DR-Submodular Function [6, 51]) The DR-submodular function F is defined over $\mathcal{X}$, i.e., $F : \mathcal{X} \to \mathbb{R}$. For $\forall \mathbf{x} \leq \mathbf{y} \in \mathcal{X}$ and $\forall i \in [n]$, if $\forall k \in \mathbb{R}_+$ such that $\mathbf{x} + k\mathbf{1}_i, \mathbf{y} + k\mathbf{1}_i \in \mathcal{X}$, then the following inequality holds:

$$F(\mathbf{x} + k\mathbf{1}_i) - F(\mathbf{x}) \geq F(\mathbf{y} + k\mathbf{1}_i) - F(\mathbf{y}) \tag{15}$$

If F is twice-differentiable, all elements of its Hessian are nonpositive.

$$\frac{\partial^2 F(\mathbf{x})}{\partial x_i \partial x_j} \leq 0 \quad \forall i, j \in [n] \tag{16}$$

Lemma 6 ([6]) *The weak DR-submodularity* $\Leftrightarrow$ *The submodularity.*

Proof Similar with the proof of Lemma 1.

(1) submodularity $\Rightarrow$ weak DR-submodularity: $\forall \mathbf{a} \leq \mathbf{b} \in \mathcal{X}, i \in [n]$ with $a_i = b_i$ and $\forall k \in \mathbb{R}_+$ such that $\mathbf{a} + k\mathbf{1}_i, \mathbf{b} + k\mathbf{1}_i \in \mathcal{X}$, denote $\mathbf{x} = \mathbf{a} + k\mathbf{1}_i$ and $\mathbf{y} = \mathbf{b}$.
(2) weak DR-submodularity $\Rightarrow$ submodularity: $\forall \mathbf{a}, \mathbf{b} \in \mathcal{X}$, denote $\mathbf{x}_0 = \mathbf{a} \wedge \mathbf{b}$, $\mathbf{y}_0 = \mathbf{b}$, $L = \{i \in [n] | a_i > b_i\} = \{e_1, \ldots, e_m\}$, $\mathbf{x}_i = \mathbf{x}_{i-1} + (a_{e_i} - b_{e_i})\mathbf{1}_{e_i}$ and $\mathbf{y}_i = \mathbf{y}_{i-1} + (a_{e_i} - b_{e_i})\mathbf{1}_{e_i}$, then $F(\mathbf{a}) - F(\mathbf{a} \wedge \mathbf{b}) = \sum_{i=1}^{m} F(\mathbf{x}_i) - F(\mathbf{x}_{i-1}) \geq \sum_{i=1}^{m} F(\mathbf{y}_i) - F(\mathbf{y}_{i-1}) = F(\mathbf{a} \vee \mathbf{b}) - F(\mathbf{b})$.

□

Without causing ambiguity, we denote $F(\mathbf{y}|\mathbf{x}) = F(\mathbf{y} \vee \mathbf{x}) - F(\mathbf{x})$. The monotonicity condition can be rewritten as $F(\epsilon\mathbf{1}_e|\mathbf{x}) \geq 0$ for $\forall \mathbf{x} \in \mathcal{X}$ and $\forall \mathbf{x} + \epsilon\mathbf{1}_e \in \mathcal{X}$. From the definition of DR-submodularity, it implies that inverse monotonicity of the gradient, i.e., $\forall \mathbf{x} \leq \mathbf{y}$, $\nabla F(\mathbf{x}) \geq \nabla F(\mathbf{y})$. In addition, weak DR-submodular functions induce diminishing returns in some special dimensions where $\mathbf{x}$ and $\mathbf{y}$ have the same component, while DR-submodular functions exhibit diminishing marginal returns in all dimensions in two vectors. In terms of the Hessian matrix, DR-submodularity imposes an additional restriction compared to weak DR-submodularity, specially that the diagonal elements of Hessian must be nonpositive. Consequently, there is a relationship between DR-submodularity and weak DR-submodularity. Because the proof is intuitive based on the property of Hessian, we omit it.

Lemma 7 ([6, 51]) *The function F is a DR-submodular function if and only if F is submodular function and coordinate-wise concave.*

It is evident that the multilinear extension of submodular set functions is DR-submodular due to the nonpositive elements of its Hessian. In light of these nonpositive elements of Hessian, several important lemmas have been widely utilized in the design of approximation algorithms for maximizing DR-submodular functions.

Lemma 8 ([6, 18, 51]) *$F : \mathcal{X} \to \mathbb{R}$ is DR-submodular, then F is concave along with all nonnegative directions, i.e., for $\forall \mathbf{x} \le \mathbf{y} \in \mathcal{X}$ and $\forall \lambda \in [0, 1]$,*

$$F(\lambda \mathbf{x} + (1-\lambda)\mathbf{y}) \ge \lambda F(\mathbf{x}) + (1-\lambda)F(\mathbf{y}) \tag{17}$$

If F is differentiable, then

$$\begin{aligned} \langle \nabla F(\mathbf{x}), \mathbf{x} \vee \mathbf{y} - \mathbf{x} \rangle &\ge F(\mathbf{x} \vee \mathbf{y}) - F(\mathbf{x}) \\ \langle \nabla F(\mathbf{x}), \mathbf{x} - \mathbf{x} \wedge \mathbf{y} \rangle &\le F(\mathbf{x}) - F(\mathbf{x} \wedge \mathbf{y}) \end{aligned} \tag{18}$$

Proof Because all elements in Hessian are nonpositive, for $\forall \mathbf{v} \in \mathbb{R}^n_{\ge 0}$, we have $\mathbf{v}^T \nabla^2 F(\mathbf{x}) \mathbf{v} \le 0$. It implies the concavity of nonnegative directions. There are two nonnegative directions $\mathbf{x} \vee \mathbf{y} - \mathbf{x}$ and $\mathbf{x} - \mathbf{y} \wedge \mathbf{x}$. Using the concavity of nonnegative directions, the Eq. (18) is clear. □

Lemma 9 ([5, 18]) *$F : \mathcal{X} \to \mathbb{R}$ is a differentiable DR-submodular function, then for $\forall \mathbf{x}, \mathbf{y} \in [0, 1]^n$*

$$\langle \nabla F(\mathbf{x}), \mathbf{y} - \mathbf{x} \rangle \ge F(\mathbf{x} \vee \mathbf{y}) + F(\mathbf{x} \wedge \mathbf{y}) - 2F(\mathbf{x}) \tag{19}$$

Proof The proof is by simply adding Eq. (18) □

Corollary 1 *For a stationary point $\mathbf{x}$ and for $\forall \mathbf{y}$, $F(\mathbf{x}) \ge \alpha F(\mathbf{x} \vee \mathbf{y}) + (1-\alpha) F(\mathbf{x} \wedge \mathbf{y})$ with $\alpha \in [0, 1/2]$.*

Proof Because $\mathbf{x}$ is a stationary point, i.e., $0 \ge \langle \nabla F(\mathbf{x}), \mathbf{y} - \mathbf{x} \rangle$, then $F(\mathbf{x} \vee \mathbf{y}) + F(\mathbf{x} \wedge \mathbf{y}) \le 2F(\mathbf{x})$. In addition, $F(\mathbf{x} \vee (\mathbf{y} \wedge \mathbf{x})) + F(\mathbf{x} \wedge (\mathbf{y} \wedge \mathbf{x})) \le 2F(\mathbf{x})$. The proof is complete by the convex combination of the above two inequalities. □

Lemma 10 ([5, 31]) *$F : \mathcal{X} \to \mathbb{R}$ is a DR-submodular function, then for $\forall \mathbf{x}, \mathbf{y} \in [\mathbf{l}, \mathbf{u}]^n$*

$$\begin{aligned} F(\mathbf{x} \vee \mathbf{y}) &\ge (1 - \|\mathbf{x}\|_\infty) F(\mathbf{y}) \\ F(\mathbf{x} \wedge \mathbf{y}) &\ge \|\mathbf{x}\|_{-\infty} F(\mathbf{y}) \end{aligned} \tag{20}$$

where $\|\mathbf{x}\|_\infty = \max_{i \in [n]} \frac{x_i - l_i}{u_i - l_i}$, $\|\mathbf{x}\|_{-\infty} = \min_{i \in [n]} \frac{x_i - l_i}{u_i - l_i}$.

Proof Firstly, for any $\alpha, \beta \in [0, 1]$, we have $\mathbf{x} \vee \mathbf{y} = (1-\alpha)\mathbf{y} + \alpha(\mathbf{y} + \frac{1}{\alpha}(\mathbf{x} \vee \mathbf{y} - \mathbf{y}))$ and $\mathbf{x} \wedge \mathbf{y} = \beta \mathbf{y} + (1-\beta)(\mathbf{y} - \frac{1}{1-\beta}(\mathbf{y} - \mathbf{x} \wedge \mathbf{y}))$. In addition, $\frac{1}{\alpha}(\mathbf{x} \vee \mathbf{y} - \mathbf{y})$ and $\frac{1}{1-\beta}(\mathbf{y} - \mathbf{x} \wedge \mathbf{y})$ are nonnegative directions.

$$F(\mathbf{x}\vee\mathbf{y})=F\left((1-\alpha)\mathbf{y}+\alpha\left(\mathbf{y}+\frac{1}{\alpha}(\mathbf{x}\vee\mathbf{y}-\mathbf{y})\right)\right)\geq(1-\alpha)F(\mathbf{y})$$
$$+\alpha F\left(\mathbf{y}+\frac{1}{\alpha}(\mathbf{x}\vee\mathbf{y}-\mathbf{y})\right)$$
$$F(\mathbf{x}\wedge\mathbf{y})=F\left(\beta\mathbf{y}+(1-\beta)\left(\mathbf{y}-\frac{1}{1-\beta}(\mathbf{y}-\mathbf{x}\wedge\mathbf{y})\right)\right)\geq\beta F(\mathbf{y})$$
$$+(1-\beta)F\left(\mathbf{y}-\frac{1}{1-\beta}(\mathbf{y}-\mathbf{x}\wedge\mathbf{y})\right)$$

To ensure that $F(\mathbf{y}+\frac{1}{\alpha}(\mathbf{x}\vee\mathbf{y}-\mathbf{y}))$ is well-defined, we restrict $y_i+\frac{1}{\alpha}(x_i-y_i)\leq u_i$ for $i\in[n]$ and $x_i\geq y_i$. If $x_i<y_i$, the inequality is obvious. It implies $\alpha\geq\frac{x_i-y_i}{u_i-y_i}$ for all $x_i\geq y_i$, i.e., $\alpha\geq\max_{i\in[n]}\frac{x_i-l_i}{u_i-l_i}$. Combining with nonnegativity, we have:

$$F(\mathbf{x}\vee\mathbf{y})\geq(1-\alpha)F(\mathbf{y})$$

Similarly, if $\beta\leq\min_{i\in[n]}\frac{x_i-l_i}{u_i-l_i}$, we have

$$F(\mathbf{x}\wedge\mathbf{y})\geq\beta F(\mathbf{y})$$

□

Lemma 11 *If $F:\mathcal{X}\rightarrow\mathbb{R}$ is L-smooth, then for $\forall\mathbf{x},\mathbf{y}\in\mathcal{X}$, we have:*

$$F(\mathbf{x})+\langle\nabla F(\mathbf{x}),\mathbf{y}-\mathbf{x}\rangle-\frac{L}{2}\|\mathbf{y}-\mathbf{x}\|^2\leq F(\mathbf{y})\leq F(\mathbf{x})+\langle\nabla F(\mathbf{x}),\mathbf{y}-\mathbf{x}\rangle+\frac{L}{2}\|\mathbf{y}-\mathbf{x}\|^2 \tag{21}$$

Proof

$$|F(\mathbf{y})-F(\mathbf{x})-\langle\nabla F(\mathbf{x}),\mathbf{y}-\mathbf{x}\rangle|=\left|\int_0^1\langle\nabla F(\mathbf{x}+t(\mathbf{y}-\mathbf{x}))-\nabla F(\mathbf{x}),\mathbf{y}-\mathbf{x}\rangle dt\right|$$
$$\leq\int_0^1\|\nabla F(\mathbf{x}+t(\mathbf{y}-\mathbf{x}))-\nabla F(\mathbf{x})\|\,\|\mathbf{y}-\mathbf{x}\|\,dt$$
$$\leq\int_0^1 tL\,\|\mathbf{y}-\mathbf{x}\|^2\,dt=\frac{L}{2}\,\|\mathbf{y}-\mathbf{x}\|^2$$

The first inequality comes from Cauthy-Schwarz inequality. The second inequality is due to the definition of L-smooth. □

3 Set Functions

In this section, we will introduce some algorithms for maximizing submodular set functions, covering both combinatorial algorithms for set functions and continuous algorithms for the multilinear extension of submodular set functions. For clarity, we give the formal statement for maximizing submodular set functions and their multilinear extension:

$$\begin{aligned} &\max f(X) && \max F(\mathbf{x}) \\ \text{s.t.} \quad & X \in \mathcal{I} \subseteq 2^N && \text{s.t.} \quad \mathbf{x} \in \mathcal{P} \end{aligned} \tag{22}$$

where $\mathcal{P}$ is a relaxation for $\mathcal{I} \subseteq 2^N$. For example, matroid constraints $\mathcal{P} = \text{conv}\{\mathbf{1}_S | S \in \mathcal{I}\} = \{\mathbf{x} \in [0,1]^n | \mathbf{x}(S) = \sum_{e \in S} x_e \leq r(S), \forall S \in N\}$, knapsack constraints $\mathcal{P} = \{\mathbf{x} \in [0,1]^n | \mathbf{A}\mathbf{x} \leq \mathbf{b}\}$.

3.1 *Monotone Cases*

Before describing algorithms, we make the following assumption:

- The submodular set function f is monotone.
- The function value of f under the empty set is 0, i.e., $f(\emptyset) = 0$.

Conventionally, a submodular function with these assumptions is referred to as a polymatroid function.

3.1.1 Combinatorial Algorithms

Cardinality Constraints As a warm-up, we first introduce the greedy algorithm for cardinality constraints, i.e., $\mathcal{I} = \{S \subseteq 2^N || S| \leq k\}$, which was proposed by Nemhauser et al. [56].

Algorithm 1 Greedy algorithm

Input: function f, the ground set N, cardinality constraints k
Output: S_k with k elements

1: $S_0 \leftarrow \emptyset$
2: **for** $i = 1$ to k **do**
3: $\quad e_{\max} = \arg\max_{e \in N} f(e|S_{i-1})$
4: $\quad S_i \leftarrow S_{i-1} \cup \{e_{\max}\}$
5: **end for**

Lemma 12 *For each iteration, the marginal gain of Algorithm 1 has the following theoretical guarantee:*

$$f(S_{i+1}) \geq \frac{k-1}{k} f(S_i) + \frac{1}{k} f(OPT) \tag{23}$$

Proof

$$\begin{aligned} f(OPT) \leq f(S_i \cup OPT) \leq f(S_i) + \sum_{e \in OPT \setminus S_i} f(e|S_i) \\ \leq f(S_i) + kf(e_{max}|S_i) = (1-k)f(S_i) + kf(S_{i+1}) \end{aligned}$$

□

Theorem 1 ([56]) *Algorithm 1 is a* $1 - e^{-1}$ *approximation algorithm for maximizing submodular set functions under cardinality constraints with the complexity* $O(nk)$.

Proof

$$\begin{aligned} f(S_k) &\geq \frac{k-1}{k} f(S_{k-1}) + \frac{1}{k} f(OPT) \\ &\geq \frac{(k-1)^2}{k^2} f(S_{k-2}) + \frac{k-1}{k^2} f(OPT) + \frac{1}{k} f(OPT) \\ &\geq \cdots \geq \frac{(k-1)^k}{k^k} f(S_0) + \left(1 - \left(1 - \frac{1}{k}\right)^k\right) f(OPT) \\ &\geq (1 - e^{-1}) f(OPT) \end{aligned}$$

In each iteration, the algorithm identifies the element with the maximal marginal gain. Generally speaking, this step has a complexity of $O(n)$. Since there are k iterations, the overall complexity of Algorithm 1 is $O(nk)$. □

Note that k is often on the same order of magnitude as n, meaning the complexity of the greedy algorithm can potentially be $O(n^2)$. To reduce the complexity of the greedy algorithm, several fast greedy algorithms have been proposed, such as [52, 53]. Next, we introduce an algorithm for another technical route, namely, threshold decrease algorithm, which has been proposed by Badanidiyuru and Vondrák [3].

Lemma 13 *For each iteration, the marginal gain of adding element e to the current solution S in Algorithm 2 has the following theoretical guarantee:*

$$f(e|S) \geq \frac{1-\epsilon}{k} \sum_{v \in OPT \setminus S} f(v|S) \tag{24}$$

Algorithm 2 Threshold decrease algorithm

Input: function f, the ground set N, cardinality constraints k, parameter ϵ
Output: S with k elements

```
1: S ← ∅, d ← max_{i∈N} f(i)
2: for θ = d; θ ≥ εd/n; θ = θ(1 − ε) do
3:     for all e ∈ N do
4:         if f(e|S) ≥ θ and |S| ≤ k then
5:             S ← S ∪ {e}
6:         end if
7:     end for
8: end for
```

Proof According to the Line 4, we have

$$f(x|S)\begin{cases}\geq \theta & \text{if } x = e\\ \leq \frac{\theta}{1-\epsilon} & \text{if } x \in OPT \setminus (S \cup \{e\})\end{cases}$$

Therefore, for each $v \in OPT \setminus S$, $f(e|S) \geq (1-\epsilon)f(v|S)$. Then, we have

$$f(e|S) \geq \sum_{v\in OPT\setminus S} \frac{(1-\epsilon)f(v|S)}{|OPT \setminus S|} \geq \sum_{v\in OPT\setminus S} \frac{(1-\epsilon)f(v|S)}{k}$$

□

Theorem 2 ([3]) *Algorithm 2 is a* $1 - e^{-1} - O(\epsilon)$ *approximation algorithm for maximizing submodular set functions under cardinality constraints with the complexity* $O(\frac{n}{\epsilon}\log\frac{n}{\epsilon})$

Proof For convenience, denote $S_m = \{e_1, \ldots, e_m\}$ where $m \leq k$. Lemma 13 implies that $f(S_m) - f(S_{m-1}) = f(e_m|S_{m-1}) \geq \sum_{v\in OPT\setminus S_{m-1}} \frac{(1-\epsilon)f(v|S_{m-1})}{k} \geq \frac{1-\epsilon}{k}[f(OPT \cup S_{m-1}) - f(S_{m-1})] \geq \frac{1-\epsilon}{k}[f(OPT) - f(S_{m-1})]$. When the algorithm ends, if $|S| = |S_k| = k$, then we have:

$$\begin{aligned} f(S_k) &\geq \frac{1-\epsilon}{k}f(OPT) + \left(1 - \frac{1-\epsilon}{k}\right)f(S_{k-1})\\ &\geq \left(1 - \left(1 - \frac{1-\epsilon}{k}\right)^k\right)f(OPT)\\ &\geq (1 - e^{-(1-\epsilon)})f(OPT) \geq (1 - e^{-1} - \epsilon)f(OPT)\end{aligned}$$

If $|S| < k$, then $f(S_k) - f(S) \leq \sum_{v\in S_k\setminus S} f(v|S) \leq (k - |S|)\frac{\epsilon d}{n} \leq \epsilon d$. It implies that $f(S) \geq (1 - e^{-1} - O(\epsilon))f(OPT)$.

The complexity can be shown to be $O(\frac{n}{\epsilon}\log\frac{n}{\epsilon})$. In each outer loop, the algorithm traverses the ground set, and the number of outer loops is determined by the precision parameter ϵ, leading to this overall complexity. □

Algorithm 2 shows faster executing time compared to the greedy algorithm, with a potential complexity of less than $O(n^2)$. By leveraging the threshold setting, it can approximate the maximal step in the greedy algorithm. Roughly speaking, the threshold decrease algorithm serves as a simplified or crude version of the greedy algorithm.

Matroid Constraints Next, we review approximation algorithms for matroid constraints, i.e., $X \in \mathcal{I}, \mathcal{M} = (N, \mathcal{I})$. Although [35] introduced the conventional greedy algorithm with a $1/2$ approximation ratio, there are also algorithms that achieve a $1 - e^{-1} - \epsilon$ approximation for matroid constraints. Here, we mainly focus on the greedy algorithm that combines non-oblivious local search, which is a $1 - e^{-1} - \epsilon$ approximation algorithms and was proposed by Filmus and Ward [34].

Algorithm 3 Non-oblivious local search algorithm

Input: function f, the matroid $\mathcal{M} = (N, \mathcal{I})$, parameter ϵ_0
Output: S

```
S ← Greedy(g̃, M = (N, I))
repeat
    for all a ∈ S and b ∈ N \ S do
        S' ← S \ {a} ∪ {b}
        if S' ∈ I and g̃(S') > (1 + ε0)g̃(S) then
            S ← S'; break
        end if
    end for
until No exchange is made
```

Algorithm 3 starts with the solution produced by greedy algorithm, which gradually expands the empty set greedily into a matroid basis based on marginal gains. Then, according to the exchange property of matroid, Algorithm 3 gradually improves the function value of the current solution by a factor of $(1 - \epsilon_0)$ at each exchange. Note that while the input of Algorithm 3 is the submodular function f, both the greedy step and the exchange step rely on the auxiliary function g, often referred to as the non-oblivious potential function. The formal statement of g is as follows:

$$\begin{aligned} g(A) &= \textstyle\sum_{B \subseteq A} m_{|A|-1,|B|-1} f(B) \\ m_{a,b} &= \textstyle\int_0^1 \frac{e^t}{e-1} \cdot t^b (1-t)^{a-b} dt \end{aligned} \tag{25}$$

From the definition of g, the non-oblivious potential function incorporates information from the objective function for all subsets of the input and adjusts the contribution with a weight based on their size. In some ways, the non-oblivious function relieve the poor local ratio often associated with oblivious functions, as oblivious local search possibly returns a locally optimal solution that is significantly inferior to the global optimum. Furthermore, the above non-oblivious function possesses several beautiful properties.

Lemma 14 *The non-oblivious potential function g is normalized, monotone, and submodular.*

Proof $g(\emptyset) = m_{-1,-1} f(\emptyset) = 0$, then g is normalized. Because f is monotone submodular, g is monotone and submodular directly from the following inequality.

$$\begin{aligned} g(A \cup \{e\}) - g(A) &= \sum_{B \subseteq A \cup \{e\}} m_{|A|,|B|-1} f(B) - \sum_{B \subseteq A} m_{|A|-1,|B|-1} f(B) \\ &= \sum_{B \subseteq A} (m_{|A|,|B|-1} - m_{|A|-1,|B|-1}) f(B) + m_{|A|,|B|} f(B \cup \{e\}) \\ &= \sum_{B \subseteq A} -m_{|A|,|B|} f(B) + m_{|A|,|B|} f(B \cup \{e\}) \\ &= \sum_{B \subseteq A} m_{|A|,|B|} f(e|B) \end{aligned}$$

□

However, evaluating $g(A)$ exactly requires querying f on all subset of A, which is known to entail an exponential number of operations for the value oracle f. Consequently, an efficient sampling procedure to estimate g becomes one of the most intuitive methods, particularly when a polynomial number of sampling is adopted. Therefore, an approximation of g can be derived through this sampling approach.

$$\tilde{g}(A) = \frac{1}{|R|} \frac{1}{\sum_{B \subseteq A} m_{|A|-1,|B|-1}} \sum_{i=1}^{|R|} f(B_i) \tag{26}$$

Next, we analyze the error between g and $\tilde{g}$ arising from the sampling process.

Lemma 15 *For any $\epsilon \geq 0$, let R be the collection of samples, then we have:*

$$Pr[|\tilde{g}(A) - g(A)| \geq \epsilon g(A)] \leq 2 \exp\left(-\Omega\left(\frac{2\epsilon^2 |R|}{\log^2 r}\right)\right) \tag{27}$$

Proof According to the Hoeffding's bound, we have

$$Pr[|\tilde{g}(A) - g(A)| \geq \epsilon g(A)] \leq 2 \exp\left(-\frac{2\epsilon^2 |R|}{(\sum_{B \subseteq A} m_{|A|-1,|B|-1})^2}\right)$$

In addition, we have

$$\sum_{B\subseteq A} m_{|A|-1,|B|-1} = \sum_{k=1}^{|A|}\binom{|A|}{k}\int_0^1 \frac{e^t}{e-1}\cdot t^{k-1}(1-t)^{|A|-k}dt \le \frac{e}{e-1}\sum_{k=1}^{|A|}\frac{1}{k}$$

Note that $\frac{e}{e-1}\sum_{k=1}^{|A|}\frac{1}{k} \le \frac{e}{e-1}\sum_{k=1}^{r}\frac{1}{k} = O(\log r)$. Therefore,

$$2\exp\left(-\frac{2\epsilon^2|R|}{(\sum_{B\subseteq A} m_{|A|-1,|B|-1})^2}\right) \le 2\exp\left(-\Omega\left(\frac{2\epsilon^2|R|}{\log^2 r}\right)\right)$$

□

The above lemma shows that errors come from the approximation $\tilde{g}$ which can be controlled. In other words, our target is to analyze the theoretical guarantees for Algorithm 3, when g is evaluated exactly. Under certain samples that satisfy Lemma 15, the theoretical guarantees of Algorithm 3 will remain closed to those derived from the exactly evaluation of g. Here, we only give conclusion; for proof details, see [34].

Theorem 3 ([34]) *Given $\epsilon > 0$, set the parameters of Algorithm 3 as follows:*

$$\epsilon_0 = \frac{\epsilon}{r\log r},\ |R| = C_1\epsilon_0^{-2}\log^2 r\log(\epsilon^{-1}r^2 n\log r) \tag{28}$$

where C_1 is a global constant, r is the rank of $\mathcal{M} = (N, \mathcal{I})$, and $n = |N|$. With probability $1-o(1)$, Algorithm 3 is a $1-e^{-1}-O(\epsilon)$ approximation algorithm with $\tilde{O}(\epsilon^{-3}r^4 n)$ complexity.

While Algorithm 3 achieves the tight approximation ratio for matroid constraints, its complexity scales with the fourth power of the rank of matroids, which is often unacceptable for many large-scale problems. To address this issue, [12] introduced a novel non-oblivious function to reduce the complexity of non-oblivious local search algorithm. The formal statement is as follows:

$$g'(S) = g(\pi_1(S),\ldots,\pi_l(S)) = \sum_{J\subseteq[l]} \alpha_{|J|} f(\pi_J(S)) \tag{29}$$

Let $S \subseteq N' = N \times [l]$, where $\pi_J(S) = \{e \in N | \exists_{i\in J}(e,i) \in S\}$ and $\alpha_1,\ldots,\alpha_l$ are positive parameters. Naturally, there is a new matroid as $\mathcal{M}' = (N', \mathcal{I}')$, where $\mathcal{I}' = \{S \in N' | \forall_{u\in N} |S \cap (u \times [l])| \le 1 \text{ and } \pi_{[l]}(S) \in \mathcal{I}\}$. In other words, the sets $S_1 = \pi_1(S),\ldots,S_l = \pi_l(S)$ are l disjoint sets, and their union $\cup_i S_i$ forms a base of the matroid $\mathcal{M}$. It is easy to see that the rank of $\mathcal{M}'$ is equal to the rank of $\mathcal{M}$. In addition, the above non-oblivious function g' shares the same advantageous properties as f.

Lemma 16 *The non-oblivious function g' is normalized, monotone, and submodular.*

Proof Because f is normalized and monotone, clearly, g' is normalized and monotone. The submodularity of g' follows immediately from the following inequality. For $\forall S \subseteq T \in N'$ and $(u, i) \notin T$

$$g'(u|S) = \sum_{J \subseteq [l], i \in J} \alpha_{|J|} \cdot f(u|\pi_J(S)) \geq \sum_{J \subseteq [l], i \in J} \alpha_{|J|} \cdot f(u|\pi_J(T)) = g'(u|T)$$

□

Note that the Eq. (29) needs to invoke 2^l value oracle queries to function f. Specially, choosing $l = 1 + \lceil 1/\epsilon \rceil$, we can upper bound 2^l by a function of ϵ, i.e., $2^l = O_\epsilon(1)$. Next, we provide details on the fast non-oblivious local search algorithm proposed by Buchbinder and Feldman [12].

Algorithm 4 Fast non-oblivious local search algorithm

Input: function f, the matroid $\mathcal{M} = (N, \mathcal{I})$, parameter ϵ
Output: $\pi_{[l]}(S)$

1: Generate the ground set N' and the matroid $\mathcal{M}' = (N', \mathcal{I}')$, construct the non-oblivious function g', set parameters $l = 1 + \lceil 1/\epsilon \rceil$, $\alpha_i = \frac{(1+1/l)^{i-1}}{\binom{l-1}{i-1}}$
2: $S, S_0 \leftarrow$Threshold Greedy$(g', \mathcal{M}')$ ([15] proved that it is 1/3 approximation algorithm with $O(n \log r)$).
3: Expand S as a base of $\mathcal{M}'$
4: **while** there exists elements $(u, i) \in S$ and $(v, j) \in N' \setminus S$ such that $S \cup (v, j) \setminus (u, i) \in \mathcal{I}'$ and $g'(v|S) - g'(u|S \setminus u) \geq \frac{\epsilon}{r} g'(S_0)$ **do**
5: $\quad S \leftarrow S \cup (v, j) \setminus (u, i) \in \mathcal{I}'$
6: **end while**

Algorithm 4 starts with a solution returned by threshold greedy, which provides a 1/3 approximation. This solution is then expanded to form a base of the new matroid $\mathcal{M}'$. Leveraging the exchange property of matroid, Algorithm 4 replaces the pair (u, i) with (v, j) if the marginal gain is enough large. It can be proved that the solution following these exchanges has the following theoretical guarantees. See [12] for proof details.

Lemma 17 *After $\tilde{O}(\epsilon^{-1} nr)$ value and independence oracle queries, the output solution S satisfies that*

$$\sum_{(v,i) \in T} g'((v, i)|S) - \sum_{(u,j) \in S} g'((u, j)|S \setminus (u, j)) \leq \epsilon' g'(OPT') \quad \forall T \in \mathcal{I}' \tag{30}$$

The above lemma illustrates that Algorithm 4 ends in a solution that resembles stationary points. Using such a feature, we can get the performance guarantee for Algorithm 4.

Theorem 4 *Algorithm 4 is a* $1 - 1/e - \epsilon$ *approximation algorithm with* $\tilde{O}_\epsilon(nr)$ *query complexity for maximizing a nonnegative monotone submodular function under a matroid constraint.*

To further accelerate non-oblivious local search, [12] developed a random algorithm that achieves the guarantees outlined in Eq. (30). This algorithm requires $\tilde{O}(\epsilon^{-1}(n+r\sqrt{n}))$ value and independence oracle queries with successful probability at least $1-\epsilon$. It implies that there is a random version of Algorithm 4 that maintains a $1-1/e-\epsilon$ approximation ratio while having a query complexity of $\tilde{O}_\epsilon(n+r\sqrt{n})$. For further details, please refer to [12].

Knapsack Constraints At last, we retrospect approximation algorithms for knapsack constraints, i.e., $\mathcal{I} = \{S \subseteq 2^N | c(S) = \sum_{e \in S} c(e) \leq b \text{ and } \forall c(e) \geq 0\}$. The first $1 - 1/e$ approximation algorithm for this problem was proposed by Sviridenko [63].

Algorithm 5 Greedy algorithm

Input: function f, the ground set N, the cost function c, the budget b
Output: S_*

$X \leftarrow \arg\max_{|S|\leq 2, c(S)\leq b} f(S)$ (enumerate all feasible solution of cardinality one or two)
$Y \leftarrow \emptyset$
for all $S \subseteq N$ with $|S| = 3$ and $c(S) \leq b$ **do**
    $V \leftarrow N \setminus S$, $S_0 \leftarrow S$, $i = 0$
    while $V \neq \emptyset$ **do**
        $e_{\max} = \arg\max_{e \in V} \frac{f(e|S_i)}{c(e)}$
        if $c(S) + c(e_{\max}) \leq b$ **then**
            $S_{i+1} \leftarrow S_i \cup \{e_{\max}\}$, $i++$
        end if
        $V \leftarrow V \setminus e_{\max}$
    end while
    if $f(S_i) \geq f(Y)$ **then**
        $Y \leftarrow S_i$
    end if
end for
$S_* \leftarrow \arg\max_{S \in \{X, Y\}} f(S)$

Algorithm 5 consists of two phase. In the first phase, the algorithm enumerates all feasible solution with at most two elements. Let X be the solution with the largest function value in the first phase. In the second phase, the algorithm considers all feasible solution with three elements, using each one as a starting solution. In each iteration, Algorithm 5 greedily expands the current solution according to the objective function value per unit cost, ensuring that the solution remains feasible. Let Y be the solution with the largest function value in second phase. The algorithm ultimately outputs the best of X and Y.

Theorem 5 ([63]) *Algorithm 5 is a* $1 - 1/e$ *approximation algorithm with* $O(n^5)$ *for maximizing monotone submodular set functions subject to knapsack constraints.*

Proof Let OPT be the optimal solution. If $|OPT| \le 3$, then $f(S_*) = f(OPT)$ because of enumeration. Then, we assume $|OPT| > 3$. First, we put OPT in descending order, i.e., $f(\{o_1, \dots, o_i\}) = \max_{o \in OPT \setminus \{o_1,\dots,o_{i-1}\}} f(\{o_1, \dots, o_{i-1}\} \cup \{o\})$. Denote $A = \{o_1, o_2, o_3\}$. For $o_k \in OPT, k \ge 4$, and $B \subseteq N \setminus \{o_1, o_2, o_3, o_k\}$, we have

$$f(A \cup B \cup \{o_k\}) - f(A \cup B) \le f(\{o_k\}) \le f(\{o_1\})$$
$$f(A \cup B \cup \{o_k\}) - f(A \cup B) \le f(\{o_k, o_1\}) - f(\{o_1\}) \le f(\{o_1, o_2\}) - f(\{o_1\})$$
$$f(A \cup B \cup \{o_k\}) - f(A \cup B) \le f(\{o_k, o_1, o_2\}) - f(\{o_1, o_2\}) \le f(\{o_1, o_2, o_3\}) - f(\{o_1, o_2\})$$

Therefore, we have $3(f(A \cup B \cup \{o_k\}) - f(A \cup B)) \le f(A)$. Define the function $g(S) = f(S) - f(A)$. It is easy to check $g(S)$ is monotone and submodular. For convenience, denote $\theta_i = \max_{e \in V \setminus S_{i-1}} \frac{f(e|S_{i-1})}{c(e)}$. Then, we have

$$\begin{aligned} g(OPT) &\le g(S_i) + \sum_{e \in OPT \setminus S_i} g(e|S_i) = g(S_i) + \sum_{e \in OPT \setminus S_i} f(e|S_i) \\ &\le g(S_i) + \left(b - \sum_{e \in A} c(e)\right) \theta_{i+1} \end{aligned}$$

Let $e_{i^*+1} \in OPT$ be the first element cannot be added in the solution S_{i^*}. If $e_{i^*+1} \notin OPT$, it has no effect on our analysis of the relationship between the current solution and the optimal solution. Therefore, we have $b' = b_{i^*+1} = \sum_{j=1}^{i^*+1} c(e_j) > b > b - c(A) = b''$. Denote $b_i = \sum_{j=1}^{i} c(e_j)$. For $j \in [b']$, we define $\rho_j = \theta_i$ if $j = b_{i-1}+1, \dots, b_i$. Then, we have $g(S_{i^*} \cup \{e_{i^*+1}\}) = \sum_{j=1}^{i^*+1} c(e_j)\theta_j = \sum_{j=1}^{b'} \rho_j$ and $g(S_i) = \sum_{j=1}^{i} c(e_j)\theta_j = \sum_{j=1}^{b_i} \rho_j$. In addition, $\min_{j=1,\dots,i^*} g(S_j) + b''\theta_{j+1} = \min_{j=1,\dots,i^*} \sum_{k=1}^{b_j} \rho_j + b''\rho_{b_j+1} = \min_{j=1,\dots,b'} \sum_{k=1}^{j-1} \rho_k + b''\rho_j$. Consequently, we have:

$$\frac{g(S_{i^*} \cup \{e_{i^*+1}\})}{g(OPT)} \ge \frac{\sum_{j=1}^{b'} \rho_j}{\min_{j=1,\dots,b'} \sum_{k=1}^{j-1} \rho_k + b''\rho_j} \ge 1 - e^{-b'/b''} \ge 1 - e^{-1}$$

Adopting the above inequality, we have

$$\begin{aligned} f(S_{i^*}) &= f(A) + g(S_{i^*}) = f(A) + g(S_{i^*} \cup \{e_{i^*+1}\}) - (g(S_{i^*} \cup \{e_{i^*+1}\}) - g(S_{i^*})) \\ &\ge (1 - e^{-1})g(OPT) + f(A) - \frac{1}{3}f(A) \ge (1 - e^{-1})f(OPT) \end{aligned}$$

□

The $O(n^5)$ complexity of Algorithm 5 is a drawback for some problems. To reduce the complexity, [47] and [33] proved that the stage of greedily expanding the current solution can start from the solution with two elements while still achieving the approximation ratio $1 - e^{-1}$.

Algorithm 6 Greedy algorithm

Input: function f, the ground set N, the cost function c, the budget b
Output: S_*

$S_* \leftarrow \emptyset$
for all $S \subseteq N$ with $|S| \leq 2$ and $c(S) \leq b$ **do**
  $V \leftarrow N \setminus S$, $S_0 \leftarrow S$, $i = 0$
  while $V \neq \emptyset$ **do**
    $e_{\max} = \arg\max_{e \in V} \frac{f(e|S_i)}{c(e)}$
    if $c(S) + c(e_{\max}) \leq b$ **then**
      $S_{i+1} \leftarrow S_i \cup \{e_{\max}\}$, $i++$
    end if
    $V \leftarrow V \setminus e_{\max}$
  end while
  if $f(S_i) \geq f(S_*)$ **then**
    $S_* \leftarrow S_i$
  end if
end for

Theorem 6 ([33, 47]) *Algorithm 6 is a $1 - 1/e$ approximation algorithm with $O(n^4)$ for maximizing monotone submodular set functions subject to knapsack constraints.*

3.1.2 Continuous Algorithm

In this subsection, we will utilize continuous optimization algorithms to maximize the monotone submodular set functions. Calinescu et al. [16] first proposed the continuous greedy algorithm based on the multilinear extension of submodular set functions. Due to the limitation of rounding techniques at the time, they were only able to prove theoretical guarantees for matroid constraints. Essentially, the continuous greedy algorithm can return a $1 - e^{-1} - \epsilon$ non-integer solution for maximizing monotone submodular set functions subject to convex down-closed constraints. Combining some theoretical guarantees rounding technique such as contention resolution schemes [18], the continuous greedy algorithm can also be adopted to maximize monotone submodular set functions under cardinality constraints and knapsack constraints. Here, we only focus on the algorithm itself and omit the rounding procedure.

Algorithm 7 starts from the $\mathbf{0}$ vector. In each iteration, the algorithm identifies a forward direction based on the suboptimization problem and takes a step of size δ along this direction. Next, we present the theoretical guarantee for $\mathbf{x}(1)$.

Algorithm 7 Continuous greedy algorithm

Input: function f, the ground set N, the feasible domain $\mathcal{P}$
Output: $\mathbf{x}(1)$
1: $\delta = \epsilon/n^3$, $\mathbf{x}(0) = \mathbf{0}$
2: **for** $t = 0$ to 1 **do**
3:     **for all** $e \in N$ **do**
4:         $w_e \leftarrow \frac{F(\mathbf{x}(t)\vee \mathbf{1}_e)-F(\mathbf{x}(t))}{1-x_e}$
5:     **end for**
6:     $\mathbf{v}(\mathbf{x}(t)) = \arg\max_{\mathbf{v}(t)\in\mathcal{P}} \langle \mathbf{w}(t), \mathbf{v}\rangle$
7:     $\mathbf{x}(t+\delta) = \mathbf{x}(t) + \delta\mathbf{v}(\mathbf{x}(t))$
8:     $t = t + \delta$
9: **end for**

Lemma 18 *The solution* $\mathbf{x}(1)$ *returned by Algorithm 7 is a feasible solution, i.e.,* $\mathbf{x}(1) \in \mathcal{P}$.

Proof Because of $\mathbf{x}(0) = \mathbf{0}$, we have $\mathbf{x}(1) = \sum_{i=1}^{n^3/\epsilon} \frac{\epsilon}{n^3}\mathbf{v}(\mathbf{x}(t_i))$ that is the convex combination of points in $\mathcal{P}$. Because $\mathcal{P}$ is convex, then $\mathbf{x}(1) \in \mathcal{P}$. □

Theorem 7 ([16]) *Algorithm 7 is a* $1 - 1/e - \epsilon$ *approximation algorithm for maximizing the multilinear extension of submodular set functions under down-closed and convex constraints.*

Proof

$$
\begin{aligned}
F(\mathbf{x}(t+\delta)) - F(\mathbf{x}(t)) &\geq \langle \nabla F(\mathbf{x}(t)), \delta\mathbf{v}(\mathbf{x}(t))\rangle - O(n^3\delta^2)F(\mathbf{x}^*) \\
&\geq \langle \nabla F(\mathbf{x}(t)), \delta(\mathbf{x}^* \vee \mathbf{x}(t) - \mathbf{x}(t))\rangle - O(n^3\delta^2)F(\mathbf{x}^*) \\
&\geq \delta(F(\mathbf{x}^* \vee \mathbf{x}(t)) - F(\mathbf{x}(t))) - O(n^3\delta^2)F(\mathbf{x}^*) \\
&\geq \delta(F(\mathbf{x}^*) - F(\mathbf{x}(t))) - O(n^3\delta^2)F(\mathbf{x}^*)
\end{aligned}
$$

The first inequality comes from Lemma 3. Because $\mathcal{P}$ is down-closed, $\mathbf{x}^* \in \mathcal{P}$ implies $\mathbf{x}^* \geq \mathbf{x}^* \vee \mathbf{x}(t) - \mathbf{x}(t) \in \mathcal{P}$. The third inequality comes from Lemma 8. Therefore, we have

$$
F(\mathbf{x}(1)) = (1 - (1-\delta)^{1/\delta})F(\mathbf{x}^*) - O(n^3\delta)F(\mathbf{x}^*) \geq (1 - e^{-1} - O(\epsilon))F(\mathbf{x}^*)
$$

□

Note that the above theoretical guarantees presented above are based on the assumption that the function value F and its derivative can be computed exactly. According to the definition of the multilinear extension, evaluating F and its derivative exactly requires querying f on all subset of N, which entails exponential complexity. To efficiently obtain $F(\mathbf{x}(t))$ and $\nabla F(\mathbf{x}(t))$ as precise as possible, sampling techniques are wildly used, and their errors can be bounded by Chernoff bounds.

Theorem 8 (Chernoff Bounds) *Let X_i, $1 \le i \le k$, be mutually independent with $\mathbb{E}[X_i] = 0$ and $|X_i| \le 1$. Set $S = X_1 + \ldots + X_k$. Then,*

$$Pr[|S| \ge a] \le 2\exp\left(\frac{-a^2}{2k}\right) \tag{31}$$

Consequently, the theoretical guarantees of the continuous greedy algorithm are $1 - e^{-1} - O(\epsilon)$ with a "high probability" depending on the number of samples.

3.2 *Non-monotone Cases*

In this subsection, we assume only that f is normalized, meaning $f(\emptyset) = 0$, and we do not require the monotonicity.

3.2.1 Combinatorial Algorithm

Cardinality Constraints First, we introduce the random greedy algorithm, proposed by Buchbinder et al. [13], to maximize non-monotone submodular set functions under cardinality constraints, i.e., $\mathcal{I} = \{S \subseteq N | |S| \le k\}$.

Algorithm 8 Random greedy algorithm

Input: function f, the ground set N, cardinality constraints k
Output: S_k with k elements

1: $S_0 \leftarrow \emptyset$
2: **for** $i = 1$ to k **do**
3: $\quad M_i = \arg\max_{M \subseteq N \setminus S_{i-1}, |M|=k} \sum_{u \in M} f(u|S_{i-1})$
4: $\quad e_i$ be a uniformly random element from M_i
5: $\quad S_i \leftarrow S_{i-1} \cup \{e_i\}$
6: **end for**

Algorithm 8 begins with the empty set. In each iteration, the algorithm expands the current solution S_{i-1} by randomly selecting an element from the k elements with the largest marginal relative to S_{i-1}.

Theorem 9 ([13]) *If f is monotone, Algorithm 8 is a $1 - e^{-1}$ approximation algorithm. If f is non-monotone, Algorithm 8 is a e^{-1} approximation algorithm. The complexity is $O(nk)$.*

Proof Here, we only put the sketch proof for non-monotone cases.

$$
\begin{aligned}
\mathbb{E}_{u_i,S_{i-1}}[f(u_i|S_{i-1})] &= \frac{\mathbb{E}_{S_{i-1}}[\sum_{u\in M_i} f(u|S_{i-1})]}{k} \\
&\geq \frac{\mathbb{E}_{S_{i-1}}[\sum_{u\in OPT\setminus S_{i-1}} f(u|S_{i-1})]}{k} \\
&\geq \frac{\mathbb{E}[f(OPT\cup S_{i-1}) - f(S_{i-1})]}{k} \\
&\geq \frac{(1-1/k)^{i-1} f(OPT) - \mathbb{E}[f(S_{i-1})]}{k}
\end{aligned}
$$

The last inequality comes from Lemma 5, because each element belongs to S_i with probability at most $1-(1-1/k)^i$. Then, we have

$$
\begin{aligned}
\mathbb{E}[f(S_k)] &= \mathbb{E}[f(S_{k-1})] + \mathbb{E}[f(u_k|S_{k-1})] \\
&\geq \mathbb{E}[f(S_{k-1})] + \frac{(1-1/k)^{k-1} f(OPT) - \mathbb{E}[f(S_{k-1})]}{k} \\
&\geq \left(1-\frac{1}{k}\right)^{k-1} f(OPT)
\end{aligned}
$$

□

Algorithm 9 0.385-approximation algorithm

Input: function f, the ground set N, cardinality constraints k, error rate $\epsilon \in (0,1)$, flip point $t_s \in [0,1]$
Output: S_L

1: $Z \leftarrow$ Fast-Local-Search $(k, f, \epsilon, L := \lceil 16k/(\epsilon(1-1/e))\rceil)$
2: **if** the last algorithm did not fail **then**
3: $\quad A \leftarrow$ Aided-Measured Discrete Stochastic Greedy(Z, k, t_s, ϵ)
4: $\quad$ **return** $\max\{f(Z), f(A)\}$
5: **else**
6: $\quad$ **return** $\emptyset$
7: **end if**

In addition, [23] developed a combinatorial algorithm for cardinality constraints with 0.385 approximation ratio and $O(nk)$ complexity. Tukan et al. [65] proposed a practical algorithm that reduces the complexity to $O(n+k^2)$ while maintaining the 0.385 approximation ratio. The following algorithms are formally stated in [65]. Please refer to [65] for proof details.

Matroid Constraints Buchbinder et al. [13] proposed a random combinatorial algorithm that is a $0.283 - O(\frac{1}{k^2})$ approximation algorithm with $O(kn)$ complexity for maximizing non-submodular set functions subject to matroid constraints $\mathcal{M} = (N, \mathcal{I})$. Sun et al. [62] proposed a deterministic algorithm with the same

Algorithm 10 Fast-local search

Input: function f, the ground set N, cardinality constraints k, an approximation factor $\epsilon \in (0, 1)$, a number of iterations
Output: A subset of N of cardinality at most k

Initialize S_0 to be a c approximation solution with probability $1 - \epsilon$
Fill S_0 with dummy elements to ensure $|S_0| = k$
for $j = 1$ to $\lceil \log_2 \frac{1}{\epsilon} \rceil$ **do**
  $S_0^j \leftarrow S_0$
  for $i = 1$ to L **do**
    $Z_i^j \leftarrow$ sample $\frac{n}{k}$ items from N uniformly at random
    $u_i^j \leftarrow \arg\max_{u' \in Z_i^j} f(S_{i-1}^j \cup u_i^j) - f(S_{i-1}^j)$
    if $f(S_{i-1}^j \cup u_i^j) - f(S_{i-1}^j) \leq 0$ **then**
      $u_i^j \leftarrow$ dummy element that does not belong to S_{i-1}^j
    end if
    $v_i^j \leftarrow \arg\min_{v' \in S_{i-1}^j} f(S_{i-1}^j) - f(S_{i-1}^j \setminus v')$
    if $f(S_{i-1}^j) \leq f(S_{i-1}^j \setminus v_i^j \cup u_i^j)$ **then**
      $S_i^j \leftarrow S_{i-1}^j \setminus v_i^j \cup u_i^j$
    else
      $S_i^j \leftarrow S_{i-1}^j$
    end if
    Pick a uniformly random integer $0 \leq i^* < L$.
    if for every integer $0 \leq t \leq k$ it holds that $\max_{S \subseteq N \setminus S_{i*}^j, |S|=t} \sum_{u \in S} f(u|S_{i*}^j) \leq \min_{S \subseteq S_{i*}^j, |S|=t} \sum_{v \in S} f(v|S_{i*}^j \setminus v) + \epsilon f(S_{i*}^j)$ **then**
      return S_{i*}^j
    end if
  end for
end for
return FAILURE

approximation ratio, employing derandomlized techniques, but requiring $O(k^2n^2)$ complexity. Here, we introduce a deterministic $0.305 - \epsilon$ approximation algorithm proposed by Chen et al. [23]. For readers interested in proofs, please refer to [23].

Unconstraints Here, we introduce algorithms for maximizing non-monotone submodular set functions without constraints, i.e., $\mathcal{I} = 2^N$. Buchbinder et al. [14] proposed a deterministic algorithm with a 1/3 approximation ratio, as well as a random algorithm that achieves a 1/2 approximation ratio.

The algorithm maintains two sets, X and Y, with initial conditions $X_0 = \emptyset$, $Y_0 = N$. In each iteration, the algorithm considers the element e_i, if the marginal gain from adding e_i into X_{i-1} is not less than that for removing it from Y_{i-1}, then e_i is added into X_{i-1}, while Y_i remains unchanged. Otherwise, X_i remains unchanged, and e_i is removed from Y_{i-1}. Eventually, the algorithm yields a solution $X_n = Y_n$. In each iteration, the relationship between a_i and b_i is as follows:

Algorithm 11 Guided stochastic greedy

Input: function f, the ground set N, positive integer $k \geq 1$, $\epsilon \in (0, 1)$, a number of iterations, a set Z, $t_s \in [0, 1]$

Output: A subset of $S_k \subseteq N$

Initialize $S_0 \leftarrow \emptyset$, $p \leftarrow 8k^{-1}\epsilon^{-2}\ln(2\epsilon^{-1})$, $s_1 \leftarrow k/(n - |Z|)$, $s_2 \leftarrow k/n$
for $i = 1$ to $\lceil kt_s \rceil$ **do**
  Let $M_i \subseteq N \setminus Z$ be a uniformly random set containing $\lceil p(n - |Z|) \rceil$ elements
  Let d_i be uniformly random scalar from the range $(0, s_1\lceil p(n - |Z|) \rceil)$
  Let u_i be an element of M_i associated with the $\lceil d_i \rceil$ the largest marginal contribution to S_{i-1}
  if $f(u_i|S_{i-1}) \geq 0$ **then**
    $S_i \leftarrow S_{i-1} \cup \{u_i\}$
  else
    $S_i \leftarrow S_{i-1}$
  end if
end for
for $i = \lceil kt_s \rceil + 1$ to k **do**
  Let $M_i \subseteq N$ be a uniformly random set containing $\lceil pn \rceil$ elements
  Let d_i be uniformly random scalar from the range $(0, s_2\lceil pn \rceil)$
  Let u_i be an element of M_i associated with the $\lceil d_i \rceil$ the largest marginal contribution to S_{i-1}
  if $f(u_i|S_{i-1}) \geq 0$ **then**
    $S_i \leftarrow S_{i-1} \cup \{u_i\}$
  else
    $S_i \leftarrow S_{i-1}$
  end if
end for
return FAILURE

Algorithm 12 $0.305 - \epsilon$ Deterministic combinatorial approximation algorithm

Input: function f, constraints $\mathcal{I}$, error $\epsilon \in (0, 1)$, an approximation results Z_0, switch point $t \in [0, 1]$

Output: A subset of $S_k \subseteq N$

Initialize $Z \leftarrow$ FAST-LS$(f, \mathcal{I}, Z_0, \epsilon)$, $l \leftarrow \frac{10}{9\epsilon}$, $G_0 \leftarrow \{\emptyset\}$
for $i = 1$ to l **do**
  Let $G_i \subseteq \emptyset$
  for $A_{i-1} \in G_{i-1}$ **do**
    if $i \leq tl$ **then**
      $G_i \leftarrow G_i \cup$ GUIDEDIG-M$(f, \mathcal{I}, Z, A_{i-1}, l)$
    else
      $G_i \leftarrow G_i \cup$ GUIDEDIG-M$(f, \mathcal{I}, \emptyset, A_{i-1}, l)$
    end if
  end for
end for

Algorithm 13 FAST-LS

Input: function f, constraints $\mathcal{I}$, error $\epsilon \in (0, 1)$, an approximation results Z_0
Output: Z

Initialize $Z \leftarrow Z_0$,
while $\exists a \in Z, e \in N \setminus Z$, s.t. $Z \setminus a \cup e \in \mathcal{I}$ and $f(e|Z) - f(a|Z \setminus a) \geq \frac{\epsilon}{k} f(Z)$ **do**
  $Z \leftarrow Z \setminus a \cup e$
end while

Algorithm 14 GUIDEDIG-M

Input: function f, constraints $\mathcal{I}$, guidance set Z, starting set G, set size l
Output: $\{(G \cup A_j) \setminus (\sum_{x \in A_j} \sigma^{-1}(x)) : i \leq j \leq l\}$

Initialize $A, A_1, \ldots, A_l \leftarrow \emptyset$,
for $i = 1$ to k **do**
  $X_i \leftarrow \{x \in N \setminus (G \cup A \cup Z) : A \cup x \in \mathcal{I}\}$
  $j_i^*, a_i^* \leftarrow \arg\max_{j \in [l], x \in X} f(x|G \cup A_j)$
  $A \leftarrow A \cup a_i^*, A_{j_i^*} \leftarrow A_{j_i^*} \cup a_i^*$
end for
$\sigma \leftarrow$ a bijection from G to A s.t. $(G \cup A_j) \setminus (\sum_{x \in A_j} \sigma^{-1}(x))$ is a basis

Algorithm 15 Deterministic double greedy

Input: function f, the ground set $N = \{e_1, \ldots, e_n\}$
Output: X_n (or equivalently Y_n)

Initialize $X_0 \leftarrow \emptyset$, $Y_0 \leftarrow N$
for $i = 1$ to n **do**
  $a_i \leftarrow f(e_i|X_{i-1})$
  $b_i \leftarrow -f(e_i|Y_{i-1} \setminus e_i)$
  if $a_i \geq b_i$ **then**
    $X_i \leftarrow X_{i-1} \cup e_i, Y_i \leftarrow Y_{i-1}$
  else
    $X_i \leftarrow X_{i-1}, Y_i \leftarrow Y_{i-1} \setminus e_i$
  end if
end for

Lemma 19 *For each* $i \in [n]$, $a_i + b_i \geq 0$

Proof It is easy to check $X_i \subseteq Y_i$. According to submodualrity,

$$a_i + b_i = f(e_i|X_{i-1}) - f(e_i|Y_{i-1}) \geq 0$$

□

Based on the above lemma, we can construct the following relationship between the current sets X_i, Y_i and the optimal solution OPT.

Lemma 20 *Let* $OPT_i = (X_i \cup OPT) \cap Y_i$, *then we have* $f(OPT_{i-1}) - f(OPT_i) \leq f(X_i) - f(X_{i-1}) + f(Y_i) - f(Y_{i-1})$

Proof Case 1: $a_i \geq b_i$, i.e., $X_i = X_{i-1} \cup e_i$ and $Y_i = Y_{i-1}$. If $e_i \in OPT$, then $f(OPT_{i-1}) - f(OPT_i) = 0 \leq f(X_i) - f(X_{i-1}) + f(Y_i) - f(Y_{i-1})$. If $e_i \notin OPT$, then $OPT_i = (X_i \cup OPT) \cap Y_i = (X_{i-1} \cup e_i \cup OPT) \cap Y_{i-1} = OPT_{i-1} \cup e_i$. It implies that $f(OPT_{i-1}) - f(OPT_i) = -f(e_i|OPT_{i-1}) \leq -f(e_i|Y_{i-1} \setminus e_i) = b_i \leq a_i = f(X_i) - f(X_{i-1}) + f(Y_i) - f(Y_{i-1})$.

Case 2: $a_i \leq b_i$, i.e., $X_i = X_{i-1}$ and $Y_i = Y_{i-1} \setminus e_i$. If $e_i \in OPT$, then $OPT_i = (X_i \cup OPT) \cap Y_i = (X_{i-1} \cup OPT) \cap (Y_{i-1} \setminus e_i) = OPT_{i-1} \setminus e_i$, then $f(OPT_{i-1}) - f(OPT_i) \leq f(e_i|X_{i-1}) \leq b_i = f(Y_i) - f(Y_{i-1}) + f(X_i) - f(X_{i-1})$. If $e_i \notin OPT$, then $f(OPT_{i-1}) - f(OPT_i) = 0 \leq b_i = f(Y_i) - f(Y_{i-1}) + f(X_i) - f(X_{i-1})$. □

Consequently, the theoretical guarantees of deterministic double greedy are as follows:

Theorem 10 ([14]) *Algorithm 15 is a* $\frac{1}{3}$ *approximation algorithm with* $O(n)$ *complexity for the unconstrained submodular maximization problem.*

Proof $OPT_0 = OPT$, $OPT_n = X_n = Y_n$.

$$
\begin{aligned}
f(OPT_0) - f(OPT_n) &= \sum_{i=1}^{n} f(OPT_{i-1}) - f(OPT_i) \\
&\leq \sum_{i=1}^{n} f(X_i) - f(X_{i-1}) + f(Y_i) - f(Y_{i-1}) \\
&\leq f(X_n) + f(Y_n)
\end{aligned}
$$

□

The above analysis is tight for deterministic double greedy, as demonstrated by Buchbinder et al. [14] using a tight example. Next, we introduce the random double greedy algorithm that is a 1/2 approximation algorithm for the unconstrained submodular maximization problem.

Algorithm 16 Random double greedy

Input: function f, the ground set $N = \{e_1, \ldots, e_n\}$
Output: X_n (or equivalently Y_n)

```
Initialize X_0 ← ∅, Y_0 ← N
for i = 1 to n do
    a_i ← f(e_i | X_{i-1})
    b_i ← −f(e_i | Y_{i-1} \ e_i)
    a'_i ← max{a_i, 0}, b'_i ← max{b_i, 0}
    X_i ← X_{i-1} ∪ e_i, Y_i ← Y_{i-1} with probability a'_i/(a'_i+b'_i) or X_i ← X_{i-1}, Y_i ← Y_{i-1} \ e_i with
    probability b'_i/(a'_i+b'_i). If a'_i + b'_i = 0, set a'_i/(a'_i+b'_i) = 1.
end for
```

The only difference between the random double greedy and the deterministic double greedy is that the decisions to add or eliminate elements are probabilistically rather than deterministically. Next, we show the difference between the random version and deterministic version concerning the relationship between the current sets X_i, Y_i and the optimal solution OPT.

Lemma 21 *For $i \in [n]$,*

$$\mathbb{E}[f(OPT_{i-1}) - f(OPT_i)] \leq \frac{1}{2}\mathbb{E}[f(X_i) - f(X_{i-1}) + f(Y_i) - f(Y_{i-1})] \tag{32}$$

Proof Case 1: $a_i \geq 0, b_i \geq 0$. If $e_i \in OPT$, then $OPT_i = OPT_{i-1}$ with probability $\frac{a_i}{a_i+b_i}$ and $OPT_i = OPT_{i-1} \setminus e_i$ with probability $\frac{b_i}{a_i+b_i}$. If $e_i \notin OPT$, then $OPT_i = OPT_{i-1} \cup e_i$ with probability $\frac{a_i}{a_i+b_i}$ and $OPT_i = OPT_{i-1}$ with probability $\frac{b_i}{a_i+b_i}$. It implies $\mathbb{E}[f(OPT_{i-1}) - f(OPT_i)] \leq \frac{b_i}{a_i+b_i}[f(OPT_{i-1}) - f(OPT_{i-1} \setminus e_i)] = \frac{a_i b_i}{a_i+b_i}$ if $e_i \in OPT$ and $\mathbb{E}[f(OPT_{i-1}) - f(OPT_i)] \leq \frac{a_i}{a_i+b_i}[f(OPT_{i-1}) - f(OPT_{i-1} \cup e_i)] \leq \frac{a_i b_i}{a_i+b_i}$ if $e_i \notin OPT$.

$$\begin{aligned}
\mathbb{E}[f(OPT_{i-1}) - f(OPT_i)] &\leq \frac{a_i b_i}{a_i + b_i} \leq \frac{1}{2}\frac{a_i^2 + b_i^2}{a_i + b_i} \\
&= \frac{1}{2}\left[\frac{a_i}{a_i + b_i} f(e_i|X_{i-1}) - \frac{b_i}{a_i + b_i}[f(e_i|Y_{i-1} \setminus e_i)]\right] \\
&= \frac{1}{2}\mathbb{E}[f(X_i) - f(X_{i-1}) + f(Y_i) - f(Y_{i-1})]
\end{aligned}$$

Case 2: $a_i \geq 0, b_i \leq 0$. It means that $X_i = X_i \cup e_i, Y_i = Y_{i-1}$ with probability 1, i.e., $\mathbb{E}[f(OPT_{i-1}) - f(OPT_i)] = -f(e_i|OPT_{i-1}) \leq -f(e_i|Y_{i-1} \setminus e_i) = b_i \leq 0 \leq \frac{a_i}{2} = \frac{1}{2}[f(X_i) - f(X_{i-1}) + f(Y_i) - f(Y_{i-1})]$.

Case 3: $a_i \leq 0, b_i \geq 0$. It means that $X_i = X_i, Y_i = Y_{i-1} \setminus e_i$ with probability 1, i.e., $\mathbb{E}[f(OPT_{i-1}) - f(OPT_i)] = f(e_i|OPT_{i-1} \setminus e_i) \leq f(e_i|X_{i-1}) = a_i \leq 0 \leq \frac{b_i}{2} = \frac{1}{2}[f(Y_i) - f(Y_{i-1}) + f(X_i) - f(X_{i-1})]$ □

Theorem 11 ([14]) *Algorithm 16 is a $\frac{1}{2}$ approximation algorithm with $O(n)$ complexity for the unconstrained submodular maximization problem.*

Proof $OPT_0 = OPT, OPT_n = X_n = Y_n$.

$$\begin{aligned}
\mathbb{E}[f(OPT_0) - f(OPT_n)] &= \sum_{i=1}^{n} \mathbb{E}[]f(OPT_{i-1}) - f(OPT_i)] \\
&\leq \frac{1}{2}\sum_{i=1}^{n} \mathbb{E}[f(X_i) - f(X_{i+1}) + f(Y_i) - f(Y_{i-1})] \\
&\leq \frac{1}{2}\mathbb{E}[f(X_n) + f(Y_n)]
\end{aligned}$$

□

Because [29] showed that no polynomial complexity algorithm can achieve $\frac{1}{2}+\epsilon$ approximation ratio under the assumption NP $\neq$ P, it follows that Algorithm 16 is a tight approximation algorithm for the unconstrained submodular maximization problem. In addition, [29] proved that the random subset $R(\frac{1}{2})$ of the ground set N, formed by independently choosing each element with a probability $1/2$, is a $\frac{1}{4}$ approximation solution for the unconstrained submodular maximization problem.

Theorem 12 ([29]) *For the unconstrained submodular maximization problem,* $\mathbb{E}[f(R(\frac{1}{2}))] \geq \frac{1}{4} f(OPT)$.

Proof The conclusion can be proved by a double application of Lemma. 4. Details see [29]. □

3.2.2 Continuous Algorithm

In this subsection, we will introduce some algorithms based on the multilinear extension of submodular set functions. Because cardinality constraints ($|S| \leq k$), matroid constraints ($\mathcal{M} = (N, \mathcal{I})$), knapsack constraints ($c(S) = \sum_{e \in S} c(e) \leq b$) are all down-closed and convex, the following algorithms concentrate on these types of constraints.

Finding stationary points is a common technique for continuous optimization. According to Lemma. 9, stationary points exhibit an interesting property, i.e., for a stationary point $\mathbf{x}$ and $\forall \mathbf{y} \in \mathcal{P}$, $2F(\mathbf{x}) \geq F(\mathbf{x} \vee \mathbf{y}) + F(\mathbf{x} \wedge \mathbf{y})$. Based on this inequality, [18] first proposed two approximation algorithms for stationary points in the context of the multilinear extension of submodular set functions. Here, we omit the proof. Readers can refer to [18] for further details.

Algorithm 17 Fractional local search

Input: function f, the polytope $\mathcal{P}$, parameter a
Output: $\mathbf{x}$

Initialize $M = \max\{f(e), f(N \setminus e) : e \in N\}$ $q = n^a$, $\delta = \frac{M}{n^{a-1}}$, $\mathbf{v}_i \leftarrow \mathbf{0}$
$\mathbf{x} \leftarrow \frac{1}{q} \sum_{i=1}^{q} \mathbf{v}_i$
while $\exists \mathbf{y} \in \mathcal{P}$ such that $(\mathbf{y} - \mathbf{x}) \cdot \nabla F(\mathbf{x}) \geq 4\delta n$ **do**
  $i \leftarrow \arg\max_{i \in [q]} F(\mathbf{x} - \mathbf{v_i} + \mathbf{y})$
  $\mathbf{x} \leftarrow \mathbf{x} - \mathbf{v_i} + \mathbf{y}$
end while

Theorem 13 ([18]) *For maximizing the multilinear extension of submodular set functions subject to down-closed convex constraints, repeated local search is a* $\frac{1}{4} - O(\frac{M}{n^{a-2}})$ *approximation algorithm, and local search is a* $0.309 - O(\frac{M}{n^{a-3}})$ *approximation algorithm when the polytope is reduced to* $\frac{3-\sqrt{5}}{2} \cdot \mathcal{P}$.

Inspired by the continuous greedy algorithm, [31] proposed a unified algorithm, namely, measured continuous greedy, whose approximation ratio is $1/e - O(\epsilon)$ for

Algorithm 18 Repeated local search for the multilinear submodular set functions

Input: function f, the polytope $\mathcal{P}$, parameter a
Output: $\mathbf{x}$

1: $\mathbf{x} \leftarrow$Fractional Local Search $(f, \mathcal{P}, a)$
2: $\mathcal{Q} \leftarrow \{\mathbf{y} \in \mathcal{P} | \mathbf{y} \leq \mathbf{1} - \mathbf{x}\}$
3: $\mathbf{y} \leftarrow$ Fractional Local Search $(f, \mathcal{Q}, a)$

non-monotone cases and $1 - 1/e - O(\epsilon)$ for monotone cases. The formal statement is as follows:

Algorithm 19 Measured continuous greedy algorithm

Input: function f, the ground set N, the feasible domain $\mathcal{I}$
Output: $\mathbf{x}(1)$

1: $\delta = \epsilon/n^3$, $\mathbf{x}(0) = \mathbf{0}$
2: **for** $t = 0$ to 1 **do**
3:     **for all** $e \in N$ **do**
4:         $w_e \leftarrow \frac{F(\mathbf{x}(t) \vee \mathbf{1}_e) - F(\mathbf{x}(t))}{1 - x_e}$
5:     **end for**
6:     $\mathbf{v}(\mathbf{x}(t)) = \arg\max_{\mathbf{v}(t) \in \mathcal{P}} \langle \mathbf{w}(t), \mathbf{v} \rangle$
7:     $\mathbf{x}(t + \delta) = \mathbf{x}(t) + \delta \mathbf{v}(\mathbf{x}(t)) \odot (\mathbf{1} - \mathbf{x}(t))$
8:     $t = t + \delta$
9: **end for**

Algorithm 19 starts from $\mathbf{0}$. The main difference from the continuous greedy algorithm lies in the forward directions. The measured continuous greedy employs a modified direction that is the hardmard product of the solution $\mathbf{v}(\mathbf{x}(t))$ returned by a suboptimization problem and $\mathbf{1} - \mathbf{x}(t)$. In each iteration, the algorithm moves along the direction $\mathbf{v}(\mathbf{x}(t)) \odot (\mathbf{1} - \mathbf{x}(t))$ with a step length of δ. The theoretical guarantee is as follows:

Theorem 14 ([31]) *Algorithm 19 is a $1 - 1/e - \epsilon$ approximation algorithm for monotone cases and a $1/e - \epsilon$ approximation algorithm for non-monotone cases.*

Proof

$$
\begin{aligned}
&F(\mathbf{x}(t+\delta)) - F(\mathbf{x}(t)) \\
&\quad \geq \langle \nabla F(\mathbf{x}(t)), \delta \mathbf{v}(\mathbf{x}(t)) \odot (\mathbf{1} - \mathbf{x}(t)) \rangle - O(n^3\delta^2) F(\mathbf{x}^*) \\
&\quad \geq \langle \nabla F(\mathbf{x}(t)), \delta \mathbf{x}^* \odot (\mathbf{1} - \mathbf{x}(t)) \rangle - O(n^3\delta^2) F(\mathbf{x}^*) \\
&\quad = \langle \nabla F(\mathbf{x}(t)), \delta (\mathbf{x}^* \vee \mathbf{x}(t) - \mathbf{x}(t)) \rangle - O(n^3\delta^2) F(\mathbf{x}^*) \\
&\quad \geq \delta (F(\mathbf{x}^* \vee \mathbf{x}(t)) - F(\mathbf{x}(t))) - O(n^3\delta^2) F(\mathbf{x}^*) \\
&\quad \geq \begin{cases} \delta (F(\mathbf{x}^*) - F(\mathbf{x}(t))) - O(n^3\delta^2) F(\mathbf{x}^*) & \text{if } F \text{ is monotone} \\ \delta ((1 - \|\mathbf{x}(t)\|_\infty) F(\mathbf{x}^*) - F(\mathbf{x}(t))) - O(n^3\delta^2) F(\mathbf{x}^*) & \text{if } F \text{ is non-monotone} \end{cases}
\end{aligned}
$$

The first inequality comes from Lemma 3. The equality is because of the integer $\mathbf{x}^*$. The third inequality comes from Lemma 8. The last inequality is due to Lemma 10. According to Line 7 in Algorithm 19, we have $\forall i \in [n]$, $x_i(t+\delta) \leq x_i(t) + \delta(1 - x_i(t))$. That means $x_i(t) \leq 1 - (1-\delta)^{t/\delta}$. Therefore, we have

$$F(\mathbf{x}(1)) \geq \begin{cases} (1-e^{-1})F(\mathbf{x}^*) - O(\epsilon) & \text{if } F \text{ is monotone} \\ e^{-1}F(\mathbf{x}^*) - O(\epsilon) & \text{if } F \text{ is non-monotone} \end{cases}$$

□

While the above algorithms have strong theoretical guarantees, they focus on only one class of algorithms. Ene and Nguyen [27] first combined two class of algorithms that are continuous greedy and double greedy. The algorithm starts from $\mathbf{0}$. Before time θ, it executes a reduced version of measured continuous greedy algorithm. After time θ, the algorithm goes back to the standard measured continuous greedy. Additionally, based on the forward direction $\mathbf{v}(\theta)$, the algorithm invokes the continuous double greedy as a subroutine to obtain a feasible solution $\mathbf{z}$. Finally, the final output is the best solution returned by either the stage of measured continuous greedy or the stage of double greedy.

Algorithm 20 Measured continuous greedy + double greedy

Input: function f, the ground set N, the feasible domain $\mathcal{I}$, parameter α, θ
Output: $\mathbf{x}$

$\delta = \epsilon/n^3$, $\mathbf{x}(0) = \mathbf{0}$
for $t = 0$ to 1 **do**
  for all $e \in N$ **do**
    $w_e \leftarrow \frac{F(\mathbf{x}(t) \vee \mathbf{1}_e) - F(\mathbf{x}(t))}{1 - x_e}$
  end for
  if $t \leq \theta$ **then**
    $\mathbf{v}(\mathbf{x}(t)) = \arg\max_{\mathbf{v}(t) \in \mathcal{P}, \|\mathbf{v}(t)\|_\infty \leq \alpha} \langle \mathbf{w}(t), \mathbf{v} \rangle$
  else
    $\mathbf{v}(\mathbf{x}(t)) = \arg\max_{\mathbf{v}(t) \in \mathcal{P}} \langle \mathbf{w}(t), \mathbf{v} \rangle$
  end if
  $\mathbf{x}(t+\delta) = \mathbf{x}(t) + \delta \mathbf{v}(\mathbf{x}(t)) \odot (\mathbf{1} - \mathbf{x}(t))$
  $t = t + \delta$
end for
$\mathbf{z} \leftarrow$ Double Greedy $(f, \mathbf{0}, \mathbf{v}(\mathbf{x}(\theta)))$
$\mathbf{x} = \arg\max_{\mathbf{x} \in \{\mathbf{x}(1), \mathbf{z}\}} F(\mathbf{x})$

Theorem 15 ([27]) *For each box* $[\mathbf{u}, \mathbf{v}]$ *constraints, the solution* $\mathbf{x}$ *returned by Algorithm 21 satisfies the following inequality*

$$F(\mathbf{x}) \geq \frac{1}{2}F(\mathbf{x}^*) + \frac{1}{4}F(\mathbf{u}) + \frac{1}{4}F(\mathbf{v}) \tag{33}$$

Algorithm 21 Double greedy

Input: function f, $\mathbf{u}$, $\mathbf{v}$

Output: $\mathbf{u}$

Initialize $\mathbf{u}_0 = \mathbf{u}$, $\mathbf{v}_0 = \mathbf{v}$

for $i = 0$ to n **do**

  $a_i \leftarrow (v_i - u_i)F(\mathbf{u}_{i-1} \vee \mathbf{1}_i) - F(\mathbf{u}_{i-1} \wedge \mathbf{1}_{[n]\setminus i})$

  $b_i \leftarrow (v_i - u_i)F(\mathbf{v}_{i-1} \wedge \mathbf{1}_{[n]\setminus i}) - F(\mathbf{v}_{i-1} \vee \mathbf{1}_i)$

  $a_i' \leftarrow \max\{a_i, 0\}$, $b_i' \leftarrow \max\{b_i, 0\}$

  $\mathbf{u}_i \leftarrow \mathbf{u}_{i-1} + \frac{a_i'}{a_i'+b_i'}\mathbf{1}_i(v_i - u_i)$, $\mathbf{v}_i \leftarrow \mathbf{v}_{i-1} - \frac{b_i'}{a_i'+b_i'}\mathbf{1}_i(v_i - u_i)$. If $a_i' + b_i' = 0$, $\mathbf{u}_i \leftarrow \mathbf{u}_{i-1} + \mathbf{1}_i(v_i - u_i)$, $\mathbf{v}_i \leftarrow \mathbf{v}_{i-1}$.

end for

Then, for maximizing the multilinear extension of submodular set functions subject to down-closed and convex constraints, Algorithm 20 is a $0.372 - O(\epsilon)$ *approximation algorithm when* $\theta = 0.18$ *and* $\alpha = 1/2$.

Algorithm 22 Aided measured continuous greedy algorithm

Input: function f, the ground set N, the feasible domain $\mathcal{I}$

Output: $\mathbf{x}$

$\mathbf{z} \leftarrow$ Fractional Local Search $(f, \mathcal{P}, \alpha)$

A random set $Z = R(\mathbf{z})$

$\delta_1 = \epsilon t_s/n^3, \delta_1 = \epsilon(1 - t_s)/n^3$ $\mathbf{x}(0) = \mathbf{0}$

for $t = 0$ to 1 **do**

  for all $e \in N$ **do**

    $w_e \leftarrow \frac{F(\mathbf{x}(t) \vee \mathbf{1}_e) - F(\mathbf{x}(t))}{1 - x_e}$

  end for

  if $t \leq t_s$ **then**

    $\mathbf{v}(\mathbf{x}(t)) = \arg\max_{\mathbf{v}(t) \in \mathcal{P}, \mathbf{v}(t) \leq \mathbf{1} - \mathbf{1}_Z} \langle \mathbf{w}(t), \mathbf{v}(t) \rangle$

    $\mathbf{x}(t + \delta_1) = \mathbf{x}(t) + \delta_1 \mathbf{v}(\mathbf{x}(t)) \odot (\mathbf{1} - \mathbf{x}(t))$

    $t = t + \delta_1$

  else

    $\mathbf{v}(\mathbf{x}(t)) = \arg\max_{\mathbf{v}(t) \in \mathcal{P}} \langle \mathbf{w}(t), \mathbf{v}(t) \rangle$

    $\mathbf{x}(t + \delta_2) = \mathbf{x}(t) + \delta_2 \mathbf{v}(\mathbf{x}(t)) \odot (\mathbf{1} - \mathbf{x}(t))$

    $t = t + \delta_2$

  end if

end for

$\mathbf{x} = \arg\max_{\mathbf{x} \in \{\mathbf{x}(1), \mathbf{z}\}} F(\mathbf{x})$

Note that the solution returned by the measured continuous greedy is not always a stationary point. Buchbinder and Feldman [10] first adopted a stationary point to guide the measured continuous greedy, which yields a 0.385 approximation algorithm for maximizing the multilinear extension of submodular set functions subject to down-closed convex constraints. Algorithm 22 first obtains a stationary point $\mathbf{z}$ from the fractional local search and randomly chooses a subset $Z = R(\mathbf{z})$. In the procedure of the measured continuous greedy, there is two stages. Before time t_s, the algorithm determines a forward direction based on a suboptimization problem within the dimension of $N \setminus Z$, while the forward direction on dimension Z is set to $\mathbf{0}$. After time t_s, the algorithm goes back to the standard measured continuous greedy. Ultimately, Algorithm 22 returns the best solution obtained from both the fractional local search and the measured continuous greedy as its output. The theoretical guarantee is as follows:

Theorem 16 ([10]) *Algorithm 22 is a* $0.385 - O(\epsilon)$ *for maximizing the multilinear extension of submodular set functions subject to down-closed convex constraints.*

Proof The proof of the measured continuous greedy stage is similar to that of standard measured continuous greedy. First, set $(t - t_s)^+ = \max\{0, t - t_s\}$, we have

$$x_i(t) \le \begin{cases} 1 - e^{-t} & \text{if } i \notin Z \\ 1 - e^{-(t-t_s)^+} & \text{if } i \in Z \end{cases}$$

$$\begin{aligned} F(\mathbf{x}(t+\delta)) - F(\mathbf{x}(t)) &\ge \langle \nabla F(\mathbf{x}(t)), \delta \mathbf{v}(\mathbf{x}(t)) \rangle - O(n^3\delta^2)F(\mathbf{x}^*) \\ &\ge \langle \nabla F(\mathbf{x}(t)), \delta \mathbf{x}^* \odot \mathbf{y} \odot (\mathbf{1} - \mathbf{x}(t)) \rangle - O(n^3\delta^2)F(\mathbf{x}^*) \\ &= \langle \nabla F(\mathbf{x}(t)), \delta((\mathbf{x}^* \wedge \mathbf{y}) \vee \mathbf{x}(t) - \mathbf{x}(t)) \rangle - O(n^3\delta^2)F(\mathbf{x}^*) \\ &\ge \delta(F((\mathbf{x}^* \wedge \mathbf{y}) \vee \mathbf{x}(t)) - F(\mathbf{x}(t))) - O(n^3\delta^2)F(\mathbf{x}^*) \end{aligned}$$

For convenience, denote $\mathbf{y} = \mathbf{1} - \mathbf{1}_Z$ if $t \le t_s$ and $\mathbf{y} = \mathbf{1}$ if $t > t_s$. The equality is because $\mathbf{x}^*$ and $\mathbf{y}$ are integer. For convenience, denote $\mathbf{m} = \mathbf{1}_M$.

$$\begin{aligned} F(\mathbf{m} \vee \mathbf{x}) \ge F_L(\mathbf{m} \vee \mathbf{x}) &= \int_0^1 f(T_\lambda(\mathbf{x} \vee \mathbf{m}))d\lambda = \int_0^1 f(T_\lambda(\mathbf{x}) \cup M)d\lambda \\ &\ge \int_{1-e^{-(t-t_s)^+}}^{1-e^{-t}} f(T_\lambda(\mathbf{x}) \cup M)d\lambda + \int_{1-e^{-t}}^{1} f(T_\lambda(\mathbf{x}) \cup M)d\lambda \\ &= \int_{1-e^{-(t-t_s)^+}}^{1-e^{-t}} f((T_\lambda(\mathbf{x}) \cap (N \setminus Z)) \cup M)d\lambda + e^{-t} f(M) \\ &\ge \int_{1-e^{-(t-t_s)^+}}^{1-e^{-t}} f(M) - f(M \cup Z)d\lambda + e^{-t} f(M) \\ &= e^{-(t-t_s)^+} f(M) - (e^{(t-t_s)^+} - e^{-t}) f(M \cup Z) \end{aligned}$$

The last inequality comes from the submodularity, i.e., $f((T_\lambda(\mathbf{x}) \cap (N \setminus Z)) \cup M) + f(M \cup Z) \ge f(M) + f(M \cup Z \cup (T_\lambda(\mathbf{x}) \cap (N \setminus Z)))$. If $t \le t_s$, $M = OPT \cap (N \setminus Z) = \mathbf{x}^* \wedge \mathbf{1}_{\hat{Z}}$. If $t > t_s$, $M = OPT = \mathbf{x}^*$. Therefore, it is easy to check

$$F(\mathbf{x}(1)) \ge e^{t_s-1}[(2-t_s-e^{-t_s})F(\mathbf{x}^*) - (1-e^{-t_s})F(\mathbf{z} \wedge \mathbf{x}^*) - (2-t_s-2e^{-t_s})F(\mathbf{z} \vee \mathbf{x}^*)]$$

Combining the theoretical guarantees of stationary points with Corollary 1, we have

$$\begin{aligned}\max\{F(\mathbf{x}(1)), \mathbf{z}\} &\geq (1-p)F(\mathbf{x}(1)) + pF(\mathbf{z}) \\ &\geq (1-p)e^{t_s-1}(2-t_s-e^{-t_s})F(\mathbf{x}^*) \\ &\quad + [p\alpha - (1-p)e^{t_s-1}(1-e^{-t_s})]F(\mathbf{z}\wedge\mathbf{x}^*) \\ &\quad + [p(1-\alpha) - (1-p)e^{t_s-1}(2-t_s-2e^{-t_s})]F(\mathbf{z}\vee\mathbf{x}^*)\end{aligned}$$

When $t_s = 0.372$ and $p = 0.23$, the approximation ratio is 0.385, and coefficients before $F(\mathbf{z}\wedge\mathbf{x}^*)$, $F(\mathbf{z}\vee\mathbf{x}^*)$ are nonnegative. □

4 Continuous Differential Functions

In this section, we introduce some algorithms for maximizing continuous DR-submodular functions. Formally, we first give the statement as follows:

$$\begin{aligned} &\max F(\mathbf{x}) \\ \text{s.t.} \quad &\mathbf{x} \in \mathcal{P} \subseteq \mathbb{R}^n \end{aligned} \tag{34}$$

where $F(\mathbf{x})$ is differentiable and $\mathcal{P}$ is convex and down-closed. Without loss of generality, we assume $F(\mathbf{0}) = 0$ and $\mathcal{P} \subseteq [0, 1]^n$.

4.1 Monotone Cases

In this subsection, we retrospect algorithms for monotone DR-submodular functions. Bian et al. [6] first proposed a Frank-Wolfe variant algorithm for maximizing monotone DR-submodular functions subject to down-closed convex constraints, achieving an approximation ratio of $1 - e^{-1} - O(\epsilon)$. Notably, this Frank-Wolfe variant algorithm can also effectively handle general convex constraints while keeping the same approximation ratio $1 - e^{-1} - O(\epsilon)$.

Algorithm 23 starts from $\mathbf{0}$. In each iteration, the algorithm solves a suboptimization problem to obtain a move forward direction. The main difference from the standard Frank-Wolfe lies in the iteration formula. In Algorithm 23, the update is given by $\mathbf{x}(t+\delta) = \mathbf{x}(t) + \delta\mathbf{v}(\mathbf{x}(t))$, whereas in the standard Frank-Wolfe, the update takes the form $\mathbf{x}(t+\delta) = \mathbf{x}(t) + \delta(\mathbf{x}(t) - \mathbf{v}(\mathbf{x}(t)))$. The formal statement is as follows:

Theorem 17 ([6, 24]) *Algorithm 23 is a* $1 - e^{-1} - \epsilon$ *approximation algorithm for maximizing monotone DR-submodular functions subject to convex constraints.*

Algorithm 23 Frank-Wolfe variant

Input: function f, the feasible domain $\mathcal{P}$, error ϵ
Output: $\mathbf{x}(1)$

```
x(0) ← 0, t ← 0, δ ← ε/(nL)
while t < 1 do
    v(x(t)) = argmax_{v(t)∈P} ⟨∇F(x(t)), v(t)⟩
    x(t + δ) = x(t) + δv(x(t))
    t = t + δ
end while
```

Proof The proof is similar with the continuous greedy algorithm. Here, we only introduce the difference.

$$\begin{aligned} F(\mathbf{x}(t+\delta)) - F(\mathbf{x}(t)) &\geq \langle \nabla F(\mathbf{x}(t)), \delta \mathbf{v}(\mathbf{x}(t))\rangle - \frac{L}{2}\left\|\mathbf{x}(t+\delta) - \mathbf{x}(t)\right\|^2 \\ &\geq \langle \nabla F(\mathbf{x}(t)), \delta \mathbf{x}^*\rangle - \frac{L\delta^2}{2}\left\|\mathbf{v}(\mathbf{x}(t))\right\|^2 \\ &\geq \langle \nabla F(\mathbf{x}(t)), \delta(\mathbf{x}^* \vee \mathbf{x}(t) - \mathbf{x}(t))\rangle - \frac{L\delta^2 n}{2} \\ &\geq \delta(F(\mathbf{x}^* \vee \mathbf{x}(t)) - F(\mathbf{x}(t))) - \frac{L\delta^2 n}{2} \\ &\geq \delta(F(\mathbf{x}^*) - F(\mathbf{x}(t))) - \frac{L\delta^2 n}{2} \end{aligned}$$

The first inequality comes from Lemma 11. The third inequality is because of the monotonicity of F, i.e., $\nabla F(\mathbf{x}(t)) \geq 0$ and $\mathbf{x}^* \geq \mathbf{x}^* \vee \mathbf{x}(t) - \mathbf{x}(t) \geq 0$. The forth inequality comes from Lemma 8. Therefore, we have

$$F(\mathbf{x}(1)) = (1 - (1-\delta)^{1/\delta})F(\mathbf{x}^*) - O(L\delta n)F(\mathbf{x}^*) \geq (1 - e^{-1} - O(\epsilon))F(\mathbf{x}^*)$$

□

It is important to note that the above algorithm achieves a tight approximation ratio for monotone DR-submodular functions subject to convex constraints, as directly inferred from the hardness results for maximizing submodular set functions. However, the solution returned by Frank-Wolfe variant algorithm is not necessarily a stationary point. Regarding stationary points, [42] first demonstrated that stationary points yield a $1/2$ approximation solution.

Theorem 18 ([42]) *For maximizing monotone DR-submodular functions, the stationary point is a* $1/2$ *approximation solution. In addition, the approximation ratio* $1/2$ *is tight.*

Proof For any stationary point $\mathbf{x}$, we have

$$F(\mathbf{x}) \geq \frac{1}{2}F(\mathbf{x} \vee \mathbf{x}^*) + \frac{1}{2}F(\mathbf{x} \wedge \mathbf{x}^*) \geq \frac{1}{2}F(\mathbf{x}^*)$$

The first inequality comes from the definition of stationary points. The second inequality is due to the monotonicity and nonnegativity. The tight example can be found in [42]. □

Building on this phenomenon and the non-oblivious local search framework for submodular set functions, [69] constructed a non-oblivious function $\tilde{F}(\mathbf{x})$ for monotone continuous DR-submodular functions. They proved that the stationary point of $\tilde{F}(\mathbf{x})$ is a $1 - e^{-1}$ approximation solution of $F(\mathbf{x})$. Here, we restate the conclusion and omit the details of proof.

Theorem 19 ([69]) *For a continuous monotone DR-submodular function F, its non-oblivious function is as follows:*

$$\tilde{F}(\mathbf{x}) = \int_0^1 \frac{e^{z-1}}{z} F(z \cdot \mathbf{x}) dz \tag{35}$$

If $\mathbf{x} \in \mathcal{P}$ is a stationary point of $\tilde{F}(\mathbf{x})$, then we have

$$F(\mathbf{x}) \geq (1 - e^{-1})F(\mathbf{x}^*) \tag{36}$$

4.2 Non-monotone Cases

In this subsection, we review algorithms for non-monotone DR-submodular functions, focusing on three types of constraints: convex constraints, convex and down-closed constraints, and box constraints. Before delving into algorithms, we introduce a negative result concerning stationary points for non-monotone DR-submodular functions.

Theorem 20 ([20]) *For non-monotone DR-submodular functions, stationary points can yield arbitrarily bad approximation ratios. In addition, these undesirable stationary points are usually on the boundary of the feasible domain.*

Proof The proof comes from a concrete example for maximizing unconstrainted regular coverage functions. Here, we just give its multilinear extension.

$$F(\mathbf{x}) = k+1-(1 - x_{2k+1})\prod_{i=1}^{k}(1 - x_i)-(1 - x_{2k+1})\left(k - \sum_{i=1}^{k} x_i\right)-\sum_{i=1}^{k} x_i - x_{2k+1}$$

where the size of ground set $|N| = 2k + 1$. There are two notable stationary points $\mathbf{x}$ and $\mathbf{y}$ such that $F(\mathbf{x}) = 1$ and $F(\mathbf{y}) = k$. This leads to an upper bound of the approximation ratio for stationary points $\frac{F(\mathbf{x})}{F(\mathbf{y})} = \frac{1}{k}$. As k increases, the approximation ratio approaches 0. It implies that the approximation ratio can degrade significantly for large k.

$$\mathbf{x} = \begin{cases} 1 & i \in [2k] \\ 0 & i = 2k+1 \end{cases} \qquad \mathbf{y} = \begin{cases} 0 & i \in [2k] \\ 1 & i = 2k+1 \end{cases}$$

□

Algorithm 24 Frank-Wolfe

Input: function f, the feasible domain $\mathcal{P}$, error ϵ
Output: $\mathbf{x}(1)$

1: $\mathbf{x}(0) \leftarrow \arg\min_{\mathbf{x}\in\mathcal{P}} \|\mathbf{x}\|_\infty$, $t \leftarrow 0$, $\delta \leftarrow \frac{\epsilon}{nL}$, $a(t) = (1+t)^2$, $b(t) = t$
2: **while** $t < 1$ **do**
3: $\quad \mathbf{v}(\mathbf{x}(t)) = \arg\max_{\mathbf{v}(t)\in\mathcal{P}} \langle \nabla F(\mathbf{x}(t)), \mathbf{v}(t)\rangle$
4: $\quad \mathbf{x}(t+\delta) = \mathbf{x}(t) + \frac{(b(t+\delta)-b(t))\sqrt{a(t)}}{a(t+\delta)}(\mathbf{v}(\mathbf{x}(t)) - \mathbf{x}(t))$
5: $\quad t = t + \delta$
6: **end while**

Convex Constraints Firstly, we consider the maximization problem of continuous non-monotone DR-submodular subject to convex constraints. Dürr et al. [26] first adopted Frank-Wolfe algorithm for this purpose, achieving an approximation ratio of $\frac{1-\min_{\mathbf{x}\in\mathcal{P}}\|\mathbf{x}\|_\infty}{3\sqrt{3}}$ with $O(e^{\sqrt{n/\epsilon}})$ complexity. However, their analysis is not tight. Subsequently, [25] provided a tighter analysis of Frank-Wolfe algorithm that is $\frac{1-\min_{\mathbf{x}\in\mathcal{P}}\|\mathbf{x}\|_\infty}{4}$ approximation algorithm. In addition, [24] reduced the sub-exponential complexity to polynomial complexity.

Algorithm 24 selects the point closest to the origin $\mathbf{0}$ as the initial solution. In each iteration, the forward direction is determined by a suboptimization problem. Along this direction, the step length is adaptive and depends on the current number of iterations. The formal statement is as follows:

Theorem 21 ([24]) *Algorithm 24 is a* $\frac{1-\min_{\mathbf{x}\in\mathcal{P}}\|\mathbf{x}\|_\infty}{4}$ *approximation algorithm with the complexity of* $O(\frac{nL}{\epsilon})$.

Proof Here, we adopt the Lyapunov framework proposed by Du [24] to illustrate the proof sketch. First, we define the Lyapunov function: $E(t) = a(t)F(\mathbf{x}(t)) - b(t)F(\mathbf{x}^*)$, where we choose $a(t) = (1+t)^2, b(t) = t$. According to Line 4 in Algorithm 24, it is easy to check $\mathbf{1} - \mathbf{x}(t) \geq \frac{\sqrt{a(0)}}{\sqrt{a(t)}}(1 - \mathbf{x}(0))$. Then, in each iteration, the marginal gain of Lyapunov function is as follows:

$$\begin{aligned}
E(t+\delta)-E(t) &= a(t+\delta)[F(\mathbf{x}(t+\delta))-F(\mathbf{x}(t))]+[a(t+\delta)-a(t)]F(\mathbf{x}(t))\\
&\quad -[b(t+\delta)-b(t)]F(\mathbf{x}^*)\\
&\geq a(t+\delta)[\langle \nabla F(\mathbf{x}(t)), \mathbf{x}(t+\delta)-\mathbf{x}(t)\rangle - \frac{L}{2}\left\|\mathbf{x}(t+\delta)-\mathbf{x}(t)\right\|^2]\\
&\quad +[a(t+\delta)-a(t)]F(\mathbf{x}(t))-[b(t+\delta)-b(t)]F(\mathbf{x}^*)\\
&\geq [b(t+\delta)-b(t)]\sqrt{a(t)}\langle \nabla F(\mathbf{x}(t)), \mathbf{v}(\mathbf{x}(t))-\mathbf{x}(t)\rangle\\
&\quad +[a(t+\delta)-a(t)]F(\mathbf{x}(t))-[b(t+\delta)-b(t)]F(\mathbf{x}^*)\\
&\quad -nL\frac{(b(t+\delta)-b(t))^2a(t)}{a(t+\delta)^2}\\
&\geq [b(t+\delta)-b(t)]\sqrt{a(t)}[F(\mathbf{x}^*\vee \mathbf{x}(t))\\
&\quad +F(\mathbf{x}^*\wedge \mathbf{x}(t))-2F(\mathbf{x}(t))]\\
&\quad +[a(t+\delta)-a(t)]F(\mathbf{x}(t))-[b(t+\delta)-b(t)]F(\mathbf{x}^*)\\
&\quad -nL\frac{(b(t+\delta)-b(t))^2a(t)}{a(t+\delta)^2}\\
&\geq [a(t+\delta)-a(t)-2\sqrt{a(t)}(b(t+\delta)-b(t))]F(\mathbf{x}(t))\\
&\quad -(b(t+\delta)-b(t))\left\|\mathbf{x}(0)\right\|_\infty F(\mathbf{x}^*)\\
&\quad -nL\frac{(b(t+\delta)-b(t))^2a(t)}{a(t+\delta)^2}\\
&\geq \delta^2F(\mathbf{x}(t))-nL\delta^2-\delta\left\|\mathbf{x}(0)\right\|_\infty F(\mathbf{x}^*)\\
&\geq -nL\delta^2-\delta\left\|\mathbf{x}(0)\right\|_\infty F(\mathbf{x}^*)
\end{aligned}$$

The first inequality comes from Lemma 11. The third inequality is because of Lemma 8. The forth inequality comes from Lemma 10. The above inequality implies that

$$\begin{aligned}
E(1)-E(0) &= a(1)F(\mathbf{x}(1))-(b(1)-b(0))F(\mathbf{x}^*)-a(0)F(\mathbf{x}(0))\\
&\geq -nL\delta-\left\|\mathbf{x}(0)\right\|_\infty F(\mathbf{x}^*)
\end{aligned}$$

Therefore, we have

$$F(\mathbf{x}(1)) \geq \frac{1-\left\|\mathbf{x}(0)\right\|_\infty}{4}F(\mathbf{x}^*)-O(\epsilon)$$

□

Note that [55] proved that there is no polynomial complexity approximation algorithm with an approximation ratio $\frac{1}{4} + \epsilon$ under the assumption P $\neq$NP. This implies that the above Frank-Wolfe algorithm is a tight approximation algorithm for maximizing non-monotone DR-submodular functions subject to convex constraints. Although stationary points have arbitrarily bad theoretical guarantees, similar to monotone cases, [70] constructed a non-oblivious function $\tilde{F}$ such that its stationary points are $1/4$ approximation solutions for maximizing the original non-monotone DR-submodular function subject to convex constraints.

Theorem 22 ([70]) *For a continuous non-monotone DR-submodular function F, define its non-oblivious function as follows:*

$$\tilde{F}(\mathbf{x}) = \int_0^1 \frac{1}{4z(1-\frac{z}{2})^3}\left[F(\frac{z}{2}\cdot(\mathbf{x}-\mathbf{x}_{\min})+\mathbf{x}_{\min}) - F(\mathbf{x}_{\min})\right]dz \tag{37}$$

where $\mathbf{x}_{\min} = \arg\max_{\mathbf{x}\in\mathcal{P}} \|\mathbf{x}\|_\infty$. If $\mathbf{x} \in \mathcal{P}$ is a stationary point of $\tilde{F}(\mathbf{x})$ subject to convex constraints, then we have

$$F(\mathbf{x}) \geq \frac{1-\|\mathbf{x}\|_\infty}{4} F(\mathbf{x}^*) \tag{38}$$

Convex and Down-Closed Constraints Building upon convex sets, we incorporate down-closedness into constraints. Bian et al. [5] extended Frank-Wolfe variant algorithm from monotone cases to non-monotone cases, which serves as a unified algorithm both monotone cases with $1 - e^{-1} - O(\epsilon)$ approximation ratio and non-monotone cases with $e^{-1} - O(\epsilon)$ approximation ratio.

Algorithm 25 starts from $\mathbf{0}$. In each iteration, the initial forward direction $\mathbf{v}(\mathbf{x}(t))$ is derived from a suboptimization problem. The algorithm then adjusts this forward direction using $\mathbf{1} - \mathbf{x}(t)$. Specially, the algorithm adopts the Hardmard product of $\mathbf{v}(\mathbf{x}(t))$ and $\mathbf{1} - \mathbf{x}(t)$ as its forward directions. The formal statement is as follows:

Algorithm 25 Frank-Wolfe variant

Input: function f, the feasible domain $\mathcal{P}$, error ϵ
Output: $\mathbf{x}(1)$

1: $\mathbf{x}(0) \leftarrow \mathbf{0}, t \leftarrow 0, \delta \leftarrow \frac{\epsilon}{nL}$
2: **while** $t < 1$ **do**
3: $\quad \mathbf{v}(\mathbf{x}(t)) = \arg\max_{\mathbf{v}(t)\in\mathcal{P}, \mathbf{v}(t)\leq \mathbf{1}-\mathbf{x}(t)} \langle \nabla F(\mathbf{x}(t)), \mathbf{v}(t)\rangle$
4: $\quad \mathbf{x}(t+\delta) = \mathbf{x}(t) + \delta \mathbf{v}(\mathbf{x}(t))$
5: $\quad t = t + \delta$
6: **end while**

Theorem 23 ([5]) *Algorithm 25 is a $1 - 1/e - \epsilon$ approximation algorithm for monotone cases and a $1/e - \epsilon$ approximation algorithm for non-monotone cases.*

Proof

$$
\begin{aligned}
&F(\mathbf{x}(t+\delta)) - F(\mathbf{x}(t)) \\
&\geq \langle \nabla F(\mathbf{x}(t)), \delta \mathbf{v}(\mathbf{x}(t)) \rangle - \frac{L}{2}\|\mathbf{x}(t+\delta) - \mathbf{x}(t)\|^2 \\
&\geq \langle \nabla F(\mathbf{x}(t)), \delta(\mathbf{x}^* \vee \mathbf{x}(t) - \mathbf{x}(t)) \rangle - \frac{nL\delta^2}{2} \\
&\geq \delta(F(\mathbf{x}^* \vee \mathbf{x}(t)) - F(\mathbf{x}(t))) - \frac{nL\delta^2}{2} \\
&\geq \begin{cases} \delta(F(\mathbf{x}^*) - F(\mathbf{x}(t))) - \frac{nL\delta^2}{2} & \text{if } F \text{ is monotone} \\ \delta((1 - \|\mathbf{x}(t)\|_\infty)F(\mathbf{x}^*) - F(\mathbf{x}(t))) - \frac{nL\delta^2}{2} & \text{if } F \text{ is non-monotone} \end{cases}
\end{aligned}
$$

The first inequality comes from Lemma 11. The third inequality comes from Lemma 8. The last inequality is due to Lemma 10. According to Line 4 in Algorithm 25, we have $\forall i \in [n]$, $x_i(t+\delta) \leq x_i(t) + \delta(1 - x_i(t))$. That means $x_i(t) \leq 1 - (1-\delta)^{t/\delta}$. Therefore, we have

$$
F(\mathbf{x}(1)) \geq \begin{cases} (1 - e^{-1})F(\mathbf{x}^*) - O(\epsilon) & \text{if } F \text{ is monotone} \\ e^{-1}F(\mathbf{x}^*) - O(\epsilon) & \text{if } F \text{ is non-monotone} \end{cases}
$$

□

Inspired by the observation that bad stationary points are probably closed to the boundary of the feasible domain, [20] employed stationary points to guide the Frank-Wolfe algorithm, ensuring a solution is far from these stationary points. Algorithm 26 first obtains a stationary point $\mathbf{z}$ by Algorithm 27. Subsequently, based on the stationary point $\mathbf{z}$, the algorithm generates a subset Z of N. During the stage of Frank-Wolfe variant, before time t_s, Algorithm 26 sets the forward direction in the dimension corresponding to Z to $\mathbf{0}$ and obtains the forward direction on $N \setminus Z$ from a suboptimization problem. After time t_s, the algorithm goes back to the standard Frank-Wolfe variant. The formal statement is as follows:

Theorem 24 ([20]) *Algorithm 26 is a* $0.385 - O(\epsilon)$ *approximation algorithm for maximizing non-monotone DR-submodular functions subject to down-closed convex constraints.*

Proof Here, we only proof the relationship between $F(\mathbf{x} \vee \mathbf{m})$ and $F(\mathbf{m})$. The reminder proof is similar with the above analysis. Details see [20].

For any $\mathbf{m} \in \mathcal{P}$ and a partition $A \cap \hat{A} = \emptyset$, $A \cup \hat{A} = [n]$, assume $\theta_A(t) = \max_{i \in A} |x_i| \leq \theta_{\hat{A}}(t) = \max_{i \in N \setminus A} |x_i|$. First, we have $\mathbf{x} \vee \mathbf{m} = \left(1 - \theta_{\hat{A}}\right)\mathbf{m} + \theta_{\hat{A}}\left[\mathbf{m} + \frac{1}{\theta_{\hat{A}}}(\mathbf{x} \vee \mathbf{m} - \mathbf{m})\right] = \left(1 - \theta_{\hat{A}}\right)\mathbf{m} + \theta_{\hat{A}}\mathbf{p}$. Then, we can define three sets $B_1 = \{i \in [n] : m_i \geq x_i\}$, $B_2 = \left\{i \in [n] : x_i > m_i, p_i \leq \frac{\theta_A}{\theta_{\hat{A}}}\right\}$, $B_3 =$

Algorithm 26 Aided Frank-Wolfe variant

Input: function f, the feasible domain $\mathcal{P}$, error ϵ, the number of iteration K, time $t_s = 0.372$
Output: $\mathbf{x}$

$\mathbf{z} \leftarrow$ Non-Convex Frank-Wolfe($f, \mathcal{P}, \epsilon, K$)
$Z \leftarrow R(\mathbf{z})$ is a random subset of N
$\mathbf{x}(0) \leftarrow \mathbf{0}, t \leftarrow 0, \delta_1 \leftarrow \frac{\epsilon t_s}{nL}, \delta_2 \leftarrow \frac{\epsilon(1-t_s)}{nL}$
while $t < t_s$ **do**
  $\mathbf{v}(\mathbf{x}(t)) = \arg\max_{\mathbf{v}(t) \in \mathcal{P}, \mathbf{v}(t) \leq (\mathbf{1}-\mathbf{x}(t)) \odot (\mathbf{1}-\mathbf{1}_Z)} \langle \nabla F(\mathbf{x}(t)), \mathbf{v}(t) \rangle$
  $\mathbf{x}(t+\delta) = \mathbf{x}(t) + \delta_1 \mathbf{v}(\mathbf{x}(t))$
  $t = t + \delta_1$
end while
while $t < 1$ **do**
  $\mathbf{v}(\mathbf{x}(t)) = \arg\max_{\mathbf{v}(t) \in \mathcal{P}, \mathbf{v}(t) \leq \mathbf{1}-\mathbf{x}(t)} \langle \nabla F(\mathbf{x}(t)), \mathbf{v}(t) \rangle$
  $\mathbf{x}(t+\delta) = \mathbf{x}(t) + \delta_2 \mathbf{v}(\mathbf{x}(t))$
  $t = t + \delta_2$
end while
return $\arg\max_{\mathbf{x} \in \{\mathbf{x}, \mathbf{z}\}} F(\mathbf{x})$

Algorithm 27 Non-convex Frank-Wolfe

Input: function f, the feasible domain $\mathcal{P}$, error ϵ, the number of iterations K
Output: $\mathbf{x}$

$\mathbf{x}(0) \leftarrow \mathbf{0}$
for $k = 0$ to K **do**
  $\mathbf{v}_k = \arg\max_{\mathbf{v}(t) \in \mathcal{P}} \langle \nabla F(\mathbf{x}_k), \mathbf{v}(t) \rangle$
  $g_k = \langle \nabla F(\mathbf{x}_k), \mathbf{v}_k - \mathbf{x}_k \rangle$
  if $g_k \leq \epsilon$ **then**
    return $\mathbf{x}_k$
  else
    $\gamma_k = \arg\max_{\gamma \in [0,1]} f(\mathbf{x}_k + \gamma(\mathbf{v}_k - \mathbf{x}_k))$
    $\mathbf{x}_{k+1} = \mathbf{x}_k + \gamma_k(\mathbf{v}_k - \mathbf{x}_k)$
  end if
end for
return $\mathbf{x}_j$ with $j = \arg\max_{i \in [K]} g_i$

$\left\{i \in [n] : x_i > m_i, 1 \geq p_i > \frac{\theta_A}{\theta_{\hat{A}}}\right\}$. Obviously, $B_3 \subseteq \hat{A}$ and $\mathbf{p}_{B_2} \geq \mathbf{m}_{B_2}$, implying that $\mathbf{p} = \mathbf{p}_{B_1} \vee \mathbf{p}_{B_2} \vee \mathbf{p}_{B_3} \vee \mathbf{m}_{B_2}$. Then, we have

$$\begin{aligned}
F(\mathbf{x}(t) \vee \mathbf{m}) &\geq \left(1 - \theta_{\hat{A}}\right) F(\mathbf{m}) + \theta_{\hat{A}} F\left(\mathbf{m} + \frac{1}{\theta_{\hat{A}}}(\mathbf{x} \vee \mathbf{m} - \mathbf{m})\right) \\
&= \left(1 - \theta_{\hat{A}}\right) F(\mathbf{m}) + \theta_{\hat{A}} F(\mathbf{p}_{B_1} \vee \mathbf{p}_{B_2} \vee \mathbf{p}_{B_3} \vee \mathbf{m}_{B_2}) \\
&\geq \left(1 - \theta_{\hat{A}}\right) F(\mathbf{m}) + \theta_{\hat{A}} \frac{\theta_{\hat{A}} - \theta_A}{\theta_{\hat{A}}} F(\mathbf{p}_{B_1} \vee \mathbf{p}_{B_3} \vee \mathbf{m}_{B_2}) \\
&\geq \left(1 - \theta_{\hat{A}}\right) F(\mathbf{m}) + \left(\theta_{\hat{A}} - \theta_A\right)(F(\mathbf{m}) - F(\mathbf{m} \vee \mathbf{1}_A)) \\
&= (1 - \theta_A) F(\mathbf{m}) - \left(\theta_{\hat{A}} - \theta_A\right) F(\mathbf{m} \vee \mathbf{1}_A),
\end{aligned}$$

where the last inequality comes from submodularity of F, i.e., $F(\mathbf{p}_{B_1} \vee \mathbf{p}_{B_3} \vee \mathbf{m}_{B_2}) + F(\mathbf{m} \vee \mathbf{1}_A) \geq F(\mathbf{m}) + F(\mathbf{m} \vee \mathbf{1}_A \vee \mathbf{p}_{B_1} \vee \mathbf{p}_{B_3} \vee \mathbf{m}_{B_2})$. □

From the perspective of improving stationary points performance, [6] developed a two-phase Frank-Wolfe algorithm that obtains two stationary points by two iterations of the non-convex Frank-Wolfe and returns the better one as the final output. Additionally, [20] proved that stationary points within a reduce feasible domain can achieve a 0.309 theoretical guarantee.

Algorithm 28 Two-phase non-convex Frank-Wolfe

Input: function f, the feasible domain $\mathcal{P}$, error ϵ, the number of iterations K
Output: $\mathbf{x}$

1: $\mathbf{x}_1 \leftarrow$ Non-Convex Frank-Wolfe $(f, \mathcal{P}, \epsilon, K)$
2: $\mathcal{Q} \leftarrow \mathcal{P} \cap \{\mathbf{y} \in \mathbf{R}^n | \mathbf{y} \leq \mathbf{1} - \mathbf{x}_1\}$
3: $\mathbf{x}_2 \leftarrow$ Non-Convex Frank-Wolfe $(f, \mathcal{Q}, \epsilon, K)$
4: **return** $\arg\max_{\mathbf{x} \in \{\mathbf{x}_1, \mathbf{x}_2\}} F(\mathbf{x})$

Algorithm 29 Reduced non-convex Frank-Wolfe

Input: function f, the feasible domain $\mathcal{P}$, error ϵ, the number of iterations K, parameter $m \in [0, 1]$
Output: $\mathbf{x}$

1: $\mathbf{x} \leftarrow$ Non-Convex Frank-Wolfe $(f, m \cdot \mathcal{P}, \epsilon, K)$

Theorem 25 ([6, 20]) *Algorithm 28 is a $1/4 - O(\epsilon)$ approximation algorithm, and Algorithm 29 is a $0.309 - O(\epsilon)$ approximation algorithm for maximizing non-monotone DR-submodular functions subject to down-closed convex constraints, when $m = \frac{3-\sqrt{5}}{2}$.*

In addition, [11] proposed a 0.401 approximation algorithm for non-monotone DR-submodular functions subject to convex and down-closed constraints. This algorithm integrates non-convex Frank-Wolfe, double greedy, Frank-Wolfe variant, and recursions techniques. Because the multilinear extension of submodular set functions is a special DR-submodular function, the algorithm proposed by Buchbinder and Feldman [11] is also a 0.401 approximation algorithm for submodular set functions. Due to space limitations, we will not detail it here; interested readers can refer to [11].

Box Constraints At last, we introduce a $1/2$ approximation algorithm for maximizing continuous DR-submodular functions subject to box constraints, i.e., $\mathcal{P} = [\mathbf{l}, \mathbf{u}]$. Bian et al. [6] proposed a continuous double greedy that includes a one-dimension optimization problem, achieving a $1/3$ approximation algorithm. On this basis, [57] proposed a $1/2$ continuous double greedy algorithm with binary search. Here, we only summarize the $1/2$ approximation algorithm.

Algorithm 30 maintains two vectors, $\mathbf{x}$ and $\mathbf{y}$. In each iteration, the algorithm identifies the component of the current dimension z_i through binary search based on $\mathbf{x}$ and $\mathbf{y}$. After n iterations, Algorithm 30 returns the vector $\mathbf{z}$ as the final output. The formal statement is as follows:

Algorithm 30 Binary search continuous double greedy

Input: function f, error ϵ
Output: $\mathbf{z}$

Initialize $\mathbf{x} = \mathbf{0}$, $\mathbf{y} = \mathbf{1}$
for $i = 0$ to n **do**
  if $\frac{\partial F}{\partial x_i}(0, \mathbf{x}_{-i}) \leq 0$ **then**
    $z_i \leftarrow 0$
  else if $\frac{\partial F}{\partial y_i}(1, \mathbf{x}_{-i}) \geq 0$ **then**
    $z_i \leftarrow 1$
  else
    while $y_i - x_i \geq \epsilon/n$ **do**
      $z_i = \frac{x_i + y_i}{2}$
      if $\frac{\partial F}{\partial x_i}(z_i, \mathbf{x}_{-i})(1 - z_i) + \frac{\partial F}{\partial y_i}(z_i, \mathbf{y}_{-i})z_i \leq 0$ **then**
        $x_i \leftarrow z_i$
      else
        $y_i \leftarrow z_i$
      end if
    end while
  end if
  $x_i \leftarrow z_i$, $y_i \leftarrow z_i$
end for

Theorem 26 ([57]) *Algorithm 30 is a* $1/2 - O(\epsilon)$ *approximation algorithm with* $O(n \log \frac{n}{\epsilon})$ *complexity for maximizing non-monotone DR-submodular functions subject to a box constraints.*

Proof The formal proof adopts ideas similar to Theorem 11. Here, we omit it. □

In addition, [11] proposed a different version of double greedy algorithm, whose approximation ratio is also $1/2$. The main idea and the proof are similar to those presented by Buchbinder et al. [14] and Niazadeh et al. [57]. For readers interested in details, please refer to [11].

5 Discussion

In this section, we partially mention some applications of DR-submodular maximization and some directions of future efforts.

5.1 Applications

In discrete optimization and non-convex continuous optimization, numerous problems are DR-submodular maximization problems. Specifically, many challenges in fields such as social network analysis and machine learning can be framed as submodular set function maximization problems or continuous DR-submodular function maximization problems. The following highlight several types of specific application scenarios.

Non-convex/concave Quadratic Programming (NQP) The goal is to maximize a quadratic function subject to linear constrains. The formal statement is as follows:

$$\begin{aligned} \max \quad & \frac{1}{2}\mathbf{x}^T\mathbf{H}\mathbf{x} + \mathbf{h}\mathbf{x} + c \\ \text{s.t.} \quad & \mathbf{A}\mathbf{x} \le \mathbf{b} \end{aligned} \tag{39}$$

where $\mathbf{H} \in \mathbb{R}^{n\times n}$, $\mathbf{A} \in \mathbb{R}^{m\times n}$, $\mathbf{h} \in \mathbb{R}^{1\times n}$, $\mathbf{b} \in \mathbb{R}^{m\times 1}$, $c \in \mathbb{R}$. If $\mathbf{H} \in \mathbb{R}^{n\times n}_{\le 0}$, the above optimization is a DR-submodular maximization problem subject to down-closed convex constraints.

Softmax Extension of Determinantal Point Process (DPPs) DPPs is a significant probabilistic model. The formal description is as follows:

$$F(\mathbf{x}) = \log\det[\mathrm{diag}(\mathbf{x})(\mathbf{L} - \mathbf{I}) + \mathbf{I}] \tag{40}$$

where diag($\mathbf{x}$) is the diagonal matrix based on the vector $\mathbf{x} \in [0, 1]^n$ and $\mathbf{I}$ is the identity matrix. It is easy to check Eq. (40) is DR-submodular.

Influence Maximization For a given social network $\mathcal{G}(V, E)$, choosing at most k seed nodes as the initial users, i.e., $S \subseteq V, |S| \le k$, such that the information spreading is as large as possible. The formal statement is as follows:

$$\begin{aligned} \max \quad & \sigma(S) \\ \text{s.t.} \quad & S \subseteq V, |S| \le k \end{aligned} \tag{41}$$

Under the independent cascade model and the linear threshold model, $\sigma(S)$ is a monotone submodular set function.

Facility Location Facility location is a conventional submodular function. The goal is to optimally serve a set of customers. It can be adopted to select a subset of data for faster training of machine learning models. The formal statement is as follows:

$$f(S) = \sum_{u\in N} \max_{s\in S} \phi(s, u) \tag{42}$$

where N is a data set, S is the selected subset where $S \subseteq N$, and $\phi(s, u)$ is the similarity between the examples.

DR-submodular maximization has numerous applications in social network, economics, and machine learning. Here, we briefly highlight a few examples. For readers interested in artificial intelligence, please refer to the survey [7].

5.2 *Future Directions*

Regarding in-approximation results for non-monotone DR-submodular maximization, [36] showed that no algorithm can achieve an approximation ratio better than 0.478 under a matroid independence constrain. On this basis, [60] refined the in-approximation result from 0.491 to 0.478 under cardinality constraints. However, the best existing approximation algorithm for non-monotone DR-submodular maximization subject to either a matroid constraint or cardinality constraints is 0.401 approximation [11]. Bridging the theoretical gap between algorithmic and in-approximation results including but not limited to the above problem remains a critical open challenge in the area of DR-submodular maximization.

From the perspective of combinatorial algorithms for non-monotone submodular set maximization, low complexity algorithms have tremendous value, especially for large-scale problems such as social networks. The existing best approximation algorithm often faces challenges when adopted to solve concrete application problems. First, these approximation algorithms typically obtain a fractional solution that needs a rounding scheme to yield an integer solution. Secondly, in each iteration, the algorithm obtains the forward direction by maximizing a linear function with millions of variables. The above two procedures are inherently time-consuming. In addition, accurately estimating the multilienar extension of set functions can also be expensive. To effectively tackle large-scale non-monotone submodular maximization problems, there is an urgent problem to design a faster and more efficient approximation algorithm. A promising approach to efficiently maximize large-scale non-monotone submodular set functions is the combinatorial algorithm, which can directly circumvent the above time-consuming procedure. Moreover, designing combinatorial algorithms with much better approximation ratios, such as 0.401, would have substantial theoretical value, helping to fill the algorithmic gap in submodular set maximization.

These open problems mentioned earlier are from a view of the conventional computation model. In recent decades, DR-submodular maximization has also been extensively studied in other interesting models, such as the streaming model, the online arrival model, the online action model, and the stochastic model, all of which has many potential applications in social network and machine learning. Under the streaming model, the design of algorithm must consider not only its approximation ratio and computational complexity but also their memory footprint. Many applications involve massive data streams, for example, images, videos, texts, and tweets, that are generated at an unprecedented volume. This rapid production means that decision-makers cannot store all data in memory. In other words, classical submodular maximization algorithms, which assume that the ground set is known, cannot handle these massive data streams. Under the action online model, the scenario involves a sequence of consecutive rounds, where the algorithm must make an action in each round. The outcome of each action is associated with a reward function revealed by the environment or an adversary. Because the reward function is unknown to the algorithm beforehand, the goals of algorithm design include maximizing the approximation ratio and minimizing the upper bound on the worst-case regret. In addition, there are two sub-settings: the full information, where the adversary provides the entire reward function, and the bandit setting, where only a reward value at a single point is returned. Under the stochastic model, each element in the ground set has a multistate governed by some specific distributions. Addressing stochastic submodular functions involves two policy strategy: a nonadaptive policy, which depends only on elements previously chosen without considering their realization of their state, and an adaptive policy, which considers both elements previously chosen and their realization. The aforementioned models present numerous fundamental problems in DR-submodular maximization worth studying.

Apart from classical optimization, DR-submodular optimization has emerged as a significant topic in algorithmic game theory in recent years. The following are partially fascinating examples. One typical example is robust DR-submodular maximization. In the robust case, two types of problems are especially relevant. The first is the robust deletion of the submodular set function, focusing on maximizing a submodular function when the selected set may have a subset removed by an adversary. The second is the max-min DR-submodular functions, which aim to maximize the minimum value of a batch DR-submodular functions. Another classical example is the social welfare maximization, where m items are distributed among n players, each with DR-submodular utility functions. This area has garnered interest due to its implications for fairness resource allocation. We are optimistic that robust DR-submodular maximization and online Nash social welfare maximization will lead to many important consequences in the future.

Acknowledgments This work is partially supported by the National Key R&D Program of China under grant (No. 2022YFA1003900) and the National Natural Science Foundation of China (Nos.12501450, 12071459, 62325210, and 62272441).

References

1. Ageev, A., Sviridenko, M.: Pipage rounding: a new method of constructing algorithms with proven performance guarantee. J. Comb. Optim. **8**, 307–328 (2004). https://doi.org/10.1023/B:JOCO.0000038913.96607.c2
2. Bach, F.: Submodular functions: from discrete to continuous domains. Math. Program. **175**(3), 419–459 (2019). https://doi.org/10.1007/s10107-018-1248-6

3. Badanidiyuru, A., Vondrák, J.: Fast algorithms for maximizing submodular functions. In: Proceedings of the 2014 Annual ACM-SIAM Symposium on Discrete Algorithms (SODA), pp. 1497–1514 (2014). https://doi.org/10.1137/1.9781611973402.110
4. Bai, W., Iyer, R., Wei, K., Bilmes, J.: Algorithms for optimizing the ratio of submodular functions. In: Proceedings of the 33rd International Conference on International Conference on Machine Learning, vol. 48. ICML'16, pp. 2751–2759 (2016)
5. Bian, A., Levy, K., Krause, A., Buhmann, J.M.: Continuous DR-submodular maximization: structure and algorithms. In: Advances in Neural Information Processing Systems, vol. 30 (2017)
6. Bian, A.A., Mirzasoleiman, B., Buhmann, J., Krause, A.: Guaranteed non-convex optimization: submodular maximization over continuous domains. In: Proceedings of the 20th International Conference on Artificial Intelligence and Statistics, vol. 54, pp. 111–120 (2017)
7. Bilmes, J.: Submodularity in machine learning and artificial intelligence (2022). https://arxiv.org/abs/2202.00132, 2202.00132
8. Bodek, K., Feldman, M.: Maximizing sums of non-monotone submodular and linear functions: understanding the unconstrained case. In: 30th Annual European Symposium on Algorithms (ESA 2022) (2022)
9. Buchbinder, N., Feldman, M.: Submodular functions maximization problems. In: Handbook of Approximation Algorithms and Metaheuristics (2018). https://api.semanticscholar.org/CorpusID:46989385
10. Buchbinder, N., Feldman, M.: Constrained submodular maximization via a nonsymmetric technique. Math. Oper. Res. **44**(3), 988–1005 (2019)
11. Buchbinder, N., Feldman, M.: Constrained submodular maximization via new bounds for dr-submodular functions. In: Proceedings of the 56th Annual ACM Symposium on Theory of Computing, STOC 2024, pp. 1820–1831 (2024). https://doi.org/10.1145/3618260.3649630
12. Buchbinder, N., Feldman, M.: Deterministic algorithm and faster algorithm for submodular maximization subject to a matroid constraint. In: 2024 IEEE 65th Annual Symposium on Foundations of Computer Science (2024)
13. Buchbinder, N., Feldman, M., Naor, J.S., Schwartz, R.: Submodular maximization with cardinality constraints. In: Proceedings of the 2014 Annual ACM-SIAM Symposium on Discrete Algorithms (SODA), pp. 1433–1452 (2014). https://doi.org/10.1137/1.9781611973402.106
14. Buchbinder, N., Feldman, M., SeffiNaor, J., Schwartz, R.: A tight linear time (1/2)-approximation for unconstrained submodular maximization. SIAM J. Comput. **44**(5), 1384–1402 (2015). https://doi.org/10.1137/130929205
15. Buchbinder, N., Feldman, M., Schwartz, R.: Comparing apples and oranges: query trade-off in submodular maximization. Math. Oper. Res. **42**(2), 308–329 (2017). https://doi.org/10.1287/moor.2016.0809
16. Calinescu, G., Chekuri, C., Pál, M., Vondrák, J.: Maximizing a monotone submodular function subject to a matroid constraint. SIAM J. Comput. **40**(6), 1740–1766 (2011). https://doi.org/10.1137/080733991
17. Chekuri, C., Vondrak, J., Zenklusen, R.: Dependent randomized rounding via exchange properties of combinatorial structures. In: 2010 IEEE 51st Annual Symposium on Foundations of Computer Science, pp. 575–584 (2010). https://doi.org/10.1109/FOCS.2010.60
18. Chekuri, C., Vondrák, J., Zenklusen, R.: Submodular function maximization via the multilinear relaxation and contention resolution schemes. SIAM J. Comput. **43**(6), 1831–1879 (2014). https://doi.org/10.1137/110839655
19. Chen, S., Yang, W., Gao, S., Jin, R.: Novel algorithms for maximum DS decomposition. Theor. Comput. Sci. **857**, 87–96 (2021). https://doi.org/10.1016/j.tcs.2020.12.041
20. Chen, S., Du, D., Yang, W., Xu, D., Gao, S.: Continuous non-monotone DR-submodular maximization with down-closed convex constraint (2023). https://arxiv.org/abs/2307.09616, 2307.09616
21. Chen, S., Du, D., Yang, W., Gao, S.: Maximizing the ratio of monotone dr-submodular functions on integer lattice. J. Oper. Res. Soc. China (2023). https://doi.org/10.1007/s40305-023-00469-1

22. Chen, S., Du, D., Yang, W., Gao, S.: Maximizing stochastic set function under a matroid constraint from decomposition. J. Comb. Optim. **48**(3) (2024). https://doi.org/10.1007/s10878-024-01193-z
23. Chen, Y., Nath, A., Peng, C., Kuhnle, A.: Guided combinatorial algorithms for submodular maximization (2024). Preprint. arXiv:240505202
24. Du, D.: Lyapunov function approach for approximation algorithm design and analysis: with applications in submodular maximization (2022). Preprint. arXiv:220512442
25. Du, D., Liu, Z., Wu, C., Xu, D., Zhou, Y.: An improved approximation algorithm for maximizing a DR-submodular function over a convex set (2022). Preprint. arXiv:220314740
26. Dürr, C., Thundefinedng, N.K., Srivastav, A., Tible, L.: Non-monotone DR-submodular maximization over general convex sets. In: Proceedings of the Twenty-Ninth International Joint Conference on Artificial Intelligence, IJCAI'20 (2021)
27. Ene, A., Nguyen, H.L.: Constrained submodular maximization: beyond 1/e. In: 2016 IEEE 57th Annual Symposium on Foundations of Computer Science (FOCS), pp. 248–257 (2016). https://doi.org/10.1109/FOCS.2016.34
28. Feige, U.: A threshold of ln n for approximating set cover. J. ACM **45**(4), 634–652 (1998). https://doi.org/10.1145/285055.285059
29. Feige, U., Mirrokni, V.S., Vondrák, J.: Maximizing non-monotone submodular functions. SIAM J. Comput. **40**(4), 1133–1153 (2011). https://doi.org/10.1137/090779346
30. Feldman, M.: Maximization problems with submodular objective functions. PhD thesis, Technion - Israel Institute of Technology (2013)
31. Feldman, M., Naor, J., Schwartz, R.: A unified continuous greedy algorithm for submodular maximization. In: 2011 IEEE 52nd Annual Symposium on Foundations of Computer Science, pp. 570–579 (2011). https://doi.org/10.1109/FOCS.2011.46
32. Feldman, M., Naor, J., Schwartz, R., Ward, J.: Improved approximations for k-exchange systems. In: Proceedings of the 19th European Conference on Algorithms, ESA'11, pp. 784–798 (2011)
33. Feldman, M., Nutov, Z., Shoham, E.: Practical budgeted submodular maximization. Algorithmica **85**, 1332–1371 (2023). https://doi.org/10.1007/s00453-022-01071-2
34. Filmus, Y., Ward, J.: Monotone submodular maximization over a matroid via non-oblivious local search. SIAM J. Comput. **43**(2), 514–542 (2014). https://doi.org/10.1137/130920277
35. Fisher, M.L., Nemhauser, G.L., Wolsey, L.A.: An analysis of approximations for maximizing submodular set functions-II. Math. Program. **8**(1), 73–87 (1978)
36. Gharan, S.O., Vondrák, J.: Submodular Maximization by Simulated Annealing, pp. 1098–1116. SIAM, Philadelphia (2011). https://doi.org/10.1137/1.9781611973082.83
37. Gillenwater, J., Kulesza, A., Taskar, B.: Near-optimal map inference for determinantal point processes. In: Advances in Neural Information Processing Systems, vol. 25 (2012)
38. Goldengorin, B.: Maximization of submodular functions: theory and enumeration algorithms. Euro. J. Oper. Res. **198**(1), 102–112 (2009). https://doi.org/10.1016/j.ejor.2008.08.022
39. Goldengorin, B., Ghosh, D.: A multilevel search algorithm for the maximization of submodular functions applied to the quadratic cost partition problem. J. Global Optim. **32**, 65–82 (2005). https://doi.org/10.1007/s10898-004-5909-z
40. Goldengorin, B., Tijssen, G.A., Tso, M.: The maximization of submodular functions: old and new proofs for the correctness of the dichotomy algorithm. Research Report 99a17, Graduate School/Research Institute SOM (Systems, Organisation and Management), University of Groningen (1999)
41. Harshaw, C., Feldman, M., Ward, J., Karbasi, A.: Submodular maximization beyond non-negativity: guarantees, fast algorithms, and applications. In: Proceedings of the 36th International Conference on Machine Learning, vol. 97, pp. 2634–2643 (2019)
42. Hassani, H., Soltanolkotabi, M., Karbasi, A.: Gradient methods for submodular maximization. In: Proceedings of the 31st International Conference on Neural Information Processing Systems, NIPS'17, pp. 5843–5853 (2017)
43. Iwata, S., Fleischer, L., Fujishige, S.: A combinatorial strongly polynomial algorithm for minimizing submodular functions. J. ACM **48**(4), 761–777 (2001). https://doi.org/10.1145/502090.502096

44. Iyer, R., Bilmes, J.: Algorithms for approximate minimization of the difference between submodular functions, with applications. In: Proceedings of the Twenty-Eighth Conference on Uncertainty in Artificial Intelligence, UAI'12, pp. 407–417 (2012)
45. Kempe, D., Kleinberg, J., Tardos, E.: Maximizing the spread of influence through a social network. In: Proceedings of the Ninth ACM SIGKDD International Conference on Knowledge Discovery and Data Mining, KDD '03, pp. 137–146 (2003). https://doi.org/10.1145/956750.956769
46. Krause, A., Singh, A., Guestrin, C.: Near-optimal sensor placements in Gaussian processes: theory, efficient algorithms and empirical studies. J. Mach. Learn. Res. **9**(3), 235–284 (2008)
47. Kulik, A., Schwartz, R., Shachnai, H.: A refined analysis of submodular greedy. Oper. Res. Lett. **49**(4), 507–514 (2021). https://doi.org/10.1016/j.orl.2021.04.006
48. Lin, H., Bilmes, J.: A class of submodular functions for document summarization. In: Proceedings of the 49th Annual Meeting of the Association for Computational Linguistics: Human Language Technologies, vol. 1, HLT '11, pp. 510–520 (2011)
49. Lovasz, L.: Submodular Functions and Convexity, pp. 235–257. Springer, Berlin (1983). https://doi.org/10.1007/978-3-642-68874-4_10
50. Lu, C., Yang, W., Gao, S.: Regularized nonmonotone submodular maximization. Optimization **73**(6), 1739–1765 (2024). https://doi.org/10.1080/02331934.2023.2173968
51. Marinacci, M., Montrucchio, L.: Ultramodular functions. Math. Oper. Res. **30**(2), 311–332 (2005)
52. Minoux, M.: Accelerated greedy algorithms for maximizing submodular set functions. In: Optimization Techniques: Proceedings of the 8th IFIP Conference on Optimization Techniques Würzburg, September 5–9, 1977, pp. 234–243 (2005)
53. Mirzasoleiman, B., Badanidiyuru, A., Karbasi, A., Vondrak, J., Krause, A.: Lazier than lazy greedy. Proc. AAAI Conf. Artif. Intell. **29**(1) (2015). https://doi.org/10.1609/aaai.v29i1.9486
54. Mitra, S., Feldman, M., Karbasi, A.: Submodular + concave. In: Ranzato, M., Beygelzimer, A., Dauphin, Y., Liang, P., Vaughan, J.W. (eds.) Advances in Neural Information Processing Systems, vol. 34, pp. 11577–11591 (2021)
55. Mualem, L., Feldman, M.: Resolving the approximability of offline and online non-monotone DR-submodular maximization over general convex sets. In: Proceedings of The 26th International Conference on Artificial Intelligence and Statistics, vol. 206, pp. 2542–2564 (2023)
56. Nemhauser, G.L., Wolsey, L.A., Fisher, M.L.: An analysis of approximations for maximizing submodular set functions-I. Math. Program. **14**(1), 265–294 (1978)
57. Niazadeh, R., Roughgarden, T., Wang, J.R.: Optimal algorithms for continuous non-monotone submodular and DR-submodular maximization. J. Mach. Learn. Res. **21**(1), 4937–4967 (2020)
58. Pedramfar, M., Quinn, C., Aggarwal, V.: A unified approach for maximizing continuous DR-submodular functions. In: Advances in Neural Information Processing Systems, vol. 36, pp. 61103–61114 (2023)
59. Perrault, P., Healey, J., Wen, Z., Valko, M.: On the approximation relationship between optimizing ratio of submodular (RS) and difference of submodular (DS) functions (2021). CoRR abs/2101.01631, 2101.01631
60. Qi, B.: On maximizing sums of non-monotone submodular and linear functions. Algorithmica **86**, 1080–1134 (2024). https://doi.org/10.1007/s00453-023-01183-3
61. Sun, X., Zhang, J., Zhang, Z.: Simple deterministic approximation for submodular multiple knapsack problem. In: 31st Annual European Symposium on Algorithms (ESA 2023), vol. 274, pp. 98:1–98:15 (2023). https://doi.org/10.4230/LIPIcs.ESA.2023.98
62. Sun, X., Zhang, J., Zhang, S., Zhang, Z.: Improved deterministic algorithms for non-monotone submodular maximization. Theor. Comput. Sci. **984**, 114293 (2024). https://doi.org/10.1016/j.tcs.2023.114293
63. Sviridenko, M.: A note on maximizing a submodular set function subject to a knapsack constraint. Oper. Res. Lett. **32**(1), 41–43 (2004). https://doi.org/10.1016/S0167-6377(03)00062-2
64. Topkis, D.M.: Supermodularity and Complementarity. Princeton University Press, Princeton (1998)

65. Tukan, M., Mualem, L., Feldman, M.: Practical 0.385-approximation for submodular maximization subject to a cardinality constraint (2024). Preprint. arXiv:240513994
66. Vondrak, J.: Optimal approximation for the submodular welfare problem in the value oracle model. In: Proceedings of the Fortieth Annual ACM Symposium on Theory of Computing, STOC '08, pp. 67–74 (2008). https://doi.org/10.1145/1374376.1374389
67. Vondrák, J.: Symmetry and approximability of submodular maximization problems. SIAM J. Comput. **42**(1), 265–304 (2013). https://doi.org/10.1137/110832318
68. Ward, J.: A (k+3)/2-approximation algorithm for monotone submodular k-set packing and general k-exchange systems. In: International Symposium on Theoretical Aspects of Computer Science (2012)
69. Zhang, Q., Deng, Z., Chen, Z., Hu, H., Yang, Y.: Stochastic continuous submodular maximization: boosting via non-oblivious function. In: Proceedings of the 39th International Conference on Machine Learning, PMLR, vol. 162, pp. 26116–26134 (2022)
70. Zhang, Q., Wan, Z., Deng, Z., Chen, Z., Sun, X., Zhang, J., Yang, Y.: Boosting gradient ascent for continuous dr-submodular maximization (2024). 2401.08330

Optimization of the Design and Control of an Optical System with Structural Inhomogeneities

G. M. Zrazhevsky, V. F. Zrazhevska, and A. M. Golodnikov

1 Introduction

In this work, we explore how to optimize the parameters of a mechanical device designed for the excitation and formation of wave motion when complete information about its structural integrity is lacking. Such devices can be utilized for the generation, transformation, and transmission of information, and more broadly, for the transmission of wave energy. Specifically, we address the problem of modulating a mirror fixed on active supports and develop a model to optimize the structure of these supports. The primary challenge is to identify the control forces and their characteristics—specifically, the application points, amplitude, and phase of oscillation—that best approximate a desired shape and phase of the mirror's oscillation. This optimization takes into account structural inhomogeneities (defects) with undefined geometric and mechanical characteristics. The mechanical properties of these defects are characterized by parameters such as material density, Young's modulus, and cylindrical stiffness. To accelerate calculations, we employ a heterogeneity model that assumes a high-order point singularity with parameters allowing us to accurately represent its physical properties [1]. We estimate the error associated with this approximation and investigate how the accuracy of defect modeling depends on the number of circular harmonics used [2].

G. M. Zrazhevsky (✉)
Taras Shevchenko National University of Kyiv, Kyiv, Ukraine
e-mail: greg.zrazhevsky@knu.ua

V. F. Zrazhevska
National Technical University of Ukraine "Igor Sikorsky Kyiv Polytechnic Institute", Kyiv, Ukraine

A. M. Golodnikov
Glushkov Institute of Cybernetics of NAS of Ukraine, Kyiv, Ukraine

B. Goldengorin (ed.), *Theory, Algorithms, and Experiments in Applied Optimization*, Springer Optimization and Its Applications 226,
https://doi.org/10.1007/978-3-031-91357-0_19

In the first part of our study, we tackle a deterministic problem by assuming that the characteristics of the defects are known. This assumption simplifies the task of determining the plate's vibration shape to a boundary value problem with distributed point quadruples, which serve as first-order models for inhomogeneities with specified characteristics. Modeling the active supports as point forces allows us to apply the Green's function method [2]. We then formulate an optimization problem to find the optimal characteristics of the exciting forces—their locations, amplitudes, and phases—that minimize the deviation of the mirror's oscillation shape from the desired form. The criterion for optimality is the minimization of the root-mean-square deviation [3].

In the second part of our study, we assume that the locations, sizes, and physical characteristics of the defects are uncertain [4]. The aim is to determine the optimal placement of the active supports and the characteristics of the oscillation excitations that ensure the device functions most effectively, specifically by minimizing the deviation of the mirror's oscillatory profile from the specified shape.

In reliability engineering, risk is typically quantified by the probability of failure—the chance that a system fails to perform its intended task. Although widely used, this measure has undesirable mathematical properties, such as discontinuity in sample distributions, which make it difficult to optimize. To address these shortcomings, an alternative risk measure called Buffered Probability of Failure (BPF) was developed [5]. BPF accounts for the degree to which the failure threshold $h = 0$ is exceeded and is more conservative than the classical probability of failure.

The Buffered Probability of Exceedance (BPOE) generalizes BPF for cases where the system's failure threshold can be any number, not just zero [6]. These risk measures are based on the properties of the Conditional Value-at-Risk (CVaR) measure [7, 8]. Notably, BPOE possesses exceptional mathematical properties; under general conditions, it is quasi-convex with respect to the random variable. This explains the use of CVaR and BPOE as risk measures in formulating the stochastic optimization problem.

2 Modeling of Defects

This section presents an approach to modeling inhomogeneities in the plate using the apparatus of generalized functions [1].

Let a defect with a cylindrical stiffness D_1 and material density ρ_1 is located in the region $P(x, y) > 0$. A cylindrical stiffness and density in region $P(x, y) < 0$ (outside of defect) are D_0, ρ_1. The plate is loaded by external forces of intensity q, which are applied in points outside of the defects zones. Suppose that $\vec{\nabla} P_{|P=0} \neq 0$, where P is simple connected and convex. Then taking into account designations: $D_1 = D_0 - \Delta D$, $\frac{\Delta D}{D_0} = \kappa_D$, $\frac{\rho_0 \omega^2}{D_0} = k^4$ the equations of harmonic oscillations of the plate can be written in the form:

$$\begin{cases} \mathbf{L}\left(-\mathbf{L}_1\right)\mathbf{L}_2 w + k^4 w = -q, & P\left(x,y\right) < 0 \\ \mathbf{L}\left(-\mathbf{L}_1\right)\mathbf{L}_2 w + \frac{\rho_1}{\rho_0(1-\kappa_D)} k^4 w = 0, & P\left(x,y\right) > 0 \end{cases}, (x,y) \in \Omega \tag{1}$$

where w—deflection of plate, ω—oscillation frequency and

$$\mathbf{L}_0 = \begin{bmatrix} \frac{\partial}{\partial x} & 0 & \frac{\partial}{\partial y} \\ 0 & \frac{\partial}{\partial y} & \frac{\partial}{\partial x} \end{bmatrix}, \mathbf{L}_1 = \begin{bmatrix} \frac{\partial}{\partial x} & \nu\frac{\partial}{\partial y} \\ \nu\frac{\partial}{\partial x} & \frac{\partial}{\partial y} \\ \frac{1-\nu}{2}\frac{\partial}{\partial y} & \frac{1-\nu}{2}\frac{\partial}{\partial x} \end{bmatrix}, \mathbf{L}_2 = \begin{bmatrix} \frac{\partial}{\partial x} \\ \frac{\partial}{\partial y} \end{bmatrix},$$
$$\mathbf{L} = \mathbf{L}_2{}^T \mathbf{L}_0 = \begin{bmatrix} \frac{\partial^2}{\partial x^2} & \frac{\partial^2}{\partial y^2} & \frac{\partial^2}{\partial x \partial y} \end{bmatrix}. \tag{2}$$

We consider a piecewise homogeneous model:

$$\begin{cases} D = D_0\left(1 - \kappa_D \theta(P)\right) \\ \rho = \rho_0\left(1 - \kappa_\rho \theta(P)\right) \end{cases} \tag{3}$$

$\theta(P) = \begin{cases} 1, & P > 0 \\ 0, & P < 0 \end{cases}$—is a function of region, $\kappa_\rho = \frac{\rho_0 - \rho_1}{\rho_0}$.

Then (1) can be rewritten in the form of a single equation, valid in the whole area:

$$\mathbf{L}\mathbf{L}_1\mathbf{L}_2 w - k^4 w = \frac{q}{D_0} + \kappa_D \mathbf{L}\theta(P)\mathbf{L}_1\mathbf{L}_2 w - k^4 \kappa_\rho \theta w, \quad (x,y) \in \Omega. \tag{4}$$

For convenience, we introduce another notation:

$$\vec{v} = \mathbf{L}_1\mathbf{L}_2 w = \left(v_x, v_y, v_{xy}\right)^T \tag{5}$$

The singular part of the right-hand side of (4) that describes the inhomogeneity can be rebuilt: $\left\langle \mathbf{L}_2^T \mathbf{L}_0 \vec{v}\theta, \Phi \right\rangle = \left\langle \vec{v}, \mathbf{L}_2^T \mathbf{L}_0 \Phi \right\rangle$ and according to the mean value theorem:

$$\left\langle \vec{v}, \mathbf{L}\Phi \right\rangle = S_P \vec{v}(X) \left(\frac{\partial^2 \Phi}{\partial x^2}, 2\frac{\partial^2 \Phi}{\partial x \partial y}, \frac{\partial^2 \Phi}{\partial y^2} \right)\Bigg|_X + \overline{\overline{O}}\left(S_P\right), \tag{6}$$

where $S_P = \mathrm{pow}(P)$.

Thus,

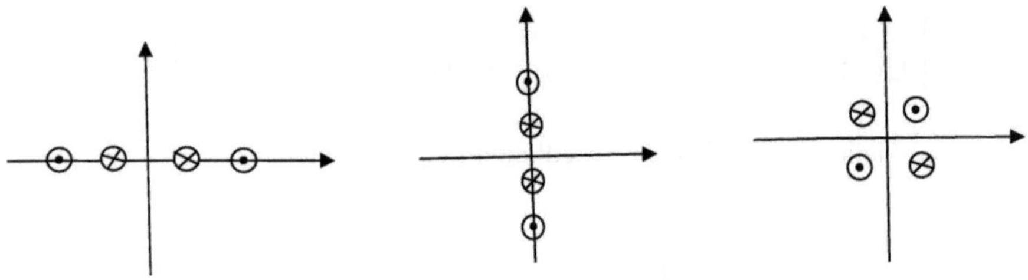

Fig. 1 Representation of inhomogeneity in the form of balanced dipoles

$$\mathbf{L}\theta(P)\vec{v} = S_P{}^2\vec{v}\left(\frac{\partial^2\delta(x-X)}{\partial x^2}\delta(y-Y),2\frac{\partial\delta(x-X)}{\partial x}\frac{\partial\delta(y-Y)}{\partial y},\frac{\partial^2\delta(y-Y)}{\partial y^2}\delta(x-X)\right) + \underline{\underline{O}}\left(S_P{}^2\right) \tag{7}$$

Volumetric loads equivalent to heterogeneity are interpreted as a superposition of double balanced dipoles (see Fig. 1).

Thus, considering (7), Eq. (4) takes the form:

$$\begin{aligned} \mathbf{L}\mathbf{L}_1\mathbf{L}_2 w - k^4 w = \tfrac{q}{D_0} + \kappa_D S_P + \Big\{ v_x(X)\tfrac{\partial^2\delta(x-X)}{\partial x^2}\delta(y-Y) \\ +2v_{xy}(X)\tfrac{\partial\delta(x-X)}{\partial x}\tfrac{\partial\delta(y-Y)}{\partial y} + v_y(X)\tfrac{\partial^2\delta(y-Y)}{\partial y^2}\delta(x-X)\Big\} \\ - k^2\kappa_\rho S_\rho w(X)\delta(x-X)\,\delta(y-Y) + \underline{\underline{O}}\left(S_P{}^2\right) \end{aligned} \tag{8}$$

where $X = (X, Y)$ is a defect center with characteristics κ_D, κ_ρ, S_ρ.

We are looking for a solution in the form: $w = w_0 + \kappa w_1$, confining ourselves to the first two terms of the expansion in a small parameter $\kappa < < 1$ ($\kappa < < 1$, $(\kappa = \underline{\underline{O}}(\kappa_\rho, \kappa_D))$)). Neglecting change in ρ ($\kappa_\rho \equiv 0$). As a result, we obtain:

$$\left\{ \begin{aligned} \mathbf{L}\mathbf{L}_1\mathbf{L}_2 w_0 - k^4 w_0 = \tfrac{q}{D_0} \\ \mathbf{L}\mathbf{L}_1\mathbf{L}_2 w_1 - k^4 w_1 = v_{0x}(X)\tfrac{\partial^2\delta(x-X)}{\partial x^2}\delta(y-Y) + \\ + 2v_{0xy}(X)\tfrac{\partial\delta(x-X)}{\partial x}\tfrac{\partial\delta(y-Y)}{\partial y} + v_{0y}(X)\tfrac{\partial^2\delta(y-Y)}{\partial y^2}\delta(x-X) \end{aligned} \right. \tag{9}$$

Equations (9) were obtained with accuracy $\underline{\underline{O}}\left(S_\rho{}^2\right)$. The first Eq. (9) describes the deflection of the plate if there is no inhomogeneity.

Let the Green's function $w_*(x, X)$ is constructed as solution of equation:

$$\mathbf{L}\mathbf{L}_1\mathbf{L}_{2x} w^*(x, X) - k^4 w^*(x, X) = \delta(x - X), x \in \Omega, X \in \Omega, X \neq x, \tag{10}$$

where $x = x(x, y)$ is the observation point, $X = X(X, Y)$ is the source point with homogeneous boundary conditions:

$$\begin{cases} w^* (x, X) = 0 \\ M^* (x, X) = 0 \end{cases}, x \in \Gamma \tag{11}$$

Then, $w_1(x, X)$ satisfying (9) with the boundary conditions (9) can be constructed as:

$$w_1 (x, X) = v_{0x} (X) \frac{\partial^2 w^* (x, X)}{\partial X^2} + 2 v_{0xy} (X) \frac{\partial^2 w^* (x, X)}{\partial X \partial Y} + v_{0y} (X) \frac{\partial^2 w^* (x, X)}{\partial Y^2} \tag{12}$$

Indeed, for example: $\frac{\partial w^*(x,X)}{\partial X} = \lim\limits_{\Delta x \to 0} \frac{w^*(x,X+\Delta X) - w^*(x,X)}{\Delta X}$. But $w_*(x, X)$ satisfies (11) for $\forall X \in \Omega$, hence, $\frac{\partial w^*(x,X)}{\partial X}$ as well as the higher derivatives (within the limits of smoothness) satisfies (11).

In the polar coordinates $(X, Y) \equiv (R, \Phi)$ the first approximation of the solution of the boundary value problem (9), (11) has the form:

$$w_1 (x, X) = \frac{1}{R^2} \left\{ R^2 v_{0r} (X) \Big(R, \Phi \Big) \frac{\partial^2 w^* (x,X)}{\partial R^2} \right.$$
$$\left. + 2 v_{0r\varphi} (X) \left(R \frac{\partial^2 w^* (x,X)}{\partial R \partial \Phi} - \frac{\partial w^* (x,X)}{\partial \Phi} \right) + v_{0\varphi} (X) \left(R \frac{\partial w^* (x,X)}{\partial R} + \frac{\partial^2 w^* (x,X)}{\partial \Phi^2} \right) \right\} \tag{13}$$

where v_{0r}, $v_{0\varphi}$, $v_{0r\varphi}$ are determined through the zero approximation in polar coordinates.

Thus, to solve the direct problem, we need:

1. Find w_0 (by solving the first Eq. (9) and satisfying the boundary conditions of the form (11).
2. Construct the Green's function w_* from (10), (11).
3. Find v_{0r}, $v_{0\varphi}$, $v_{0r\varphi}$ from (5).
4. Implement the formula (13).
5. As a result, we get the solution for a plate with defect with parameters R, Φ, S_P, κ_D, as a first approximation: $w = w_0 + S_P \frac{\kappa_D}{1-\kappa_D} w_1$.

3 Optimization Problem. Deterministic Setting

Let $I + 1$ forces with intensities $F_k = u_k + i v_k$ $(k = \overline{0, I})$ be applied to the plate at points $\xi_k \left(r_{\xi_k}, \varphi_{\xi_k} \right)$ $k = \overline{1, I}$, $\xi_0(0)$. In addition, the plate contains J defects $\left(r_{d_j}, \varphi_{d_j}, \kappa_j, s_j \right)$, $j = \overline{1, J}$. For a given profile function $W(r, \varphi)$, it is required to determine ξ_k, F_k from the optimization problem:

$$I\left(\vec{F}\right) = \frac{1}{2\pi}\int_0^1 rdr\int_{-\pi}^{\pi}\left|w\left(r,\varphi\right) - W\left(r,\varphi\right)\right|^2 d\varphi \rightarrow \min \tag{14}$$

Let

$$w\left(r,\varphi\right) = w_0\left(r,\varphi\right) + w_1\left(r,\varphi\right) \tag{15}$$

where $w_0(r, \varphi)$ is the optimal solution to the problem for a homogeneous plate with appropriate controls ξ_k^0, F_k^0, which minimize

$$I_0\left(\vec{F}\right) = \frac{1}{2\pi}\int_0^1 rdr\int_{-\pi}^{\pi}\left|w_0\left(r,\varphi\right) - W\left(r,\varphi\right)\right|^2 d\varphi. \tag{16}$$

We are looking for a solution in the form: $w_0\left(r,\varphi\right) = F_0^0 w_{00}^* + \sum_{k=1}^{I} F_k^0 w^* \left(r, r_{\xi_k}^0, \varphi - \varphi_{\xi_k}^0\right)$, where w_{00}^*, w^* are Green's functions, defined in the problem (10), (11). Implementation of this stage is very similar to methodology presented in [2]. In what follows, w_0 is considered as known. Moreover, $w_1(r, \varphi)$ is defined in (13).

$$\begin{aligned} w_1\left(r,\varphi\right) &= \sum_{j=1}^{J}\varepsilon_j\left\{v_{0x}\left(X_j\right)\frac{\partial^2 w^*(x;X_j)}{\partial X^2} + 2v_{0xy}\left(X_j\right)\frac{\partial^2 w^*(x;X_j)}{\partial X\partial Y} + v_{0y}\left(X_j\right)\frac{\partial^2 w^*(x;X_j)}{\partial Y^2}\right\} \\ &\equiv \sum_{j=1}^{J}\frac{\varepsilon_j}{R_j^2}\left\{R_j^2 v_{0r}\left(X_j\right)\frac{\partial^2 w^*(x;X_j)}{\partial R_j^2} + 2v_{0r\varphi}\left(X_j\right)\left(R_j\frac{\partial^2 w^*(x;X_j)}{\partial R\partial\Phi} - \frac{\partial w^*(x;X_j)}{\partial\Phi}\right)\right. \\ &\quad \left. + v_{0\varphi}\left(X_j\right)\left(R_j\frac{\partial w^*(x;X_j)}{\partial R} + \frac{\partial^2 w^*(x;X_j)}{\partial\Phi^2}\right)\right\} \end{aligned} \tag{17}$$

where $\varepsilon_j = \frac{\kappa_j}{1-\kappa_j} S_j$, with κ_j as inhomogeneity measure, and S_j is the area of the j-th inhomogeneity. We assume that $\sum_j^J \varepsilon_j << 1$, i.e. $\|w_1\| < < \|w_0\|$.

4 Remarks on the Implementation of the Optimization Problem

In a preliminary study of problem (14), we can take the known part of the control variables of the problem, namely, the points of application of forces: r_{ξ_k}, φ_{ξ_k} $k = \overline{1, I}$. In this case, the optimization problem becomes quadratic and has an exact solution.

If we take $r_{\xi_k} = r_{\xi_k}^0$, $\varphi_{\xi_k} = \varphi_{\xi_k}^0$ (that is the optimal positions of the forces for a homogeneous plate), then we can obtain the first approximation of the problem of the complete problem (14). So $I(R) = (w - W, w - W)$, and $w = w_0 + w_1$, where $\|w_1\| < < \|w_0\|$. The solution of the problem can be represented as: $w_0 = \sum_{i=0}^{I} g_i F_i$, $w_1 = \sum_{i=0}^{I} s_i F_i$, where g_i, s_i are some functions defined from (13). Problem (14) is the classical optimal approximation problem in $L^2(\Omega)$ with basis $\{g_i, s_i\}_{i=0}^{I}$. The Gram system for this task has the form:

$$\sum_{i=0}^{I} ((g_i + s_i)\, F_i, (g_s + s_s)) = (W, g_s + s_s)\,, s = \overline{0, I}. \tag{18}$$

Wherein, in the absence of inhomogeneities, (14) has the form of a system of linear equations:

$$\mathbf{K}^0 \vec{F}^0 = \vec{b}^0, \mathbf{K}^0 = \{K_{is}\}_{i,s=0}^{I}, \vec{b}^{0T} = \left\{ b_s{}^0 \right\}_{s=0}^{I}, K_{is}^0 = (g_i, g_s)\,, b_s^o = (W, g_s) \tag{19}$$

The solution of this problem is denoted by $\vec{F}^{0opt}$.

With this in mind, (18) is a system of linear equations of the form: $\left(\mathbf{K}^0 + \mathbf{K}^1\right) \vec{F} = \vec{b}^0 + \vec{b}^1$,

Where

$$\begin{cases} K_{is}^1 = (g_i, s_s) + (s_i, g_s) + (s_i, s_s) \\ b_s^1 = (W, s_s) \end{cases} \tag{20}$$

Let $\vec{F} = \vec{F}^0 + \vec{F}^1$, where $\vec{F}^0$ is defined in (19) and $\left\| \vec{F}^1 \right\| << \left\| \vec{F}^0 \right\|$. Then $\left(\mathbf{K}^0 + \mathbf{K}^1\right) \left(\vec{F}^0 + \vec{F}^1 \right) = \vec{b}^0 + \vec{b}^1$. Neglecting a small member $\mathbf{K}^1 \vec{F}^1$, with known $\vec{F}^0 = \vec{F}^{0opt}$, solving (18) reduces to a system of linear equations to find $\vec{F}^1$:

$$\mathbf{K}^0 \vec{F}^1 = \vec{b}^1 - \mathbf{K}^1 \vec{F}^0 \tag{21}$$

with the solution $\vec{F}^1 = \left(\mathbf{K}^0\right)^{-1} \left(\vec{b}^1 - \mathbf{K}^1 \vec{F}^0 \right)$.

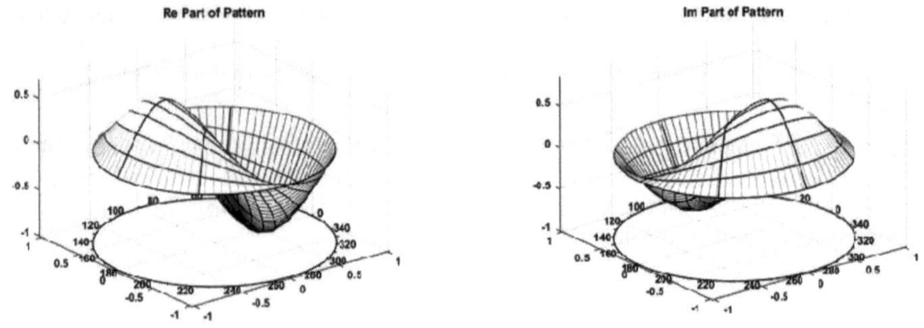

Fig. 2 Target surface shape

5 Examples of Implementation

As examples, the optimal excitation parameters of the plate with defects were found to achieve the shape of vibration defined by the expression:

$$W(k, r, \phi) = (-I_0(k)/J_0(k)J_0(kr) + I_0(kr)) + (-I_1(k)/J_1(k)J_1(kr) + I_1(kr))\, e^{i\phi}, \quad (22)$$

where $J_{0(1)}(.)$, $I_{0(1)}(.)$—Bessel functions.

The pattern (22) is presented in Fig. 2. As was shown [3], this form provides a wave traveling in the circular direction. In dimensionless form, the first resonant frequencies are $k_{01} = 2.232$, $k_{02} = 5.455$, $k_{11} = 0$, $k_{12} = 3.734$.

The calculations were performed for the low-frequency region ($k = 1.2324$—quasi-static excitation), for the midrange frequency region ($k = 3.2324$—between the first and the second resonant frequencies), and for the high frequencies ($k = 5.91$—higher than the second resonant frequency) of excitation. The number of harmonics taken into account was determined automatically.

In each group, at a fixed excitation frequency, in addition to the central force, examples with different numbers of forces were considered. Forces were grouped by two forces located centrally symmetrically. Group radii, orientation angles (angles between axes of force groups and a fixed direction), amplitudes, and phases of forces are taken as control parameters. As the location of the active supports, the optimal locations for a defect-free plate are taken. The optimal values of the excitation forces for a defect-free and defective plate are determined independently. Thus, the location of the supports, which is a design parameter, is determined based on the assumption that the plate is ideal. The presence of defects in the plate is proposed to be compensated by the selection of exciting forces.

We placed three groups of inhomogeneities on the plate. The centers of the groups were set randomly in accordance with the uniform distribution. Inhomogeneities in groups were placed randomly relative to the centers of groups in accordance with the normal distribution law with $\sigma = 0.03, 0.04, 0.05$. The total

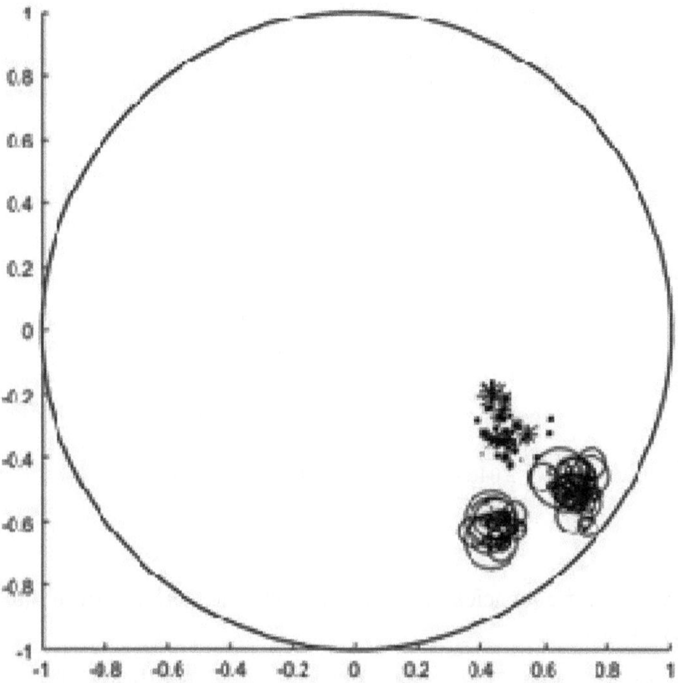

Fig. 3 The characteristic distribution of defects on the plate

number of inhomogeneities in the groups was randomly selected (evenly distributed) within $N_{\min} = 150$, $N_{\max} = 250$. Measures of defect heterogeneity κ_D were chosen in accordance with the law of uniform distribution within $\kappa_{D\min} = -0.5$, $\kappa_{D\max} = 0.5$. Positive values of k_D correspond to weakening of the material in the defect area (circles in Fig. 3), negative values to toughening (asterisks in Fig. 3). Effective defect radii are uniformly distributed random variables with limits $r_{D\min} = 0.01$, $r_{D\max} = 0.08$. Some typical examples of defect distributions are shown in Fig. 3.

The computational procedure for each example consists of the following steps:

- Determining the optimal locations of the active supports and calculating the optimal excitation forces for a defect-free plate.
- Introducing a random set of defects into the plate.
- Calculating the optimal excitation forces for the defective plate, using the optimal arrangement of active supports determined for the defect-free plate.
- Generating the waveforms for both the defect-free and defective plates.

The calculations were carried out for two and three groups of forces, each of which includes two forces located symmetrically. The frequency of excitation of oscillations was chosen in three frequency ranges:

Table 1 k—vibration excitation frequency, key _ def—presence or absence of defects (1/0), $I(R)$—the value of the root-mean-square deviation (formula (18), (20)), NF—the number of control forces, R_g—the optimal radii of force groups, SF—the root-mean-square value of forces amplitudes, MF—the maximum amplitude of forces

k	key _ def	$I(R)$	NF	R_g	SF	MF
1.2324	0	1.82e - 04	5	0.443, 0.443	1.74	1.37
1.2324	1	3.2e - 03	5	0.443, 0.443	2.05	7.3
1.2324	0	5.77e - 05	7	0.545, 0.545, 0.545	1	1.18
1.2324	1	3.1e - 03	7	0.545, 0.545, 0.545	2.17	5.94
3.2324	0	4.08e - 04	5	0.447, 0.447	0.094	0.136
3.2324	1	6.85e - 03	5	0.447, 0.447	0.128	0.152
3.2324	0	1.18e - 05	7	0.546, 0.546, 0.546	0.072	0.089
3.2324	1	1.46e - 03	7	0.546, 0.546, 0.546	0.086	0.133
5.91	0	1.32e - 02	5	0.351, 0.351	0.437	0.628
5.91	1	1.07e - 01	5	0.351, 0.351	0.518	0.595
5.91	0	2.04e - 03	7	0.551, 0.551, 0.551	0.791	0.987
5.91	1	1.94e - 02	7	0.551, 0.551, 0.551	0.886	1.05

The excitation frequencies of the oscillations were selected from three different frequency ranges:

- In the low-frequency (Quasi-static) range $k = 1.2324$ (below the first resonant frequency $k_{01} = 2.232$), the reference waveform corresponds to the first asymmetric waveform.
- In mid-frequency range $k = 3.2324$—above the value of the first resonant frequency, but below the second resonant frequency $k_{12} = 3.734$—the reference waveform corresponds to the first asymmetric form of natural vibrations.
- In high-frequency range $k = 5.91$—above the value of the frequency of the second symmetric resonance $k_{02} = 5.455$.

The results of the calculations are presented in Table 1 and Fig. 4 (in the high frequency range for 5 number of forces).

6 Stochastic Model

Stochastic model for modulating mirror fixed on active supports takes into consideration multiple defects (inhomogeneities) with uncertain characteristics located on the mirror. We suppose that number of defects N_{def} is random, and each defect is characterized by a vector of stochastic parameters (geometric and mechanical characteristics) $\vec{\ell}_{\text{def}}$, and a vector of random locations $\vec{\chi}_{\text{def}}$. Stochastic vector $\vec{\theta}^{T} = \left\{N_{\text{def}}, \overrightarrow{mathrm\ell}_{\text{def}}, \chi_{\text{def}}\right\}$ transfers deterministic function of mirror oscillation form $w\left(\vec{F}, r, \varphi\right)$ into random function $w\left(\vec{\theta}, \vec{F}, r, \varphi\right)$. Therefore,

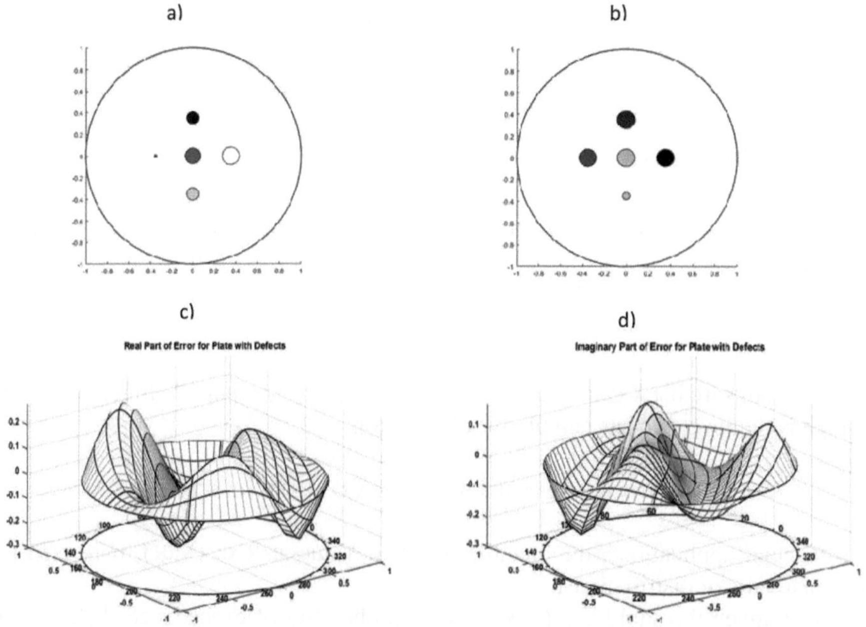

Fig. 4 The oscillations of the plate with five forces in the high frequency range ($k = 5.91$). (**a** and **b**) positions of forces, their normed amplitudes and phases, (**c** and **d**) real and imaginary parts of approximation error in the case of the plate with defects

deterministic mean-square deviation $I\left(\overrightarrow{F}\right)$ in (14) is transformed to the following random function:

$$L\left(\overrightarrow{\theta}, \overrightarrow{F}\right) = \frac{1}{2\pi}\int_0^1 rdr \int_{-\pi}^{\pi} \left|w\left(\overrightarrow{\theta}, \overrightarrow{F}, r, \varphi\right) - W\left(r, \varphi\right)\right|^2 d\varphi. \tag{23}$$

Let η be a measure of risk. We build the following risk functional $H\left(\overrightarrow{F}\right)$ with respect to mean-square deviation of the deflection of the plate from the given oscillation form $W(r, \varphi)$:

$$H\left(\overrightarrow{F}\right) = \eta\left(\int \left|w\left(\overrightarrow{\theta}, \overrightarrow{F}, r, \varphi\right) - W\left(r, \varphi\right)\right|^2 drd\varphi\right). \tag{24}$$

where η is a measure of risk, and $\overrightarrow{F} = (F_0, \ldots, F_I)$. Information about measure of risk η is given in the next section.

Thus, the control problem is reduced to the problem of minimization of the risk functional (24) on a given domain of control space $\overrightarrow{F} \in U$:

$$\min_{\vec{F} \in U} H\left(\vec{F}\right) \tag{25}$$

7 Risk Measures

Let F_X be the cumulative distribution function (CDF) of a random variable X modeling losses:

$$F_X(x) = P\{X \le x\}$$

The random variable X can be alternatively described by the quantile function $q_\alpha(X)$, which is the inverse of its cumulative distribution:

$$F_X^{-1}(\alpha) = q_\alpha(X) = \inf\{x : \alpha \le F_X(x)\} \;\; \text{for} \;\; \alpha \in (0, 1)\,.$$

In financial applications, the quantile is called Value-at-Risk (VaR). It is used to estimate tails of distributions.

For a given threshold $x \in (-\infty, \infty)$ the probability of exceedance (POE) with threshold x equals:

$$p_X(x) = P\{X > x\} = 1 - F_X(x).$$

Since POE describes the upper tail of the distribution, it is commonly used in reliability engineering for quantifying of failure probability, which is the chance that a system fails to carry out its intended task. We use POE to quantify the probability that the approximation error would exceed a certain threshold. Both POE and VaR are "optimistic" risk functions which are based on the lower bound of outcomes in the tail. It should be noted that these characteristics do not provide information about large losses that may occur with low probability. Besides, POE and VaR are difficult to optimize for discrete distributions, because they are non-convex, non-smooth, and they have multiple local extremes.

More attractive properties have Conditional Value-at-Risk (CVaR) and Buffered Probability of Exceedance (BPOE). These measures of risk are conservative counterparts of VaR and POE. By definition, CVaR for a distribution, which has no jump at the point $q_\alpha(X)$, is equal to the expected loss exceeding this quantile, see [5, 6]:

$$CVaR_\alpha(X) = E\left[X : X \ge q_\alpha(X)\right].$$

In general case, when $F_X(x)$ has jump at the point $q_\alpha(X)$, the definition of CVaR is more complicated:

$$CVaR_{\alpha}(X) = \min_{u}\left(u + \frac{E[X-u]^{+}}{1-\alpha}\right),$$

where $[\cdot]^{+} = \max\{\cdot\ ,0\}$.

Compared to VaR, CVaR possesses significantly better mathematical properties. It is convex with respect to random variables and can be optimized using linear programming and convex optimization algorithms. BPOE equals one minus the inverse of the superquantile [5]:

$$\text{bPOE}_x(X) = \begin{cases} 0 & \text{if } x \geq \sup X; \\ 1 - \text{CVaR}^{-1}(x; X) & \text{if } EX < x < \sup X; \\ 1 & \text{otherwise.} \end{cases}$$

Buffered Probability of Exceedance (BPOE) describes the probability that the average of data points in the upper tail of a distribution equals a specified threshold, whereas the Probability of Exceedance (POE) only indicates the likelihood that a specified threshold will be surpassed [6]. In the calculation of POE, the threshold is set by the user, while in BPOE, the threshold is strictly defined as the average of the data points in the tail. BPOE acts as an upper bound for POE because it includes all outcomes exceeding the threshold, as well as some outcomes below it—the so-called buffer.

When BPOE and Conditional Value-at-Risk (CVaR) are used solely as constraints in optimization problems, the results are equivalent. However, the problems of minimizing BPOE and CVaR are quite different [8]. The risk measure BPOE has been applied in numerous optimization problems, as demonstrated in [6–9].

8 Numerical Approach for Solving the Problem

We tackle problem (24) using a three-step methodology:

Step 1: For a given vibration form, wave number, and number of forces, we solve the deterministic optimization problem without accounting for defects. This provides the optimal characteristics of the forces, including their application points and the real and imaginary parts of their amplitudes.

Step 2: We perturb the optimal deterministic solution by introducing a random number of defects with random parameters. The locations, geometric properties, and mechanical characteristics of these defects are simulated as independent random variables uniformly distributed within specified limits. This process yields a sample of deviations in the mirror's deflection and phase from the desired oscillation form.

Step 3: We minimize the buffered Probability of Exceedance (BPOE) as a measure of risk.

To demonstrate this approach, we consider the following examples.

Example 1
Parameters: wave number is 1.23; number of forces is 5.

Step 1. Deterministic problem. For given form of the plate surface $W(r,\phi) = \text{Re}\, W(r,\phi) + i\, \text{Im}\, W(r,\phi)$ and these parameters we determined the optimal characteristics of forces, which minimize the value of the deterministic function (16) ($\min\limits_{\vec{F}} I_0\left(\vec{F}\right)$). We determined the following characteristics of optimal forces $\vec{F}^* = \left(u_0^* + i w_0^*, \ldots, u_4^* + i w_4^*\right)$ where:

1. Real parts of optimal forces $u_j^*,\ \ j = 0, \ldots, 4$ are:

u_0^*	u_1^*	u_2^*	u_3^*	u_4^*
−8.49E–03	−1.74E+00	1.67E+00	−3.80E–02	−3.80E–02

2. Imaginary parts of optimal forces $w_j^*,\ \ j = 0, \ldots, 4$ are:

w_0^*	w_1^*	w_2^*	w_3^*	w_4^*
3.55E−15	−1.71E–08	1.71E–08	-1.71E+00	1.71E+00

$F_0{}^* = u_0^* + i w_0^*$ —the value of the force applied at the center of the plate ($R_0 = 0$). The other forces form two groups with two forces, having a common radius and a centrally symmetrical arrangement:

$R_1 = R_2 = 0.443,\ \Phi_1 = 0,\ \Phi_2 = \pi,$

$R_3 = R_4 = 0.443,\ \Phi_3 = \pi/2,\ \Phi_4 = 3\pi/2$

Optimal value of mean-square deviation of the deflection of the plate (and the deflection phase) from the given oscillation form in deterministic case is $Q\left(\vec{F}^*\right) = 1.826\text{E–}3$.

Step 2. Perturbation of the optimal deterministic solution of problem (16).

We generated $N_{\text{Sample}} = 200$ independent random realizations of defects $\vec{\theta}_1, \ldots, \vec{\theta}_{N_{\text{Sample}}}$. Defects were modeled by small inhomogeneities with changed elastic characteristics. Using iterative technique, developed in [2], we modeled finite-size defects in a Kirchhoff plate by point quadruples. Then we perturbed the optimal deterministic solution of optimization problem (14).

As a result, we obtained N_{Sample} values $L_1\left(\vec{\theta}_1, \vec{F}^*\right), \ldots, L_{N_{\text{Sample}}}\left(\vec{\theta}_{N_{\text{Sample}}}, \vec{F}^*\right)$, which we consider as independent realizations of the random value $L\left(\vec{\theta}, \vec{F}^*\right)$ in (23). Using these values, we determined $\text{CVaR}_{\alpha_i}\left(L\left(\vec{\theta}, \vec{F}^*\right)\right)$, where $\alpha_i = \frac{i}{N_{\text{Sample}}}$, $i = 1, \ldots, N_{\text{Sample}}$. Let

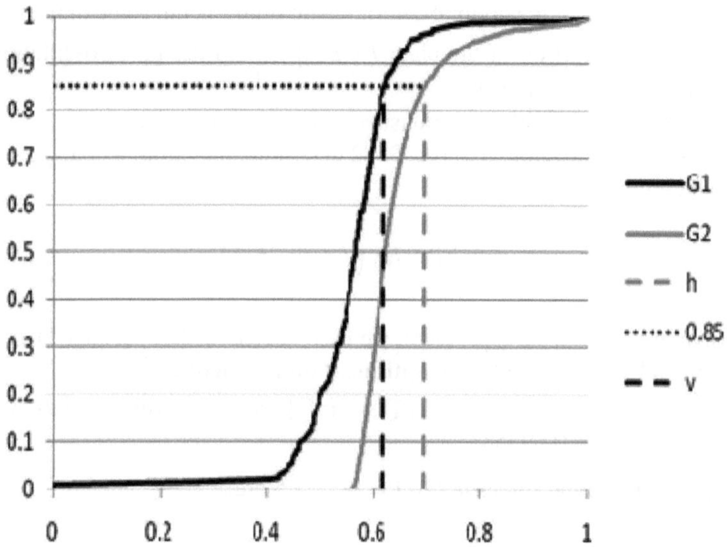

Fig. 5 Distribution G1 of random value $I\left(\vec{\theta}, \vec{F}^*\right)$ vs. distribution G2 of random value $\mathrm{CVaR}\left(L\left(\vec{\theta}, \vec{F}^*\right)\right)$ before optimization. Wave number is 1.23, number of forces

$\mathrm{CVaR}\left(L\left(\vec{\theta}, \vec{F}^*\right)\right)$ denote discrete random value taking equally probable values $\mathrm{CVaR}_{\alpha_i}\left(L\left(\vec{\theta}, \vec{F}^*\right)\right)$, $i = 1, \ldots, N_{\mathrm{Sample}}$.

Cumulative distribution functions of the random value $L\left(\vec{\theta}, \vec{F}^*\right)$, denoted by G1, and random value $\mathrm{CVaR}\left(L\left(\vec{\theta}, \vec{F}^*\right)\right)$, denoted by G2, are presented in Fig. 5.

The black solid line in Fig. 5 corresponds to the distribution of random value $L\left(\vec{\theta}, \vec{F}^*\right)$, and the gray solid line corresponds to distribution of random value $\mathrm{CVaR}\left(L\left(\vec{\theta}, \vec{F}^*\right)\right)$. The black dotted line indicates 0.85-quantile of the distribution function of the random value $L\left(\vec{\theta}, \vec{F}^*\right)$. It corresponds to the threshold $v = 0.6193943$. The gray dotted line indicates 0.85-quantile of the distribution function $\mathrm{CVaR}\left(L\left(\vec{\theta}, \vec{F}^*\right)\right)$. It corresponds to the threshold $h = 0.693582$. The threshold v was chosen such that $E\left(L\left(\vec{\theta}, \vec{F}^*\right) | L\left(\vec{\theta}, \vec{F}^*\right) \geq v\right) = \mathrm{CVaR}_{0.85}\left(L\left(\vec{\theta}, \vec{F}^*\right)\right) = h$. Interception between the CDF G1 and the threshold v and interception between the CDF G2 and the threshold h are located on the black dotted line, which corresponds to 0.85 level of probability. Thus, before optimization, the value of POE, corresponding to the threshold v, and the value of BPOE, corresponding to the threshold h, are the same and are equal to 0.15 (1–0.85).

After perturbation of the optimal deterministic solution, the value of mean-square deviation of the deflection of the plate $I\left(\overrightarrow{F}^{*}\right)$ from the given oscillation form increased from 1.826E-3 to 0.15.

Step 3. Minimize BPOE as a measure of risk (24) **in** (25).

Problem 2. For threshold $h = 0.693582$ determine the optimal characteristics of forces, which minimize BPOE.

$$\min_{\overrightarrow{F} \in U} bPOE_h\left(L\left(\overrightarrow{\theta}, \overrightarrow{F}\right)\right) \tag{26}$$

To solve this optimization problem, we used AORDA PSG package [10].

Optimal solution of stochastic optimization problem 2.

Let $\overrightarrow{F}^{**} = \left(u_1^{**} + iw_1^{**}, \ldots, u_5^{**} + iw_5^{**}\right)$ denote optimal solution of the problem (7) with $\mathbf{h} = 0.693582$.

1. Real parts of optimal forces $u_j^{**}, \quad j = 1, \ldots, 5$ are:

u_1^{**}	u_2^{**}	u_3^{**}	u_4^{**}	u_5^{**}
−6.40E–03	−1.74E+00	1.67E+00	−3.59E–02	6.31E–02

2. Imaginary parts of optimal forces $w_j^{**}, \quad j = 1, \ldots, 5$ are:

w_1^{**}	w_2^{**}	w_3^{**}	w_4^{**}	w_5^{**}
1.01E–01	2.09E–03	2.09E–03	−1.61E+00	1.81E+00

Similar to Step 2, we used optimal solution $\overrightarrow{F}^{**}$ of the problem (26) to determine independent realizations $L_1\left(\overrightarrow{\theta}_1, \overrightarrow{F}^{**}\right), \ldots, L_{N_{\text{Sample}}}\left(\overrightarrow{\theta}_{N_{\text{Sample}}}, \overrightarrow{F}^{**}\right)$ of new discrete random value $L\left(\overrightarrow{\theta}, \overrightarrow{F}^{**}\right)$ and independent realizations $\text{CVaR}_{\alpha_i}\left(L\left(\overrightarrow{\theta}, \overrightarrow{F}^{**}\right)\right)$, $i = 1, \ldots, N_{\text{Sample}}$, of new discrete random value $\text{CVaR}\left(L\left(\overrightarrow{\theta}, \overrightarrow{F}^{**}\right)\right)$. Tails of cumulative distribution functions of the random value $L\left(\overrightarrow{\theta}, \overrightarrow{F}^{**}\right)$, denoted by G1_1, and the random value $\text{CVaR}\left(L\left(\overrightarrow{\theta}, \overrightarrow{F}^{**}\right)\right)$, denoted by G2_1, are presented in Fig. 6 together with tails of CDFs G1 and G2.Characters G1, G2, h, and v in the legend of Fig. 6 have the same meaning as in the legend of Fig. 5. The black solid line, denoted by G1_1, corresponds to distribution of random value $L\left(\overrightarrow{\theta}, \overrightarrow{F}^{**}\right)$, obtained after BPOE minimization. The gray solid line, denoted by G2_1, corresponds to the distribution of the random value $\text{CVaR}\left(L\left(\overrightarrow{\theta}, \overrightarrow{F}^{**}\right)\right)$, obtained after

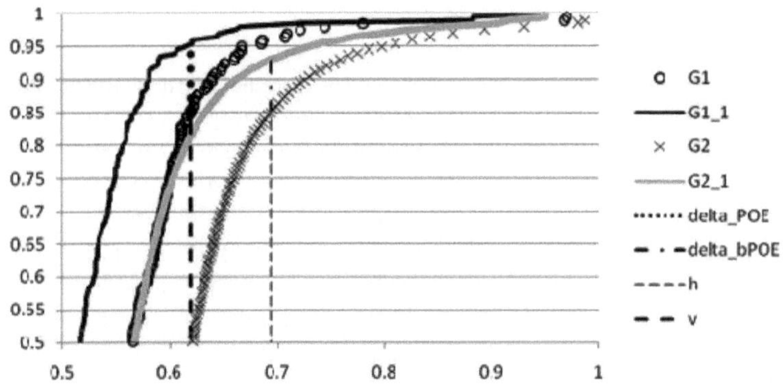

Fig. 6 Comparison of cumulative distribution functions before and after minimization of BPOE with threshold h is 0.693582. Wave number is 1.23, number of forces is 5

BPOE minimization. The black dotted line segment "delta_POE" demonstrates how much the value of POE, corresponding to the threshold v, diminished after BPOE minimization. The line segment "delta_BPOE" demonstrates how much the value of BPOE, corresponding to the threshold h, diminished after BPOE minimization. In the considered case delta_BPOE $= 0.07 - 0.15 = -0.08$, and delta_POE $= 0.04 - 0.15 = -0.11$. This example demonstrates that minimization of BPOE reduces not only BPOE (by 53.33%), but also POE (by 73.33%).

9 Results of Calculations

Using approach described in Sect. 7, we solved the problem (26) with the following parameters: confidence level: 0.8, 0.85, 0.9, 0.95, 0.99; wave number: 1.23, 3.23, 5.91, number of forces: 5, 7. The results obtained after BPOE optimization are summarized in Table 2.

10 Conclusions

The article presents fundamental formulas necessary for calculating control actions, modeled as point active supports, to achieve the optimal wave form of mono harmonic forced oscillations in a circular plate. The optimization procedure minimizes the root-mean-square deviation between the plate's wave profile and a specified target profile. Control parameters include the locations of the active supports and the amplitudes and phases of the applied forces. Since the optimization problem is non-convex, additional studies were conducted to validate the results. The article analyzes bothanideallyhomogeneousplateandonewithrandomlylocated-

Table 2 Results obtained after BPOE optimization

Confidence level	Wave number	Number of forces	Threshold h	Threshold v	delta BPOE	delta POE
0.85	1.23	5	0.693582	0.6193943	−53.33%	−73.33%
0.9	1.23	5	0.726248	0.6393482	−48.47%	−64.82%
0.85	5.91	7	0.893926	0.4931682	−42.30%	−35.40%
0.9	5.91	7	1.057949	0.6432257	−43.28%	−50.68%
0.95	1.23	5	0.7977	0.6664913	−41.46%	−60.50%
0.99	1.23	5	0.987562	0.9695511	−100.00%	−100.00%
0.8	5.91	7	0.785515	0.4161413	−40.08%	−28.05%
0.95	5.91	7	1.381567	0.6470629	−41.93%	−10.23%
0.99	5.91	7	2.312896	1.452711	−14.75%	−54.65%
0.85	3.23	7	0.452861	0.1950224	−4.33%	−83.13%
0.99	3.23	7	1.540914	1.263241	−100%	−75.49%
0.9	1.23	7	1.272189	1.02435	−16.01%	−15.99%
0.95	1.23	7	1.445576	1.150952	−100%	−94.28%
0.99	1.23	7	1.886962	1.502909	−21.36%	−17.36%

inhomogeneities simulated as small defects. The developed theory has practical applications. An approximate solution to the formal boundary value problems was sought using truncated Fourier series along the circumferential coordinate. The inhomogeneities were modeled as point sources with high-orders singularities.

The theory presented serves as the foundation for numerical implementation. Test calculations have been performed, demonstrating the correctness of both the problem formulation and its solution. These calculations demonstrate that for both a homogeneous plate and a plate with defects, it is possible to select the optimal locations and control parameters of active supports to achieve a high-quality approximation of a given plate vibration form. The control forces and the quality of the approximation are significantly influenced by the presence of inhomogeneities and the oscillation frequency. However, the optimal placement of the supports is largely unaffected by these factors. Instead, the supports should be positioned based on the specified vibration form.

In the numerical example considered, to achieve the best results, the supports (excluding the central fixed support) should be placed symmetrically along a single concentric circle, with the radius heavily dependent on the number of supports. Naturally, the quality of the approximation improves as the number of forces increases. It should be noted that these conclusions are partly a consequence of the symmetry of the chosen vibration form, which was selected as a waveform propagating in the circumferential direction.

We have demonstrated how risk measures such as CVaR and BPOE can be applied to structural reliability problems. A corresponding stochastic model was developed. Building upon the solutions of the deterministic optimization problems, we developed an algorithm to solve the stochastic optimization problem. In the first step, for a given vibration form, wave number, and number of forces, we solved the deterministic optimization problem without considering defects and obtained the optimal force characteristics. In the second step, we introduced defects by perturbing the optimal deterministic solution. A random number of defects with random parameters were generated, with their locations, geometric, and mechanical characteristics simulated as independent random variables uniformly distributed within specified limits. This process produced a sample of deviations in the mirror's deflection (and phase deflection) from the desired oscillation form. In the third step, we minimized the risk of deviation in the plate's deflection (and phase deflection) from the specified oscillation form using the BPOE and CVaR risk measures, and determined the optimal forces. Our numerical experiments demonstrated that minimizing BPOE leads to minimizing the Probability of Exceedance (POE). Considering that POE optimization is a challenging task, optimizing BPOE presents a practical alternative for structural reliability problems.

The results of the article can be applied to the design of actively controllable reflective elements in multichannel formation or transmission energy systems.

References

1. Zrazhevsky, G.M., Zrazhevska, V.F., Golodnikov, A.N.: Modeling finite inhomogeneities with discrete features. J. Comput. Appl. Math. **1**(135), 138–144 (2021). https://doi.org/10.17721/2706-9699.2021.1.18
2. Zrazhevsky, G., Golodnikov, A., Uryasev, S.: Mathematical methods to find optimal control of oscillations of a hinged beam (deterministic case). Cybern. Syst. Anal. **55**, 1009–1026 (2019). https://doi.org/10.1007/s10559-019-00211-x
3. Zrazhevsky, G.M., Zrazhevska, V.F., Golodnikov, A.N.: Developing a model for a modulating Mirror fixed on active supports. Deterministic problem. Cybern. Syst. Anal. **58**, 702–712 (2022). https://doi.org/10.1007/s10559-022-00503-9
4. Zrazhevsky, G.M., Zrazhevska, V.F., Golodnikov, A.N.: Developing a model for a modulating mirror fixed on active supports. Stochastic model. Cybern. Syst. Anal. **59**, 101–107 (2023). https://doi.org/10.1007/s10559-023-00546-6
5. Rockafellar, T., Royset, J.O.: On buffered failure probability in design and optimization of structures. Reliab. Eng. Syst. Saf. **95**, 499–510 (2010). https://doi.org/10.1016/j.ress.2010.01.001
6. Mafusalov, A., Uryasev, S.: Buffered probability of exceedance: mathematical properties and optimization. SIAM. J. Optim. **28**, 1077–1103 (2018). https://doi.org/10.1137/15M1042644
7. Rockafellar, R.T., Uryasev, S.: Optimization of conditional value-at-risk. J. Risk. **2**(3), 21–41 (2000). https://doi.org/10.21314/JOR.2000.038
8. Shang, D., Kuzmenko, V., Uryasev, S.: Cash flow matching with risks controlled by buffered probability of exceedance and conditional value-at-risk. Ann. Oper. Res. **260**(1–2), 501–514 (2018). https://doi.org/10.1007/s10479-016-2354-6
9. Zrazhevsky, G.M., Golodnikov, A.N., Uryasev, S.P., Zrazhevsky, A.G.: Application of buffered probability of exceedance in reliability optimization problems. Cybern. Syst. Anal. **56**, 476–484 (2020). https://doi.org/10.1007/s10559-020-00263-4
10. AORDA Portfolio Safeguard (PSG). http://www.aorda.com

MIX
Papier aus verantwortungsvollen Quellen
Paper from responsible sources
FSC® C105338

If you have any concerns about our products,
you can contact us on
ProductSafety@springernature.com

In case Publisher is established outside the EU,
the EU authorized representative is:
Springer Nature Customer Service Center GmbH
Europaplatz 3, 69115 Heidelberg, Germany

Printed by Libri Plureos GmbH
in Hamburg, Germany